THE ART OF EXPERIMENTAL PHYSICS

THE ART OF
EXPERIMENTAL PHYSICS

Daryl W. Preston
California State University, Hayward

Eric R. Dietz
California State University, Chico

With software by

R. H. Good and Daryl W. Preston
California State University, Hayward

JOHN WILEY & SONS
NEW YORK/CHICHESTER/BRISBANE/TORONTO/SINGAPORE

Library of Congress Cataloging-in-Publication Data

Preston, Daryl W.
    The art of experimental physics / Daryl W. Preston, Eric R. Dietz
; with software by R.H. Good and Daryl W. Preston.
      p.   cm.
    Includes bibliographical references.
    ISBN 0-471-84748-8
    1. Physics--Experiments--Methodology. 2. Physics--Methodology.
I. Dietz, Eric R. II. Title.
QC33.P74 1991
530′.078--dc20   90-33872

Printed in the United States of America

10 9 8 7 6 5 4 3

Printed and bound by Malloy Lithographing, Inc..

TO THE STUDENT

Before you begin to study the text of this book and to do its experiments, you should read the description of its special features provided here. Doing so will acquaint you with the book's organization and content and will help you to get the most out of your assignments.

BACKGROUND AND ORIENTATION: SETTING THE STAGE

In the first part of the book, the "Introduction," you will find the following features:

- Two sections entitled "Physics—A Human Endeavor" and "Responsibility of the Experimentalist" and twenty-five Nobel prize citations quoted in the text discussion, some of which are accompanied by a description of the background and milieu in which a particular discovery was made. These features will give you valuable insight into the importance of the human element in physics as well as the course of its historical development.
- A section on "Objectives of the Physics Laboratory," which describes for your easy reference the basic aims of the laboratory.
- A section entitled "Journals of Physics" as well as more than fifty references to twelve different journals in the text. These features provide an in-depth introduction to the journals of physics and an understanding of how results are published.

- A section entitled "Spectroscopy: An Important Class of Experiments," which presents a general discussion designed to familiarize you with these types of experiments.
- Sections on the "Laboratory Notebook," "Error Analysis," "Significant Figures," "Graphical Analysis," and "Curve Fitting," which describe in practical terms the skills you must master to maintain a notebook and to analyze data.

The Role of the Computer in the Laboratory: Learning How to Use It

The computer is a unique laboratory device in that it may be used to control the experiment, collect and analyze the data, and present the experimentalist with a labeled graph of the finished results complete with error analysis. Therefore, it is essential that you learn how to use the computer in the laboratory. With this goal in mind, the second part of this book, "Experiments" and two of its appendices includes these valuable features:

- Experiment 6, "Introduction to Computer-Assisted Experimentation." In this experiment the emphasis is on interfacing experimental apparatus to the computer, that is, on the hardware. John Wiley & Sons, Inc., the publisher of this text, has provided your instructor with the software for Experiment 6.

- Several experiments with optional sections on computer-assisted experimentation. (Experiment 6 is a prerequisite for these sections.)
- Appendix F on writing computer programs, which is included to assist you in writing programs in BASIC and in assembly language. (It assumes that the reader has had prior experience in writing programs.)
- Appendix G, "A Software Tutorial for the IBM PC." You will find this appendix useful if you have had experience with microcomputers but not with the IBM PC.

The Experiments

An important goal of an experimental physics course is to provide students with the opportunities to use a variety of instruments in carrying out many different measurements. To further this goal, the twenty-two experiments of this book include the following features:

- *List of Apparatus*. The list of apparatus included at the beginning of each experiment will alert you to the equipment you will need.
- *Statement of Objectives*. This feature will enable you to focus your attention on the major goals of each experiment.
- *Key Concepts*. By studying the list of key concepts provided at the beginning of each experiment, you will become familiar with the terms you should use in describing the experiment in your notebook.
- *References*. The list of references included with each experiment will enable you to broaden your knowledge of useful books on physics and, especially, journals of physics.
- *Introduction*. The theoretical background you will need to understand each experiment is presented in a carefully written introduction.
- *Experiment*. The text of each experiment includes suggested measurements and is structured to guide you as you carry out your experimental work.
- *Computer-Assisted Experimentation*. These optional sections are included in several experiments to help you increase your knowledge of computer usage in the laboratory.
- *Exercises*. Exercises are included to provide direction and to enhance your understanding of the experiment.

We hope that this summary of the features of this book will prove useful and will contribute to a challenging and successful study of its contents.

DARYL W. PRESTON
ERIC R. DIETZ

TO THE INSTRUCTOR

SOFTWARE FOR EXPERIMENT 6

The publisher, John Wiley & Sons, Inc., will provide you with the software for Experiment 6, "Introduction to Computer-Assisted Experimentation." The software is written for both the Apple II ™ and the IBM PC ™. The Apple II software is written for two ADC/DAC cards: the 8-bit Mountain card and the 12-bit Sunset-Vernier card. The IBM PC software is also written for two ADC/DAC cards: the 8-bit ML-16 Multi-Lab card and the 12-bit IBM DACA card. The instructions for each card are at the end of Experiment 6.

INSTRUCTOR'S MANUAL

There is an Instructor's Manual available that contains the following information:

- List of equipment suppliers.
- Data for some of the experiments.
- Answers to some of the exercises posed in the first part of the book, the "Introduction," the second part of the book, the "Experi-

ments," and in the appendixes at the end of the book.

SUGGESTIONS FOR USING THE BOOK

We offer these suggestions for using this book:

- Reserve laboratory time for experimentation and data analysis.
- Provide that the reading of the text and the reference material be done outside of the laboratory.
- Assign the exercises in the "Introduction" as homework.
- Assign the exercises in the experiment each student is performing as homework. For some of the experiments this includes the exercises in the appendixes and the sections on "Introduction to Laser Physics" and "Introduction to Magnetic Resonance."
- Require one report per term that is written according to the directions in the *Style Manual* of the American Institute of Physics.

D. W. P.
E. R. D.

ACKNOWLEDGMENTS

This book grew out of the experiences one of us (DWP) had, starting in 1981, in teaching an upper-division experimental physics course. After teaching the course for three years using a patchwork array of materials, it was apparent that there existed an acute need for an experimental physics text. The recognition led to this book, which was officially started in 1984.

When a project is labored over for a number of years it is difficult to recall all of the students and colleagues who have contributed to the project. Our apologies to any individuals that we fail to acknowledge.

We are indebted to Sumner Davis, University of California, Berkeley, for suggesting that Key Concepts be included with each experiment. John Powell, Reed College, did an extraordinary job of reviewing the entire manuscript. His many suggestions improved the book.

We acknowledge Daedalon Corporation for providing a drawing of their Rutherford scattering apparatus along with information concerning its operation. Charles Leming, Henderson State University, and Shane Thompson kindly provided data regarding the development of the cellulose nitrate film.

The notes for the advanced lab at the Massachusetts Institute of Technology provided inspiration, along with some of the information and techniques used in several of the experiments, most notably "Zeeman Effect," "Hall Effect," "Ionization of Gases," and "Faraday Effect."

One of us (DWP) is indebted to the following individuals: Marvin M. Abraham, Oak Ridge National Laboratory, and Wesley P. Unruh, Los Alamos National Laboratory, who suggested many possible electron spin resonance samples. Also, Dr. Unruh suggested using the SLAC/LBL discovery of the ψ particle as an example in the Error Analysis section of the "Introduction." James H. Breen, Los Alamos National Laboratory, granted permission for use of the front cover photograph. John R. Burke, San Francisco State University, reviewed parts of the manuscript, and he not only tolerated our endless discussions, but also was an enthusiastic participator to the end. Richard K. Cooper, Los Alamos National Laboratory, provided constructive criticism of parts of the manuscript. R. H. Good, California State University, Hayward, wrote Appendixes F and G, and the software for Experiment 6. Professor Good also reviewed parts of the manuscript and was always willing to discuss details of the project. His very significant contributions to this book are gratefully acknowledged. Ken Grove, University of California, Berkeley, provided bench space and equipment for the construction and testing of circuits. Ken Grove, Mason Bolton, John Davis, and LeVern Garner provided suggestions and constructive criticism of the circuits. Andree Nel, Hewlett-Packard Corporation, made useful suggestions pertaining to Experiment 3. Tom Palmer, Instrument Shop, California State University,

Hayward, constructed the Compton scattering apparatus, and provided mechanical assistance in the development of many of the experiments. F. N. H. Robinson, Clarendon Laboratory, University of Oxford, generously designed, constructed, and tested the NMR spectrometer shown in Figure 15.2. The advice and assistance of Dr. Robinson and Larry Wald, University of California, Berkeley, pertaining to the NMR experiment are gratefully acknowledged. Eugene Y. Wong, University of California, Los Angeles, generously spent most of a day demonstrating and discussing the Ebert-mount spectrograph with DWP. In addition, Professor Wong made available shop drawings of the spectrograph. Finally, we thank the students at California State University, Hayward, who performed many of the experiments and also contributed to their improvement.

D. W. P.
E. R. D.

GUIDE TO THE EXPERIMENTS

Chapters in a textbook are usually linearly arranged; that is, the reader starts with Chapter 1 and proceeds through the text in the order that the chapters are numbered. The experiments in this book are not so arranged. The level of the physics does vary with individual experiments, and we recommend that each student perform experiments for which he or she has the appropriate background.

As an aid to both students and instructors, the table that follows summarizes each experiment, providing the following information:

• Prerequisites, denoted by (p), and/or corequisites, denoted by (c).

• A description of the level of the physics.

• A typical textbook where most of the physics is presented. (The reference number refers to the number at the beginning of the experiment.)

• The approximate number of three-hour laboratory periods required to complete the experiment. (This number is dependent on the student performing the experiment, and on the advice and assistance offered by the instructor. Also some experiments have optional parts, and this approximate number of lab periods does not include the time required to perform such optional measurements.)

D. W. P.
E. R. D.

GUIDE TO THE EXPERIMENTS

Experiment	Prerequisite and/or corequisite	Level of Physics	Text	Number of Lab Periods
1. Coaxial transmission line	None	Telegrapher's equation for both voltage and current waves, and their harmonic space and time solutions with exponential damping	Scott, ref. 1	3
2. Waveguide	None	Maxwell's equations and their solutions in rectangular coordinates	Wangsness, ref. 4	5
3. Optical fiber	None	Maxwell's equations and their solutions in cylindrical coordinates, cylindrical Bessel functions	Marcuse, ref. 3	4

GUIDE TO THE EXPERIMENTS (*Continued*)

Experiment	Prerequisite and/or corequisite	Level of Physics	Text	Number of Lab Periods
4. Laser spectra	Introduction to Laser Physics (c)	Principle of detailed balance, Einstein coefficients, electric dipole selection rules, standing waves, and eigenfrequencies	Eisberg and Resnick, ref. 3	3
5. Laser cavity modes	Introduction to Laser Physics (c)	Same as Experiment 4		3
6. Computer-assisted experimentation	Appendix C, GPIB; Appendix F, Writing Programs; Appendix G, Software Tutorial: (optional, c)	Hall voltage, Poisson and interval distributions, electronics: transistor amplifier, op-amps; monostable multivibrator; software provided	Evans, ref. 6, Horowitz and Hill, ref. 5	5
7. Electron physics	None	Free electron wave function, dynamics of electrons, electrostatics, Fermi–Dirac distribution	Kittel, ref. 2	3
8. Blackbody radiation	None	Boltzmann factor, standing electromagnetic waves	Reif, ref. 2	3
9. Photoelectric	Experiment 6 (optional, p)	Fermi–Dirac distribution, contact potential, qualitative description of the free electron model of a metal and the energy bands of a semiconductor	Eisberg and Resnick, ref. 1	4
10. Diffraction of x rays and microwaves by periodic structures	Experiment 6 (optional, p)	Interference of waves, Bragg law, periodicity in crystal lattices, reciprocal lattice, x-ray spectra	Kittel, ref. 1	3
11. Franck–Hertz	Experiment 6 (optional, p)	Quantum theory description of the energy levels of mercury, ordinary and metastable states, mean free path	Eisberg and Resnick, ref. 1	3
12. Structure of hydrogen	Appendix D, Spectrograph (c)	First-order perturbation theory	Richtmyer, ref. 3	5
13. Molecular structure	Appendix D, Spectrograph (c)	Detailed discussion of rotational, vibrational, and electronic structure based on quantum theory; selection rules, bands, and P, Q, and R branches	Herzberg, ref. 2	5
14. Zeeman effect	Experiment 6 (optional, p)	Schrödinger's equation and quantum states of electrons in atoms, *LS* coupling and Landé *g* factor, dipole matrix elements and selection rules	Eisberg and Resnick, ref. 1	3
15. NMR	Introduction to Magnetic Resonance (c); Appendix A, Modulation Spectroscopy (optional, c)	Transition probability, selection rules, Maxwell–Boltzmann statistics, Dirac notation	Eisberg and Resnick, ref. 5	4

GUIDE TO THE EXPERIMENTS (*Continued*)

Experiment	Prerequisite and/or corequisite	Level of Physics	Text	Number of Lab Periods
16. ESR	Introduction to Magnetic Resonance (c); Appendix A, Modulation Spectroscopy (optional, c); Experiment 2 (p)	Hamiltonian of a multielectron atom, term states, crystalline electric field, spin Hamiltonian, Dirac notation	*Rep. Prog. Phys.*, refs. 2 and 3	4
17. Electrical conductivity and the Hall effect	None	Semiclassical theory of electrical conduction, elementary description of bands in metals and semiconductors, Boltzmann factor and Lorentz force	Kittel, ref. 1	3
18. Gamma rays	Appendix B, Scintillation Counter (c)	Compton scattering, photoelectric effect, pair production, relativistic kinematics,	Eisberg and Resnick, ref. 1	3
19. Compton scattering	Appendix B, Scintillation Counter (c)	Compton scattering, relativistic kinematics	Eisberg and Resnick, ref. 1,	3
20. Ionization of gases by α particles	None	Dynamics of electrons, electrostatics, energy loss calculations	Bleuler and Goldsmith, ref. 2	3
21. Rutherford scattering	None	Cross section and solid angle, dynamics of charged particles in a Coulomb field	Tipler, ref. 2	3
22. The Faraday effect	None	Electromagnetic waves, polarization and phase, classical polarization response of atomic electrons, complex dielectric constant, Zeeman effect	Portis, ref. 6	3

CONTENTS

xv

INTRODUCTION

PHYSICS: A HUMAN ENDEAVOR

Taken as a story of human achievement, and human blindness, the discoveries in the sciences are among the great epics.

Robert Oppenheimer

A basic quest of humans, regardless of culture and nationality, is to know where they are in the universe, how they came to be, and what the future holds for them. Physics is a highly developed manifestation of the desire to determine our origin and our future. For instance, quantum mechanics, in principle, allows us to understand the macromolecules that make up our bodies, and the general theory of relativity is fundamental to our understanding of stellar and galactic evolution.

The theories of physics are, in general, abstract and mathematical, and the physical systems that they describe are often far removed from everyday experiences. For these reasons nonscientists often consider physics as devoid of the characteristics referred to as human, namely, feelings, sentiments, and emotions. What they do not recognize is that the laws of physics are an expression of the relations between humans and the universe.

Although some people may view physics as lacking human characteristics, all agree that the observation of natural phenomena is often an awe-inspiring event. Examples are observation of a lunar or solar eclipse, or simply the observation of a moonlit mountain covered with snow. There is also a deeper beauty in nature that scientists see which is usually not observed by others.

Commonplace as such experiments (nuclear magnetic resonance) have become in our laboratories, I have not yet lost a feeling of wonder, and of delight, that this delicate motion (precession of nuclear spins in a magnetic field) should reside in all the ordinary things around us, revealing itself only to him who looks for it. I remember, in the winter of our first experiments, just seven years ago, looking on snow with new eyes. There the snow lay around my doorstep—great heaps of protons quietly precessing in the earth's magnetic field. To see the world for a moment as something rich and strange is the private reward of many a discovery.

Edward M. Purcell, Nobel Lecture, 1952

Nature—from crystals to flowers—is full of ever recurring characteristic shapes and symmetries. There must be a fundamental reason for the typical properties of materials and forms that we observe in the flow of natural events. This reason is found in quantum mechanics: The wave nature of electrons forces them into typical patterns, the shapes of standing waves in the spherically symmetric Coulomb field. These shapes are the fundamental patterns of Nature, which are the basis of all the shapes we observe . . . .

Victor F. Weisskopf, Oersted Medal recipient, 1976

The human element of scientific investigation is rarely brought to the attention of the general public. The form of scientific publications is an example of this kind of omission. A publication typically includes an abstract, introduction, results, and conclusion, and suggests a straight-arrow path from the conception of the project to its completion.

> We have a habit in writing articles published in scientific journals to make the work as finished as possible, to cover the tracks, to not worry about the blind alleys or to describe how you had the wrong idea first, and so on. So there isn't any place to publish, in a dignified manner, what you actually did in order to get to do the work, although, there has been in these days, some interest in this kind of thing. Since winning the [Nobel] prize is a personal thing, I thought I could be excused in this particular situation, if I were to talk personally about my relationship to quantum electrodynamics, rather than to discuss the subject itself in a refined and finished fashion.
>
> *Richard P. Feynman, Nobel Lecture, 1965*

The form of a scientific publication suggests that scientists have a method that always leads to the final result. Indeed, there is a method, called the scientific method, which may be summarized as the following sequence of steps:

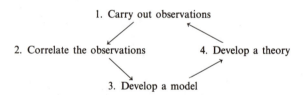

These four steps are shown in Figure I.1. It is not always possible to correlate all of the observations, and one observation is not correlated in the figure.

It is hindsight that allows us to claim that a scientific method exists. From a historical perspective it is possible to identify the existence of this method, but rarely can we discern it clearly from the work of an individual scientist. Planetary astronomy provides a clean-cut example. Tycho Brahe carried out careful observations. Johannes Kepler correlated the observations and developed a model for planetary motions (usually stated as Kepler's three laws). Sir Isacc Newton discovered laws of motion and a gravitational force law that he used to predict Kepler's laws. Newton also

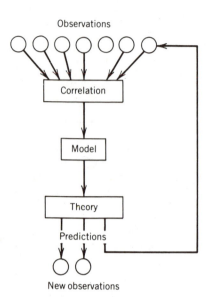

FIGURE I.1 Schematic diagram of the scientific method.

predicted new observations—for example, the orbital speed of artificial satellites.

In addition to the form of a scientific publication, the scientific method does not suggest a human side to science. In reality, the path to the completion of a scientific project is usually long and arduous and, as indicated by Feynman, is filled with wrong ideas, blind alleys, and the like. Therefore, it is not surprising that scientists experience the complete spectrum of human emotions, ranging from joyous excitement to severe depression, in the pursuit of scientific investigations.

> It took almost a full year before a workable scheme (experimental method to measure the fine structure of hydrogen) was clear in my mind.
>
> *Willis E. Lamb, Nobel Lecture, 1955*

> I worked on this problem (quantum electrodynamics) about eight years until the final publication in 1947.
>
> *Richard P. Feynman, Nobel Lecture, 1965*

> My first introduction to superconductivity came in the 1930's .... My first abortive attempt to construct a theory, in 1940's .... It was not until 1950, as a result of the discovery of the isotope effect, that I again began to become interested in superconductivity, ...
>
> *John Bardeen, Nobel Lecture, 1972*

> For a few days I was beside myself with joyous excitement.
>
> *Albert Einstein's response to his predictions that explained anomalies in the orbit of Mercury*

Finally, there is the tragic case of Ludwig Boltzmann, who was born in 1844 and committed suicide in 1906. Today, Boltzmann's work in physics is highly regarded, primarily because of his contributions to kinetic theory and for the statistical interpretation that he gave to classical thermodynamics. A large part of his lifework was related to the atomic theory of matter. Until the 1890s, it was generally agreed among physicists that matter was composed of atoms. But near the end of the nineteenth century, various paradoxes (e.g., specific heats and reversibility) were seen as serious defects of the atomic theory of matter, and Boltzmann came to be referred to as the last defender of atomism. The despondency that led to his suicide may have resulted from the rejection of his work by the physics community. It is one of the most tragic ironies in the history of science that Boltzmann ended his life just before the existence of atoms was finally established by experiments on Brownian motion.

During any experiment you are likely to experience some of the emotions just described, ranging from the disappointment of equipment failure to immense joy when the wiring, oscillator, amplifier, and computer respond in unison to yield a beautiful spectrum.

REFERENCES

1. *Nobel Lectures. Physics*, Vols. 1–4, Elsevier, New York. The four volumes cover from 1901 (the first year the Nobel prize was awarded) to 1970. Unfortunately, later volumes did not exist when this book was written.

2. V. Weisskopf, *Physics Today*, **23** (June 1976). This article is entitled "Is Physics Human?" and is based on the author's response as recipient of the 1976 Oersted Medal of the American Association of Physics Teachers.

3. J. Bardeen, *Physics Today*, **41** (July 1973). This paper is a reprint of Professor Bardeen's 1972 Nobel Lecture.

4. C. Gillispie (Ed. in chief), *Dictionary of Scientific Biography*, Vol. 2, Charles Scribner's Sons, New York, 1970. The biography of Ludwig Boltzmann is on pp. 260–268.

OBJECTIVES OF THE PHYSICS LABORATORY

The basic aims of the laboratory are to have the student do the following:

1. Gain an understanding of some basic physical concepts and theories.
2. Realize that completely functioning experimental apparatus are rarely encountered, and learn how to recognize and correct an equipment malfunction.
3. Gain familiarity with a variety of instruments and learn to make reliable measurements.
4. Learn how precisely a measurement can be made with a given instrument and the size of the measurement error. (See the section on "Error Analysis," page 7.)
5. Learn how to do calculations so that the results have the appropriate number of significant figures. (See the section on "Significant Figures," page 16.)
6. Learn how to analyze data by calculations and by plotting graphs that illustrate functional relations. (See the sections on "Graphical Analysis" and "Curve Fitting," pages 18 and 23.)
7. Learn how to keep an accurate and complete laboratory notebook. (See the following discussion on the "Laboratory Notebook.")
8. Ultimately, learn how best to approach a new laboratory problem.

LABORATORY NOTEBOOK

In general, loose-leaf paper is not appropriate for recording data or doing laboratory calculations. It is recommended that you record data and do calculations directly in your laboratory notebook.

General questions to be considered when writing a lab notebook are "If I pick up this notebook in a year or two, is there enough information in it for me to understand what was done, why, and what the results and conclusions were? Could I reproduce the experiment if I wanted to?"

With these concepts in mind, it is suggested that for each experiment in your lab notebook you record or perform the following eight procedures:

1. First write the title, date, partner, and page numbers in the upper right-hand corner of the appropriate notebook page.

2. Next: state a general purpose in one or two sentences. Throughout the experiment indicate the purpose of each new set of measurements or calculations. (In this instance, the purpose may simply be a statement of exactly what is being measured if the "why" is obvious.)

3. Sketch the apparatus, free hand, but with the parts labeled.

4. Record all original data directly in the laboratory notebook, *not* on scratch paper. The original data readings are the most important pieces of information you have, and their loss should not be risked by recording them on scratch paper. Copying the data wastes valuable time and risks mistakes. Be sure to indicate clearly what is being measured and in what units. You may cross out data that appear to be useless or wrong, but do not erase them—they may turn out to be valuable.

5. Make certain that measured quantities include a figure of uncertainty or "error." (See the following section on "Error Analysis.")

6. Write all calculated values in your notebook with the method of calculation clearly indicated. They will usually appear near the data and may be presented in the form of a table. Each calculated result should include appropriate significant figures. (Refer to the section on "Significant Figures," page 16.)

7. To graph the data follow the guidelines listed in the section on "Graphical Analysis."

8. To record your results and conclusion, tell briefly what you did and how it came out. For example, if you measured a physical constant, how does it compare with the "accepted" value in the light of your estimated errors?

The format of a notebook is not rigid, but it should follow the order in which you worked. As you perform the experiment, you should carry out error, data, and graphical analyses. Such analyses should not be postponed.

ERROR ANALYSIS

REFERENCES

1. H. Young, *Statistical Treatment of Experimental Data*, McGraw-Hill, New York, 1952.
2. J. Taylor, *An Introduction to Error Analysis: The Study of Uncertainties in Physical Measurements*, University Science Books, Mill Valley, CA, 1982.
3. N. Barford, *Experimental Measurements: Precision, Error, and Truth*, Addison-Wesley, Reading, MA, 1967.

The verification of a physical law or the determination of a physical quantity involves measurements. A reading taken from the scale on a voltmeter, a stopclock, or a meter stick, for example, may be directly related by a chain of analysis to the quantity or law under study. Any uncertainty in these readings would result in an uncertainty in the final result. A measurement alone, without a quantitative statement as to the uncertainty involved, is of limited usefulness. It is therefore essential that any course in basic laboratory technique include a discussion of the nature of the uncertainty in individual measurements and the manner in which uncertainties in two or more measurements are propagated to determine the uncertainty in the quantity or law being investigated. Such uncertainties are called experimental errors and their analysis is called error analysis.

We begin our discussion of error analysis with a story. It is a true story; we would not tell any other kind.

MEASUREMENTS, ERRORS, AND A NOBEL PRIZE

In 1974 a SLAC–LBL (Stanford Linear Accelerator–Lawrence Berkeley Laboratory) team was doing a somewhat routine study of the total cross section for positron–electron annihilation as a function of energy. In this experiment positrons and electrons were accelerated to a center-of-mass energy of a few giga electron-volts (GeV) or less and then forced to collide with each other. The team varied the total energy in 200-MeV steps and measured the cross section for each energy.

The measured cross sections were routinely about 23 ± 3 nb (nanobarns) until the center-of-mass energy reached 3.2 GeV, where a value of approximately 30 nb was measured. (The barn, b, originally a tongue-in-cheek name for the "large" area, 10^{-28} m^2, has become the standard unit for expressing nuclear cross sections. It typifies the size of a nucleus.) As you will learn in the discussion on error analysis, 30 nb is more than two standard deviations from the mean value of 23 nb, and the probability of measuring a value that is more than two standard deviations from the mean value is less than 5 percent. Now, a careless physicist might ignore a single data point that is more than two standard deviations from the mean, and perhaps miss a major discovery.

The SLAC–LBL team did not ignore this single data point; instead, they carried out detailed measurements of the cross section from 3.1 to 3.3 GeV, and they discovered a new massive hadron that did not fit into the three-quark classification scheme. This new particle, which they called the ψ particle, has a rest mass of 3.105 ± 0.003 GeV (1974 value), and it is now known to be a bound state of a charmed quark and its antiquark.

The same particle was simultaneously discovered by a MIT–BNL (Massachusetts Institute of Technology–Brookhaven National Laboratory) team and they called it the J particle. It is now called the ψ/J particle.

Historical Note

The 1976 Nobel prize was divided equally between

Burton Richter, the United States, Stanford Linear Accelerator, Stanford, California, and

Samuel C. C. Ting, the United States, Massachusetts Institute of Technology, Cambridge, Massachusetts

For their pioneering work in the discovery of a heavy particle of a new kind.

The *moral of this story*: There may be a Nobel prize lurking beneath what at first appears to be a spurious data point. Thus, explore (take more measurements) before you ignore (the spurious measurement).

REFERENCES

1. J. E. Augustin, A. M. Boyarski, M. Breidenbach, F. Bulos, J. T. Dakin, G. J. Feldman, G. E. Fischer, D. Fryberger, G. Hanson, B. Jean-Marie, R. R. Larsen, V. Luth, H. L. Lynch, D. Lyon, C. C. Morehouse, J. M. Paterson, M. L. Perl, B. Richter, P. Rapidis, R. F. Schwitters, W. M. Tanenbaum, F. Vannucci, G. S. Abrams, D. Briggs, W. Chinowsky, C. E. Friedberg, G. Goldhaber, R. J. Hollebeek, J. A. Kadyk, B. Lulu, F. Pierre, G. H. Trilling, J. S. Whitaker, J. Wiss, J. E. Zipse, *Phys. Rev. Lett.* **33**, 1406 (1974).

2. B. Richter, *Adventures in Experimental Physics* **5**, 143 (1976). The name of this journal has been changed to *Adventures in Sciences*.

3. S. C. C. Ting, *Adventures in Experimental Physics* **5**, 115 (1976).

4. J. J. Aubert, U. Becker, P. J. Biggs, J. Burger, M. Chen, G. Everhart, P. Goldhagen, J. Leong, T. McCorriston, T. G. Rhodes, M. Rohde, S. C. C. Ting, S. L. Wu, Y. Y. Lee, *Phys. Rev. Lett.* **33**, 1404 (1974).

TYPES OF EXPERIMENTAL ERRORS

In the collection of data two types of experimental errors, systematic errors and random errors, usually contribute to the error in the measured quantity.

Systematic errors are due to identifiable causes and can, in principle, be eliminated. Errors of this type result in measured values that are consistently too high or consistently too low. Systematic errors may be of four kinds:

1. *Instrumental.* For example, a poorly calibrated instrument such as a thermometer that reads $102\,°C$ when immersed in boiling water and $2\,°C$ when immersed in ice water at atmospheric pressure. Such a thermometer would result in measured values that are consistently too high.

2. *Observational.* For example, parallax in reading a meter scale.

3. *Environmental.* For example, an electrical power "brown out" that causes measured currents to be consistently too low.

4. *Theoretical.* Due to simplifications of the model system or approximations in the equations describing it—for example, if a frictional force is acting during the experiment but this force is not included in the theory, then the theoretical and experimental results will consistently disagree.

In principle an experimentalist wants to identify and to eliminate systematic errors.

Random errors are positive and negative fluctuations that cause about one-half of the measurements to be too high and one-half to be too low. Sources of random errors cannot always be identified. Possible sources of random errors are as follows:

1. *Observational.* For example, errors in judgment of an observer when reading the scale of a measuring device to the smallest division.

2. *Environmental.* For example, unpredictable fluctuations in line voltage, temperature, or mechanical vibrations of equipment.

Random errors, unlike systematic errors, can often be quantified by statistical analysis; therefore, the effects of random errors on the quantity or physical law under investigation can often be determined.

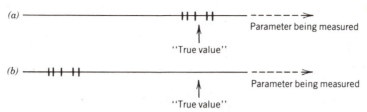

FIGURE I.2 Set of measurements with (a) random errors only and (b) systematic and random errors. Each mark indicates the result of a measurement.

The distinction between random errors and systematic errors can be illustrated with the following example. Suppose the measurement of a physical quantity is repeated five times under the same conditions. If there are only random errors, then the five measured values will be spread about the "true value"; some will be too high and others will be too low, as shown in Figure I.2a. If in addition to the random errors there is also a systematic error, then the five measured values will be spread, not about the true value, but about some displaced value as shown in Figure I.2b.

STATISTICAL ANALYSIS OF RANDOM ERRORS

If a physical quantity, such as a length measured with a meter stick or a time interval measured with a stopclock, is measured many times, then a distribution of readings is obtained because of random errors. For such a set of data the average or mean value $\bar{x}$ is defined by

$$\bar{x} \equiv \frac{1}{n} \sum_{i=1}^{n} x_i \qquad (1)$$

where x_i is the ith measured value and n is the total number of measurements. The n measured values will be distributed about the mean value as shown in Figure I.3. (In many instances $\bar{x}$ approaches the "true value" if n is very large and there are no systematic errors.) A small spread of

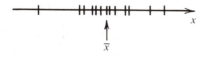

FIGURE I.3 The measured values are distributed about the mean value $\bar{x}$.

measured values about the mean value implies high *precision*.

Now that we have determined the "best value" for the measurement, that is, $\bar{x}$, we need to estimate the uncertainty or error in this value. We start with defining one way in which the spread of data about the mean value can be characterized.

The standard deviation s is defined as

$$s \equiv \sqrt{\frac{1}{n-1} \sum_{i=1}^{n} (x_i - \bar{x})^2} \qquad \text{(units of } x_i\text{)} \quad (2)$$

If the standard deviation is small then the spread in the measured values about the mean is small; hence, the precision in the measurements is high. Note that the standard deviation is always positive and that it has the same units as the measured values.

The error or uncertainty in the mean value, $\bar{x}$, is the **standard deviation of the mean**, s_m, which is defined to be

$$s_m \equiv \frac{s}{n^{1/2}} \qquad (3)$$

where s is the standard deviation and n is the total number of measurements.

The result to be reported is then

$$\bar{x} \pm s_m \qquad (4)$$

The interpretation of equation 4 is that the measured value lies in the range from $\bar{x} - s_m$ to $\bar{x} + s_m$ with a specified probability.

Examples

1. The measurement of the speed of light in a vacuum is an example of the determination of a fundamental constant with high precision. In 1974 the speed of light and its error were

reported as $299\ 792\ 459.0 \pm 0.8$ m/s. Such precise measurements have resulted in an accepted value of $299\ 792\ 458$ m/s.

2. In 1916 Einstein predicted, using his general theory of relativity, that light from a star would be bent through an angle of $1.75''$ as it passed near the sun. A careful classical calculation predicts the angle would be $0.87''$. In 1919 the bending of light was measured to be $2.0 \pm 0.3''$.* Clearly, the measurement was consistent with general relativity and inconsistent with the classical result.

The optical measurement of the bending of starlight by the sun is difficult for the following reasons. The experiment involves taking a photograph of the star field near the sun during a total eclipse and then taking a photograph of the same star field several months later. The two photographic plates are then compared to determine the gravitational displacement of the star positions. The main difficulty with these experiments is the conditions under which the first photographs are taken. The eclipse photographs must be taken at a temporary field observatory in some remote region of earth. During the eclipse the temperature drops suddenly, causing a contraction of the telescope and an abrupt change in atmospheric turbulence. Another difficulty is that the second photographs are made several months later, and during the time between exposures the previously exposed plates or the telescope or both may have changed. Systematic errors are hard to avoid in such an experiment.

More recent experiments have used radio telescopes to observe the bending of radiowaves by the sun. The measurements are made when the sun is within about $10°$ of a distant radio source. Unlike the optical experiment, the radio experiment does not require a solar eclipse. In 1976, the mean gravitational deflection of radiowaves by the sun was reported as 1.007 ± 0.009 times the value predicted by general relativity.† A major source of error is the diffraction of the radiowaves by the solar corona.

To more fully discuss the spread or distribution of measured values about the mean value it is useful to consider the Gauss distribution.

* Dyson, Eddington, and Davidson, *Philos. Trans. R. Soc. A* **220**, 291 (1920).
† Fomalont and Sramek, *Phys. Rev. Lett.* **36**, 1475 (1976).

GAUSS DISTRIBUTION

Figure I.4a shows the distribution of measured values about the mean value for two sets of data, where each set is n repeated measurements of the same physical quantity. The x axis has been divided into increments of width Δx and each dot indicates a measured value. The dots are spread vertically for clarity. The data on the left are more closely clustered about the mean; hence, they imply higher precision. In Figure I.4b the number of measured values, $N(x)$, in an increment Δx centered on x is plotted vertically. Note that the curve on the left, which corresponds to the data of higher precision, is more sharply peaked than the curve on the right. The smooth curves, which are nonsymmetrical in Figure I.4b, are drawn to illustrate the approximate dependence of the number of measured values $N(x)$ on x. If the number of measurements n becomes very large, then the measured values are symmetrically distributed about the mean value, as shown in Figure I.4c. For very large n the **standard deviation** is denoted by σ. Each curve in Figure I.4c, $N(x)$ versus x, represents the frequency with which the value x is obtained as the result of any *single* measurement. Ideally, the analytical expression for such curves is

$$N(x) = \frac{n}{(2\pi)^{1/2}\sigma} \exp\left[-\frac{(x - \bar{x})^2}{2\sigma^2}\right] \tag{5}$$

where n is the very large number of measurements, $\bar{x}$ is the mean value, and σ is the standard deviation. Equation 5 is the **Gauss (or normal) distribution**.

For a very large number of measurements n, the normal or Gauss distribution is the theoretical distribution of measured values x about the mean value $\bar{x}$. If the measurements are carried out with high precision, then σ will be small and the Gauss distribution will be sharply peaked at the mean value $\bar{x}$. See Figure I.5.

If both sides of equation 5 are divided by n and defining $N(x)/n$ to be $P(x)$, we have

$$P(x) = \frac{1}{(2\pi)^{1/2}\sigma} \exp\left[-\frac{(x - \bar{x})^2}{2\sigma^2}\right] \tag{6}$$

The probability of obtaining the value x as a result of any *single* measurement is given by $P(x)$. Note that the most probable value resulting from any single measurement is the mean value $\bar{x}$.

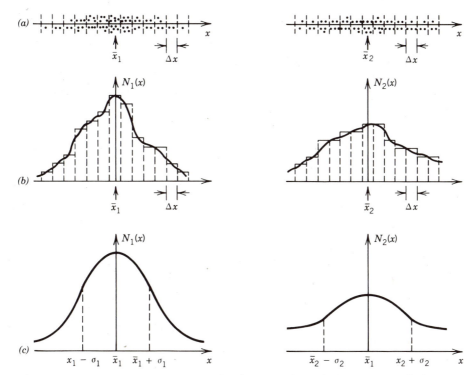

FIGURE 1.4 Two sets of measurements for the same physical quantity. (*a*) Each dot indicates the result of a measurement. The dots are spread vertically for clarity. (*b*) $N(x)$ is the number of measured values in an increment Δx centered on x. (*c*) For very large n the distribution of measured values about the mean value is the normal or Gauss distribution, and $\bar{x}_1$ and $\bar{x}_2$ approach the same value, the "true value." The data of higher precision have a smaller standard deviation, that is, $\sigma_1 < \sigma_2$.

The Gauss distribution has the property that 68 percent of the measurements will fall within the range from

$$\bar{x} - \sigma \quad \text{to} \quad \bar{x} + \sigma \qquad (7)$$

and 95 percent of the measurements will fall

within the range from

$$\bar{x} - 2\sigma \quad \text{to} \quad \bar{x} + 2\sigma \qquad (8)$$

EXERCISE 1

The probability of a measurement being within one standard deviation of the mean is given by

$$P\,(\text{within }\sigma) = \int_{\bar{x}-\sigma}^{\bar{x}+\sigma} P(x)\,dx$$

where $P(x)$ is given by equation 6. Defining $z \equiv (x - \bar{x})/\sigma$, show that the above integral may be written as

$$P\,(\text{within }\sigma) = \frac{1}{(2\pi)^{1/2}} \int_{-1}^{1} e^{-z^2/2}\,dz$$

The integral cannot be evaluated analytically, but it is readily evaluated with a hand calculator. Use your calculator to show that $P\,(\text{within }\sigma) = 0.68$.

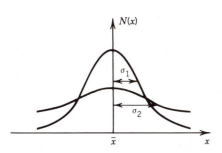

FIGURE 1.5 Gauss distributions for the same $\bar{x}$ and different values of σ. σ_1 is smaller than σ_2 and implies data of higher precision.

Summary of How to Treat Random Errors

The *best ways* to treat random errors can be summarized as follows:

1. Repeat the measurement n times. The measurement could be, for example, a reading of an ammeter to measure electrical current or a reading of a stopclock to measure a time interval.

2. Calculate the mean value $\bar{x}$, the standard deviation s, and the standard deviation of the mean s_m. The result to be reported is $\bar{x} \pm s_m$.

3. In the limit when $n \to \infty$, then $\bar{x} \to$ "true value," $s \to \sigma$ and $s_m \to \sigma_m$, where $\sigma_m = \sigma/\sqrt{n}$.

A word of caution is in order pertaining to the distribution of measured values about the mean value. The distribution of measurements is often describable by the Gauss distribution; however, not all distributions follow the Gauss distribution. In the preceding discussion we have assumed that the measurements are describable by the Gauss distribution, as often is the case.

REVIEW EXERCISES

For each exercise calculate: (a) the mean value, (b) the standard deviation s, (c) the standard deviation of the mean s_m, (d) the percentage error, and (e) the result to be reported.

1. The data are a set of measurements of the length of a sheet of paper, made with a 30-cm rule.

   $\ell_1 = 27.94$ cm $\ell_2 = 27.96$ cm $\ell_3 = 27.99$ cm

   $\ell_4 = 27.97$ cm $\ell_5 = 28.00$ cm $\ell_6 = 27.93$ cm

   $\ell_7 = 27.96$ cm $\ell_8 = 27.98$ cm

   Answers

   (a) $\bar{\ell} = 27.97$ cm
   (b) $s = 0.02$ cm
   (c) $s_m = 0.01$ cm
   (d) $\frac{s_m}{\bar{\ell}} \times 100$ percent $= 0.04$ percent
   (e) $\bar{\ell} \pm s_m = 27.97 \pm 0.01$ cm

We round off the calculated values, keeping an appropriate number of significant figures. (See the section on "Significant Figures," page 16.)

2. In Experiment 2 you are asked to slide a probe along a longitudinal slot in a waveguide and thereby to measure the electric field strength of the electromagnetic field in the waveguide. Points of minimum and maximum field alternate for each quarter-wavelength the probe is moved. For a certain frequency of the field, the alternating minima and maxima were observed at the following probe positions (in cm): 2.84, 3.59, 4.31, 5.09, 5.86, 6.57, 7.36, 8.14, 8.85, 9.55. There are several ways to analyze the data to determine the wavelength λ. Two ways are suggested:

   (a) Subtract the first measurement from the second, the third from the fourth, and so forth, thus obtaining five values of $\lambda/4$.
   (b) Subtract the first from the sixth, the second from the seventh, and so forth, thus obtaining five values of $5\lambda/4$.

   For both (a) and (b) calculate the mean value of λ, the standard deviation, and so forth, to show that method a yields a fractional error that is about five times greater than method b. This result depends on your keeping the appropriate number of significant figures in both calculations.

In general, in this text we will not use the above-described best ways to determine random error because we seldom have the time and resources to make so many measurements, and because a simpler method of estimation of random error, which is described next, is usually adequate.

ESTIMATION OF RANDOM ERROR

We will estimate measurement errors in a somewhat subjective way based on judgment and experience. For example, we know that the ERROR IN A GIVEN INSTRUMENT IS LIKELY TO BE ABOUT THE SAME SIZE AS THE SMALLEST SEPARATE DIVISION ON THAT INSTRUMENT; but instruments vary widely in the reliability of that smallest division, and some judgment must come into play. If we measure the position of a mark to be 92.4 cm, using a meter

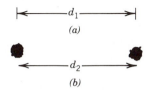

FIGURE I.6 The estimated error in d_2 is likely to be larger than that of d_1 because of the uncertainty in estimating the centers of the globs.

stick whose smallest division is a millimeter, then we might write the result as 92.4 ± 0.1 cm.

Figure I.6 illustrates one way judgment must be exercised when one estimates measurement errors. The distance d_1 is the separation of the sharply defined vertical lines, and d_2 is the distance from center to center of the two "globs." Even though we may measure d_1 and d_2 with the same ruler (hence, the smallest separate division is the same), the error in d_2 is likely to be larger than that of d_1 because of the uncertainty in locating the centers of the globs.

PROPAGATION OF ERRORS

The propagation of errors is simply a method to determine the error in a value where the value is calculated by using two or more measured values with known estimated errors. This method will be discussed separately for the addition and subtraction of measurements, and for the multiplication and division of measurements.

Suppose that x, y, and z are three measured values and the estimated errors are δx, δy, and δz. The results of the three measurements would be reported in the form

$$x \pm \delta x \qquad y \pm \delta y \qquad z \pm \delta z \qquad (9)$$

where each estimated error may be the smallest division of the measuring instrument.

If w is some known function of the measured values, $w(x, y, z)$, then we may calculate w and its error δw. To determine δw, we start by calculating the differential dw:

$$dw = \frac{\partial w}{\partial x} dx + \frac{\partial w}{\partial y} dy + \frac{\partial w}{\partial z} dz \qquad (10)$$

We approximate the differentials in equation 10

with deltas, that is,

$$\delta w = \frac{\partial w}{\partial x} \delta x + \frac{\partial w}{\partial y} \delta y + \frac{\partial w}{\partial z} \delta z$$

(incorrect as the error in w)   (11)

We assumed that the estimated errors δx, δy, and δz are random errors, and statistical theory shows that δw is the square root of the sum of squares:

$$\delta w = \sqrt{\left(\frac{\partial w}{\partial x} \delta x\right)^2 + \left(\frac{\partial w}{\partial y} \delta y\right)^2 + \left(\frac{\partial w}{\partial z} \delta z\right)^2} \qquad (12)$$

Equation 12 is the **basic formula** for error propagation.

Addition and Subtraction of Measurements

Suppose that

$$w = ax + by + cz \qquad (13)$$

where a, b, and c are known positive or negative constants, and x, y, and z are measured values with known estimated errors δx, δy, and δz.

We obtain δw by using equation 12

$$\delta w = \sqrt{(a\,\delta x)^2 + (b\,\delta y)^2 + (c\,\delta z)^2} \qquad (14)$$

Note that δw is independent of the sign of the constants, since each constant is squared in equation 14. If one of the estimated errors is significantly larger than the others, then we may ignore the others; for example, if $b\,\delta y$ is the largest, then

$$\delta w \simeq \sqrt{(b\,\delta y)^2} = b\,\delta y \qquad (15)$$

Equation 15 is a quick way to approximate the error in w, and it is suggested that this method be used while working in the laboratory. The final error analysis should be carried out by using equation 14.

Examples

1. Suppose that three measured lengths and their estimated errors are

$$\ell_1 \pm \delta\ell_1 = 23.5 \pm 0.1 \text{ cm}$$

$$\ell_2 \pm \delta\ell_2 = 17.8 \pm 0.2 \text{ cm}$$

$$\ell_3 \pm \delta\ell_3 = 93.9 \pm 0.2 \text{ cm}$$

If the quantity to be calculated, L, is defined to be

$$L = \ell_1 + 2\ell_2 - \ell_3$$

then

$$L \pm \delta L = -34.8 \pm 0.5 \text{ cm}$$

where δL was calculated by using equation 14.

2. Suppose two time intervals and their errors are

$$t_1 \pm \delta t_1 = 0.743 \pm 0.005 \text{ s}$$

$$t_2 \pm \delta t_2 = 0.384 \pm 0.005 \text{ s}$$

If the total time t is defined to be

$$t = 2t_1 + 5t_2$$

then

$$t \pm \delta t = 3.406 \pm 0.027 \text{ s}$$

Multiplication and Division of Measurements

Suppose that

$$w = kx^a y^b z^c \qquad (16)$$

where k, a, b, and c are constants, positive or negative. To find δw, we apply the basic formula for error propagation, equation 12:

$$\delta w = [(kax^{a-1}y^b z^c)^2 (\delta x)^2 + (kx^a by^{b-1} z^c)^2 (\delta y)^2$$
$$+ (kx^a y^b cz^{c-1})^2 (\delta z)^2]^{1/2} \qquad (17)$$

Equation 17 becomes more useful if we divide both sides by w, where on the right side we replace w with $kx^a y^b z^c$:

$$\frac{\delta w}{w} = \frac{1}{kx^a y^b z^c} [(kax^{a-1}y^b z^c)^2 (\delta x)^2$$
$$+ (kx^a by^{b-1} z^c)^2 (\delta y)^2$$
$$+ (kx^a y^b cz^{c-1})^2 (\delta z)^2]^{1/2} \qquad (18)$$

By squaring $1/kx^a y^b z^c$, moving it under the square root, and simplifying, we obtain

$$\frac{\delta w}{w} = \sqrt{\left(\frac{a\,\delta x}{x}\right)^2 + \left(\frac{b\,\delta y}{y}\right)^2 + \left(\frac{c\,\delta z}{z}\right)^2} \qquad (19)$$

Finally,

$$\delta w = w \sqrt{\left(\frac{a\,\delta x}{x}\right)^2 + \left(\frac{b\,\delta y}{y}\right)^2 + \left(\frac{c\,\delta z}{z}\right)^2} \qquad (20)$$

In equation 20 note that each term of the form $\delta q/q$ is a dimensionless fraction error multiplied by the appropriate exponent. We may estimate the error in the result, δw, by ignoring all but the largest fractional error in the measurements. For example, if $b\,\delta y/y$ is the largest, then

$$\delta w \simeq w \frac{b\,\delta y}{y} \qquad (21)$$

Equation 21 is a quick approximation that is useful while one carries out laboratory work. However, the final error analysis should be carried out by using equation 20.

Examples

1. Suppose the gravitational force, F, where $F = Gm_1 m_2/r^2$, is to be calculated and the measured values and their errors are

$$m_1 \pm \delta m_1 = 19.7 \pm 0.2 \text{ kg}$$

$$m_2 \pm \delta m_2 = 9.4 \pm 0.2 \text{ kg}$$

$$r \pm \delta r = 0.641 \pm 0.009 \text{ m}$$

$$G = 6.67 \times 10^{-11} \text{ Nm}^2 \text{ kg}^{-2}$$

Then,

$$F = 3.0 \times 10^{-8} \text{ N}$$

and

$$\delta F = F \sqrt{\left(\frac{\delta m_1}{m_1}\right)^2 + \left(\frac{\delta m_2}{m_2}\right)^2 + \left(-2\frac{\delta r}{r}\right)^2}$$
$$= 1.10 \times 10^{-9} \text{ N}$$

The result to be reported is

$$F \pm \delta F = (3.0 \pm 0.1) \times 10^{-8} \text{ N}$$

2. Suppose the index of refraction of glass, n_g, is to be calculated by using Snell's law of refraction:

$$n_g = n_1 \frac{\sin \theta_1}{\sin \theta_2}$$

where the measured values and their errors are

$$n_1 = \text{index of air} = 1.000$$

$$\theta_1 \pm \delta\theta_1 = 61 \pm 2°$$

$$\theta_2 \pm \delta\theta_2 = 36 \pm 1°$$

Then

$$n_g = 1.5$$

and to calculate δn_g it is best to start with the basic formula for error propagation, since n_g is not expressible as sums or products of directly measured quantities as in equations 13 or 16. Equation 12 in this case becomes

$$\delta n_g = \sqrt{\left(\frac{\partial n_g}{\partial \theta_1}\delta\theta_1\right)^2 + \left(\frac{\partial n_g}{\partial \theta_2}\delta\theta_2\right)^2}$$

$$= \sqrt{\left(\frac{n_1 \cos\theta_1}{\sin\theta_2}\delta\theta_1\right)^2 + \left(-\frac{n_1 \sin\theta_1 \cos\theta_2}{\sin^2\theta_2}\delta\theta_2\right)^2}$$

where $\delta\theta_1$ and $\delta\theta_2$ must be in (dimensionless) radians. Dividing both sides by n_g, where on the right side we write n_g as $n_1 \sin\theta_1/\sin\theta_2$, and simplifying we obtain

$$\frac{\delta n_g}{n_g} = \sqrt{\left(\frac{\cos\theta_1}{\sin\theta_1}\delta\theta_1\right)^2 + \left(\frac{\cos\theta_2}{\sin\theta_2}\delta\theta_2\right)^2}$$

Evaluating the right side and multiplying by n_g, we find

$$\delta n_g = 0.0490 \times 1.5 = 0.07$$

Finally, after rounding off to an appropriate number of significant figures, we determine that

$$n_g \pm \delta n_g = 1.5 \pm 0.1$$

SIGNIFICANT FIGURES

The significant figures in a number are all of the figures that are obtained directly from the measuring process and that exclude those zeros that are included solely for the purpose of locating the decimal point. This definition can be illustrated with several examples (see table). In this table, the number 2, for example, implies that the measured value is between 1.5 and 2.5, and the precision is about 0.5/2 or 25 percent.

A measurement and its experimental error should have their last significant digits in the same location (relative to the decimal point). Examples are 54.1 ± 0.1, 121 ± 4, 8.764 ± 0.002, and $(7.63 \pm 0.10) \times 10^3$.

HANDLING OF SIGNIFICANT FIGURES IN CALCULATIONS

Properly, the correct number of significant figures to which a result should be quoted is obtained via error analysis. However, error analysis takes time, and frequently in actual laboratory practice it is postponed. In such a situation, one should retain enough significant figures that round-off error is no danger, but not so many as to constitute a burden. Here is an example:

$$0.91 \times 1.23 = 1.1 \quad \textbf{(WRONG)}$$

In this instance, the numbers 0.91 and 1.23 are

Measured Value	Number of Significant Figures	Remarks
2	1	Implies = 25 percent precision
2.0	2	Implies = 2.5 percent precision
2.00	3	Implies = 0.25 percent precision
0.136	3	Leading zero is not necessary, but it does make the reader notice the decimal point.
2.483	4	
2.483×10^3	4	
310	2 or 3	Ambiguous. The zero may be significant or it may be present only to show the location of the decimal point.
3.10×10^2	3	No ambiguity
3.1×10^2	2	

known to about 1 percent, whereas the result, 1.1, is defined to about 10 percent. In this extreme case, the accuracy of the result is reduced by almost a factor of 10, because of round-off error. Now, a factor of 10 in accuracy is usually precious and expensive, and it must not be thrown away by careless data analysis.

$$0.91 \times 1.23 = 1.1193 \qquad \textbf{(WRONG)}$$

The extra digits, which are not really significant, are just a burden, and they also carry the incorrect implication of a result of absurd accuracy.

$$0.91 \times 1.23 = 1.12 \qquad \text{(Okay)}$$
$$0.91 \times 1.23 = 1.119 \qquad \text{(Less good, but still acceptable)}$$

In multiplication or division it is often acceptable to keep the same number of significant figures in the product or quotient as are in the least precise factor. Examples are as follows:

$$2.6 \times 31.7 = 82.42 = 82$$
$$5.3 \div 748 = 0.007085 = .0071$$

The handling of significant figures in addition and subtraction can be illustrated with these examples:

$\begin{array}{r} 51.4 \\ -\ 1.67 \\ \hline 49.73 \to 49.7 \end{array}$	$\begin{array}{r} 7146 \\ -\ 12.8 \\ \hline 7133.2 \to 7133 \end{array}$
$\begin{array}{r} 20.8 \\ 18.72 \\ +\ .851 \\ \hline 40.371 \to 40.4 \end{array}$	$\begin{array}{r} 1.4693 \\ 10.18 \\ +\ 1.062 \\ \hline 12.7113 \to 12.71 \end{array}$

Note that the answer is appropriately rounded off in each example.

REVIEW EXERCISES

In each exercise, measured values and their errors are given. A quantity to be calculated is defined in terms of the measured quantities. Calculate: (a) the defined quantity, (b) the error in the defined quantity, and (c) the percentage error in the defined quantity.

1. The quantity to be calculated is L where $L = \ell_1 - a\ell_2/\ell_3$. The constant a is precisely 2 cm, and measured values and their errors are

$$\ell_1 \pm \delta\ell_1 = 17.4 \pm 0.2 \text{ cm}$$
$$\ell_2 \pm \delta\ell_2 = 9.76 \pm 0.05 \text{ cm}$$
$$\ell_3 \pm \delta\ell_3 = 11 \pm 1 \text{ cm}$$

Answers

(a) $L = 15.6$ cm
(b) $\delta L = 0.3$ cm
(c) $\dfrac{\delta L}{L} \times 100$ percent = 2 percent

2. The quantity to be calculated is K, the kinetic energy, where $K = \frac{1}{2}mv^2$. The measured values and their errors are

$$m \pm \delta m = (1.25 \pm 0.05) \times 10^{-1} \text{ kg}$$
$$v \pm \delta v = 0.87 \pm 0.01 \text{ m/s}$$

Answers

(a) $K = 4.73 \times 10^{-2}$ J
(b) $\delta K = 0.22 \times 10^{-2}$ J
(c) $\dfrac{\delta K}{K} \times 100$ percent = 5 percent

3. The quantity to be calculated is T, the period for underdamped harmonic motion, where

$$T = 2\pi \left/ \sqrt{\frac{k}{m} - \left(\frac{b}{2m}\right)^2} \right. \qquad \text{(s)}$$

and the measured values and their errors are

$$k \pm \delta k = 0.11 \pm 0.01 \text{ N/m}$$
$$m \pm \delta m = 0.500 \pm 0.005 \text{ kg}$$
$$b \pm \delta b = 0.062 \pm 0.008 \text{ kg/s}$$

Answers

(a) $T = 14$ s
(b) $\delta T = 1$ s
(c) $\dfrac{\delta T}{T} \times 100$ percent = 7 percent

4. The quantity to be calculated is a_x, the component of acceleration down a frictionless incline, where $a_x = g \sin\theta$ and the measured values and their errors are

$$g \pm \delta g = 978 \pm 3 \text{ cm/s}^2$$
$$\theta \pm \delta\theta = 0.859 \pm 0.002°$$

Answers

(a) $a_x = 14.7$ cm/s^2
(b) $\delta a_x = 0.1$ cm/s^2
(c) $\dfrac{\delta a_x}{a_x} \times 100$ percent = 0.7 percent

GRAPHICAL ANALYSIS

A purpose of many experiments is to find the relationship between measured variables. A good way to accomplish this task is to plot a graph of the data and then analyze the graph. These guidelines should be followed in plotting your data:

1. Use a sharp pencil or pen. A broad-tipped pencil or pen will introduce unnecessary inaccuracies.

2. Draw your graph on a full page of graph paper. A compressed graph will reduce the accuracy of your graphical analysis.

3. Give the graph a concise title.

4. The dependent variable should be plotted along the vertical (y) axis and the independent variable should be plotted along the horizontal (x) axis.

5. Label axes and include units.

6. Select a scale for each axis and start each axis at zero, if possible.

7. Use error bars to indicate errors in measurements, for example,

Data point ‖↕ Error range

8. Draw a smooth curve through the data points. If the errors are random, then about one-third of the points will not lie within their error range of the best curve.

The microcomputer is a powerful tool for data analysis. Commercial software is available that handles data and instructs the microcomputer to carry out graphical analysis. See your instructor about the availability of this software for your laboratory.

As an example consider the study of the speed of an object (dependent variable) as a function of time (independent variable). The data are as follows:

Speed (m/s)	Time (s)
0.45 ± 0.06	1
0.81 ± 0.06	2
0.91 ± 0.06	3
1.01 ± 0.06	4
1.36 ± 0.06	5
1.56 ± 0.06	6
1.65 ± 0.06	7
1.85 ± 0.06	8
2.17 ± 0.06	9

Using the above guidelines, the data are graphed in Figure I.7.

The graphed data show that the speed v is a linear function of the time t. The general equation for a straight line is

$$y = mx + b \tag{22}$$

where m is the slope of the line and b, the vertical intercept, is the value of y when $x = 0$. Let $v = y$,

18

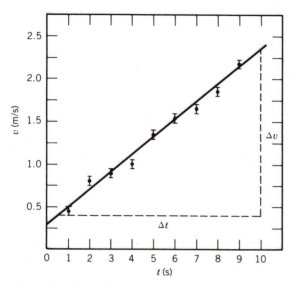

FIGURE I.7 Speed versus time. The graphed data, v versus t, show a linear relation.

$x = t$, $a = m$, and $v_0 = b$; then,

$$v = at + v_0 \quad \text{(m/s)} \qquad (23)$$

This is the form of the equation for the line drawn through the data, where v_0 is the value of the velocity at $t = 0$ and a is the slope of the line that is the acceleration of the object. From the graph we see that $v_0 = 0.32$ m/s. To determine the slope select two points on the line, but not data points, which are well separated, then

$$a = \text{slope} = \frac{\Delta v}{\Delta t} = \frac{2.35 - 0.40 \text{ (m/s)}}{10.0 - 0.5 \text{ (s)}}$$

$$= \frac{1.95 \text{ (m/s)}}{9.5 \text{ (s)}} = 0.20 \text{ m/s}^2 \qquad (24)$$

The equation for the line is

$$v = 0.20t + 0.32 \quad \text{(m/s)} \qquad (25)$$

The data plotted in Figure I.7 are analyzed in the section on "Curve Fitting," page 23, as an example of linear regression.

As a second example, let us consider the study of the distance traveled by an object as a function of time. The data are as follows:

Distance (m)	Time (s)
0.20 ± 0.05	1
0.43 ± 0.05	2
0.81 ± 0.05	3
1.57 ± 0.10	4
2.43 ± 0.10	5
3.81 ± 0.10	6
4.80 ± 0.20	7
6.39 ± 0.20	8

The data are graphed, using the above guidelines, in Figure I.8.

In this instance a straight line through the data points would not be acceptable. An inspection of the graph suggests that d is proportional to t^n, where $n > 1$; for example, d may be a quadratic function of time and, hence, $n = 2$.

Suppose that we know the theoretical relation between d and t is

$$d = \tfrac{1}{2}at^2 \quad \text{(m)} \qquad (26)$$

where a is the object's acceleration. Often it is useful to know if the data agree with the theory. If the data follow the above theoretical relation, then a graph of d versus t^2 should result in a straight line.

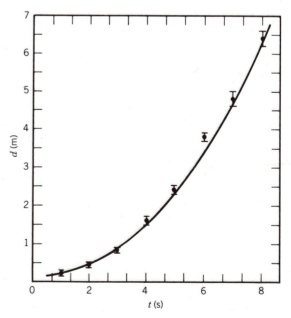

FIGURE I.8 Distance versus time. The graphed data, d versus t, show a nonlinear relation.

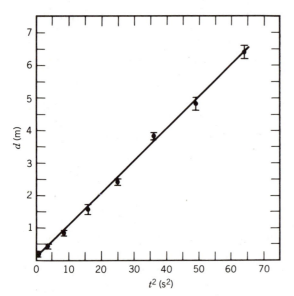

FIGURE I.9 Plotting d versus t^2 yields a linear relation.

The graph in Figure I.9 indicates that d is a linear function of t^2 and, hence, that the data agree with the theoretical relation. The equation for the straight line is

$$d = mt^2 + d_0 \qquad \text{(m)} \qquad (27)$$

where m is the slope and d_0 is the vertical intercept.

PLOTTING DATA ON SEMILOG PAPER

Often the relationship between the measured variables is not linear. For example, consider the intensity of light I transmitted through a sample of thickness x, shown in Figure I.10, where I_0 is the incident intensity of the light.

Lambert's law states the theoretical relationship between the dependent variable I and the independent variable x:

$$I = I_0 e^{-\mu x} \qquad \text{(W/cm}^2\text{)} \qquad (28)$$

FIGURE I.10 I_0 is the incident light intensity, x is the sample thickness, and I is the transmitted intensity.

where μ is the absorption coefficient, a constant that depends on the wavelength of light and the absorbing properties of the sample. Suppose I is measured as a function of x, and the data are plotted as is shown in Figure I.11.

From the smooth curve it would be difficult to determine the relationship between I and x, that is, it would be difficult to conclude the data obey Lambert's law.

A good way to determine the experimental relationship between I and x is to use semilog paper. Semilog paper has a logarithmic y axis (it automatically takes logarithms of data plotted) and a regularly spaced x axis. The data are plotted on semilog paper in Figure I.12. Note that there is never a zero on the logarithmic axis, and that when reading values off of a logarithmic axis you read the logarithm of the value and not the value, for example, log 9 and not 9.

The smooth curve drawn through the data is a straight line with a negative slope and the intensity at the point on the vertical axis intercepted by the curve is I_0. Lambert's law does agree with this result as can be seen by taking the logarithm of Lambert's law:

$$\begin{aligned}
\log I &= \log(I_0 e^{-\mu x}) \\
&= \log e^{-\mu x} + \log I_0 \\
&= -\mu x \log e + \log I_0 \\
&= -0.434\mu x + \log I_0 \qquad \text{(unitless) (29)}
\end{aligned}$$

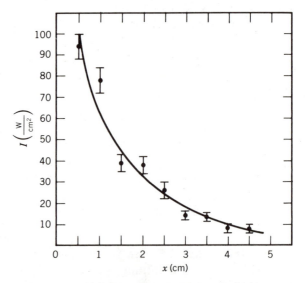

FIGURE I.11 Light intensity versus sample thickness, showing a nonlinear relation. From the graph it is not clear if the data obey Lambert's law or not.

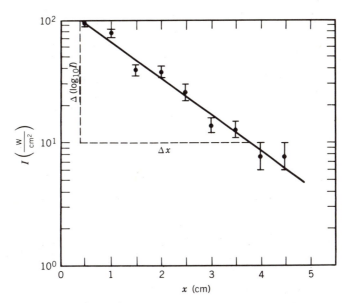

FIGURE I.12 Light intensity versus sample thickness. The linear relation obtained on semilog paper shows that the data obey Lambert's law.

Again, the general equation of a straight line is of the form:

$$y = mx + b \qquad (30)$$

Now let $y = \log I$, $m = -0.434\mu$, and $b = \log I_0$. Then, if $\log I$ is plotted vertically and x is plotted horizontally, the curve will be a straight line with slope -0.434μ and vertical intercept $\log I_0$. Using semilog paper, I is plotted on the logarithmic axis; the vertical intercept on this axis is I_0. Note that the slope of the line drawn through the data points may be used to calculate μ:

$$\text{slope} = \frac{\Delta(\log I)}{\Delta x} = \frac{\log 10 - \log 100}{(3.80 - 0.40)\ \text{cm}} = -0.294\ \text{cm}^{-1} \qquad (31)$$

From Lambert's law the theoretical slope is

$$\text{slope} = -0.434\mu$$

By equating theoretical and experimental slopes, we find that

$$-0.434\mu = -0.294\ \text{cm}^{-1}$$

and

$$\mu = +0.678\ \text{cm}^{-1}$$

EXERCISE 2

Suppose the functional relation between the dependent variable y and the independent variable x is given by

$$y = a e^{-x} + b$$

where a and b are nonzero constants. Explain why a graph of y versus x on semilog paper would not give a straight line.

PLOTTING DATA ON LOG–LOG PAPER

Log–log paper is used to obtain a straight line plot when y and x satisfy a power-law relation:

$$y = cx^n \qquad (32)$$

where c and n are constants. For example, the semimajor axis R of the orbit of a planet is related to its period (time for one revolution around the sun) T:

$$R^3 = KT^2 \quad \text{or} \quad R = K^{1/3}T^{2/3} \qquad (33)$$

where K is a constant. R is nonlinearly related to T.

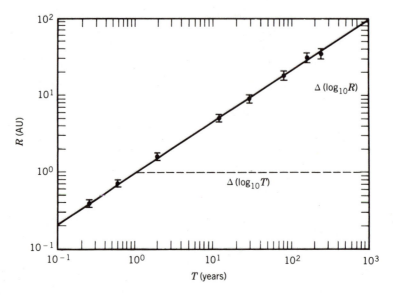

FIGURE I.13 Planets: Semimajor axis versus period. The linear relation
on log–log paper indicates R and T obey a power law of
the form of equation 32.

A straight-line plot is obtained in the following
way. Take logarithms

$$\log R = \log(K^{1/3}T^{2/3})$$
$$= \log T^{2/3} + \log K^{1/3}$$
$$= 2/3 \log T + \log K^{1/3} \qquad (34)$$

Let $y = \log R$, $x = \log T$, $m = \frac{2}{3}$, and $b = \log K^{1/3}$.
Then a plot of $\log R$ versus $\log T$ would be a
straight line. Log–log graph paper automatically
takes the logarithm of the plotted data. A log–log
graph is shown in Figure I.13.

The units used are years and astronomical
units (AU), where 1 AU is the semimajor axis of
earth's orbit. (The errors shown in the graph are
fictitious.) The slope of the log–log plot is

$$\text{slope} = \frac{\Delta(\log R)}{\Delta(\log T)} = \frac{\log 10^2 - \log 10^0}{\log 10^3 - \log 10^0}$$
$$= \frac{2 - 0}{3 - 0} = \frac{2}{3} \qquad (35)$$

Note that the slope of the log–log plot is the
exponent of the power law relation. For example,
the power law relation $y = cx^n$ plotted on log–log
paper has a slope equal to n. Hence, a log–log
plot is a good way to determine the exponent in a
power law relation.

Another way to obtain a straight-line plot is to
plot y versus x^n or R versus $T^{2/3}$ on regular graph
paper (see Figure I.14).

A problem with plotting R versus $T^{2/3}$ is that
values of R less than about 1 AU cannot be
plotted with much accuracy.

In units of years and astronomical units the
constant K is one, and an inspection of the curve
in the figure shows a slope of approximately one.

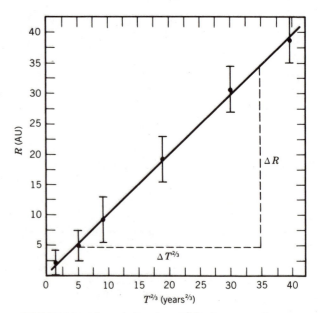

FIGURE I.14 Planets: R versus $T^{2/3}$, showing a linear
relation. This graph requires knowing the
exponent in the power-law relation.

CURVE FITTING

Just as it is important to learn to use a microcomputer to do graphical analysis, it is also important to use the microcomputer to carry out curve fitting. Commercial software is available for curve fitting. See your instructor about the availability of this software for your use.

Given n data points (x_i, y_i), we want to find the equation for the "best" curve for this set of data. If the data points are linearly related, then the process is called **linear regression**. In general, data points are not linearly related and the process of obtaining the equation for the best curve is called **nonlinear regression**. The technique to be used in determining the best-fitting curve is the **method of least squares**.

Before we consider linear and nonlinear regression, we use the method of least squares to determine the best estimate of a quantity x.

Suppose a physical quantity is measured n times, x_i, $i = 1, 2, \ldots, n$. An example is the measurement of a single period of a pendulum n times, where for each measurement the length, mass, and amplitude are constant. The method of least squares states that the best estimate for the result of the n measurements is that which minimizes the sum of the squares of the deviations of the measurements from their best estimate x, that is, we minimize

$$\sum_{i=1}^{n} (x - x_i)^2 \tag{36}$$

where x is the unknown best estimate. Minimizing expression 36 and solving for x, we find that

$$\frac{d}{dx} \sum_{i=1}^{n} (x - x_i)^2 = 0$$

$$2nx - 2 \sum_{i=1}^{n} x_i = 0 \tag{37}$$

$$x = \frac{1}{n} \sum_{i=1}^{n} x_i \equiv \bar{x}$$

Hence, the best estimate is the average or mean value, $\bar{x}$.

Note that minimizing the sum of the squared deviations is equivalent to maximizing the probability $P(x_1, x_2, \ldots, x_n)$ of obtaining our set of measurements $x_1, x_2, \ldots, x_n$. We assume that the data points (x_i) are distributed according to the Gauss distribution; then the probability of obtaining a measurement within an interval dx of x_i is

$$P(x_i) = \frac{1}{\sigma(2\pi)^{1/2}} \exp\left[-\frac{(x - x_i)^2}{2\sigma^2} \right] dx \tag{38}$$

where

$$x = \text{best estimate for } x_i \tag{39}$$

and σ is the theoretical standard deviation.

The probability of obtaining our set of measurements is

$$P(x_1, x_2, \ldots, x_n) = P(x_1)P(x_2) \cdots P(x_n)$$

$$= \left(\frac{dx}{\sigma\sqrt{2\pi}} \right)^n$$

$$\times \exp\left[-\sum_{i=1}^{n} \frac{(x - x_i)^2}{2\sigma^2} \right] \tag{40}$$

If we minimize the exponent in equation 40, then $P(x_1, \ldots, x_n)$ will be a maximum. The sum in the exponent is called the **least-squares sum**,

$$\sum_{i=1}^{n} \frac{(x - x_i)^2}{2\sigma} \qquad (41)$$

and minimizing it is equivalent to minimizing $\Sigma (x - x_i)^2$, since σ is (assumed) a constant.

Note: We assume the data points follow the Gauss distribution, and the method of least squares is used to find the most probable value.

METHOD OF LEAST SQUARES AND LINEAR REGRESSION

Given n data points (x_i, y_i) (for example, x_i could be the time and y_i the average speed of a falling object), we would like to find the equation for the best straight line. Typical data points (x_i, y_i) and the equation of the line, which we want to determine, are shown in Figure I.15. We make the following assumptions:

1. The measured values (x_i, y_i) are distributed according to the Gauss distribution (this is usually so if the errors are random).
2. The errors in x_i, δx_i, are negligible in comparison to the errors in y_i, δy_i (then we only consider the distribution of the values y_i).
3. The errors in y are all the same: $\delta y_1 = \delta y_2 = \cdots = \delta y_n$ (then the standard deviation σ_y is constant).

We approximate the set of n measurements (x_i, y_i)

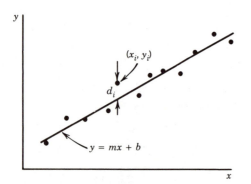

FIGURE I.15 Minimizing the least-squares sum gives the equation for the best straight line.

by a linear relation:

$$y(x) = a_0 + a_1 x \qquad (42)$$

The probability of obtaining the observed value y_i is

$$P(y_i) \propto \frac{1}{\sigma_y} \exp\left[-\frac{[y_i - y(x_i)]^2}{2\sigma_y^2} \right] \qquad (43)$$

where

$$y(x_i) = \text{best estimate for } y_i = a_0 + a_1 x_i \qquad (44)$$

and σ_y is the theoretical standard deviation. The probability $P(y_1, \ldots, y_n)$ of obtaining the set of measurements is

$$P(y_1, \ldots, y_n) = P(y_1)P(y_2) \cdots P(y_n)$$

$$\propto \frac{1}{(\sigma_y)^n} \exp\left[-\sum_{i=1}^{n} \frac{(y_i - a_0 - a_1 x_i)^2}{2\sigma_y^2} \right] \qquad (45)$$

We want this probability to be a maximum; hence, the exponent (least-squares sum) must be a minimum. Minimizing the least-squares sum gives the equation for the best straight line.

In Figure I.15, d_i is the vertical distance from each point (x_i, y_i) to the line $y = a_0 + a_1 x$. We wish to find values of a_0 and a_1 such that we minimize the function $M(a_0, a_1)$ defined to be

$$M(a_0, a_1) = \sum_{i=1}^{n} \frac{d_i^2}{2\sigma_y^2} = \sum_{i=1}^{n} \frac{[y_i - (a_0 + a_1 x_i)]^2}{2\sigma_y^2} \qquad (46)$$

which is the exponent in equation 45. Expanding the squared term and ignoring the (assumed) constant σ_y, we find that

$$M = \Sigma (y_i)^2 - 2a_1 \Sigma x_i y_i - 2a_0 \Sigma y_i$$

$$+ a_1^2 \Sigma x_i^2 + 2a_0 a_1 \Sigma x_i + n a_0^2 \qquad (47)$$

where Σ is understood as a sum over the index i. Next we set

$$\frac{dM}{da_0} = 0 \quad \text{and} \quad \frac{dM}{da_1} = 0 \qquad (48)$$

to find a_0 and a_1 corresponding to the minimum

value of M. This results in two simultaneous equations:

$$\frac{dM}{da_0} = -2 \Sigma y_i + 2a_1 \Sigma x_i + 2na_0 = 0$$

$$\frac{dM}{da_1} = -2 \Sigma x_i y_i + 2a_1 \Sigma x_i^2 + 2a_0 \Sigma x_i = 0 \quad (49)$$

which when solved for a_0 (intercept) and a_1 (slope) yield

$$a_0 = \frac{(\Sigma x_i^2) \Sigma y_i - (\Sigma x_i)(\Sigma x_i y_i)}{n \Sigma x_i^2 - (\Sigma x_i)^2} \quad (50)$$

$$a_1 = \frac{n \Sigma x_i y_i - (\Sigma x_i)(\Sigma y_i)}{n \Sigma x_i^2 - (\Sigma x_i)^2} \quad (51)$$

The equation for the best-fitting line is obtained by substituting equations 50 and 51 into equation 42.

We ask this question: "What are the uncertainties in a_0 and a_1?" Each y_i has an uncertainty (assumed the same for all y_i) and, hence, a_0 and a_1 will both have uncertainties. These uncertainties are the standard deviations of the means, s_{ma_0} and s_{ma_1}. To calculate s_{ma_0} and s_{ma_1}, we need the standard deviation s_y.

We ask the question: "What is the statistical uncertainty in the measurements $y_1, y_2, \ldots, y_n$?" In this case the standard deviation s_y is

$$s_y = \sqrt{\frac{1}{n-2} \sum_{i=1}^{n} (y_i - a_0 - a_1 x_i)^2} \quad (52)$$

The standard deviation of the mean s_{my} is

$$s_{my} = \frac{s_y}{n^{1/2}} \quad (53)$$

For each y_i the result to be reported is

$$y_i \pm s_{my} \qquad i = 1, 2, \ldots, n \quad (54)$$

The reason for the factor of $n - 2$ in the denominator of equation 52 is that the calculation of a_0 and a_1 reduces the number of independent data points (x_i, y_i) from n to $n - 2$; the denominator in the equation for the standard deviation is the number of independent data points.

Remark: It is important to check whether the estimated errors, δy_i, recorded during data taking are consistent with the calculated statistical error s_{my}. A standard deviation of the mean s_{my}, which is much larger than the estimated errors, δy_i, would indicate estimated errors that are unaccounted for. Experimental errors, δy_i, which are much larger than s_{my} suggest a too conservative error estimate, that is, the δy_i should have been estimated as smaller values.

EXERCISE 3

A physicist plans to calibrate her equipment by determining an average value for some parameter x. She does this by measuring four values of x and estimates the error δx. Suppose that the values of $x \pm \delta x$ are 2.741 ± 0.010, 2.832 ± 0.010, 2.678 ± 0.010, 2.763 ± 0.010. Calculate the mean, $\bar{x}$, and the standard deviation of the mean, s_m. Is her estimated error too large, too small, or reasonable? Explain.

We now consider the errors in a_0 and a_1, s_{ma_0} and s_{ma_1}. Equations 50 and 51 give a_0 and a_1 as functions of the measured values (x_i, y_i) where the statistical error for each y_i is given in equation 53. Since a_0 and a_1 are known functions of y_i and the errors in y_i are known, the errors in a_0 and a_1 may be determined by error propagation. The basic formula for error propagation, equation 12, may be written as

$$\delta Q = \sqrt{\sum_{j=1}^{n} \left(\frac{\partial Q}{\partial b_j}\right)^2 (\delta b_j)^2} \quad (55)$$

where the measured values are $b_j \pm \delta b_j$, $j = 1, 2, \ldots, n$, and δQ is the error in the calculated quantity $Q(b_1, b_2, \ldots, b_n)$. Replacing δQ and δb_j with standard deviations of the mean s_{mQ} and s_{mb_j} and squaring, we have

$$s_{mQ}^2 = \sum_{j=1}^{n} \left(\frac{\partial Q}{\partial b_j}\right)^2 s_{mb_j}^2 \quad (56)$$

Applying equation 56, s_{ma_0} is

$$s_{ma_0}^2 = \sum_{j=1}^{n} \left(\frac{\partial a_0}{\partial y_j}\right)^2 s_{my}^2 \quad (57)$$

where the partial derivative $\partial a_0 / \partial y_i$ is calculated by using equation 50:

$$\frac{\partial a_0}{\partial y_j} = \frac{\sum_i x_i^2 - \left(\sum_i x_i\right) x_j}{n \sum_i x_i^2 - \left(\sum_i x_i\right)^2} \quad (58)$$

Then, after some algebra, equation 57 becomes

$$s_{ma_0}^2 = \frac{s_{my}^2 \sum x_i^2}{n \sum x_i^2 - (\sum x_i)^2} \quad (59)$$

The result to be reported is

$$a_0 \pm s_{ma_0} \quad (60)$$

The calculation of $s_{ma_1}^2$ is similar to the calculation of $s_{ma_0}^2$. The result is

$$s_{ma_1}^2 = \frac{n s_{my}^2}{n \sum x_i^2 - (\sum x_i)^2} \quad (61)$$

and we report

$$a_1 \pm s_{ma_1} \quad (62)$$

Example

The method of least squares and linear regression is applied to the speed versus time data given in the section on "Graphical Analysis," p. 18, and plotted in Figure I.7.

The vertical intercept a_0 is calculated using equation 50, where $n = 9$, and the result is

$$a_0 = 0.305 \text{ m/s}$$

The slope a_1 is obtained from equation 51:

$$a_1 = 0.201 \text{ m/s}^2$$

When a_0 and a_1 are known, equations 52 and 53 may be used to calculate s_{my}:

$$s_{my} = 0.025 \text{ m/s}$$

where, in this case, the dependent variable y is the speed v. For each v_i the result to be reported is

$$v_i \pm s_{mv} = v_i \pm 0.025$$

Note that s_{mv} is smaller than the estimated errors $\delta v_i = 0.06$ m/s (see data on p. 18), which suggests the estimated errors were too conservative or too large.

When s_{mv} is known, the uncertainties in a_0 (s_{ma_0}) and a_1 (s_{ma_1}) may be calculated by using equations 59 and 61. The results are

$$s_{ma_0} = 0.018 \text{ m/s}$$

$$s_{ma_1} = 0.003 \text{ m/s}^2$$

Thus,

$$a_0 \pm s_{ma_0} = 0.305 \pm 0.018 \text{ m/s}$$

$$a_1 \pm s_{ma_1} = 0.201 \pm 0.003 \text{ m/s}^2$$

METHOD OF LEAST SQUARES AND NONLINEAR REGRESSION

Given n data points (x_i, y_i), $i = 1, 2, \ldots, n$, that are nonlinearly related, we want to determine the polynomial in x that gives the best fit to the set of n measurements:

$$y(x) = a_0 + a_1 x + a_2 x^2 + \cdots + a_m x^m \quad (63)$$

If a plot of the data or theoretical considerations suggest a quadratic function of x, then we consider only the first three terms in equation 63. We make the same three assumptions as in the method of least squares and linear regression.

As before, equation 43, the probability of obtaining the observed value y_i is

$$P(y_i) \propto \frac{1}{\sigma_y} \exp\left\{-\frac{[y_i - y(x_i)]^2}{2\sigma_y^2}\right\} \quad (64)$$

where

$$y(x_i) = \text{the best estimate for } y_i$$
$$= a_0 + a_1 x_i + \cdots + a_m x_i^m \quad (65)$$

and σ_y is the theoretical standard deviation. The probability of obtaining the set of measurements is

$$P(y_1, y_2, \ldots, y_n) = P(y_1)P(y_2) \cdots P(y_n)$$
$$\propto \frac{1}{(\sigma_y)^n} \exp\left[-\sum_{i=1}^{n} \frac{(y_i - a_0 - a_1 x_i - \cdots - a_m x_i^m)^2}{2\sigma_y^2}\right]$$
$$(66)$$

Again we want the probability to be a maximum; hence, the exponent (least-squares sum) must be a minimum and minimizing the least-squares sum gives the equation for the best fitting curve.

We wish to find the values of $a_0, a_1, \ldots, a_m$

such that we minimize the function M defined to be

$$M = \sum_{i=1}^{n} \frac{(y_i - a_0 - a_1 x_i - \cdots - a_m x_i^m)^2}{2\sigma_y} \quad (67)$$

Taking the partial derivative of M with respect to a_k and setting it equal to zero yields

$$\frac{\partial M}{\partial a_k} = 2 \sum_{i=1}^{n} (y_i - a_0 - a_1 x_i - \cdots - a_m x_i^m) x_i^k = 0 \quad (68)$$

where $k = 0, 1, 2, \ldots, m$. Equation 68 is a set of $m+1$ equations in the $m+1$ variables $a_0, a_1, \ldots, a_m$ which determines the best-fitting curve.

CHI-SQUARE TEST OF FIT

If a measurement is repeated many times then the distribution of measured values is expected to follow a theoretical distribution precisely in the limit that the number of measurements approaches infinity. The Gauss and Poisson distributions are two of many theoretical distributions used in physics, corresponding to different kinds of experiments. (The Poisson distribution is discussed in Experiment 6.)

Suppose we have repeated a measurement n times. We ask the question, "How do we determine whether the measurements follow the expected theoretical distribution?" The chi-square,

or χ^2, test provides the answer to this question. χ^2 is a number, without units, defined by

$$\chi^2 \equiv \sum_{k=1}^{m} \frac{(O_k - E_k)^2}{E_k} \quad (69)$$

where m is the number of bins, O_k is the number of observed or measured values in the kth bin, and E_k is the number of expected values in the kth bin. The n measured values are divided into bins or ranges of values, where the bins must be chosen so that each bin contains several measured values. By assuming that the measurements follow an expected theoretical distribution, such as Gauss or Poisson distribution, we can calculate the expected number E_k of measurements in each bin k:

$$E_k = nP_k \quad (70)$$

where P_k is the probability that any measurement falls in bin k. Figure I.16 shows a Gauss distribution with 6 bins and probabilities $P_1 - P_6$, where $P_1 = P_6 = 0.02$, $P_2 = P_5 = 0.14$, and $P_3 = P_4 = 0.34$ for the Gauss distribution.

The interpretation of χ^2, calculated from equation 69, is as follows:

1. If $\chi^2 = 0$, then the measured values follow the theoretical distribution exactly.
2. If $\chi^2 \leq m - c$, then the agreement between the distribution of measured values and the theoretical distribution is good, where m is the number of bins and c is the number of parameters that had to be calculated from the data to

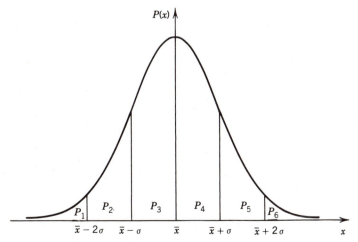

FIGURE I.16 A Gauss distribution with six bins and probabilities P_1 through P_6.

compute the expected number E_k. In statistical calculations $m - c$ is the number of degrees of freedom.

3. If $\chi^2 \gg m - c$, then the agreement is bad.

A more precise interpretation of χ^2 is obtained from a table of values of χ^2.

Example

A distance is measured 20 times. The measured values of x (in cm) are given in Table I.1. The mean value, calculated from equation 1, is $\bar{x} = 16.70$ cm. From equation 2 the standard deviation is $s = 0.16$ cm. To simplify the determination of P_k, we choose the bin boundaries at $\bar{x} - s$, $\bar{x}$, and $\bar{x} + s$, giving four bins as shown in Table I.2. The probability P_k is shown in Figure I.17

TABLE I.1 TWENTY MEASUREMENTS OF THE DISTANCE x

16.7	16.9	16.8	16.7	16.8	16.7	16.6
17.0	16.7	16.7	16.9	16.5	16.3	16.7
16.8	16.7	16.6	16.4	16.7	16.7	

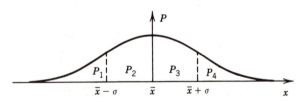

FIGURE I.17 A Gauss distribution with four bins and probabilities P_1 through P_4.

and the expected number E_k is calculated from equation 70 with $n = 20$. If a measured value falls on a bin boundary, then the observed number is determined by alloting 0.5 to each bin. χ^2 is calculated from equation 69, where $m = 4$. The result is $\chi^2 = 0.11$. To calculate E_k, two parameters, $\bar{x}$ and s, had to be determined from the data. In addition,

$$n = \sum_{k=1}^{4} O_k \qquad (71)$$

is a constraint. Hence, $c = 3$ and $m - c = 1$. Since $\chi^2 < 1$, the agreement is good.

The probability obtained from a table of χ^2 values is that, on repeating the series of measurements, larger deviations from the expected values would be observed. In this example the probability, obtained from tables (see reference 1), is between 0.90 and 0.95 that a set of measurements with two degrees of freedom will have $\chi^2 > 0.11$. In other words, if the set of measurements was repeated 100 times then we would expect that 90 to 95 cases would yield values of χ^2 greater than 0.11.

In interpreting the value of P obtained from tables, we may say that if

$$0.1 < P < 0.9 \qquad (72)$$

then the assumed distribution very probably corresponds to the observed one, while if

$$P < 0.02 \quad \text{or} \quad P > 0.98 \qquad (73)$$

then the assumed distribution is very unlikely.

TABLE I.2 DIVIDING THE 20 MEASURED VALUES OF x INTO FOUR BINS FOR A χ^2 CALCULATION

Bin Number, k	1	2	3	4
Range of x in each bin	$x < \bar{x} - s$ or $x < 16.54$	$\bar{x} - s < x < \bar{x}$ or $16.54 < x < 16.70$	$\bar{x} < x < \bar{x} + s$ or $16.70 < x < 16.86$	$\bar{x} + s < x$ or $16.86 < x$
Probability P_k	0.16	0.34	0.34	0.16
Expected number $E_k = nP_k$	3.2	6.8	6.8	3.2
Observed number O_k	3	6.5	7.5	3

JOURNALS OF PHYSICS

HOW TO PUBLISH YOUR RESULTS: SUMMARY OF TWELVE REFERENCED JOURNALS, STYLE MANUAL OF THE AMERICAN INSTITUTE OF PHYSICS

Often the results of research are worthy of publication in a scientific journal. There is a wide spectrum of publishable results, ranging from a cosmetic addendum on previously published work to original, Nobel prize quality, work. The first step in publishing a paper is to decide whether the work is worthy of publication. Your results in this course are more likely to be publishable if you try a different approach and/or go beyond what is asked in a given experiment. You are urged to discuss you results with your instructor and to consider publishing them.

The second step in publishing a paper is to select the appropriate journal. To do so, it is necessary to become familiar with the various physics journals. An important goal in an upper-division laboratory course is gaining familiarity with the journals. Many experiments in this book reference one or more journal articles. Some of the journals referenced are:

1. *American Journal of Physics* (*Am. J. Phys.*). This journal is devoted to the instructional (usually college and university level) and cultural aspects of physical science.

2. *The Physics Teacher*. This journal is dedicated primarily to the teaching of introductory physics at the high school, college, and university level.

3. *Physical Review* (*Phys. Rev.*). Prior to 1970, *Physical Review* was a single journal; in 1970 it split into four journals, *Physical Review A, B, C,* and *D*, with subtitles *General Physics, Condensed Matter* (the subtitle was *Solid State* prior to 1978), *Nuclear Physics,* and *Particles and Fields*, respectively. Papers published in *Physical Review* generally must contain new results, although confirmation of previously published results of unusual importance can be considered as new.

4. *The Journal of Undergraduate Research in Physics.* This journal is for publication of the results of undergraduate students.

5. *Reviews of Modern Physics* (*Rev. Mod. Phys.*). Tutorial articles in rapidly developing fields of physics as well as comprehensive scholarly reviews of significant topics are published in this journal.

6. *Astrophysical Journal* (*Astrophys. J.*). This journal publishes papers that generally contain new results in astronomy, astrophysics, and closely related fields.

7. *Physica.* The level of the papers published in *Physica* is analogous to those published in *Physical Review.* Like *Physical Review, Physica* is divided into journals *A, B,* and so on, but the subtitles are not the same as those of *Physical Review.*

8. *Physics Education* (*Phys. Educ.*). Like *The Physics Teacher*, this journal is dedicated to high school, college, and university level teaching.

9. *Scientific American* (*Sci. Am.*). *Scientific American* publishes timely articles on all aspects of science, and it is intended for both scientists and nonscientists.

10. *Physics Today.* *Physics Today* publishes educational, research, political, and calendar of events information. In addition, it publishes articles of historical, cultural, and scientific interest.

11. *Physical Review Letters* (*Phys. Rev. Lett.*). *Physical Review Letters* is dedicated to publishing short communications dealing with important new discoveries or topics of high current interest in rapidly changing fields of research.

12. *Computers in Physics.* Papers published in this journal focus on computer-based research and computer applications in experimental, computational, and educational areas of physics and allied fields.

You are strongly urged to read the referenced journal articles in each experiment that you perform.

The third step in publishing a paper is to become knowledgeable about the style required by the journal. Most journals have a section entitled "Information for Contributors," which specifies the general style and other details required of a manuscript. The articles in many journals, for example *American Journal of Physics*, *Physical Review*, and *The Journal of Undergraduate Research in Physics*, are written in a common style. The required style of papers submitted to these journals is specified in the *Style Manual of the AIP* (American Institute of Physics) (available from AIP for a few dollars). The general requirements of this style manual are summarized in each January issue of the *American Journal of Physics*.

The first chapter of the *Style Manual of the AIP* is a summary of the entire manual, and it is reproduced below.

It is strongly recommended that at least one experiment be written as a formal report in the style of the *Style Manual of the AIP*.

CHAPTER 1 OF THE STYLE MANUAL OF THE AMERICAN INSTITUTE OF PHYSICS

I. Summary information for journal contributors

A. MANUSCRIPT PREPARATION

The information below is a summary of information given in detailed form in later sections of the manual. Authors should note in addition the particular requirements stated in the information for contributors published in the journal for which they are writing, and the form and style of published articles.

1. General instructions

Manuscripts must be in English (American spelling), typed double spaced throughout, on white bond paper preferably 21.5×28 cm ($8\frac{1}{2} \times 11$ in.) in size. Type on one side of the page only. Leave wide margins at both sides and at top and bottom. Indent paragraphs. Number all pages consecutively, beginning with the title-abstract page. Submit the original manuscript with production-quality figures and one or more duplicate copies (including clear copies of figures), as required by the editor of the journal to which the manuscript is submitted. Illegible manuscripts will be returned.

Include the following material, in the order shown:

(a) title, with the first word capitalized,
(b) author's name, in a form used for all publications,
(c) author's affiliation, including an adequate postal address,
(d) abstract, preferably on the first page with the title,
(e) appropriate Physics and Astronomy Classification Scheme indexing codes (see Appendix I),
(f) text,
(g) acknowledgments,
(h) if necessary, appendices,
(i) collected footnotes (including references) in order of citation,
(j) tables, each with a caption,
(k) collected figure captions,
(l) figures,
(m) if necessary, supplementary material for deposit in AIP's Physics Auxiliary Publication Service (see Appendix J).

2. Abstract

An abstract must accompany each article. It should be a concise summary of the results, conclusions, and/or other significant items in the paper. Together with the title, it must be adequate as an index to all the subjects treated in the paper, and will be used as a base for indexing. It must be complete and intelligible in itself, since it may appear separately in abstract journals. Do not refer by number to a footnote at the end of the paper; define all nonstandard symbols and abbreviations (see Appendices C and D). Type the abstract double spaced, preferably as one paragraph of running text. It should be about 5% of the length of the paper, but never more than 500 words.

3. Mathematics

Type as much of the mathematical material as possible. Handwritten material must be neatly lettered in black ink. When confusion is possible, distinguish between similar looking letters, numbers, and special symbols when they first occur: l (ell) and 1 (one), k (kay) and κ (kappa), $\langle$ (Dirac bra) and $<$ (less than), $\sum$ (summation) and Σ (sigma), $\propto$ (proportional to) and α (alpha), etc. (see Table V). Write the identification in black pencil above the symbol or in the left margin; identify special symbols by their numbers in Appendix F. Identify handwritten script, German, and sans serif letters the first time they appear.

Notation should be clear, as simple as possible, and consistent with standard usage. Avoid multilevel accents (dots above tildes, etc.) and complicated subscripts and superscripts. Type superscripts directly above subscripts (A_i^q), except when a special order is required; note exceptional cases in black pencil in the left margin.

Display all numbered and complicated unnumbered equations on separate lines set off from the text above and below; an equation is "complicated" if it is longer than, say, a word of five syllables, or contains a built-up fraction, an integral or summation sign with limits, or multilevel indices. For equation numbers use arabic numerals in parentheses, placed flush with the right margin; number consecutively throughout the text.

4. Footnotes and references

Type all footnotes (including references) in order of citation as a separate, double-spaced list at the end of the manuscript, before tables and figures. Start with the footnotes to the title, authors' names, and authors' affiliations; for these, the sequence of symbols [a], [b], [c], etc., is used in some journals, while others use the sequence *, †, ‡, §, ||, etc. Check a recent issue of the journal to which the paper is submitted for the correct form. Acknowledgments of financial support should be made in the acknowledgments section of the paper, not as footnotes to the title or to an author's name.

For footnotes in the body of the paper use superscript arabic numerals running consecutively throughout the text: [1], [2], [3], etc. The names of authors in references should be given in the form in which they appear on the title page of the cited work, with the surname last. For journal references use the standard abbreviations for journal names given in Appendix G; give the volume number, the first page number (the last page number may also be included), and the year of publication. For model footnotes, see Table II.

5. Tables

Tabular material more than four or five lines long should be removed from running text and presented as a separate table. Type each table double spaced on a separate page after the list of footnotes and before the collected figure captions. Number tables with roman numerals in order of appearance in the text; be sure to cite every table in the text. Each table must have a caption, complete and intelligible in itself without reference to the text. Column headings should be clear and concise, with appropriate units. Type a double horizontal line below the caption, a single line below the headings, and another double line at the end of the table. For footnotes to a table use the sequence of letters [a], [b], [c], etc., with a new sequence for each table. Place the footnotes themselves below the double line at the end of the table. For a model table, see Table III.

6. Figures and figure captions

Number figures with arabic numerals in order of appearance in the text; be sure to cite every figure in the text. Give every figure a caption, complete and intelligible in itself without reference to the text. Type the list of captions double spaced on a separate page or pages at the end of the manuscript. Place the figures themselves in sequence after the collected captions. Write the figure number and authors' names at the bottom of each figure; if it is necessary to write on the back of a photograph, write very lightly with a soft pencil. To protect figures against damage in transit, make them no larger than 21.5×28 cm ($8\frac{1}{2} \times 11$ in.); mail them flat, well protected by heavy cardboard. Never roll or fold photographs.

In general, figures should be planned for reduction to the journal column width (7.5 cm, or 3 in.). Line drawings should be made with India ink on Bristol board, heavy smooth paper, or high-quality tracing cloth. Use white material only. Draw lines solid and black. Draw symbols and letters so that the smallest ones will be not less than 1.5 mm ($\frac{1}{16}$ in.) tall after reduction; the largest lettering should not be out of proportion. Avoid gross disparities in the thicknesses of lines and in the sizes of symbols and letters. Do not handletter; use a mechanical device instead. Submit original line drawings or, preferably, high-quality glossy prints. For a model line drawing, see Fig. 3.

Submit continuous-tone photographs, which require halftone reproduction, only when line drawings will not serve. The need for color should first be discussed with the editor.

7. Physics Auxiliary Publication Service

Material which is part of and supplementary to a paper but of too limited interest to warrant full publication in the journal should be prepared for deposit in AIP's Physics Auxiliary Publication Service and submitted to the editor along with the paper. See Appendix J.

B. PROCEDURES AND CORRESPONDENCE

1. Correspondence before acceptance

Submit manuscripts directly to the journal editor. Specify in the covering letter which author and address, if there are several, is to be used in correspondence. Also include a signed statement transferring copyright to the journal owner, i.e., to AIP or one of its member societies (see Appendix K).

All manuscripts submitted to journals published by AIP or its member societies are subject to anonymous peer review. The editor chooses the referees and makes the final decision on acceptance. Most manuscripts are returned to authors for revisions recommended by the editor and the referees. Thus it will typically take some months for a paper to be finally accepted. At that point (or earlier) most editors send the author a Publication Charge Certification form for certifying whether or not the publication charge will be honored and for entering orders for reprints. The signed form should be returned to AIP, not to the editor. If the publication charge is not honored or the form not returned, publication of the paper may be delayed, since most journals have only a limited number of pages for unsupported papers.

2. Correspondence after acceptance

After a paper has been accepted, send correspondence about all editorial matters to the office indicated in the notice of acceptance. In all correspondence, reference must be made to the title, author, journal, and scheduled date of publication.

3. Proof

Proof is sent directly from the printer to the author. Check and return proof promptly to the Editorial Supervisor at AIP, not to the printer or editor. Extensive changes from the original are costly and may delay publication while being reviewed by the editor. In keeping with common publication practice, authors will be charged for excessive alterations in proof.

PHYSICS AND ASTRONOMY CLASSIFICATION SCHEME (PACS)

A large number of articles are published each year in physics journals. The PACS indexing scheme is the method used to index articles according to subject matter. When an article is submitted to a journal for publication the author includes a PACS number on the title page of the article. The author should list all of the PACS numbers that apply to the article. Hence, in the index of a journal a single article is often listed separately under more than one subject category.

SUMMARY OF PHYSICS AND ASTRONOMY CLASSIFICATION SCHEME (PACS)

COMMUNICATION, EDUCATION, HISTORY, AND PHILOSOPHY

01.30.B	Publications of Lectures
01.30.C	Conference Proceedings
01.30.E	Monographs and Collections
01.30.K	Handbooks, Dictionaries, Tables and Data Compilations
01.30.La	Physics Laboratory Manuals; Secondary Schools
01.30.Lb	Physics Laboratory Manuals; Undergraduate
01.30.M	Textbooks for Graduates and Researchers
01.30.Pe	Errors in Physics Classroom Materials
01.30.Pp	Textbooks for Undergraduates
01.30.Ps	Textbooks for Students in Grades 9–12
01.30.Pt	Textbooks for Students in Grades K–8
01.30.Q	Books of General Interest to Physics Teachers
01.30.R	Surveys and Tutorial Papers; Resource Letters
01.30.T	Bibliographies
01.40.D	Course Design and Evaluation
01.40.Ea	Science in Elementary School
01.40.Eb	Science in Secondary School
01.40.Ga	Curricula and Evaluation
01.40.Gb	Teaching Methods and Strategies
01.40.H	Learning Theory and Science Teaching
01.40.J	Preservice Teacher Training
01.40.K	Inservice Teacher Training
01.50.Fa	Audio and Visual Aids, Films, Electronic Video Devices
01.50.Fb	Other Instructional Aids; Posters, Cartoons, Art, etc.
01.50.Ha	Instructional Computer Use, Classroom
01.50.Hb	Instructional Computer Use, Laboratory
01.50.J	Computer Software and Software Reviews
01.50.K	Testing Theory and Techniques
01.50.M	Demonstration Experiments and Apparatus
01.50.P	Laboratory Experiments and Apparatus
01.50.Q	Laboratory Course Design, Organization, and Evaluation
01.50.R	Physics Games and Contests
01.50.T	Buildings and Facilities
01.55	General Physics
01.60	Biographical, Historical, and Personal Notes
01.65	History of Science
01.70	Philosophy of Science
01.75	Science and Society
01.76	Physics of Games and Sports
01.77	Careers in Physics and Science
01.90	Other Topics of General Interest

MATHEMATICAL METHODS IN PHYSICS

02.10	Algebra, Set Theory, and Graph Theory
02.40	Geometry, Differential Geometry, and Topology
02.50	Probability Theory, Stochastic Processes, and Statistics
02.60	Numerical Approximation and Analysis
02.70	Computational Techniques
02.90	Other Topics in Mathematical Methods in Physics

CLASSICAL AND QUANTUM PHYSICS: MECHANICS AND FIELDS

03.20	Classical Mechanics of Discrete Systems
03.30	Special Relativity
03.40	Classical Mechanics of Continuous Media
03.65	Quantum Mechanics

GENERAL RELATIVITY AND GRAVITATION

04.01A	Topics on General Relativity and Gravitation

STATISTICAL PHYSICS AND THERMODYNAMICS

05.20.Dd	Kinetic Theory
05.70.Ce	Thermodynamic Functions and Equations of State

MEASUREMENT SCIENCE, GENERAL LABORATORY TECHNIQUES, AND INSTRUMENTATION SYSTEMS

06.20.Dk	Measurement and Error Theory
06.20.Fn	Units
06.20.Hq	Measurement Standards; Calibration
06.20.Jr	Determination of Fundamental Constants
06.30.Bp	Measurement of Spatial Dimensions
06.30.Dr	Mass and Density Measurement
06.30.Ft	Time and Frequency Measurement
06.30.Gv	Velocity, Acceleration, and Rotation Measurement
06.30.Lz	Measurememt of Basic Electromagnetic Quantities
06.50	Data Handling and Computation
06.60	Laboratory Techniques and Safety
06.70.Ep	Testing Equipment
06.70.Hs	Display, Recording, and Indicating Instruments
06.70.Mx	Transducers
06.70.Td	Servo and Control Devices
06.90	Other Topics in Measurement Science, General Laboratory Techniques, and Instrumentation Systems

SPECIFIC INSTRUMENTATION AND TECHNIQUES OF GENERAL USE IN PHYSICS

07.10	Mechanical Instruments and Measurement Methods
07.20	Thermal Instruments and Techniques
07.30.F	Hygrometry
07.30.T	Vacuum Production and Techniques
07.35	High-Pressure Production and Techniques
07.50	Electrical Instruments and Techniques
07.55	Magnetic Instruments and Techniques
07.60	Optical Instruments and Techniques
07.62	Detection of Radiation
07.65	Optical Spectroscopy and Spectrometers

51.30	Thermal Properties of Gases
51.40	Acoustical Properties of Gases
51.50	Electrical Phenomena in Gases
51.60	Magnetic Phenomena in Gases
51.70	Optical Phenomena in Gases
51.90	Other Topics in the Physics of Fluids

THE PHYSICS OF PLASMAS AND ELECTRIC DISCHARGES

52.01A	Topics in Plasmas and Electric Discharges

MECHANICAL AND ACOUSTICAL PROPERTIES OF CONDENSED MATTER

62.10	Mechanical Properties of Liquids
62.20	Mechanical Properties of Solids
62.20D	Elastic Constants
62.30	Mechanical and Elastic Waves
62.60	Acoustical Properties of Liquids
62.65	Acoustical Properties of Solids
62.90	Other Topics in Mechanical and Acoustical Properties of Condensed Matter
63.01A	Lattice Dynamics and Crystal Statistics
64.01A	Equations of State, Phase Equilibria, and Phase Transitions

THERMAL PROPERTIES OF CONDENSED MATTER

65.20	Heat Capacities of Liquids
65.40	Heat Capacities of Solids
65.50	Thermodynamic Properties and Entropy
65.70	Thermodynamic Expansion and Thermomechanical Effects
65.90	Other Topics in Thermal Properties of Condensed Matter
66.00	Transport Properties of Condensed Matter
67.00	Quantum Fluids and Solids; Liquid and Solid Helium
68.00	Surfaces and Interfaces; Thin Films and Whiskers

ELECTRONIC AND MAGNETIC PROPERTIES OF CONDENSED MATTER

71.01A	Electron States
72.01A	Electronic Transport in Condensed Matter
73.01A	Electronic Structure and Electrical Properties of Surfaces, Interfaces, and Thin Films
74.01A	Superconductivity
75.01A	Magnetic Properties and Materials
76.01A	Magnetic Resonances and Relaxations in Condensed Matter; Mossbauer Effect
77.01A	Dielectric Properties and Materials

OPTICAL PROPERTIES AND CONDENSED MATTER SPECTROSCOPY AND OTHER INTERACTIONS OF MATTER WITH PARTICLES AND RADIATION

78.20	Optical Properties and Materials
78.40	Visible and Ultraviolet Spectra
78.65	Optical Properties of Thin Films
78.70	Other Interactions of Matter with Particles and Radiation
79.01A	Electron and Ion Emmission by Liquids and Solids; Impact Phenomena

GENERAL TOPICS IN ELECTRICAL AND MAGNETIC DEVICES

85.10	Electron Tubes
85.20	Conductors, Inductors, and Switches
85.25	Superconducting Devices
85.30	Semiconductor Devices
85.40	Integrated Circuits
85.50	Dielectric Devices
85.60	Photoelectric and Optoelectronic Devices and Systems
85.70	Magnetic Devices
85.80	Electrochemical, Thermoelectromagnetic, and Other Devices
85.90	Other Topics in Electrical and Magnetic Devices

AREAS OF GENERAL INTEREST TO PHYSICISTS

87.01A	Biophysics, Medical Physics, and Biomedical Engineering
89.02	Materials Sciences
89.03	Physical Chemistry
89.04	Chemical Thermodynamics
89.20	Industrial and Technological Research and Development
89.30	Energy Resources
89.31	Direct Energy Conversion and Energy Storage
89.40	Transportation
89.50	Urban Planning and Development
89.60	Environmental and Ecological Studies
89.70	Information Science
89.80	Computer Science and Technology
89.81	Solid Earth Physics
89.82	Hydrospheric and Atmospheric Geophysics
89.83	Geophysical Observations, Instrumentation, and Techniques
89.84	Aeronomy
89.90	Other Areas of General Interest to Physicists

FUNDAMENTAL ASTRONOMY/ASTROPHYSICS; INSTRUMENTATION, TECHNIQUES, AND ASTRONOMICAL OBSERVATIONS

95.10	Fundamental Astronomy
95.20	Historical and Archaeoastronomy
95.30	Fundamental Aspects of Astrophysics
95.40	Artificial Earth Satellites
95.45	Observatories
95.55	Astronomical and Space Research Instrumentation
95.75	Observation and Reduction Techniques
95.80	Catalogues, Atlases, etc.
95.85	Astronomical Observations
95.90	Other Topics in Astronomy and Astrophysics
96.01A	General Topics on the Solar System
97.01A	General Topics on Stars
98.01A	General Topics on Stellar Systems; Galactic and Extragalactic Objects and Systems; The Universe

Source: "Physics and Astronomy Classification Scheme," *The Physics Teacher* **27**, 2 (1989), American Institute of Physics, New York, copyright ©, 1989. Reproduced with permission.

RESPONSIBILITY OF THE EXPERIMENTALIST

When doing an experiment it is important for the experimentalist not to accept the data as correct without adequately questioning it. This is especially true if the experimentalist is not sure what to expect and/or if the data are automatically or semiautomatically recorded. The experimentalist should feel a sense of responsibility for the numbers (data) obtained.

It is suggested that you follow the guidelines listed below as a means of accepting the responsibility for your data.

1. If the data are semiautomatically recorded, then examine the "raw" data closely to determine whether the equipment malfunctioned.

2. After recording the data in your notebook examine the numbers for observational mistakes. Examination of the numbers may also reveal an equipment malfunction that was not previously detected.

3. Before making use of the techniques for propagating errors through the various intermediate steps and into the final result, use *common sense* and ask yourself whether a given experimental result or error is reasonable. If it is not, there is an excellent chance that either the equipment malfunctioned or you made an arithmetic or observational mistake. (Note the distinction between mistake and error.)

4. There are a variety of honest mistakes that an experimentalist can make in recording or in analyzing data, and we briefly discuss some of them. One possible mistake is to assume that an observed signal is due to the sample, when it is actually due to the instrumentation. A technique that may prevent an experimentalist from falling into this trap is to devise and carry out tests for which the experiment or instrumentation should give known results. For example, often a standard sample or a standard test signal may be available for which the experiment or instrumentation should give a known result. Your are urged to devise tests of this sort in order to avoid this kind of mistake. Another mistake is to reach the wrong conclusion in analyzing data. It may be difficult to avoid this pitfall. Perhaps the best advice is to urge caution, especially if you have some doubts. Publishing a mistake and then later publishing a retraction is embarrassing and may damage one's career.

5. Unfortunately, some researchers have intentionally falsified data. Not long ago, a researcher at a medical school was accused of, and the individual subsequently admitted to, falsifying data. The initial investigation indicated that the integrity of six published papers and a large number of published abstracts were in question. As a result of the data fabrication, the National Institutes of Health (NIH), which provided the research funding, recommended that the individual

be barred from receiving federal research funds for the next 10 years, and that the laboratory repay the $122,371 that the NIH provided for the project. The investigators did note that part of the blame lies with the way research is done in the modern laboratory: "A hurried pace and emphasis on productivity, coupled with limited interaction with senior scientists, has contributed to the disappointing events."

Do not let anyone, either directly or indirectly, pressure you into falsifying data. You are responsible for your data.

SPECTROSCOPY: AN IMPORTANT CLASS OF EXPERIMENTS

Spectroscopic experiments are important because so many experiments involve spectroscopy. Note in the Table of Contents, under Fundamental Experiments, that there are nine spectroscopic experiments listed. We would like to present the general ideas of such an experiment.

Historical Note

It is perhaps worth pointing out that many Nobel prizes have been awarded for spectroscopic work; for example, one half of the 1981 prize was awarded jointly to

Nicolaas Bloembergen, the United States, Harvard University, Cambridge, Massachusetts, and

Arthur L. Schawlow, the United States, Stanford University, Stanford, California

For their contribution to the development of laser spectroscopy

and the other one half was awarded to

Kai M. Siegbahn, Sweden, Uppsala University, Uppsala, Sweden

For his contribution to the development of high resolution electron spectroscopy.

A typical spectroscopic experiment includes a source of particles, a sample, and an analyzer. (We include radiation as particles. Radiation, like matter, has both wave and particle properties, and in some experiments it is adequate to treat radiation as waves, while in others it is treated as particles or photons. For example, the wave model of radiation is used in microwave spectroscopy and the particle model is used in gamma-ray spectroscopy.) A block diagram of the apparatus is shown in Figure I.18. Loosely speaking, this diagram represents a spectrometer. A microwave spectrometer is shown in Figure 2.15, Experiment 2. The source of microwaves is the klystron, the sample inserts into the cavity, and the analyzer or detector is the diode. Often the analyzer is connected to a chart recorder, multichannel analyzer, computer, or some other device for recording the signal.

Both elementary particles and nuclei are used as the incident particles in spectrometers; however, electrons and radiation are easily made available and many experiments use one of these as the incident particle. In a typical spectroscopic

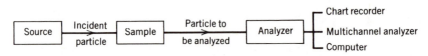

FIGURE I.18 Block diagram of the apparatus used in a spectroscopic experiment.

38

experiment particles in a more or less well-defined state (energy, momentum, polarization, etc.) are incident on a sample and, after the interaction between the incident particle and the sample, particles escape from the sample. The particles that escape carry information about the quantum states of the sample. The information that the experiment yields about these states depends on how thoroughly the escaping particles are analyzed.

One electronics technique of analyzing the escaping particles, called modulation spectroscopy, which allows us to examine, with increased signal-to-noise ratio, the information obtained from the sample is discussed in Appendix A.

Table I.3 is a summary of the spectroscopic experiments that appear in this text.

TABLE I.3 SUMMARY OF SPECTROSCOPIC EXPERIMENTS

Experiment	Source	Incident Particle	Sample	Analyzed Particle	Analyzer
9. Photoelectron spectroscopy	Hg lamp	Photon (visible)	Photocathode (metal or semiconductor)	Electron	Ammeter
10. Bragg spectroscopy	X-ray tube/klystron	X-ray/microwave	NaCl, KCl, Ni/Al spheres	X-ray/microwave	Film/crystal diode
11. Electron spectroscopy	Heated filament	Electron	Mercury atoms	Electron	Ammeter
12. Optical spectroscopy	Cathode	Electron	Hydrogen and deuterium atoms	Photon (visible)	Spectrograph
13. Optical spectroscopy	Cathode	Electron	Nitrogen molecules	Photons (visible)	Spectrograph
14. Optical spectroscopy	Hg, Na, or He discharge tube	Photon (visible)	Hg, Na, or He atoms	Photons (visible)	Spectrograph or spectrometer
15. Radiowave spectroscopy	Oscillator	Radiowaves	Distilled water, mineral oil, Teflon	Radiowaves	Diode in an RC circuit
16. Microwave spectroscopy	Klystron	Microwaves	DPPH, Cr^{3+} in alum	Microwaves	Crystal diode
20. Gamma ray spectroscopy	Radionuclide	Photon (gamma ray)	Aluminum	Photon	Scintillation counter

EXPERIMENTS

LABORATORY INSTRUMENTATION

Transmission of Radiation

---

1. COAXIAL TRANSMISSION LINE. VELOCITY OF PROPAGATION, IMPEDANCE MATCHING

APPARATUS

Oscilloscope
Coaxial line (about 20 m in length)
Coaxial line (about 1 m in length)
Various line terminations
Pulse-generating circuit (see Figure 1.7a)

OBJECTIVES

To study the amplitude of a reflected pulse as a function of load resistance.

To measure the velocity of a pulse propagating on a coaxial transmission line.

To determine the capacitance of a 1-m coaxial line and use the result to calculate a theoretical velocity of a pulse propagating on a coaxial line.

To observe the decrease in amplitude of a multi-reflected pulse and determine the attenuation constant of the coaxial line.

To observe the time dependence of a reflected pulse for loads of various impedance.

To gain an understanding of impedance, impedance matching, and reflection coefficients.

KEY CONCEPTS

TEM mode	Phase constant
Inductance per unit length	Characteristic impedance
Resistance per unit length	Reflection coefficient

Capacitance per unit length   Load impedance
Siemens per unit length   Source impedance
Propagation constant   Impedance matching
Attenuation constant

REFERENCES

1. W. T. Scott, *The Physics of Electricity and Magnetism*, 2d ed., Wiley, New York, 1966. Coaxial lines are discussed on pages 530–542. Admittance, conductance, and impedance are discussed on various pages. See the index.
2. R. K. Wangsness, *Electromagnetic Fields*, Wiley, New York, 1979. Coaxial lines and TEM modes are a part of Chapter 26.
3. R. B. Adler, L. J. Chu, and R. M. Fano, *Electromagnetic Energy Transmission and Radiation*, Wiley, New York, 1966. Coaxial lines are discussed on pages 524–530. Admittance, conductance, and impedance are discussed on various pages. See the index.
4. A. M. Portis, *Electromagnetic Fields: Sources and Media*, Wiley, New York, 1978. Transmission lines are discussed on various pages. See the index.
5. D. Halliday and R. Resnick, *Physics*, Part 2, 3d ed., Wiley, New York, 1978. Coaxial lines are discussed on pages 899–902. Capacitance and inductance per unit length of coaxial line are discussed on pages 654 and 804.
6. *American Institute of Physics Handbook*, McGraw-Hill, New York, 1957. Characteristics of standard radio-frequency cables are listed. Included are the nominal capacitance per foot and approximate impedance.

INTRODUCTION

A coaxial line is a two-conductor transmission line consisting of a center conductor, a dielectric spacer, and a concentric outer conductor. The electric and magnetic field configurations, or modes, are most commonly transverse electromagnetic (TEM) field modes. In a TEM mode both the electric and magnetic fields are entirely transverse to the direction of propagation. A longitudinal section of a coaxial line is shown in Figure 1.1, along with the (radial) electric field **E** and the (concentric circular) magnetic field **B** of a TEM mode.

A coaxial line is often used to transmit energy from a generator to a load. We ask, "How does the line effect the transmission of energy?"

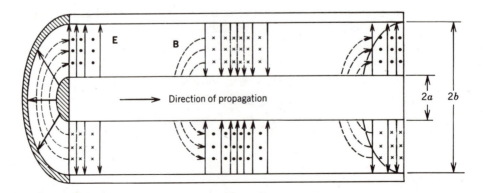

FIGURE 1.1 A TEM mode propagating on a longitudinal section of a coaxial transmission line.

1. A current in the center conductor sets up a magnetic field encircling the conductor. Hence, the line has inductance L'.

2. The line has resistance R', since the conductors making up the line have resistance. R' depends on the "skin depth" and will thus be frequency dependent.

3. There will be a voltage between and charges on the conductors. Hence, the line has capacitance C'.

4. The dielectric between the conductors is not perfect; therefore, it is necessary to associate a conductivity with the dielectric. Hence, the line has a conductance G'.

We shall see that these four parameters determine the impedance of the line, and the velocity, phase, and attenuation of waves propagating on the line. In this experiment, the propagation of voltage and current waves, rather than electric and magnetic field waves, are examined.

The value per unit length of line can be calculated for each parameter. For a coaxial line with an inner conductor of radius a and an outer conductor of inside radius b, the calculated values per unit length of line are

$$L = \frac{\mu}{2\pi} \ln \frac{b}{a} \qquad \text{(H/m)} \tag{1}$$

$$R = \frac{1}{2} \sqrt{\frac{\mu_c \nu}{\pi \sigma_c}} \left(\frac{1}{b} + \frac{1}{a} \right) \qquad \text{(}\Omega\text{/m)} \tag{2}$$

$$C = \frac{2\pi\varepsilon}{\ln(b/a)} \qquad \text{(F/m)} \tag{3}$$

$$G = \frac{\sigma}{\varepsilon} C = \frac{2\pi\sigma}{\ln(b/a)} \qquad \text{(S*/m)} \tag{4}$$

where

μ is the magnetic permeability of the dielectric (H/m),
ε is the electric permittivity of the dielectric (F/m),
σ is the conductivity of the dielectric (S/m),
μ_c is the magnetic permeability of the conductors (H/m),
σ_c is the conductivity of the conductors (S/m), and
ν is the frequency of the waves.

EXERCISE 1

Derive the expressions for the inductance and capacitance per unit length, equations 1 and 3, for a coaxial line.

For the resistance and inductance, all sections of the line are in series; therefore, for a line of length ℓ the total resistance and inductance are $R\ell$ and $L\ell$. For the capacitance and conductance, all sections are in parallel; therefore, for a line of length ℓ the total capacitance and conductance are $C\ell$ and $G\ell$. At very low frequencies the effect of the inductance, capacitance, resistance, and conductance of the line can be taken into account by either

*S stands for siemens after Ernst Werner von Siemens (1816–1892). The siemens was formerly called the mho. One siemens is equal to one ampere/volt.

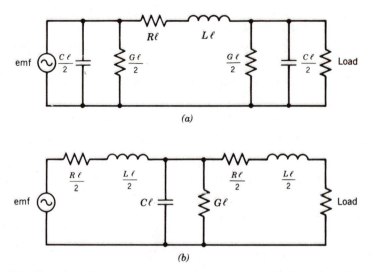

(a)

(b)

FIGURE 1.2 Both circuits are equivalent to a coaxial line when the propagating fields have very low frequencies.

equivalent circuit shown in Figure 1.2. At very high frequencies the impedance presented to the generator is

(i) the capacitive reactance for the circuit in Figure 1.2a

$$\frac{1}{2\pi v C\ell /2} \xrightarrow[v \to \infty]{} 0$$

(ii) the inductive reactance for the circuit in Figure 1.2b

$$\frac{2\pi v L\ell}{2} \xrightarrow[v \to \infty]{} \infty$$

A way of making the impedances the same and obtaining a circuit that is the exact equivalent of the coaxial line is to divide the total inductance, capacitance, resistance, and conductance into an infinite number of infinitesimal elements. The exact equivalent circuit of a coaxial line is shown in Figure 1.3.

The general equations for the current in each conductor and the voltage between the conductors may be obtained by considering a line element of length Δz (Figure 1.4a). Figure 1.4b shows the corresponding length of the coaxial line. If we denote by q the charge per unit length of the conductor, we can write, using superior bars to denote average values in

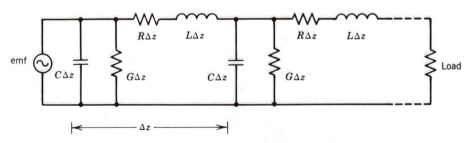

FIGURE 1.3 Equivalent circuit of a coaxial line.

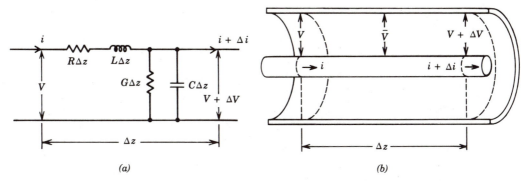

FIGURE 1.4 Current along and voltage across a section of (a) an equivalent circuit of a coaxial line and (b) a coaxial line.

the length Δz,

$$\bar{V} = \frac{\bar{q}\Delta z}{C\Delta z} = \frac{\bar{q}}{C} \qquad \text{(C/F)} \tag{5}$$

The net rate of decrease of the charge in Δz is given by the net current leaving Δz along the wire plus the leakage current $\bar{V}G\Delta z$, so that

$$-\Delta i = \frac{\partial(\bar{q}\Delta z)}{\partial t} + \bar{V}G\Delta z \qquad \text{(C/s)} \tag{6}$$

Substituting from equation 5 into 6, dividing by Δz, and taking limits, we obtain

$$\frac{\partial i}{\partial z} = -C\frac{\partial V}{\partial t} - GV \qquad \text{(C/s · m)} \tag{7}$$

with $\bar{V}$ replaced by V.

The total voltage drop in Δz is given by

$$-\Delta V = L\Delta z\frac{\partial i}{\partial t} + iR\Delta z \qquad \text{(V)} \tag{8}$$

Dividing by Δz and taking limits, we obtain

$$\frac{\partial V}{\partial z} = -L\frac{\partial i}{\partial t} - iR \qquad \text{(V/m)} \tag{9}$$

Differentiating equation 9 with respect to z yields

$$\frac{\partial^2 V}{\partial z^2} = -L\frac{\partial}{\partial z}\left(\frac{\partial i}{\partial t}\right) - R\frac{\partial i}{\partial z}$$

$$= -L\frac{\partial}{\partial t}\left(\frac{\partial i}{\partial z}\right) - R\frac{\partial i}{\partial z} \qquad \text{(V/m}^2\text{)} \tag{10}$$

If equation 7 is used to replace $\partial i/\partial z$, then equation 10 becomes

$$\frac{\partial^2 V}{\partial z^2} = -L\frac{\partial}{\partial t}\left(-C\frac{\partial V}{\partial t} - GV\right) - R\left(-C\frac{\partial V}{\partial t} - GV\right) \qquad \text{(V/m}^2\text{)} \tag{11}$$

Equation 11 may be written

$$\frac{\partial^2 V}{\partial z^2} = RGV + (RC + LG)\frac{\partial V}{\partial t} + LC\frac{\partial^2 V}{\partial t^2} \qquad (\text{V/m}^2) \qquad (12)$$

Equation 12 is the general equation for the transmission of electric signals along a wire. It is called the *telegraph equation* because it describes the transmission of telegraph signals in conductors.

The equation for the current is of the same form as equation 12, and can be obtained by differentiating equation 7 with respect to z.

We can obtain solutions of equation 12 for the case of a harmonically varying wave with a time dependence given by

$$V(z, t) = V(z)\, e^{j\omega t} \qquad (\text{V}) \qquad (13)$$

where $j^2 = -1$. For this case, equation 12 becomes

$$\frac{d^2 V(z)}{dz^2} = RGV(z) + (RC + LG)j\omega V(z) - LC\omega^2 V(z) \qquad (\text{V/m}^2) \qquad (14)$$

where the time dependence cancels.

The net effect of the series resistance and inductance can be expressed by the **series impedance** Z per unit length:

$$Z = R + j\omega L \qquad (\Omega/\text{m}) \qquad (15)$$

The net effect of the shunt conductance and capacitance can be expressed by the **shunt admittance** Y per unit length:

$$Y = G + j\omega C \qquad (\text{S/m}) \qquad (16)$$

Writing equation 14 in terms of Y and Z we have

$$\frac{d^2 V(z)}{dz^2} = ZYV(z) \qquad (\text{V/m}^2) \qquad (17)$$

The equation for $i(z)$ is obtained in a similar manner:

$$\frac{d^2 i(z)}{dz^2} = ZYi(z) \qquad (\text{A/m}^2) \qquad (18)$$

If we try a solution for equation 17 of the form $e^{\gamma z}$, we find

$$V(z) = V_1\, e^{\gamma z} + V_2\, e^{-\gamma z} \qquad (\text{V}) \qquad (19)$$

where V_1 and V_2 are arbitrary constants that can be found from initial or boundary conditions, and γ, called the **propagation constant**, is given by

$$\gamma = \sqrt{ZY} = \sqrt{(R + j\omega L)(G + j\omega C)} \qquad (\text{m}^{-1}) \qquad (20)$$

The propagation constant is complex with a real part α called the **attenuation constant** and an imaginary part β called the **phase constant**:

$$\alpha = \text{Re}(\gamma) \qquad \beta = \text{Im}(\gamma) \qquad (\text{m}^{-1}) \qquad (21)$$

The solution $V(z, t)$ is obtained by multiplying equation 19 by the harmonic time dependence $e^{j\omega t}$:

$$V(z, t) = V_1 e^{\alpha z} e^{j(\omega t + \beta z)} + V_2 e^{-\alpha z} e^{j(\omega t - \beta z)}$$

$$\equiv V_r(z, t) + V_i(z, t) \qquad (V) \qquad (22)$$

The term involving $\omega t + \beta z$ represents a reflected wave $V_r(z, t)$, reflected from the load, traveling in the negative z direction along the transmission line. The factor $e^{\alpha z}$ indicates that this wave decreases in magnitude as it travels in the negative z direction. The term involving $\omega t - \beta z$ represents an incident wave $V_i(z, t)$, incident on the load, traveling in the positive z direction, and the factor $e^{-\alpha z}$ indicates that this wave decreases in magnitude as it travels in the positive z direction. The total voltage $V(z, t)$ at any point z along the line is the superposition of the two traveling waves.

The total current $i(z, t)$ may be obtained by substituting equation 22 into equation 7 and then integrating over z:

$$i(z, t) = -\frac{V_1}{(Z/Y)^{1/2}} e^{\alpha z} e^{j(\omega t + \beta z)} + \frac{V_2}{(Z/Y)^{1/2}} e^{-\alpha z} e^{j(\omega t - \beta z)}$$

$$\equiv i_r(z, t) + i_i(z, t) \qquad (A) \qquad (23)$$

where $i_r(z, t)$ and $i_i(z, t)$ represent reflected and incident current waves that are traveling in the negative and positive directions, respectively. The total current $i(z, t)$ is a superposition of these two traveling waves.

The incident wave, reflected wave, and total wave are shown in Figure 1.5 at three separate points along a section of coaxial line. The load that reflects the reflected wave is to the right and the source of emf that produces the waves is to the left. The outer conductor is shown grounded, which is usually the case, and, hence, current is not shown in the outer conductor. As you will see, V_1, the constant term in the amplitude of the reflected wave, is less than or equal to V_2, the constant term in the amplitude of the incident wave. It is usually less, and letters of different size are used to indicate this in Figure 1.5.

For the waves specified by equations 22 and 23, β equals $2\pi/\lambda$, where λ is the wavelength, and ω/β is the wave velocity.

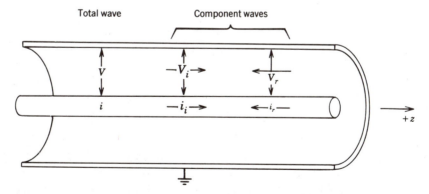

FIGURE 1.5 The incident, reflected, and total waves are shown at three separate points along a coaxial line.

EXERCISE 2

If R and G are small or if the frequency is large so that $\omega L \gg R$ and $\omega C \gg G$, then factor $\omega\sqrt{LC}$ from the right-hand side of equation 20, followed by a binomial expansion, and ignoring small terms in the expansion raised to the second and higher powers, show that

$$\alpha = \text{Re}(\gamma) \simeq \frac{\sqrt{LC}}{2}\left(\frac{G}{C} + \frac{R}{L}\right) \qquad (\text{m}^{-1})$$

$$\beta = \text{Im}(\gamma) \simeq \omega\sqrt{LC} \qquad (\text{m}^{-1})$$

Hence, show that the wave velocity v is given by

$$v = \frac{\omega}{\beta} \simeq \frac{1}{(LC)^{1/2}} = \frac{1}{(\mu\varepsilon)^{1/2}} \qquad (\text{m/s})$$

where the last equality was obtained by using equations 1 and 3. If there is no material between the conductors, then $\mu = \mu_0$, $\varepsilon = \varepsilon_0$, and $v = c$, the velocity of light.

When we confine our attention to a single wave traveling in the positive z direction with voltage $V_i(z, t)$ and current $i_i(z, t)$, the **characteristic impedance** Z_c of the line is defined to be the ratio of the voltage across the line to the current through the line for such a wave:

$$Z_c = \frac{V_i(z, t)}{i_i(z, t)} = \sqrt{\frac{Z}{Y}} \qquad (\Omega) \tag{24}$$

Using equations 15 and 16,

$$Z_c = \sqrt{\frac{R + j\omega L}{G + j\omega C}} \qquad (\Omega) \tag{25}$$

Note that if R and G are negligibly small (an ideal, lossless line), then

$$Z_c = \sqrt{\frac{L}{C}} = \frac{1}{2\pi}\sqrt{\frac{\mu}{\varepsilon}}\ln\frac{b}{a} \qquad (\Omega) \tag{26}$$

where equations 1 and 3 were used in the last equality.

EXERCISE 3

If R and G are small or if the frequency is large so that $\omega L \gg R$ and $\omega C \gg G$, then expand equation 25, similar to the expansion carried out in Exercise 2, to show that

$$Z_c \simeq \sqrt{\frac{L}{C}}\left[1 + j\left(\frac{G}{2\omega C} - \frac{R}{2\omega L}\right)\right] \qquad (\Omega) \tag{27}$$

Reflection Coefficient at the Output

Consider a transmission line of length ℓ and characteristic impedance Z_c, which is fed by an emf ϵ with a (source) impedance Z_0 and terminated with a load of impedance Z_L (Figure 1.6). At the output end where $z = \ell$, the load impedance Z_L is defined to be

$$Z_L = \frac{V(\ell, t)}{i(\ell, t)} \qquad (\Omega) \tag{28}$$

If equations 23 and 22 are substituted, then equation 28 becomes

$$Z_L = Z_c\left(\frac{V_1 e^{\alpha\ell + j\beta\ell} + V_2 e^{-\alpha\ell - j\beta\ell}}{-V_1 e^{\alpha\ell + j\beta\ell} + V_2 e^{-\alpha\ell - j\beta\ell}}\right) \qquad (\Omega) \tag{29}$$

Solving equation 29 for V_2, we have

$$V_2 = e^{2\alpha\ell + j2\beta\ell} V_1 \frac{Z_L + Z_c}{Z_L - Z_c} \quad \text{(V)} \tag{30}$$

The ratio of the reflected wave to the incident wave at $z = \ell$ is given by

$$\frac{V_r(\ell, t)}{V_i(\ell, t)} = \frac{V_1 e^{\alpha\ell} e^{j(\omega t + \beta\ell)}}{V_2 e^{-\alpha\ell} e^{j(\omega t - \beta\ell)}} = \frac{Z_L - Z_c}{Z_L + Z_c} \tag{31}$$

where equation 30 was used. The ratio of the reflected to the incident current at $z = \ell$ is given by

$$\frac{i_r(\ell, t)}{i_i(\ell, t)} = -\frac{(V_1/Z_c) e^{\alpha\ell} e^{j(\omega t + \beta\ell)}}{(V_2/Z_c) e^{-\alpha\ell} e^{j(\omega t - \beta\ell)}}$$

$$= -\frac{Z_L - Z_c}{Z_L + Z_c} \tag{32}$$

The **output reflection coefficient** Γ_L is defined to be

$$\Gamma_L = \frac{Z_L - Z_c}{Z_L + Z_c} \tag{33}$$

Note that if Γ_L is positive the current changes sign on reflection and the voltage does not, and if Γ_L is negative then the voltage changes sign and the current does not. For an open-circuited line $\Gamma_L = +1$, and for a short-circuited line $\Gamma_L = -1$, and therefore $-1 \leqq \Gamma_L \leqq +1$.

Impedance matching of the load to the line occurs when $Z_L = Z_c$; then $\Gamma_L = 0$ and there are no reflected waves.

EXERCISE 4

(a) If the line is open-circuited, then show that the current through the load is zero and the voltage across the load is twice the incident voltage. (b) If the line is short-circuited, then show that the voltage across the load is zero and the current through the load is twice the incident current.

Reflection Coefficient at the Input

Let τ be the time for a wave to propagate from one end of the line to the other; that is, $\tau = \ell/v$. If we turn on the emf at $t = 0$, then for $t < 2\tau$ only the incident wave (V_i, i_i) from the emf exists at the input end of the line. For such a case, we apply Kirchhoff's loop theorem to the input of the coaxial line shown in Figure 1.6 to obtain

$$\epsilon(t) - Z_0 i_i(0, t) = V_i(0, t) \quad \text{(V)} \tag{34}$$

where

$$V_i(0, t) = Z_c i_i(0, t) \quad \text{(V)} \tag{35}$$

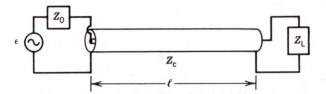

FIGURE 1.6 The source of emf with impedance Z_0 is shown connected to a coaxial line of characteristic impedance Z_c and the line is terminated with a load of impedance Z_L.

Solving for V_i and i_i, we find that

$$V_i(0, t) = \frac{Z_c}{Z_0 + Z_c} \epsilon(t) \qquad \text{(V)}$$

$$i_i(0, t) = \frac{\epsilon(t)}{Z_0 + Z_c} \qquad \text{(A)}$$

(36)

For $t \geq 2\tau$, a second incident wave (V_i', i_i'), which has reflected off of the load, arrives at the input. There are now three waves at the input of the coaxial line: (1) (V_i, i_i), a wave incident from the generator; (2) (V_i', i_i'), a wave reflected from the load; (3) (V_r', i_r'), the reflection of (V_i', i_i') at the input. Applying Kirchhoff's loop theorem to the input of the coaxial line shown in Figure 1.6, we obtain

$$\epsilon(t) - Z_0[i_i(0, t) + i_i'(0, t) + i_r'(0, t)] = V_i(0, t) + V_i'(0, t) + V_r'(0, t) \qquad \text{(V)} \qquad (37)$$

Also

$$i_i = \frac{V_i}{Z_c} \qquad i_i' = -\frac{V_i'}{Z_c} \qquad i_r' = \frac{V_r'}{Z_c} \qquad \text{(A)}$$

(38)

where i_i' and V_i' are out of phase by π following the reflection at $z = \ell$. Substituting equation 38 into 37, and using equaton 36 to eliminate V_i and i_i, we find

$$V_r' = \frac{Z_0 - Z_c}{Z_0 + Z_c} V_i' \quad \text{(V)} \qquad i_r' = -\frac{Z_0 - Z_c}{Z_0 + Z_c} i_i' \quad \text{(A)}$$

(39)

If we define

$$\Gamma_0 = \frac{Z_0 - Z_c}{Z_0 + Z_c}$$

(40)

to be the **reflection coefficient** at the **input** to the line, then equations 39 become

$$V_r' = \Gamma_0 V_i' \quad \text{(V)} \qquad i_r' = -\Gamma_0 i_i' \quad \text{(A)}$$

(41)

where $-1 \leq \Gamma_0 \leq 1$.

Impedance matching of the emf to the line occurs when $Z_0 = Z_c$ and $V_r' = 0$.

EXERCISE 5

If the emf is matched to the line, that is, $Z_0 = Z_c$, but the load is not matched to the line, that is, $Z_L \neq Z_c$, then show that the voltage across the load and the current through the load are given by

$$i = \frac{2V_i}{Z_L + Z_c} \quad (A) \qquad V = \frac{2Z_L}{Z_L + Z_c} V_i \quad (V)$$

where V_i is the voltage incident on the load.

EXPERIMENT

Connect the circuit shown in Figure 1.7a with $R = Z_c$. Set the square wave generator to 10^4 Hz and observe the voltage pulses on the oscilloscope. Connect the circuit shown in Figure 1.7b with $Z_L = \infty$. The generator pulse period is much greater than τ, the time for the pulse to travel the length of the coaxial line. The oscilloscope will display the pulse incident on the transmission line from the generator, followed by a pulse that has reflected from the load and returned to the input of the line.

Undesirable reflected signals at the input can be reduced by making the leads connecting the resistor R, the oscilloscope, and the coaxial line as short as possible. One way to do this is by using a male BNC and a female banana connector. The male BNC connects to the oscilloscope and the resistor R and the coax connects directly to the banana receptacles.

Perhaps it is worth comparing the circuits shown in Figures 1.6 and 1.7. In Figure 1.6 the impedance Z_0 is the impedance as seen from the transmission line side of the input. In Figure 1.7b the impedance seen from the transmission line side of the input is the total impedance of three parallel impedances: (1) the oscilloscope impedance, (2) the resistance R, and (3) the impedance of the 510 Ω, the 5 pF, and the generator in series. The typical impedance of a generator is 600 Ω; hence, the impedance of the series combination is given by

$$Z_s = \sqrt{R_s^2 + (\omega C')^{-2}} = \sqrt{1110^2 + (\omega 5 \times 10^{-12})^{-2}} \quad (\Omega) \tag{42}$$

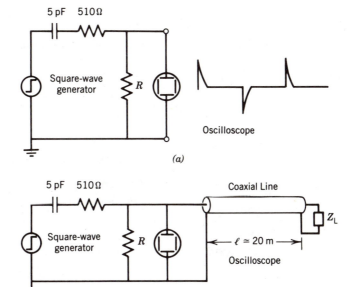

FIGURE 1.7 (a) Circuit to generate pulses. The pulses are sketched to the right. (b) Circuit to study pulses propagating on a coaxial line.

The oscilloscope impedance is about 1 MΩ, and since the three impedances are in parallel the reciprocal of Z_0 is given by

$$\frac{1}{Z_0} = \frac{1}{R} + \frac{1}{Z_s} + \frac{1}{1 \text{ M}\Omega} \qquad (1/\Omega) \qquad (43)$$

Note that when $R = Z_c$, which is much less than Z_s and 1 MΩ, then $Z_0 \simeq R$ and the input is impedance matched to the transmission line.

With $R = Z_c$, observe and compare the pulse incident on the load with the pulse reflected from the load when the load impedance Z_L is (a) open circuit, (b) short circuit, and (c) a resistor that matches the line impedance.

EXERCISE 6

In cases a and b, do you expect the reflected and incident pulses to have the same or opposite signs? Explain.

With $Z_L = \infty$ and $R = \infty$, observe the multireflected pulse.

EXERCISE 7

From your observation of the multireflected pulse determine the attenuation constant α.

With $R = Z_c$, use a 1-kΩ potentiometer as a load and measure the voltage of the reflected pulse as a function of the load resistance. Plot a graph of the reflected voltage versus load resistance. Compare your graph with theory by plotting a theoretical curve of the reflected voltage versus load resistance. See equation 31.

With $R = Z_c$ and $Z_L = \infty$ observe the pulses on the oscilloscope and determine an experimental value for the velocity of a pulse. The theoretical value of the velocity is $1/\sqrt{\mu\varepsilon}$, where $\mu \simeq \mu_0$. To determine ε, and, hence, a theoretical value for the velocity, connect the circuit shown in Figure 1.8. Measure the RC' time constant, use your measured value to determine the capacitance C', and then calculate the capacitance per unit length C, which is C'/ℓ. When C is known, use equation 3 to solve for ε and, hence, the theoretical velocity of the pulse.

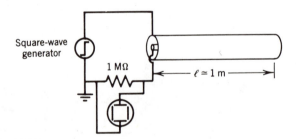

FIGURE 1.8 Circuit to measure the RC' time constant.

EXERCISE 8

Compare the experimental and theoretical values of the velocity of the pulse.

The 1-MΩ resistance shown in Figure 1.8 is of the order of the oscilloscope impedance, and therefore the resistance in the RC time constant will be an appropriate combination of the oscilloscope impedance and the 1-MΩ resistance. The scope impedance may be specified on the scope; if not, look it up in the manufacturer's operating manual.

Reconnect the circuit in Figure 1.7b with $R = Z_c$. Terminate the 20-m line with an inductor and observe the reflected pulse. Replace the inductor with a capacitor and observe the reflected pulse.

EXERCISE 9

Explain the shape (time dependence) of the reflected pulse for each termination.

2. WAVEGUIDE. MICROWAVE PRODUCTION AND DETECTION, IMPEDANCE MATCHING, CAVITY Q

APPARATUS

Oscilloscope

Klystron

Klystron cooling fan (if recommended by the klystron manufacturer)

Klystron power supply

Isolator

Variable attenuator (0–50 dB recommended)

Wavemeter

Diode and mount

Slotted section of waveguide with a movable diode

Slide screw tuner

Transmission cavity (TE$_{102}$ suggested) (see Figure 2.5).

OBJECTIVES

To measure various parameters of the reflex klystron and become familiar with its operation.

To study the characteristics of a microwave diode detector.

To measure the voltage standing-wave ratio (VSWR) and determine the reflection coefficient.

To match the characteristic impedance of the waveguide to a load using a slide screw tuner.

To measure the resonant frequency and quality factor or Q of a microwave cavity.

At the end of the experiment you will have constructed a microwave spectrometer that will be used in Experiment 16, "Electron Spin Resonance: Microwave Spectroscopy."

KEY CONCEPTS

TE, TM, and TEM modes	Guide wavelength
Propagation constant	Cutoff wavelength
Attenuation constant	TE_{mnp} cavity modes
Phase constant	Klystron mode
Characteristic impedance	Square-law detector
Boundary conditions	Linear detector
TE_{mn} modes	VSWR (voltage standing-wave ratio)
Cutoff frequency	Cavity Q

REFERENCES

1. D. Halliday and R. Resnick, *Physics*, Part 2, 3d ed., Wiley, New York, 1978. Rectangular waveguide is discussed on pages 902–903. Resonant cavity is discussed on page 926.

2. C. P. Poole, *Electron Spin Resonance: A Comprehensive Treatise on Experimental Techniques*, 2d ed., Wiley, New York, 1983. Rectangular waveguides are discussed on pages 39–51. Klystrons are discussed on pages 71–75. Rectangular resonant cavities are discussed on pages 128–135. Most microwave circuit components, such as slide screw tuner, isolator, and diode detector, are discussed.

3. A. Portis, *Electromagnetic Fields-Sources and Media*, Wiley, New York, 1978. Rectangular waveguide is discussed on pages 513–517.

4. R. Wangsness, *Electromagnetic Fields*, Wiley, New York, 1979. Rectangular waveguide is discussed on pages 489–497, and resonant cavities are discussed on pages 501–504. Boundary conditions are summarized on page 397.

5. W. Scott, *The Physics of Electricity and Magnetism*, 2d ed., Wiley, New York, 1966. Waveguides are discussed on pages 592–602. The Q of an *RLC* circuit is discussed on pages 467–470.

6. *American Institute of Physics Handbook*, McGraw-Hill, New York, 1957. Waveguides and resonant cavities are discussed on pages 5–59 to 5–68.

INTRODUCTION

The first two sections, *Waveguide* and *Resonant Cavity*, involve the solutions of Maxwell's equations at the level of a junior course in electromagnetism. The experiment may be carried out without the theoretical background provided by these two sections. If you have not completed such a course in electromagnetism it is suggested that the waveguide section, the resonant cavity section, and Exercise 11 be omitted. Consult with your instructor about omitting this material.

Waveguide

A waveguide is a hollow single conductor. Figure 2.1 shows a section of rectangular waveguide. The electric and magnetic field configurations, or modes, in a waveguide are transverse electric field (TE) modes or transverse magnetic (TM) modes. The electric field of a TE mode is transverse to the direction of propagation, while the magnetic field of a TE mode has a component in the direction of propagation. For a TM mode the magnetic field is transverse to the direction of propagation, while the electric field has a component in the direction of propagation. Waveguides will not transmit a TEM mode (one in which both fields are entirely transverse).

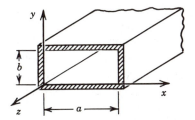

FIGURE 2.1 Cross-sectional view of waveguide. Inside dimensions are a and b, and the waveguide axis is along the z axis.

We determine the electric and magnetic fields in a metallic waveguide by solving Maxwell's equations for the fields in the metal walls and for the fields in the dielectric inside the guide. We then require the fields in the two regions to satisfy the boundary conditions, that is, Maxwell's equations at the boundary between the two regions.

Important point: There exist an infinite number of solutions to Maxwell's equations. We obtain the fields of a particular problem by requiring the solutions of Maxwell's equations to satisfy the appropriate boundary conditions.

The procedure to determine the electric and magnetic fields in a rectangular waveguide for TE modes is outlined below.

1. Start with Maxwell's equations:

$$\mathbf{\nabla} \times \mathbf{H} = \mathbf{J} + \frac{\partial \mathbf{D}}{\partial t} \qquad (\text{A/m}^2) \tag{1}$$

$$\mathbf{\nabla} \times \mathbf{E} = -\frac{\partial \mathbf{B}}{\partial t} \qquad (\text{V/m}^2) \tag{2}$$

$$\mathbf{\nabla} \cdot \mathbf{D} = \rho \qquad (\text{C/m}^3) \tag{3}$$

$$\mathbf{\nabla} \cdot \mathbf{B} = 0 \qquad (\text{Wb/m}^3) \tag{4}$$

where we eliminate $\mathbf{J}$, $\mathbf{D}$, and $\mathbf{B}$ with the following equations:

$$\mathbf{J} = \sigma \mathbf{E} \qquad (\text{A/m}^2) \tag{5}$$

$$\mathbf{D} = \varepsilon \mathbf{E} \qquad (\text{C/m}^2) \tag{6}$$

$$\mathbf{B} = \mu \mathbf{H} \qquad (\text{Wb/m}^2) \tag{7}$$

and σ, ε, and μ are the conductivity, permittivity, and permeability. For a conductor, $\sigma \gg 1$, and for a dielectric, $\sigma \ll 1$.

2. Select the mode type. In this case the mode type is TE, so $E_z = 0$ and $H_z \neq 0$.

3. Find solutions that have harmonic variation with respect to time and attenuation with respect to z. We assume

$$\mathbf{E}(x, y, z, t) = \mathbf{E}_0(x, y)\, e^{j\omega t - \gamma z} \qquad (\text{V/m}) \tag{8}$$

$$\mathbf{H}(x, y, z, t) = \mathbf{H}_0(x, y)\, e^{j\omega t - \gamma z} \qquad (\text{A/m}) \tag{9}$$

where $j^2 = -1$ and γ is an unknown **propagation constant** $= \alpha + j\beta$; α is an **attenuation**

constant and β is a **phase constant**. The angular frequency ω is an independent variable, controlled by a knob on the source of waves (the klystron). γ depends on ω as well as σ, ε, μ, and the mode of propagation.

4. Find equations for the other four field components (E_x, E_y, H_x, H_y) in terms of H_z. We first substitute equations 8 and 9 into equations 1 and 2. Then, in terms of components, equations 1 and 2 are six equations that may be solved for the four field components in terms of derivatives of H_z. If

$$k^2 \equiv \gamma^2 + \mu\varepsilon\omega^2 - j\omega\mu\sigma \qquad (\text{rad/m})^2 \tag{10}$$

then the four fields are given by

$$E_x = -\frac{j\mu\omega}{k^2}\frac{\partial H_z}{\partial y} \qquad (\text{V/m}) \tag{11}$$

$$E_y = \frac{j\mu\omega}{k^2}\frac{\partial H_z}{\partial x} \qquad (\text{V/m}) \tag{12}$$

$$H_x = -\frac{\gamma}{k^2}\frac{\partial H_z}{\partial x} \qquad (\text{A/m}) \tag{13}$$

$$H_y = -\frac{\gamma}{k^2}\frac{\partial H_z}{\partial y} \qquad (\text{A/m}) \tag{14}$$

Another equation that may be obtained from the six component equations is the ratio of E_x/H_y or $-E_y/H_x$, which is defined to be the **characteristic impedance** Z_c of the waveguide:

$$Z_c \equiv \frac{E_x}{H_y} = -\frac{E_y}{H_x} = \frac{j\mu\omega}{\gamma} \qquad (\Omega) \tag{15}$$

EXERCISE 1

Derive equations 11–15.

5. Develop the wave equation for H_z. We obtain the wave equation for H_z by substituting equations 11 and 12 into the z component of equation 2. The resulting equation is

$$\frac{\partial^2 H_z}{\partial x^2} + \frac{\partial^2 H_z}{\partial y^2} + k^2 H_z = 0 \qquad (\text{A/m}^3) \tag{16}$$

The time t does not appear explicitly in equation 16, since we have assumed harmonic variation with time.

EXERCISE 2

Carry out the necessary steps to obtain equation 16.

6. Solve equation 16 for H_z in both regions and require the fields in the two regions to satisfy the boundary conditions. A general boundary is shown in Figure 2.2, where **n** is a unit vector perpendicular to the boundary. Each region is characterized by ε, μ, and σ. The

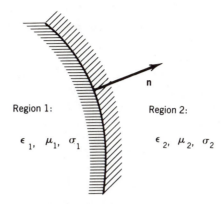

FIGURE 2.2 Boundary between two regions, where **n** is a unit vector perpendicular to the boundary.

fields must satisfy Maxwell's equations at the boundary and this yields the general boundary conditions:

$$(\mathbf{D}_2 - \mathbf{D}_1) \cdot \mathbf{n} = \Sigma \qquad (\text{C/m}^2) \tag{17}$$

$$(\mathbf{E}_2 - \mathbf{E}_1) \times \mathbf{n} = 0 \qquad (\text{V/m}) \tag{18}$$

$$(\mathbf{B}_2 - \mathbf{B}_1) \cdot \mathbf{n} = 0 \qquad (\text{Wb/m}^2) \tag{19}$$

$$(\mathbf{H}_2 - \mathbf{H}_1) \times \mathbf{n} = -\mathbf{K} \qquad (\text{A/m}) \tag{20}$$

where Σ is the surface charge density and $\mathbf{K}$ is the linear current density at the surface.

H_z in the Metal

We will determine the z dependence of H_z by assuming that H_z does not depend on x and y: that is, we assume a plane wave. Then equation 16 becomes

$$k^2 H_z = 0 \qquad (\text{A/m}^3) \tag{21}$$

or

$$k^2 \equiv \gamma^2 + \mu\varepsilon\omega^2 - j\omega\mu\sigma = 0 \qquad (\text{rad/m})^2 \tag{22}$$

Hence, the propagation constant γ inside the metal is given by

$$\gamma = \sqrt{\omega\mu(j\sigma - \varepsilon\omega)} \qquad (\text{rad/m}) \tag{23}$$

σ, ε, and μ are frequency dependent and at the microwave frequency of 10^{10} Hz the dimensionless ratio $\sigma/\varepsilon\mu$ is about 10^8. So to a good approximation we may write

$$\gamma \simeq \sqrt{j\mu\omega\sigma} = (1+j)\sqrt{\frac{\mu\omega\sigma}{2}} \qquad (\text{rad/m}) \tag{24}$$

where we used $\sqrt{j} = \sqrt{2j/2} = \sqrt{(1 + 2j - 1)/2} = \sqrt{(1+j)^2/2}$. The z dependence of H_z, according to equations 9 and 24, is given by

$$H_z = H_0 \exp\left[j\left(\omega t - \sqrt{\frac{\mu\omega\sigma}{2}}\,z\right)\right] \exp\left(-\sqrt{\frac{\mu\omega\sigma}{2}}\,z\right) \qquad (\text{A/m}) \tag{25}$$

We will assume an ideal conductor, that is, $\sigma = \infty$; hence, H_z and, from equations 11–14, the other four components are zero inside the conductor. This point is worth emphasizing: The fields in the metal are zero.

If we let the metal be region 2 in Figure 2.2, then for a perfect conductor we obtain the boundary conditions by setting $\mathbf{D}_2$, $\mathbf{E}_2$, $\mathbf{B}_2$, and $\mathbf{H}_2$ to zero in equations 17–20.

In general, the surface charge density Σ and the surface current density $\mathbf{K}$ are unknowns; therefore, we use the boundary conditions given by equations 18 and 19. For a perfect conductor these two boundary conditions become

$$\mathbf{E} \times \mathbf{n}|_S = 0 \qquad \text{(V/m)} \qquad (26)$$

$$\mathbf{B} \cdot \mathbf{n}|_S = \mu \mathbf{H} \cdot \mathbf{n}|_S = 0 \qquad \text{(Wb/m}^2) \qquad (27)$$

where S implies the surface and the subscripts to denote the fields in the dielectric have been dropped.

For the waveguide $\mathbf{n}$ is perpendicular to the guide axis z (see Figure 2.1). Carrying out the cross-product in equation 26 yields $E_z|_S = 0$. The dot product in equation 27 yields $H_x|_S = 0$ for $\mathbf{n}$ in the x direction and $H_y|_S = 0$ for $\mathbf{n}$ in the y direction. Using equations 13 and 14, this boundary condition may be written $\partial H_z/\partial x|_S = 0$ and $\partial H_z/\partial y|_S = 0$. (Recall that equations 13 and 14 are for the TE mode only.)

Thus, the boundary conditions for the TE mode with one surface a perfect conductor are

$$E_z = 0 \text{ everywhere, including on S} \qquad \text{(V/m)} \qquad (28)$$

$$\left.\frac{\partial H_z}{\partial \xi}\right|_S = 0 \qquad \text{(A/m}^2) \qquad (29)$$

where ξ is a coordinate normal to the surface; in this case it is x or y.

H_z in the Dielectric

We assume the dielectric is air with $\mu \simeq \mu_0$, $\varepsilon \simeq \varepsilon_0$, and $\sigma \simeq 0$, and then $k^2 \simeq \gamma^2 + \mu_0\varepsilon_0\omega^2$. We solve equation 16 by assuming that we may separate variables:

$$H_z(x, y, z) = X(x)Y(y)Z(z) \qquad \text{(A/m)} \qquad (30)$$

where, from equation 9, $Z(z) \propto e^{-\gamma z}$. We then substitute equation 30 into equation 16, obtaining ordinary differential equations involving $X(x)$ and $Y(y)$. We solve these ordinary differential equations and then require H_z to satisfy equation 29. This procedure yields H_z and k:

$$H_z(x, y, z, t) = H_0 \cos\frac{m\pi x}{a} \cos\frac{n\pi y}{b} e^{j\omega t - \gamma z} \qquad \text{(A/m)} \qquad (31)$$

$$k^2 = \left(\frac{n\pi}{b}\right)^2 + \left(\frac{m\pi}{a}\right)^2 \qquad \text{(rad/m)}^2 \qquad (32)$$

where m and n are integers $(0, 1, 2, 3, \ldots)$, but m and n cannot both be zero, that is, there is no TE$_{00}$ mode, and a and b are the waveguide dimensions shown in Figure 2.1.

7. Substitute H_z back into the equations of step 4 to obtain a set of equations expressing each field component as a function of space and time. This completes the solution of the problem.

The four field components that follow from step 7 are

$$H_y(x, y, z, t) = \frac{\gamma H_0}{k^2} \frac{n\pi}{b} \cos \frac{m\pi x}{a} \sin \frac{n\pi y}{b} e^{j\omega t - \gamma z} \quad \text{(A/m)} \quad (33)$$

$$H_x(x, y, z, t) = \frac{\gamma H_0}{k^2} \frac{m\pi}{a} \sin \frac{m\pi x}{a} \cos \frac{n\pi y}{b} e^{j\omega t - \gamma z} \quad \text{(A/m)} \quad (34)$$

$$E_y(x, y, z, t) = \frac{\gamma Z_c H_0}{k^2} \frac{m\pi}{a} \sin \frac{m\pi x}{a} \cos \frac{n\pi y}{b} e^{j\omega t - \gamma z} \quad \text{(V/m)} \quad (35)$$

$$E_x(x, y, z, t) = -\frac{\gamma Z_c H_0}{k^2} \frac{n\pi}{b} \cos \frac{m\pi x}{a} \sin \frac{n\pi y}{b} e^{j\omega t - \gamma z} \quad \text{(V/m)} \quad (36)$$

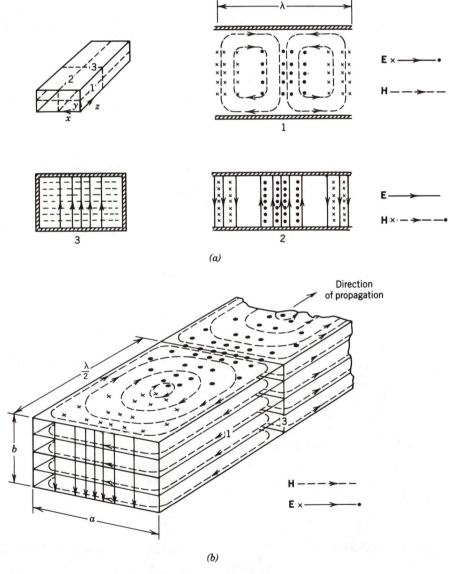

FIGURE 2.3 (a) The field lines are shown in three planes of the waveguide, where the planes are specified in the upper left figure. (b) The field lines are shown in three dimensions.

Since m and n are integers $(0, 1, 2, 3, \ldots)$, but both m and n are not zero, the TE modes are designated as TE_{mn}, and, in principle, an infinite number of TE modes may exist simultaneously in the waveguide. The particular mode or modes that are actually present depend on the guide dimension and the method of exciting the guide. The field configurations for the TE_{10} mode are shown in Figure 2.3. Figure 2.3a shows the field lines in three planes and Figure 2.3b shows the field lines in three dimensions.

EXERCISE 3

Use equations 31 and 33–36 to write out the field components for the TE_{10} mode. Sketch E_y, H_x, and H_z versus x. Do your sketches agree with the fields shown in Figure 2.3?

Combining equations 22 and 32 and solving for the propagation constant γ, we find

$$\gamma = \sqrt{\left(\frac{n\pi}{b}\right)^2 + \left(\frac{m\pi}{a}\right)^2 - \mu_0 \varepsilon_0 \omega^2} \qquad \text{(rad/m)} \qquad (37)$$

where for the dielectric we assumed $\sigma \simeq 0$. Note from equation 37 that γ may be positive, zero, or imaginary, depending on the angular frequency ω:

γ	ω	Frequency
Positive	$\omega^2 < \dfrac{1}{\mu_0 \varepsilon_0}\left[\left(\dfrac{n\pi}{b}\right)^2 + \left(\dfrac{m\pi}{a}\right)^2\right]$	Low
Zero	$\omega^2 = \dfrac{1}{\mu_0 \varepsilon_0}\left[\left(\dfrac{n\pi}{b}\right)^2 + \left(\dfrac{m\pi}{a}\right)^2\right]$	Intermediate
Imaginary	$\omega^2 > \dfrac{1}{\mu_0 \varepsilon_0}\left[\left(\dfrac{n\pi}{b}\right)^2 + \left(\dfrac{m\pi}{a}\right)^2\right]$	High

At sufficiently low frequencies γ is positive and the wave is attenuated according to equation 31. Under this condition it is said that the wave or mode is not propagated.

At sufficiently high frequencies γ is imaginary and therefore the wave or mode is propagated without attenuation.

At some intermediate frequency $\gamma = 0$. This frequency is called the **cutoff frequency** for the mode under consideration. At frequencies higher than cutoff this mode propagates, while at lower frequencies it is attenuated.

The cutoff frequency v_c for the TE_{mn} mode is given by

$$v_c = \frac{1}{2(\mu_0 \varepsilon_0)^{1/2}} \sqrt{\left(\frac{n}{b}\right)^2 + \left(\frac{m}{a}\right)^2} \qquad \text{(Hz)} \qquad (38)$$

EXERCISE 4

Measure the inside dimensions, a and b, of your waveguide. For a waveguide that is air filled, what are the cutoff frequencies for the TE_{10}, TE_{20}, and TE_{11} modes? What is the frequency range such that only the TE_{10} mode propagates?

Note that if we let the guide dimensions a and b approach infinity, then we have an unbounded medium of the same dielectric material that fills the guide. In such an unbounded medium equation 37 becomes

$$\gamma = \sqrt{-\omega^2\mu_0\varepsilon_0} = j\omega\sqrt{\mu_0\varepsilon_0} = j\frac{\omega}{c} = j\frac{2\pi}{\lambda_0} \qquad \text{(rad/m)} \qquad (39)$$

where c is the phase velocity $= \omega/k_0 = v\lambda_0$, and λ_0 is the **wavelength in the unbounded medium**.

In the waveguide at frequencies higher than cutoff the equation analogous to equation 39 is

$$\gamma = \sqrt{\left(\frac{n\pi}{b}\right)^2 + \left(\frac{m\pi}{a}\right)^2 - \omega^2\mu_0\varepsilon_0} = j\frac{2\pi}{\lambda} \qquad \text{(rad/m)} \qquad (40)$$

where λ is the **wavelength in the guide**. From equation 40, λ may be written

$$\lambda = 2\pi \left/ \sqrt{\omega^2\mu_0\varepsilon_0 - \left(\frac{n\pi}{b}\right)^2 - \left(\frac{m\pi}{a}\right)^2} \right. \qquad \text{(m)} \qquad (41)$$

If we use $\omega^2\mu_0\varepsilon_0 = \omega^2/c^2 = (2\pi/\lambda_0)^2$, the wavelength of the TE_{mn} mode becomes

$$\lambda = \lambda_0 \left/ \sqrt{1 - \left(\frac{n\lambda_0}{2b}\right)^2 - \left(\frac{m\lambda_0}{2a}\right)^2} \right. \qquad \text{(m)} \qquad (42)$$

The phase velocity of the TE_{mn} mode is given by

$$v = \frac{\omega\lambda}{2\pi} = \omega\lambda_0 \left/ 2\pi\sqrt{1 - \left(\frac{n\lambda_0}{2b}\right)^2 - \left(\frac{m\lambda_0}{2a}\right)^2} \right.$$

$$= c \left/ \sqrt{1 - \left(\frac{n\lambda_0}{2b}\right)^2 - \left(\frac{m\lambda_0}{2a}\right)^2} \right. \qquad \text{(m/s)} \qquad (43)$$

EXERCISE 5

The **cutoff wavelength**, λ_{oc}, is the wavelength in an unbounded medium at the cutoff frequency v_c. Show that the cutoff wavelength of the TE_{10} mode in an air-filled waveguide is $2a$. Knowing the dimension a, calculate λ_{oc}.

Resonant Cavity

A transmission cavity is shown in Figure 2.4. The cavity may be constructed by soldering a flange on each end of a section of waveguide of length d. The end plates may be constructed from brass shim stock with an iris diameter of approximately $a/4$, where a is the guide dimension shown in Figure 2.1.

As before, we assume that the metal is a perfect conductor and the air inside the cavity is a perfect dielectric. We also assume that the end plates are perpendicular to the z axis.

Radiation initially propagating in the guide in the positive z direction is partially reflected and partially transmitted at the front plate of the cavity. The wave transmitted at the front plate is partially reflected at the back plate, and this wave will be partially reflected at the front plate, and so on. Hence, there exist oppositely directed traveling waves in the cavity

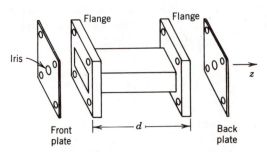

FIGURE 2.4 A microwave transmission cavity.

which superimpose to produce a standing wave. The z dependence of the five fields given by equations 31 and 33–36 will be that of standing waves:

$$A \sin\sqrt{\mu_0\varepsilon_0\omega^2 - k^2}\, z + C \cos\sqrt{\mu_0\varepsilon_0\omega^2 - k^2}\, z \tag{44}$$

where k^2 is given by equation 32.

The boundary conditions that the **E** and **H** fields must satisfy at the end plates are given by equations 26 and 27, where **n** is directed along the z axis. If the end plates are at $z = 0$ and $z = d$, then for the electric field components equation 26 yields $C = 0$ and $\sqrt{\mu_0\varepsilon_0\omega^2 - k^2}\,d = p\pi$, where $p = 0, 1, 2, 3, \ldots$. For the component H_z equation 27 yields $C = 0$ and $\sqrt{\mu_0\varepsilon_0\omega^2 - k^2}\,d = p\pi$, where $p = 0, 1, 2, 3, \ldots$. Once we know H_z, we can obtain H_x and H_y from equations 13 and 14.

Inside the cavity, the fields given by equations 31 and 33–36 become

$$H_z \propto H_0 \cos\frac{m\pi x}{a} \cos\frac{n\pi y}{b} \sin\frac{p\pi z}{d} e^{j\omega t} \quad \text{(A/m)} \tag{45}$$

$$H_y \propto -H_0 \frac{\gamma}{k^2} \frac{n\pi}{b} \frac{p\pi}{d} \cos\frac{m\pi x}{a} \sin\frac{n\pi y}{b} \cos\frac{p\pi z}{d} e^{j\omega t} \quad \text{(A/m)} \tag{46}$$

$$H_x \propto -H_0 \frac{\gamma}{k^2} \frac{m\pi}{a} \frac{p\pi}{d} \sin\frac{m\pi x}{a} \cos\frac{n\pi y}{b} \cos\frac{p\pi z}{d} e^{j\omega t} \quad \text{(A/m)} \tag{47}$$

$$E_y \propto H_0 Z_c \frac{\gamma}{k^2} \frac{m\pi}{a} \sin\frac{m\pi x}{a} \cos\frac{n\pi y}{b} \sin\frac{p\pi z}{d} e^{j\omega t} \quad \text{(V/m)} \tag{48}$$

$$E_x \propto -H_0 Z_c \frac{\gamma}{k^2} \frac{n\pi}{b} \cos\frac{m\pi x}{a} \sin\frac{n\pi y}{b} \sin\frac{p\pi z}{d} e^{j\omega t} \quad \text{(V/m)} \tag{49}$$

where

$$\sqrt{\mu_0\varepsilon_0\omega^2 - k^2} = \frac{p\pi}{d} \qquad p = 1, 2, 3, \ldots \quad \text{(rad/m)} \tag{50}$$

and $p \neq 0$, since in this case one has the trivial solution $H_z = 0$. Note that the fields depend on the integers m, n, and p, and the mode is designated as TE_{mnp}. The integers m, n, and p are the number of half-wavelength variations in the standing wave pattern in the x, y, and z directions, respectively. The electromagnetic field configurations in a TE_{102} mode resonant cavity are shown in Figure 2.5.

The resonant frequency of the cavity is, according to equation 50, given by

$$\omega_{mnp}^2 = \frac{\pi^2}{\mu_0\varepsilon_0}\left[\left(\frac{m}{a}\right)^2 + \left(\frac{n}{b}\right)^2 + \left(\frac{p}{d}\right)^2\right] \quad \text{(rad/s)}^2 \tag{51}$$

where equation 32 was used and the approximation $\varepsilon_{\text{air}} \simeq \varepsilon_0$ was made.

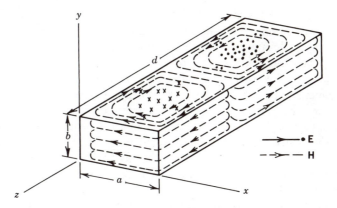

FIGURE 2.5 Field lines of a TE_{102} mode resonant cavity. The
irises are not shown for clarity.

The superposition of waves in the cavity, which produces standing waves, gives rise to larger field amplitudes than exist in the waveguide. The constant of proportionality in equations 45–49 includes the square root of the quality factor Q of the cavity. The quality factor Q is a measure of the field strength in the cavity; a high Q implies strong fields. The Q of a cavity is defined and discussed later in this experiment. For now we simply point out that the Q of a microwave transmission cavity is typically a few hundred. A Q of the order of 10^5 is possible with cavities constructed from superconductors.

EXERCISE 6

Assuming that your cavity is a TE_{102}, measure d, and, knowing the inside dimensions of your waveguide, calculate the resonant frequency ω_{102}.

Reflex Klystron

A qualitative description of the operation of a reflex klystron will be given in this section. The arrangement of operating voltages for a typical tube is shown in Figure 2.6. Electrons are thermionically emitted from the cathode (see Experiment 7, Electron Physics:

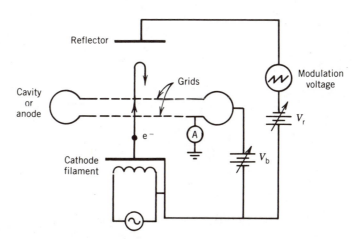

FIGURE 2.6 Schematic diagram of a reflex klystron with applied
voltages. The partial path of a single electron is shown.

Thermionic Emission and Charge-to-Mass Ratio). The beam voltage V_b accelerates the electrons as they travel between the cathode and the cavity or anode. The reflector voltage V_r decelerates the electrons to rest and then accelerates them back toward the cavity, where they pass through a second time.

When the klystron is oscillating, currents in the cavity walls establish standing electromagnetic waves in the cavity at a definite frequency determined primarily by the cavity dimensions. The cavity is constructed such that the electric field in the cavity points instantaneously toward the reflector or cathode and it oscillates at the resonant frequency of the cavity. Hence, as the electrons pass between the grids they are either accelerated or decelerated by the electric field of the cavity. If an electron is decelerated by the cavity field, then the electron gives energy to the field. If it is accelerated, then it removes energy from the cavity field.

Electrons are randomly emitted by the cathode and, hence, electrons arrive at the cavity from the cathode at random times. On the average, as many of these electrons are accelerated as decelerated and there is no net exchange of energy with the cavity field. However, the electron beam is velocity modulated by the cavity field; that is, the accelerated electrons leave the cavity with a higher velocity than the decelerated electrons. After traversing the cavity, the velocity-modulated beam travels toward the reflector, where the fast electrons move away from the slower ones behind them and catch up with the slower ones in front of them. The net effect is the formation of groups or bunches of electrons.

Figure 2.7a shows the path versus time for five electrons, where d_0 is the distance from the cavity center to the reflector, and Figure 2.7b shows the electric field across the cavity versus time. Note that as the electrons initially pass through the cavity, electrons 1 and 3 are neither accelerated nor decelerated, 2 and 5 are accelerated, and 4 is decelerated. Velocity modulation results in electrons 2, 3, and 4 returning to the cavity in a bunch.

For a fixed-beam voltage V_b the transit time from the cavity toward the reflector and back to the cavity is controlled by the reflector voltage V_r. Electrons that have transit times that are in the vicinity of $(q - \frac{1}{4})T$, where T is the period of the cavity fields and $q = 1, 2, 3, \dots$, will be returned to the cavity in a bunch and will be decelerated by the electric field of the cavity. Hence, the electrons give up energy to the cavity and oscillation of the tube continues. In Figure 2.7b note that $q = 2$.

If the transit time is such that the bunches arrive at a time when they are accelerated by

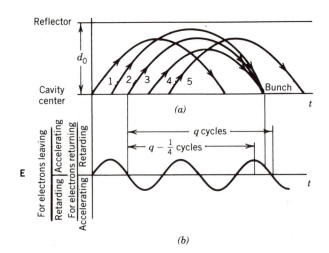

FIGURE 2.7 Velocity modulation and electron bunching. (a) The paths of 5 electrons and their transit times are shown. (b) The microwave electric field across the grids of the cavity.

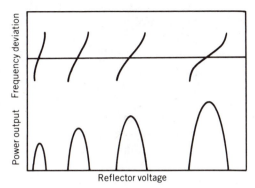

FIGURE 2.8 Klystron frequency deviation from a central value and klystron power output as a function of reflector voltage.

the cavity field, then energy is removed from the cavity and oscillations will tend to stop. To generate oscillations at a given frequency the reflector voltage must be adjusted until the transit time is in the vicinity of $(q - \frac{1}{4})T$. One way to determine if the desired transit time has been obtained is to observe the beam current as the reflector voltage is varied. In most tubes the beam current increases when oscillations occur.

Figure 2.8 shows the effect of reflector voltage on both the power output and frequency deviation from a central value. Each of the curves shown in Figure 2.8 corresponds to a different transit time. Note in the figure that for a given transit time the frequency output varies with the reflector voltage. The frequency of the cavity is determined primarily by the cavity dimensions; however, a small change in frequency may be obtained by changing the reflector voltage. This adjustment is known as **electronic tuning**. The electronic tuning range is the deviation from the cavity resonant frequency v_0 that may be obtained through

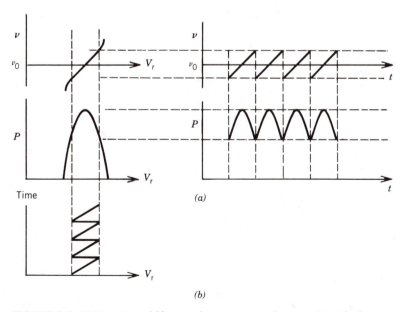

FIGURE 2.9 (a) Variation of klystron frequency v and power P as (b) the reflector voltage V_r is periodically modulated by a sawtooth voltage.

variations in the reflector voltage. Typical electronic tuning ranges are about 0.5 percent of the resonant frequency ν_0. For an X-band klystron with $\nu_0 = 10^{10}$ Hz, the electronic tuning is about $0.005\nu_0 = 50$ MHz.

A convenient way to electronically tune a klystron is by applying a periodic voltage to the reflector. Voltage modulation is applied to the reflector as shown in Figure 2.6. The effect of sawtooth voltage modulation on the frequency and power is shown in Figure 2.9.

The primary method of altering the frequency of a klystron is by **mechanical tuning**. The dimensions and, hence, resonant frequency ν_0 of a klystron cavity may be changed by adjusting a tuning screw, that is, by mechanical tuning. A typical klystron has a mechanical tuning range between 5 and 50 percent of ν_0. For example, the Varian X-13 klystron may be mechanically tuned from 8100 to 12 400 MHz.

To turn on a klystron, follow these steps in the order given:

1. Turn on the cooling fan if required by the manufacturer.
2. Turn on the filament and the reflector voltage (V_r).
3. Wait 5 minutes, then turn on the beam voltage (V_b) to a predetermined value (recommended by your instructor or the manufacturer).

Reverse these steps to turn off the klystron.

EXERCISE 7

What happens to electrons after being reflected back to the cavity and then decelerated by the cavity electric field? *Hint*: Note the ammeter in Figure 2.6.

EXPERIMENTS

Reflex Klystron Characteristics

Connect the components as shown in Figure 2.10 (K, klystron; I, isolator; A, attenuator; WM, wavemeter; and D, diode). The physical principles of the isolator are discussed in Experiment 22, The Faraday Effect. An attenuator is a circuit element that reduces the amplitude of the microwave field. If the microwave power incident on the attenuator is P_0, then the power P that exits the attenuator is reduced. This reduction in power is specified

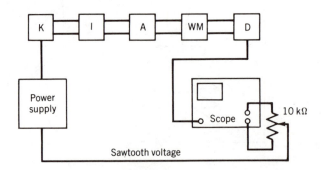

FIGURE 2.10 Microwave components to study the characteristics of both the klystron and the diode detector.

by a quantity called the attentuation A, which is defined by

$$A = 10 \log \frac{P_0}{P} \quad \text{(dB)} \quad (52)$$

Solving equation 52 for P yields

$$P = 10^{-A/10} P_0 \quad \text{(W)} \quad (53)$$

A calibrated attenuator has a dial that can be set for a predetermined attenuation A. The dial controls the insertion depth of a carbon-coated resistance card into the waveguide. The carbon coating absorbs microwave power.

A wavemeter is an adjustable, calibrated resonant cavity. An X-band wavemeter is typically adjustable from 8.0 to 12.0 kMHz. If the wavemeter is tuned to the frequency of the klystron, then a small fraction of the power incident on the wavemeter is absorbed. Hence, the wavemeter may be used to measure the klystron frequency. A klystron mode, discussed below, is shown on the oscilloscope trace in Figure 2.15, the central dip in the mode is created by the power absorbed by the wavemeter.

The diode will be discussed in the next section. The diode will have a longer lifetime if the incident microwave power is low. Therefore, the oscilloscope sensitivity and the attenuator setting should be set near maximum.

Turn on the klystron as recommended in the introduction to this experiment. Observe the beam current (the meter is on the front panel of the klystron power supply) as you manually vary the reflector voltage. When the reflector voltage does not create bunching of the electrons, then the klystron does not oscillate. For particular values of the reflector voltage bunching occurs and the klystron has a power output. The beam current is a few milliamperes larger when the tube is oscillating.

EXERCISE 8

As you vary the reflector voltage how many times does the klystron go into oscillations? Each such oscillation is called a **klystron mode**.

Modulate the reflector voltage with the sawtooth voltage from the oscilloscope. If your klystron power supply has a modulation voltage amplitude knob, then the 10-kΩ potentiometer shown in Figure 2.10 is not needed. Display one or more klystron modes on the scope face. Use the wavemeter to measure the center frequency of a mode and its frequency width.

EXERCISE 9

What is the electronic tuning range of the klystron?

EXERCISE 10

If your klystron can be mechanically tuned, what is its mechanical tuning range? To answer this question, vary the tuning screw of the klystron while adjusting the wavemeter in order to track the klystron frequency. When the screw becomes tight you have reached the high-frequency end of the klystron's range.

EXERCISE 11

From your answers to Exercises 4 and 10, which waveguide mode(s) are propagating?

Diode Detector

A commonly used microwave detector is the crystal diode that consists of a crystal of silicon in contact with a fine tungsten wire, called a "cat's whisker." A typical characteristic curve of such a diode is shown in Figure 2.11. The rectification properties of a crystal diode are defined in terms of the equivalent circuit shown in Figure 2.12. The incident microwave power produces the voltage drop V across the diode D and causes the current I to flow through the variable resistor R. At microwatt powers the rectified current is proportional to the microwave power or the square of the microwave voltage. As a result we say the diode is a **square-law detector** in the microwatt power region. In the milliwatt power region the rectified current is proportional to the square root of the microwave power or proportional to the microwave voltage, and the diode is said to be a **linear detector**. The transition from square law to linear behavior is very gradual, and the crossover point between the two regions is typically near the power range of 10^{-5} to 10^{-4} W. In Figure 2.11 square-law detection occurs when the applied voltage is about 0.0 to 0.3 V and linear detection starts when the applied voltage is about 0.5 V.

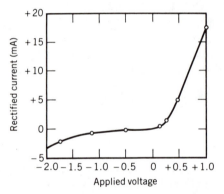

FIGURE 2.11 Characteristic curve of a crystal diode.

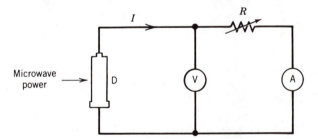

FIGURE 2.12 Equivalent circuit to define the rectification properties of a diode D.

EXERCISE 12

Examine the position of the diode in the waveguide and note that the diode is a short distance (approximately 1 cm) from the end of the waveguide. Measure the distance from the diode to the end of the waveguide. What fraction of the guide wavelength is the measured distance? Why is the diode so positioned in the waveguide?

Using the circuit shown in Figure 2.10 measure the dc voltage across the diode as a function of the incident microwave power. Periodically tune the reflector voltage so that the klystron is operating at the peak of a klystron mode, otherwise it may drift off of the peak. You do not need to know the absolute power, but rather the relative power for each change of the attenuator setting. As in equations 52 and 53, label the power incident on the attenuator as P_0 and the power transmitted as P. From equation 53, an attenuation of 3 dB gives $P = P_0/2$; hence, it is convenient to measure the diode voltage as the microwave power is varied in 3-dB increments. Start with P equal to $P_0/2$.

The voltage observed on the oscilloscope is proportional to the diode current, which in turn is proportional to the nth power of the microwave power, where $\frac{1}{2} \lesssim n \lesssim 1$.

Plot your data, observed voltage versus microwave power, by using graph paper, which best displays the power-law relation.

EXERCISE 13

From your graph determine the exponent n for low and high microwave powers. Do you find $n \simeq 1$ for low powers and $n \simeq \frac{1}{2}$ for high powers?

Voltage Standing-Wave Ratio

If a transmission line is not matched to the load, that is, if line and load impedances are not equal, then, in addition to the incident wave traveling in the positive z direction, there will be a reflected wave traveling in the negative z direction. For a TE_{10} mode the total electric field at some position z is, from equations 35 and 36,

$$E_y(x, z, t) = \frac{\gamma Z_c H_0}{k^2} \frac{\pi}{a} \sin \frac{\pi x}{a} e^{j\omega t - \gamma z} + \Gamma \frac{\gamma Z_c H_0}{k^2} \frac{\pi}{a} \sin \frac{\pi x}{a} e^{j\omega t + \gamma z} \quad \text{(V/m)} \quad (54)$$

where Γ is the reflection coefficient (see Experiment 1, Coaxial Transmission Line):

$$\Gamma \equiv \frac{\text{Amplitude of the reflected wave}}{\text{Amplitude of the incident wave}} = \frac{E_r}{E_i} \quad (55)$$

For convenience, let $x = a/2$ (center of the waveguide). Then equation 54 may be written

$$E_y(z, t) = E_i e^{j\omega t - \gamma z} + E_r e^{j\omega t + \gamma z} \quad \text{(V/m)} \quad (56)$$

where $E_r = \Gamma E_i$.

Along the waveguide the electric field goes through maxima and minima when one moves from points where E_i and E_r add constructively to positions where they partially cancel. Thus, along the waveguide there are points of maximum electric field

$$E_{max} = |E_i| + |E_r| \quad \text{(V/m)} \quad (57)$$

and points of minimum electric field

$$E_{min} = |E_i| - |E_r| \quad \text{(V/m)} \quad (58)$$

The voltage standing-wave ratio, VSWR, is defined by

$$VSWR \equiv \frac{E_{max}}{E_{min}} = \frac{|E_i| + |E_r|}{|E_i| - |E_r|} = \frac{1 + |\Gamma|}{1 - |\Gamma|} \quad (59)$$

which may be rearranged to yield

$$|\Gamma| = \frac{\text{VSWR} - 1}{\text{VSWR} + 1} \tag{60}$$

If the load impedance Z_L equals the line impedance Z_c, then there is no reflected wave and the VSWR is one. Thus, the VSWR is a measure of the impedance mismatch.

In this experiment you will determine the VSWR and $|\Gamma|$ by measuring E_{max} and E_{min} using a slotted section of waveguide. Connect the components as shown in Figure 2.13,

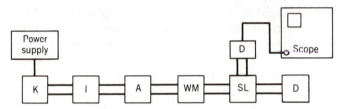

FIGURE 2.13 Microwave components to measure the VSWR.

where SL is the slotted line. Adjust the reflector voltage until the power output is a maximum, that is, the center of a klystron mode. The probe mounted on the slotted line will extract some power from the line to supply the diode. The probe will also set up reflections in the line. Ideally, we want to sample the field without altering it. You should insert the probe as little as is feasible to minimize disturbing the electromagnetic field configuration. With minimum probe insertion the power reaching the diode mounted on the probe is low and, hence, the diode is a square-law detector. For such a detector the VSWR is given by

$$\text{VSWR} = \frac{E_{max}}{E_{min}} = \frac{(V_{ob,max})^{1/2}}{(V_{ob,min})^{1/2}} \tag{61}$$

where V_{ob} is the voltage observed on the oscilloscope. Determine the VSWR and $|\Gamma|$.

EXERCISE 14

Knowing the distance between the positions where $V_{ob,max}$ and $V_{ob,min}$ occurred, what is the wavelength of the em waves in the guide? Is this wavelength consistent with your answers for Exercises 10 and 11 and equation 41?

Resonant Cavity

Using the arrangement of components shown in Figure 2.13, connect the oscilloscope to the diode at the end and tune the klystron to the resonant frequency of the cavity and leave it so tuned. Connect the circuit shown in Figure 2.14, where SST is the slide screw tuner and

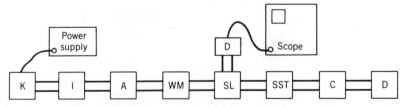

FIGURE 2.14 Microwave components to study impedance matching and the resonance frequency of a transmission cavity. The assembled components form a microwave spectrometer.

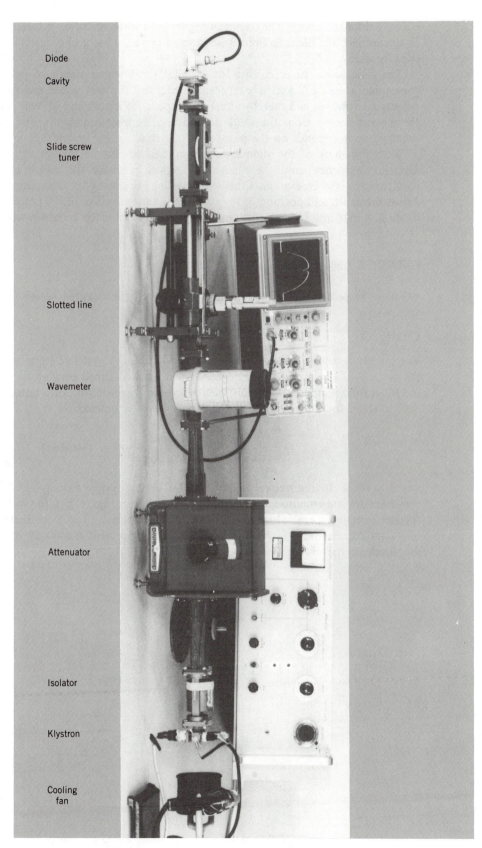

FIGURE 2.15 A microwave spectrometer.

C is the resonant cavity. The circuit in Figure 2.14 is the microwave spectrometer to be used in Experiment 16, Electron Spin Resonance. Figure 2.15 is a photograph of the microwave spectrometer.

The slide screw tuner consists of a metal shaft whose depth of penetration into the waveguide at $x = a/2$ and whose position z along the waveguide are variable. The shaft reflects microwaves and thereby alters the impedance match and VSWR. The magnitude of the wave reflected from the shaft depends on its insertion depth, and the phase of the reflected wave depends on the position of the shaft.

With the shaft of the slide screw tuner backed out of the waveguide, fine tune the klystron frequency until the microwave power reaching the diode at the end of the spectrometer is a maximum. Determine the magnitude of the reflection coefficient. Then adjust the depth and position of the slide screw tuner shaft until the power reaching the end diode is a maximum. Again determine the magnitude of the reflection coefficient.

EXERCISE 15

What is the percentage change in the amplitude of the reflected waves?

Cavity Q

The quality factor or Q of a slightly damped oscillatory system determines the response of the system to an external driving force. The slightly damped oscillatory system could be a mechanical oscillator, *RLC* circuit, microwave cavity resonator, or an atomic electron in an excited state, for example. The Q for such systems is defined by

$$Q = \omega \frac{\text{energy stored}}{\text{average power dissipated}} \quad \text{(unitless)} \tag{62}$$

where ω is the driving frequency and $\omega = 2\pi v$. For a system having one degree of freedom the characteristic amplitude versus ω/ω_r is sketched in Figure 2.16 for two values of Q and the same driving force, where ω_r is the resonant frequency. For a mechanical oscillator the characteristic amplitude is the amplitude of the displacement and for an *RLC* circuit it is the amplitude of the charge on the capacitor.

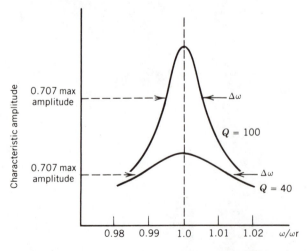

FIGURE 2.16 Curves to determine the Q of a damped oscillatory system.

The characteristic amplitude reaches a maximum at a frequency slightly less than the resonance value. The magnitude of this frequency shift increases as the Q decreases. For the Q values and horizontal scale shown in Figure 2.16 the shift is negligible. For example, for $Q = 40$ the amplitude peaks at $\omega/\omega_r = 0.9994$. Can you show that this is correct by considering the amplitude of an underdamped harmonic oscillator?

For an RLC circuit at resonance ($\omega = \omega_r$) equation 62 becomes

$$Q = \frac{\omega_r L}{R} \quad \text{(unitless)} \tag{63}$$

and for an underdamped mechanical oscillator at resonance

$$Q = \frac{\omega_r m}{b} \quad \text{(unitless)} \tag{64}$$

where m and b are the mass and damping constant.

A method of describing the sharpness of the response of the system to the driving force is to find two values of ω for which the amplitude is $1/\sqrt{2}$ or 0.707 times its maximum value (called half-power points) and to record their difference $\Delta\omega$. For an RLC circuit $\Delta\omega = R/L$ and for a mechanical oscillator $\Delta\omega = b/m$. For such systems, and for microwave cavities, Q may be written as

$$Q = \frac{\omega_r}{\Delta\omega} = \frac{\text{frequency at resonance}}{\text{full-width at half-maximum power}} \tag{65}$$

Using the circuit shown in Figure 2.14, display a klystron mode on the oscilloscope with the center frequency of the mode set to the resonant frequency of the cavity. Adjust the slide screw tuner until the mode resembles that shown in Figure 2.17. For the square-law region of the diode the observed voltage will be proportional to the microwave power.

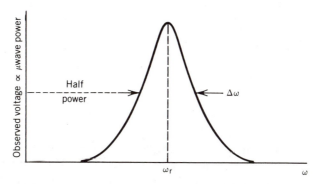

FIGURE 2.17 Microwave signal passed by a transmission cavity as a function of klystron angular frequency ω, where ω_r is the cavity resonance frequency.

EXERCISE 16

What is the frequency width $\Delta\omega$ and the Q of the cavity?

3. OPTICAL FIBER. NUMERICAL APERTURE, ATTENUATION

APPARATUS

~20-m optical fiber (50-μm diameter core, multimode, graded index, 850 nm)
Optical fiber receiver
Helium–neon laser
Digital voltmeter
Rotatable fiber mount (see Figure 3.12)
Mandrel wrap filter (see Figure 3.17)
Cladding mode stripper (see Figure 3.11)
Converging lens having a numerical aperture greater than that of the optical fiber

OBJECTIVES

To measure two important parameters of an optical fiber: numerical aperture and attenuation constant.

To gain an understanding of the optical fiber transmission line from the point of view of both geometrical (ray) optics and physical (wave) optics.

KEY CONCEPTS

Total internal reflection
Core
Cladding
Step index
Graded index
Attenuation constant
Numerical aperture
Critical angle

Meridional ray
Monomode fiber
Multimode fiber
Maximum acceptance angle
Phase constant
Cladding mode stripper
Mandrel wrap filter

REFERENCES

1. A. Portis, *Electromagnetic fields — Sources and Media*, Wiley, New York, 1978. Optical fibers are discussed on pages 523–525.
2. D. Marcuse, *Theory of Dielectric Optical Waveguides*, Academic, New York, 1974. The guidance of em waves by optical fibers is presented in Chapter 2.
3. D. Marcuse, *Light Transmission Optics*, Van Nostrand-Reinhold, Princeton, NJ, 1972. The first three sections of Chapter 8 contain a rigorous treatment of the application of Maxwell's equations to an optical fiber.
4. W. Allan, *Fibre optics, Theory and Practice*, Plenum, New York, 1973. A treatment of fibers based on geometrical optics is presented in Chapter 2.
5. Hewlett Packard *Fiber Optics Handbook*, prepared by Christian Hentschel, Hewlett Packard GmbH, Boeblingen Instruments Division, Federal Republic of Germany, October 1983. A good introduction and reference guide to fiber optic technology and measurement techniques.

6. D. Marcuse, *Principles of Optical Fiber Measurements*, Academic, New York, 1981. A variety of measurements are discussed, including measurement of the numerical aperture and the attentuation constant.

INTRODUCTION

Fiber optics is the branch of physics pertaining to the passage of light through thin filaments of transparent material. Such filaments are referred to as optical fibers, optical waveguides, dielectric waveguides, or simply fibers. The materials used in the manufacture of fibers are dielectrics and include optical glasses, fused silica, and certain plastics. A cross section of a fiber is shown in Figure 3.1a. The index of refraction of the core is greater than that of the cladding. The basic phenomenon involved in fiber optics is the guiding of light along the core by total reflection from the walls of the core. Figure 3.1b shows the index of refraction of the core n_1 and the cladding n_2 for a step-index fiber. Figure 3.1c shows indices for a graded-index fiber.

The principal characteristics of an optical fiber are its diameter (in μm), attenuation (in dB/km), and numerical aperture (NA).

In 1966, glass fibers were first suggested as transmission lines. They offer advantages over conventional transmission lines such as coaxial cables and waveguides. Some of the advantages are (1) lower attenuation, (2) larger bandwidth, (3) freedom from electromagnetic interference, and (4) low weight. (See Experiments 1 and 2, Coaxial Transmission Line and Waveguide.)

We first present a treatment of fibers based on geometrical optics. We will then outline the solution of Maxwell's equations and obtain the electromagnetic fields guided by the fiber.

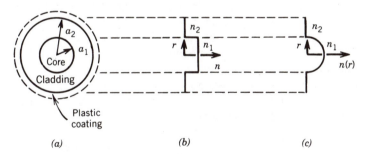

FIGURE 3.1 (a) Cross section of a fiber. Index of refraction n as a
function of fiber radius for (b) a step index fiber and
(c) a graded index fiber.

Geometrical Optics

If the dimensions of the fiber are large compared with the wavelength of light, then the guiding of light may be described by geometrical optics. A ray of light will be totally reflected at the boundary between two dielectric media when the ray is incident within the denser medium and the angle of incidence is greater than a critical angle. The critical angle depends on the refractive indices of the media and can be calculated from Snell's law. In Figure 3.2a incident, reflected, and refracted rays are shown for the case where the incident beam is in the region of higher index. The angle of incidence θ_1 and the angle of refraction θ_2 are related by Snell's law:

$$n_2 \sin \theta_2 = n_1 \sin \theta_1 \quad \text{(dimensionless)} \tag{1}$$

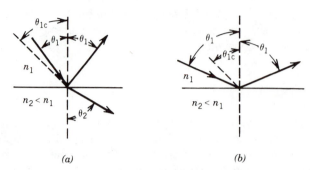

FIGURE 3.2 (a) Incident, reflected, and refracted rays.
(b) Total internal reflection.

Since $n_1 > n_2$, then $\sin \theta_2 > \sin \theta_1$, so that $\sin \theta_2$ equals one for a value of θ_1 less than $90°$. For a larger value of θ_1, equation 1 will yield a value for $\sin \theta_2$ greater than one, which is not possible. Physically, there is no refracted ray for such values of θ_1; that is, the incident ray is totally reflected. Total reflection is shown in Figure 3.2b. The critical angle for total reflection θ_{1c} is obtained by setting $\sin \theta_2$ to unity, which gives

$$\theta_{1c} = \arcsin \frac{n_2}{n_1} \qquad \text{(degrees)} \tag{2}$$

If θ_1 is greater than θ_{1c}, then total reflection occurs.

The reflectivity of a boundary between two dielectrics is not 100 percent. However, glass–glass boundaries have reflectivities as high as 99.95 percent. Such boundaries are more efficient reflection systems than metals; for example, the reflectivity of aluminum is about 90 percent. Hence, metallic waveguides have greater losses than dielectric waveguides.

We assume the rays guided by the fiber are meridional rays. A meridional ray is one whose path through the fiber is confined to a single plane. For a fiber in the form of a circular cylinder, such a plane contains the cylindrical axis. Figure 3.3 shows the passage of a meridional ray into a longitudinal-sectioned straight step-index fiber with end faces normal to the fiber axis. We now show that total internal reflection occurs in the fiber only if ϕ_1 is less than some angle ϕ_M. By applying Snell's law at the boundary between the media of indices n_0 and n_1, we find that

$$n_0 \sin \phi_1 = n_1 \sin \phi_2$$
$$= n_1 \cos \theta_1 \tag{3}$$

For total internal reflection to occur in the fiber we must have

$$\sin \theta_1 > \frac{n_2}{n_1} \tag{4}$$

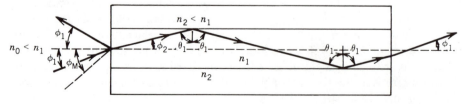

FIGURE 3.3 A meridional ray is shown entering, passing through, and exiting a step index fiber.

By using the identity $1 = \cos^2 \theta + \sin^2 \theta$ and equation 4, we have

$$\cos \theta_1 = \sqrt{1 - \sin^2 \theta_1} < \sqrt{1 - \frac{n_2^2}{n_1^2}} \qquad (5)$$

Then equations 3 and 5 yield

$$\sin \phi_1 < \frac{1}{n_0} \sqrt{n_1^2 - n_2^2} \qquad (6)$$

or

$$\phi_1 < \arcsin\left(\frac{1}{n_0} \sqrt{n_1^2 - n_2^2}\right) \equiv \phi_M \qquad \text{(degrees)} \qquad (7)$$

Equation 7 gives the condition on ϕ_1 for total internal reflection to occur in the fiber. ϕ_M is the maximum acceptance angle of the fiber. Note in Figure 3.3 that the ray emerging from the fiber makes an angle ϕ_1 with respect to the axis.

Figure 3.4 shows the passage of two parallel rays through a step-index fiber where the number of reflections differs by one. If the number of reflections is even, then the ray emerges parallel to its original direction. If the number of reflections is odd, then the ray emerges at an angle of $2\phi_1$ to its original direction. Therefore, a straight fiber will accept and propagate a cone of light incident on its end face, provided the conical semiangle is less than ϕ_M.

Normally the fiber is immersed in air or a vacuum and $n_0 \simeq 1.0$. Then, using equation 7, we may write

$$\sin \phi_M = \sqrt{n_1^2 - n_2^2} \qquad (8)$$

Equation 8 is a measure of the light-gathering power of the fiber and the right side is defined to be the numerical aperture (NA) of the fiber, by analogy with lens optics. Thus,

$$\text{NA} \equiv \sqrt{n_1^2 - n_2^2} \qquad (9)$$

A graded-index fiber has a refractive index profile $n(r)$ as shown in Figure 3.1c. The refractive index is a maximum at the center of the core and it (ideally) decreases uniformly until the core–cladding boundary, where it is equal to the cladding index. The variation of the refractive index with radial distance usually is described by a power law:

$$n(r) = \sqrt{n_1^2 - \text{NA}^2\left(\frac{r}{a_1}\right)^\alpha} \qquad r \le a_1 \qquad (10)$$

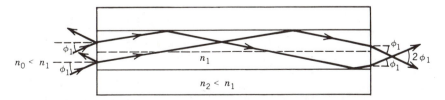

FIGURE 3.4 A ray with an even number of reflections emerges parallel to its original direction, and one with an odd number of reflections emerges at an angle of $2\phi_1$ to its original direction.

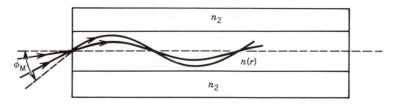

FIGURE 3.5 Path of two rays in a graded-index fiber.

where n_1 is the index at the center of the core and α is the exponent of the power law. Note that $n(0) = n_1$ and $n(a_1) = n_2$. When $\alpha = \infty$, the profile is that of a step index. For $\alpha = 2$, the profile is parabolic.

The NA for a graded-index fiber is defined identically to the step-index fiber:

$$\text{NA} \equiv \sqrt{n_1 - n_2} \tag{11}$$

The path of light rays in a graded-index fiber are shown in Figure 3.5 for two angles of incidence. The dependence of the core refractive index on r causes the light to travel on wavelike paths. The optimum value of α is 2, since in this instance there is no dispersion. That is, any two rays, such as those shown in Figure 3.5, require the same time to travel through a fiber.

EXERCISE 1

Knowing the manufacturer's value of NA for your fiber, what is the acceptance angle ϕ_M?

One of the important properties of optical fibers is that they continue to guide light if the fiber follows a curved path, provided the radius of curvature is not too small. Figure 3.6 shows rays in a step-index fiber of core radius a_1 and bent to a radius R, with a meridional ray passing through it. Note that the angle of incidence in the straight section is θ_1, for the outer curved surface it is θ_1', and for the inner curved surface it is θ_1'', where $\theta_1' < \theta_1 < \theta_1''$. If R is too small, then θ_1' becomes less than the critical angle θ_{1c} and total internal reflection does not occur at the outer curved surface; hence, light passes from the core to the cladding. (This is the principle of the **mandrel wrapped filter**; see Figure 3.17.)

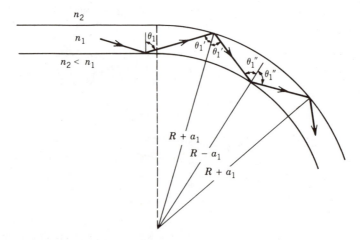

FIGURE 3.6 Path of a ray in a curved fiber.

Guidance of Electromagnetic Waves

The guidance of em waves by optical fibers is treated in detail in this section. Without loss of continuity it is possible to proceed to the experiments, omitting this section. If this section is omitted, then Exercise 5, under Attenuation of a Fiber, should also be omitted. Junior level electromagnetism and knowledge of Bessel's function and associated special functions are prerequisites for this section.

The electric and magnetic field configurations, or modes, transmitted by coaxial cables are most commonly TEM modes (see Experiment 1). Metallic waveguides will transmit TE or TM modes, but will not transmit TEM modes (see Experiment 2). In general both the **E** and **H** fields of the modes that propagate in a fiber have components in the direction of propagation; hence, in general the modes are not TE, TM, or TEM. Such modes are sometimes designated as HE modes. The term *mode* refers to a single electromagnetic wave satisfying Maxwell's equations and the boundary conditions of the fiber.

The procedure to determine the **E** and **H** fields in a fiber is to solve Maxwell's equations for the fields in both the core and the cladding, and then to require that the fields satisfy the boundary conditions at the core–cladding boundary. The steps to determine the fields in a fiber are summarized below.

1. Start with Maxwell's equations:

$$\nabla \times \mathbf{H} = \mathbf{J} + \frac{\partial \mathbf{D}}{\partial t} \qquad (\text{A/m}^2) \tag{12}$$

$$\nabla \times \mathbf{E} = -\frac{\partial \mathbf{B}}{\partial t} \qquad (\text{V/m}^2) \tag{13}$$

$$\nabla \cdot \mathbf{D} = 0 \qquad (\text{C/m}^3) \tag{14}$$

$$\nabla \cdot \mathbf{B} = 0 \qquad (\text{Wb/m}^3) \tag{15}$$

where we eliminate **J**, **D**, and **B** with the following equations:

$$\mathbf{J} = \sigma \mathbf{E} \qquad (\text{A/m}^2) \tag{16}$$

$$\mathbf{D} = \varepsilon \mathbf{E} \qquad (\text{C/m}^2) \tag{17}$$

$$\mathbf{B} = \mu \mathbf{H} \qquad (\text{Wb/m}^2) \tag{18}$$

where σ, ε, and μ are the conductivity, permittivity, and permeability. For both the core and cladding we assume $\sigma = 0$, $\rho = 0$, and $\mu = \mu_0$. The permittivity of the core and cladding will designated by ε_1 and ε_2, respectively.

2. Obtain the wave equation for **E** and **H**. Taking the curl of equation 13,

$$\nabla \times (\nabla \times \mathbf{E}) = -\mu_0 \frac{\partial}{\partial t}(\nabla \times \mathbf{H}) = -\mu_0 \varepsilon \frac{\partial^2 \mathbf{E}}{\partial t^2} \qquad (\text{V/m}^3) \tag{19}$$

where equation 18 was used to eliminate **B** and equation 12 was used to eliminate **H**. A vector identity is

$$\nabla \times (\nabla \times \mathbf{C}) = \nabla(\nabla \cdot \mathbf{C}) - \nabla^2 \mathbf{C} \tag{20}$$

Care must be exercised when equation 20 is applied in curvilinear coordinates. This is because the Laplacian ∇^2 operates not only on the components of **C** along coordinate directions but on the unit coordinate vectors as well. Using the vector identity, equation 19 becomes

$$-\mu_0\varepsilon\frac{\partial^2 \mathbf{E}}{\partial t^2} = \nabla\left(\nabla\cdot\frac{\mathbf{D}}{\varepsilon}\right) - \nabla^2\mathbf{E} = -\nabla^2\mathbf{E} \qquad (\text{V/m}^3) \tag{21}$$

where we substituted $\mathbf{E} = \mathbf{D}/\varepsilon$, used $\nabla\cdot\mathbf{D} = 0$, and assumed ε is not space dependent. Rewriting the wave equation, equation 21; we have

$$\nabla^2\mathbf{E} = \frac{1}{v^2}\frac{\partial^2 \mathbf{E}}{\partial t^2} \qquad (\text{V/m}^3) \tag{22}$$

where $v = 1/\sqrt{\mu_0\varepsilon}$ is the phase velocity and is expressed as $v = c/n$, where $c = 1/\sqrt{\mu_0\varepsilon_0}$ is the speed of light in vacuum and $n = \sqrt{\varepsilon/\varepsilon_0}$ is the index of refraction for the core or cladding.

The wave equation 22 is approximately satisfied by $\mathbf{E}$, even when ε varies in space, provided that its variation is slight over the distance of the light wavelength.

3. We seek solutions for E_i and H_i, the components of $\mathbf{E}$ and $\mathbf{H}$, where $i = \rho, \phi, z$, that have harmonic variation with respect to t and z, and periodic variations in ϕ. We assume

$$E_i(\rho, \phi, z, t) = AF(\rho)\, e^{jm\phi}\, e^{j(\omega t - \beta z)} \qquad (\text{V/m}) \tag{23}$$

$$H_i(\rho, \phi, z, t) = BF(\rho)\, e^{jm\phi}\, e^{j(\omega t - \beta z)} \qquad (\text{A/m}) \tag{24}$$

where we assumed that the fiber is cylindrical, with its axis in the z direction, $j^2 = -1$, A and B are amplitude coefficients, m is a positive or negative integer, $F(\rho)$ is an unknown function of ρ, ω is the angular frequency determined by the radiation source, and β is the unknown **phase constant** in the z direction or the unknown z component of the wave vector. We will see that β depends on ε_1, ε_2, ω, and the mode of propagation. Note that the physical E_i and H_i are the real parts of equations 23 and 24.

4. Solve the wave equation for E_z and H_z in both the core and cladding. Ideally, we would like all of the energy to be guided by the core and, hence, the cladding fields would be zero. Thus, we require that the fields be finite for $\rho = 0$ (center of the core) and that the fields vanish for large ρ in the cladding.

The wave equation in cylindrical coordinates for E_z is given by

$$\frac{\partial^2 E_z}{\partial\rho^2} + \frac{1}{\rho}\frac{\partial E_z}{\partial\rho} + \frac{1}{\rho^2}\frac{\partial^2 E_z}{\partial\phi^2} + \frac{\partial^2 E_z}{\partial z^2} - \frac{1}{v^2}\frac{\partial^2 E_z}{\partial t^2} = 0 \qquad (\text{V/m}^3) \tag{25}$$

Substituting equation 23 into 25 yields

$$\frac{d^2 F(\rho)}{d\rho^2} + \frac{1}{\rho}\frac{dF(\rho)}{d\rho} + \left(k^2 - \frac{m^2}{\rho^2}\right)F(\rho) = 0 \qquad (1/\text{m}^2) \tag{26}$$

where k is defined to be

$$k^2 \equiv -\beta^2 + \mu_0\varepsilon\omega^2 \qquad (\text{rad/m})^2 \tag{27}$$

Note that $\mu_0\varepsilon\omega^2 = \omega^2/v^2 = (2\pi/\lambda)^2$, where λ is the wavelength.

Equation 26 is a form of Bessel's equation. To obtain the standard form of Bessel's equation we define the dimensionless variable r:

$$r = k\rho \qquad (\text{dimensionless}) \tag{28}$$

In terms of r, equation 26 becomes the standard form of Bessel's equation:

$$r^2 \frac{d^2 F(r)}{dr^2} + r \frac{dF(r)}{dr} + (r^2 - m^2)F(r) = 0 \qquad \text{(dimensionless)} \qquad (29)$$

The solutions of equation 29 are cylindrical Bessel functions, and there are two independent solutions since the differential equation is of second order.

(i) $J_m(k\rho)$ and $J_{-m}(k\rho)$ are Bessel functions of the first kind of order m. However, if m is an integer, which it is in this case, then they are not linearly independent. $J_m(k\rho)$ is given by

$$J_m(k\rho) = \sum_{s=0}^{\infty} (-1)^s \frac{(k\rho/2)^{m+2s}}{s!\,\Gamma(m+s+1)} \qquad \text{(dimensionless)} \qquad (30)$$

and

$$J_{-m}(k\rho) = (-1)^m J_m(k\rho) \qquad \text{(m an integer)} \qquad (31)$$

Since m is an integer, $J_m(k\rho)$ and $J_{-m}(k\rho)$ are not independent solutions.

(ii) $N_m(k\rho)$ is the Bessel function of the second kind of order m, and it is called the Neumann function. It is defined as a linear combination of the Bessel functions of the first kind:

$$N_m(k\rho) \equiv \frac{J_m(k\rho) \cos m\pi - J_{-m}(k\rho)}{\sin m\pi} \qquad \text{(dimensionless)} \qquad (32)$$

We may take $J_m(k\rho)$ and $N_m(k\rho)$ as the two independent solutions.

(iii) $H_m^{(1)}(k\rho)$ and $H_m^{(2)}(k\rho)$ are Bessel functions of the third kind of order m. They are also called Hankel functions. They are defined to be

$$H_m^{(1)}(k\rho) \equiv J_m(k\rho) + jN_m(k\rho) \qquad \text{(dimensionless)} \qquad (33)$$

$$H_m^{(2)}(k\rho) \equiv J_m(k\rho) - jN_m(k\rho) \qquad (34)$$

Acceptable solutions to equation 29 inside the core are those functions that are finite at $\rho = 0$, and in the cladding the acceptable solutions must vanish for large values of $k\rho$. From equation 30, $J_m(k\rho)$ remains finite as ρ approaches zero. An important feature of $N_m(k\rho)$ is that it diverges as ρ approachs zero. The Hankel functions also diverge since $N_m(k\rho)$ diverges. Therefore, $J_m(k\rho)$ is the acceptable solution in the core.

We now consider the acceptable solution in the cladding. For imaginary values of k, $k \equiv j\kappa$, where κ is real, and large ρ, the Hankel functions are proportional to

$$H_m^{(1)} \propto e^{-\kappa\rho} \qquad (35)$$

$$H_m^{(2)} \propto e^{+\kappa\rho} \qquad (36)$$

Equation 36 is rejected as a solution in the cladding since it does not vanish for large ρ. $H_m^{(1)}$ will be the acceptable solution in the cladding.

Recognizing that the defined parameter k has different values in the core and cladding, we define the parameter in the core as k_1:

$$k_1^2 \equiv -\beta^2 + \mu_0 \varepsilon_1 \omega^2 \qquad (\rho < a) \qquad \text{(rad/m)}^2 \qquad (37)$$

The requirement that the solutions in the cladding vanish exponentially for large ρ dictates that k be imaginary. Thus, we define the parameter in the cladding as κ_2:

$$\kappa_2^2 \equiv +\beta^2 - \mu_0 \varepsilon_2 \omega^2 \qquad (\rho > a) \qquad (\text{rad/m})^2 \qquad (38)$$

Then the solutions for the z component of the fields are

$$\left. \begin{array}{l} E_z(\rho, \phi, z, t) = A J_m(k_1 \rho)\, e^{jm\phi}\, e^{j(\omega t - \beta z)} \\[2mm] H_z(\rho, \phi, z, t) = B J_m(k_1 \rho)\, e^{jm\phi}\, e^{j(\omega t - \beta z)} \end{array} \right\} \rho < a \qquad \begin{array}{l} (\text{V/m}) \\[2mm] (\text{A/m}) \end{array} \qquad (39)$$

$$\left. \begin{array}{l} E_z(\rho, \phi, z, t) = C H_m^{(1)}(j\kappa_2 \rho)\, e^{jm\phi}\, e^{j(\omega t - \beta z)} \\[2mm] H_z(\rho, \phi, z, t) = D H_m^{(1)}(j\kappa_2 \rho)\, e^{jm\phi}\, e^{j(\omega t - \beta z)} \end{array} \right\} \rho > a \qquad \begin{array}{l} (\text{V/m}) \\[2mm] (\text{A/m}) \end{array} \qquad (40)$$

where A, B, C, and D are the amplitude coefficients and β is the unknown phase constant. These five constants are determined from the boundary conditions.

5. The other components of $\mathbf{E}$ and $\mathbf{H}$ can be found from equations 12 and 13. We outline how to solve for these components: E_ρ, E_ϕ, H_ρ, and H_ϕ.

(a) Write equations 12 and 13 in terms of Cartesian components and, as before, assume harmonic t and z dependence of the fields.
(b) From (a) obtain six equations and solve them for E_x, E_y, H_x, and H_y in terms of the derivatives of E_z and H_z.
(c) Finally, transform the four Cartesian field components obtained in (b) to cylindrical components.

The resulting fields inside the core ($\rho < a$) are

$$E_z = A J_m(k_1 \rho)\, e^{jm\phi}\, e^{j(\omega t - \beta z)} \qquad (\text{V/m}) \qquad (41)$$

$$H_z = B J_m(k_1 \rho)\, e^{jm\phi}\, e^{j(\omega t - \beta z)} \qquad (\text{A/m}) \qquad (42)$$

$$E_\rho = \frac{-1}{k_1^2} \left[j\beta k_1 A J_m'(k_1 \rho) - \omega \mu_0 \frac{m}{\rho} B J_m(k_1 \rho) \right] e^{jm\phi}\, e^{j(\omega t - \beta z)} \qquad (\text{V/m}) \qquad (43)$$

$$E_\phi = \frac{1}{k_1^2} \left[\beta \frac{m}{\rho} A J_m(k_1 \rho) + j\omega \mu_0 k_1 B J_m'(k_1 \rho) \right] e^{jm\phi}\, e^{j(\omega t - \beta z)} \qquad (\text{V/m}) \qquad (44)$$

$$H_\rho = \frac{-1}{k_1^2} \left[\omega \varepsilon_1 \frac{m}{\rho} A J_m(k_1 \rho) + j\beta k_1 B J_m'(k_1 \rho) \right] e^{jm\phi}\, e^{j(\omega t - \beta z)} \qquad (\text{A/m}) \qquad (45)$$

$$H_\phi = \frac{-1}{k_1^2} \left[j\omega \varepsilon_1 k_1 A J_m'(k_1 \rho) - \beta \frac{m}{\rho} B J_m(k_1 \rho) \right] e^{jm\phi}\, e^{j(\omega t - \beta z)} \qquad (\text{A/m}) \qquad (46)$$

where

$$\beta^2 = -k_1^2 + \mu_0 \varepsilon_1 \omega^2 \qquad (\rho < a) \qquad (\text{rad/m})^2 \qquad (47)$$

The prime indicates differentiation of the Bessel function with respect to the argument $k_1 \rho$ and not ρ.

The resulting fields inside the cladding ($\rho > a$) are

$$E_z = C H_m^{(1)}(j\kappa_2 \rho)\, e^{jm\phi}\, e^{j(\omega t - \beta z)} \qquad (\text{V/m}) \qquad (48)$$

$$H_z = D H_m^{(1)}(j\kappa_2 \rho)\, e^{jm\phi}\, e^{j(\omega t - \beta z)} \qquad (\text{A/m}) \qquad (49)$$

$$E_\rho = \frac{-1}{\kappa_2^2}\left[\beta\kappa_2 CH_m^{(1)\prime}(j\kappa_2\rho) + \omega\mu_0\frac{m}{\rho}DH_m^{(1)}(j\kappa_2\rho)\right]e^{jm\phi}\,e^{j(\omega t - \beta z)} \qquad \text{(V/m)} \qquad (50)$$

$$E_\phi = \frac{-1}{\kappa_2^2}\left[\beta\frac{m}{\rho}CH_m^{(1)}(j\kappa_2\rho) - \omega\mu_0\kappa_2 DH_m^{(1)\prime}(j\kappa_2\rho)\right]e^{jm\phi}\,e^{j(\omega t - \beta z)} \qquad \text{(V/m)} \qquad (51)$$

$$H_\rho = \frac{1}{\kappa_2^2}\left[\omega\varepsilon_2\frac{m}{\rho}CH_m^{(1)}(j\kappa_2\rho) - \beta\kappa_2 DH_m^{(1)\prime}(j\kappa_2\rho)\right]e^{jm\phi}\,e^{j(\omega t - \beta z)} \qquad \text{(A/m)} \qquad (52)$$

$$H_\phi = \frac{-1}{\kappa_2^2}\left[\omega\varepsilon_2\kappa_2 CH_m^{(1)\prime}(j\kappa_2\rho) + \beta\frac{m}{\rho}DH_m^{(1)}(j\kappa_2\rho)\right]e^{jm\phi}\,e^{j(\omega t - \beta z)} \qquad \text{(A/m)} \qquad (53)$$

where

$$\beta^2 = \kappa_2^2 + \mu_0\varepsilon_2\omega^2 \qquad (\rho > a) \qquad (\text{rad/m})^2 \qquad (54)$$

In this case the prime indicates differentiation of the Bessel function with respect to $j\kappa_2\rho$.

6. We require the fields to satisfy the boundary conditions at the core–cladding boundary. General boundary conditions are given by equations 17–20 in Experiment 2, where for the fiber the subscripts 1 and 2 designate the core and cladding, respectively. We now make two simplifying assumptions: (1) We assume the fields have no azimuthal variation, that is, $m = 0$, and (2) we seek solutions of the TE mode type; that is, we set the amplitude coefficients A and C to zero. The fields in the core become

$$H_z = BJ_0(k_1\rho)\,e^{j(\omega t - \beta z)} \qquad \text{(A/m)} \qquad (55)$$

$$E_\phi = \frac{j\omega\mu_0}{k_1}BJ_0'(k_1\rho)\,e^{j(\omega t - \beta z)} \qquad (\rho < a) \qquad \text{(V/m)} \qquad (56)$$

$$H_\rho = \frac{-j\beta}{k_1}BJ_0'(k_1\rho)\,e^{j(\omega t - \beta z)} \qquad \text{(A/m)} \qquad (57)$$

and the fields in the cladding are given by

$$H_z = DH_0^{(1)}(j\kappa_2\rho)\,e^{j(\omega t - \beta z)} \qquad \text{(A/m)} \qquad (58)$$

$$E_\phi = \frac{\omega\mu_0}{\kappa_2}DH_0^{(1)\prime}(j\kappa_2\rho)\,e^{j(\omega t - \beta z)} \qquad (\rho > a) \qquad \text{(V/m)} \qquad (59)$$

$$H_\rho = \frac{-\beta}{\kappa_2}DH_0^{(1)\prime}(j\kappa_2\rho)\,e^{j(\omega t - \beta z)} \qquad \text{(A/m)} \qquad (60)$$

The boundary condition on **H** is given by equation 20 in Experiment 2. We will assume the surface current density **K** is zero and therefore the tangential component of **H** must be continuous. Hence, for the fiber,

$$BJ_0(k_1 a_1) = DH_0^{(1)}(j\kappa_2 a_1) \qquad \text{(A/m)} \qquad (61)$$

The boundary condition on **E** is given by equation 18 in Experiment 2, that is, the tangential component of **E** must be continuous. Thus,

$$\frac{jBJ_1(k_1 a_1)}{k_1} = \frac{DH_1^{(1)}(j\kappa_2 a_1)}{\kappa_2} \qquad \text{(V/m)/(rad/m)} \qquad (62)$$

where a property of Bessel functions was used: $J_1(x) = dJ_0(x)/dx$ and $H_1^{(1)}(x) = dH_0^{(1)}(x)/dx$.

TABLE 3.1 FIRST, SECOND, AND THIRD ZEROS OF THE ZEROTH-, FIRST-, AND SECOND-ORDER BESSEL FUNCTIONS OF THE FIRST KIND

n	$J_0(x) = 0$	$J_1(x) = 0$	$J_2(x) = 0$
1	$x = 2.405$	$x = 3.832$	$x = 5.136$
2	$x = 5.520$	$x = 7.016$	$x = 8.417$
3	$x = 8.654$	$x = 10.174$	$x = 11.620$

Taking the ratio of equations 61 and 62, dividing by j, and multiplying by $1/a_1$, we have

$$\frac{J_1(k_1 a_1)}{k_1 a_1 J_0(k_1 a_1)} = \frac{H_1^{(1)}(j\kappa_2 a_1)}{j\kappa_2 a_1 H_0^{(1)}(j\kappa_2 a_1)} \qquad \text{(dimensionless)} \qquad (63)$$

and, from equations 47 and 54,

$$-k_1^2 + \mu_0 \varepsilon_1 \omega^2 = \kappa_2^2 + \mu_0 \varepsilon_2 \omega^2 \qquad \text{(rad/m)}^2 \qquad (64)$$

Equations 63 and 64 are the two equations needed to determine k_1 and κ_2, and, hence, the phase constant β.

The propagation constant will be determined graphically. Some of the zeroes of $J_m(x)$ are given in Table 3.1, where n labels the roots ($n = 1$ is the first root, etc.). $J_0(x)$ versus x and $J_1(x)$ versus x are graphed in Figure 3.7a, and $J_1(x)/xJ_0(x)$ versus x is graphed in Figure 3.7b.

The asymptotic form of $H_m^{(1)}(x)$ is given by

$$H_m^{(1)}(x) = \frac{2}{\pi j^{m+1}} \sqrt{\frac{\pi}{2x}} e^{-x} \left[1 + 0\left(\frac{1}{x}\right) \right] \qquad \text{(dimensionless)} \qquad (65)$$

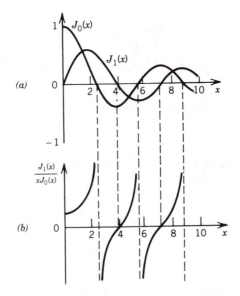

FIGURE 3.7 (a) Zeroth- and first-order Bessel functions (of the first kind) versus x. (b) The ratio $J_1(x)/xJ_0(x)$ versus x is zero when $J_1(x)$ is zero and undefined when $J_0(x)$ is zero.

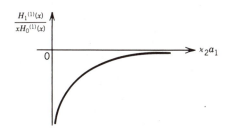

FIGURE 3.8 The ratio $H_1^{(1)}(x)/xH_0^{(1)}(x)$ approaches $-\infty$ and 0 as $\kappa_2 a_1$ approaches 0 and $+\infty$, respectively.

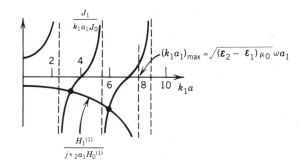

FIGURE 3.9 The intersection points are the solutions to equation 63.

where $0(1/x)$ denotes terms involving reciprocal powers of x that are very small. Using equation 65 we may write

$$\frac{H_1^{(1)}(x)}{xH_0^{(1)}(x)} \simeq \frac{1}{jx} \qquad \text{(dimensionless)} \tag{66}$$

For $x = j\kappa_2 a_1$ the right side of equation 66 becomes $-1/\kappa_2 a_1$. Figure 3.8 is a graph of the left side of equation 66, for $x = j\kappa_2 a_1$ versus $\kappa_2 a_1$.

We may superimpose the graphs in Figures 3.7b and 3.8 by using equation 64 and replacing $\kappa_2 a_1$ with $[(\varepsilon_1 - \varepsilon_2)\mu_0\omega^2 a_1^2 - k_1^2 a_1^2]^{1/2}$. The superimposed graphs are shown in Figure 3.9. The intersection points are marked with circles and they are the solutions to equation 63. The phase constant β is determined from the intersection points. This completes the solution for the fields in the fiber.

In Figure 3.9 the frequency is assumed high enough that two modes (two values of β) exist. Note that $(k_1 a_1)_{max}$ is determined by the frequency ω. If the maximum value of $k_1 a_1$ is less than the first root of $J_0(2.41)$, then the two curves do not intersect and radiation does not propagate in the fiber. Hence, the lowest cutoff frequency for TE_{0n} (n specifies the root number) modes is given by

$$\omega_{01} = \frac{2.405}{[(\varepsilon_1 - \varepsilon_2)\mu_0]^{1/2} a_1} \qquad \text{(rad/s)} \tag{67}$$

EXERCISE 2

Calculate ω_{01} for your fiber.

The fields of the TE_{01} mode inside the core are shown in Figure 3.10. Figure 3.10a shows

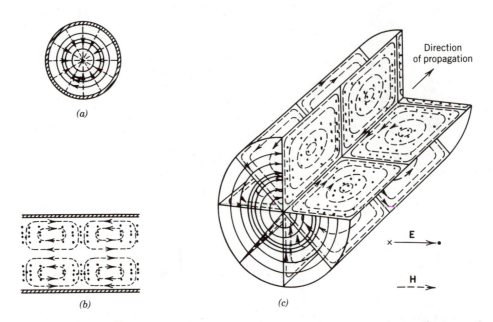

FIGURE 3.10 Fields of the TE_{01} mode inside the core. (a) Cross-sectional view. (b) Longitudinal section. (c) Three-dimensional perspective.

a cross section, Figure 3.10b a longitudinal section, and Figure 3.10c a three-dimensional perspective.

The solutions of Maxwell's equations allow only a limited number N of modes to propagate in a fiber. N can be calculated using an important fiber parameter called the V number. For a graded index fiber with parabolic profile the number of modes that will propagate is given by

$$N = \frac{V^2}{4} \tag{68}$$

where the V number is

$$V \equiv \frac{4\pi a_1}{\lambda} \sqrt{n_1^2 - n_2^2} = \frac{4\pi a_1}{\lambda} \text{ NA} \tag{69}$$

and a_1 is the core radius, λ is the wavelength in vacuum, n_1 is the refractive index at the core center, and n_2 is the refractive index of the cladding.

For a step-index fiber N is given by

$$N = \frac{V^2}{2} \tag{70}$$

where V is given by equation 69.

EXERCISE 3

How many modes can propagate in your fiber? To answer this question use the manufacturer's values of core diameter and NA.

EXERCISE 4

Using the value of NA for your fiber, calculate the core diameter required for only one mode to propagate. Such a monomode fiber has a step-index profile; hence, you should use equation 70.

EXPERIMENTS

Numerical Aperture (NA) of a Fiber

To effectively couple power into an optical fiber it is necessary to know the NA (see equation 11). The apparatus used to measure the NA is shown in Figure 3.11. One end of the 20-m fiber is attached to a rotatable mount, where the end of the fiber is situated at the rotational center of the mount. An exploded view of the mount is shown in Figure 3.12, where the rotational center of the mount is designated by the letter c. The angle of incidence of the laser beam can be varied by rotating the mount. Figure 3.13 shows a laser light plane wave incident at an angle ϕ. If the fiber end is rotated about the point c we can measure the voltage with the digital voltmeter in Figure 3.11 as a function of ϕ.

Typically the receiver shown in Figure 3.11 consists of a shielded integrated photodetector and dc amplifier (see the manufacturer's circuit diagram for your receiver). The photodetector converts the incident light signal from the fiber to a current proportional to the received power. Hence, the voltage read on the voltmeter is proportional to the received power.

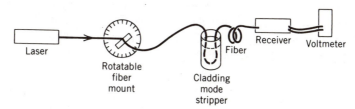

FIGURE 3.11 Apparatus to measure the numerical aperture.

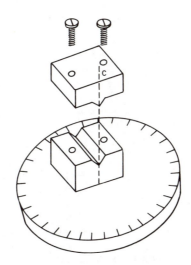

FIGURE 3.12 Rotatable fiber mount.

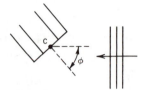

FIGURE 3.13 Plane wave incident at an angle ϕ on the end of a fiber.

At the fiber input the laser light is launched into both the cladding and the core. Unless light propagating in the cladding is removed, both the cladding and the core deliver power to the receiver. The radiation modes can be removed from the cladding by the cladding mode stripper. The cladding mode stripper consists of about 10 cm of fiber, with the jacket coating removed, immersed in a liquid that has a refractive index greater than the cladding index, such as oil or glycerine. Since the oil or glycerine has a larger refractive index than the cladding, internal reflection will not occur at the cladding–liquid boundary and radiation will escape into the liquid. In a 100-m or longer length of fiber a cladding mode stripper is not necessary because the attenuation of the cladding modes is greater than the attenuation of the core modes due to the smaller refractive index of the cladding. Hence, little power is delivered to the load via the cladding.

The jacket covering can be removed with a razor blade or dissolved with methylene chloride. Methylene chloride is toxic and it is recommended that rubber gloves be worn and a chemical hood be utilized.

It is often desirable to minimize the expenditure of resources by carrying out an approximate measurement of a parameter. Before doing a careful measurement of the NA you are asked to do a measure of it in the following way.

Note in Figure 3.4 that light entering at an angle ϕ_1 results in a cone of light with an angle $2\phi_1$ exiting the fiber. Connect the apparatus shown in Figure 3.14, where the NA of the lens is larger than that of the fiber. Figure 3.15 shows the lens, fiber core, and input

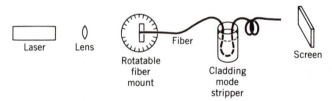

FIGURE 3.14 Apparatus to "quickly" measure the numerical aperture.

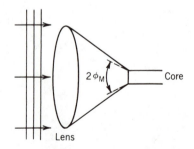

FIGURE 3.15 Radiation inside the cone of angle $2\phi_M$ enters the core.

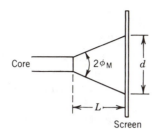

FIGURE 3.16 Output radiation is a cone of angle $2\phi_M$.

radiation. Figure 3.16 shows the output radiation from the end of the fiber. Measure L and d and determine the NA. Compare your result with the manufacturer's value.

Connect the apparatus shown in Figure 3.11. Measure the voltage as a function of angle for both positive and negative angles (this eliminates errors resulting from the axis of the fiber not being parallel to the laser beam at $\phi = 0$). The definition of the experimentally determined NA is the sine of the angle at which the power has fallen to 5 percent of the peak power at $\phi = 0$. Plot the voltage versus $\sin \phi$ on semilog paper for both positive and negative values of $\sin \phi$. Determine the NA and compare your value with the manufacturer's value.

A flat end face perpendicular to the fiber axis and free of blemishes is required for any fiber application. Several manufacturers sell kits for cutting and polishing fiber ends to an optically flat finish. For the purposes of this lab it is adequate to scribe the fiber with a carbide or diamond blade and to stress the fiber by pulling uniformly in opposite directions across the scribe mark. The initial fracture produced by the scribe mark will propagate due to the stress, producing an adequately flat end face. One way to test for irregularities in the end face is to examine the radiation pattern on a screen, as shown in Figure 3.16. It is recommended that some length of fiber be sacrificed in order to practice this technique of producing an adequate end face.

Stray light from overhead lights, windows, and so on, will enter the fiber along with the laser light. Arrange an aluminum foil shield around the input end of the fiber to reduce the amount of stray light entering the fiber. With the shield in place, block the laser light and measure the voltage produced by the stray light. Subtract the voltage due to the stray light before plotting your data to determine the NA.

Attenuation of a Fiber

The attenuation of a light signal in a fiber is one of the most important fiber parameters. It determines how long a fiber can be before the light signal that is launched into the input end can no longer be detected at the output end. In this part of the experiment you will determine the **attenuation constant** of a fiber.

Attenuation of light signals in silica fibers has three causes: (1) Rayleigh scattering, (2) absorption, and (3) bending losses. Rayleigh scattering is the scattering of light by particles that have linear dimensions considerably smaller than the wavelength of light. The scattered intensity is strongly dependent on the wavelength λ; it is proportional to $1/\lambda^4$. The sky appears blue because of Rayleigh scattering. Rayleigh scattering applies to any particles having a refractive index different from that of the surrounding medium, and in a silica fiber the scattering is caused by microscopic nonuniformity of glass and its refractive index. A ray of light is partially scattered into many directions; thus, some electromagnetic energy is lost.

Absorption of light depends on unwanted material in the fiber. Water (OH-ion) is the dominant absorber in most fibers, causing peaks in optical power loss at 1.25 and 1.39 μm. Above 1.7 μm, glass starts absorbing light energy due to a molecular resonance of SiO_2.

Bending losses occur when the angle of incidence at the core–cladding boundary is less than the critical angle.

The solutions obtained for equation 22 are fields that are not attenuated along the z axis, the direction of propagation. These nonattenuated fields are given in equations 41–46 and 48–53 for $\rho < a$ and $\rho > a$, respectively. In this experiment we are considering the attenuation of waves and we need a more general wave equation than equation 22; that is, we need a wave equation whose solutions are attenuated waves in the direction of propagation. If we assume the conductivity σ is not zero, we obtain the desired wave equation:

$$\nabla^2 \mathbf{E} = \frac{1}{v^2} \frac{\partial^2 \mathbf{E}}{\partial t^2} + \mu_0 \sigma \frac{\partial \mathbf{E}}{\partial t} \qquad (\text{V/m}^3) \tag{71}$$

EXERCISE 5

Derive equation 71 by following the steps that led to equation 22 but do not assume that the conductivity σ is zero.

By defining the constant α' by

$$\alpha' \equiv \frac{v^2 \mu_0 \sigma}{2} \qquad (1/\text{s}) \tag{72}$$

we may write equation 71 as

$$\nabla^2 \mathbf{E} = \frac{1}{v^2} \left(\frac{\partial^2 \mathbf{E}}{\partial t^2} + 2\alpha' \frac{\partial \mathbf{E}}{\partial t} \right) \qquad (\text{V/m}^3) \tag{73}$$

We try a solution of equation 73 of the form

$$\mathbf{E}(\rho, \phi, z, t) = \mathbf{f}(\rho, \phi, z, t) \, e^{-\alpha' t} \qquad (\text{V/m}) \tag{74}$$

Taking the first and second derivatives of equation 74, substituting into equation 73, and simplifying yields

$$\nabla^2 \mathbf{f} = \frac{1}{v^2} \left(\frac{\partial^2 \mathbf{f}}{\partial t^2} - \alpha'^2 \mathbf{f} \right) \qquad (\text{V/m}^3) \tag{75}$$

For glass α' is small ($\sim 2 \times 10^{-2} \, \text{s}^{-1}$), and in the approximation of small α' equation 75 becomes the wave equation without attenuation:

$$\nabla^2 \mathbf{f} \simeq \frac{1}{v^2} \frac{\partial^2 \mathbf{f}}{\partial t^2} \qquad (\text{V/m}^3) \tag{76}$$

The solutions to equation 76 are equations 41–46 for $\rho < a$ and equations 48–53 for $\rho > a$.

It is customary to write the attenuation in terms of the coordinate z rather than the time t by the substitution $t = z/v$. With this substitution the exponential attenuation becomes

$$e^{-\alpha' t} = e^{-(v^2 \mu_0 \sigma / 2)(z/v)} \equiv e^{-\alpha z} \tag{77}$$

where α is called the **attenuation constant**. The attenuated wave solutions of equation 71 are

$$\mathbf{E}(\rho, \phi, z, t) = \mathbf{f}(\rho, \phi, z, t) \, e^{-\alpha z} \qquad (\text{V/m}) \tag{78}$$

There are similar attenuated waves for the fields $\mathbf{H}(\rho, \phi, z, t)$.

The **phase constant** β was previously introduced in the solution of equation 22. The **propagation constant** γ is defined to be

$$\gamma \equiv \alpha + i\beta \qquad (1/\text{km}) \tag{79}$$

where $\mathrm{Re}(\gamma) = \alpha$ and $\mathrm{Im}(\gamma) = \beta$, and typically units are 1/km. Often the fields given by equation 78 are expressed in terms of γ.

The decrease in power as the light signal propagates along the fiber axis (assumed in the z direction) is given by

$$P(z) = P(0)e^{-2\alpha z} \quad \text{(W)} \tag{80}$$

where $P(z)$ is the power at a distance z from the input to the fiber, $P(0)$ is the power at the fiber input, and 2α is the **power attenuation constant** (1/km). The factor of 2 is included in the definition of the power attenuation constant because traditionally α is the attenuation constant for the electric field and the power is proportional to the square of the electric field.

Equation 80 is often written in terms of an attenuation constant expressed in decibels per kilometer (dB/km). The attenuation (measured in decibels) is defined by

$$A \equiv 10 \log \left[\frac{P(0)}{P(z)} \right] \quad \text{(dB)} \tag{81}$$

From equation 80, the expression for A becomes

$$A = 10(2\alpha z) \log e = 10(2\alpha z)0.434 \quad \text{(dB)} \tag{82}$$

We define the attenuation constant ξ by

$$2\xi \equiv 10(2\alpha)0.434 \quad \text{(dB/km)} \tag{83}$$

Rewriting equation 80 in terms of 2ξ yields

$$P(z) = P(0)(e^{1/0.434})^{-2\xi z/10\,\text{dB}} = P(0)10^{-2\xi z/10\,\text{dB}} \quad \text{(W)} \tag{84}$$

A graded index fiber with a 50-μm-diameter core typically allows about 500 different core modes, where each mode has a different attenuation constant. Hence, equation 80 is not valid for the total power propagating in the fiber. In addition, fiber imperfections, such as microbending and ellipticity, give rise to mode coupling; that is, the transfer of power among modes. Thus, equation 80 is also not valid for any one mode when mode coupling exists. Studies of the behavior of coupled modes in multimode fibers have shown that mode mixing causes the distribution of power over the various modes to reach a steady state beyond a certain mixing length of fiber. For modern low mode-mixing fibers the mixing length is several kilometers. In the steady-state configuration light power does decay according to equation 80.

In the steady state the attenuation constant can be interpreted as a steady-state value. The steady-state distribution of light power is called the equilibrium mode distribution, EMD.

Meaningful attenuation measurements in multimode fibers require the establishment of an EMD. A **mode mixer** (also called mode scrambler or mode filter) is a device that produces an EMD over a short length of fiber. The mode mixer suggested in this experiment is a **mandrel wrap filter** (see Figure 3.17). A mandrel wrap filter for a 50-μm graded-index fiber typically consists of a 12.7-mm- (0.5-in.-) diameter rod with five turns of coated fiber.

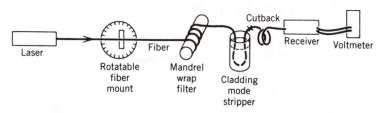

FIGURE 3.17 Apparatus to measure the attenuation constant.

The fiber tension should be such that it does not move by itself. This filter causes low-order modes to be converted to high-order modes and vice versa, until EMD is reached. Highest-order modes are forced to leave the core and are converted to cladding modes, or they completely leave the fiber. Because of this effect, a mandrel wrap filter always introduces a loss of a few decibels. A mode mixer should always be followed by a cladding mode stripper to remove the unwanted cladding modes.

Two methods are commonly used to perform attenuation measurements. Both methods measure the power output from a short length of fiber (~ 2 m) and from a long length of fiber (~ 20 m). The two measurements allow you to eliminate $P(0)$ in equation 80 and to solve for α. The cutback method is the most accurate. After measuring the power that reaches the end of the long fiber, the fiber is cut just beyond the cladding mode stripper (see Figure 3.17). The power is then measured at the end of the ~ 2-m fiber. The only problem with this method is its destructive nature. The second method is to measure the power propagated through a short length of fiber and through a separate long length of fiber. The problem with the second method is the difficulty in maintaining identical conditions of launching the light into each fiber.

The apparatus is shown in Figure 3.17. Using one of the two methods, measure the two voltages corresponding to the power delivered to the end of the two fibers. Use equation 80 to eliminate $P(0)$ and solve for α and its error. Also calculate ξ and its error. Compare your value with the manufacturer's value.

EXERCISE 6

What is the maximum length of fiber such that the signal output is barely detectable? To answer this question look up the minimum detectable signal specified by the receiver manufacturer and use your value of α. You may assume the conditions of launching the light into the fiber are the same as in your experimental setup.

Lasers

INTRODUCTION TO LASER PHYSICS

The 1964 Nobel prize in Physics was divided, one half being awarded to
    Charles Hard Townes, the United States
and the other one half was awarded jointly to
    Nikolai G. Basov, USSR, and Aleksandre M. Prochorov, USSR
    For fundamental work in the field of quantum electronics, which has led to the construction of oscillators and amplifiers based on the maser-laser-principle.

OBJECTIVES

To understand the physics involved in a helium–neon laser and, hence, to master the theory required for Experiments 4 and 5:

No. 4. Helium, Neon, and Helium–Neon Laser Spectra
No. 5. Laser Cavity Modes

KEY CONCEPTS

Stimulated absorption	LS coupling
Spontaneous emission	Electric dipole selection rules
Stimulated emission	Axial modes
Incoherent radiation	TEM modes
Coherent radiation	Spectral width
Einstein coefficients	Atomic lineshape
Population inversion	Loss coefficient
Forbidden transitions	Gain coefficient
Metastable states	

REFERENCES

1. B. A. Lengyel, *Lasers*, 2d ed., Wiley, New York, 1971. The first three chapters contain a good general description and a discussion of the theory of lasers. Helium–neon lasers are discussed on pages 302–308. Laser cavity modes and cavities with spherical mirrors are discussed on pages 75–103.

2. B. A. Lengyel, *Introduction to Laser Physics*, Wiley, New York, 1966. The first two chapters contain background material and a general description and a discussion of the theory of lasers. Helium–neon lasers are discussed on pages 191–201. Laser cavity modes are discussed on pages 67–83.

3. R. Eisberg and R. Resnick, *Quantum Physics of Atoms, Molecules, Solids, Nuclei, and Particles*, 2d ed., Wiley, New York, 1985. The laser and transition rates are discussed on pages 392–397. Transition rates and selection rules are discussed on pages 288–295. LS coupling is discussed on pages 356–361.

4. A. Yariv, *Quantum Electronics*, 2d ed., Wiley, New York, 1975. Excellent book, but advanced for most undergraduate students.

5. A. L. Bloom, *Gas Lasers*, Wiley, New York, 1968. Basic principles of laser operation are presented in Chapter 1; however, some knowledge of laser physics is assumed.

6. H. G. Heard, *Laser Parameter Measurements Handbook*, Wiley, New York, 1968. The radiation pattern of a laser is discussed for a variety of cavity configurations, pages 26–31.

7. R. A. Phillips and R. D. Gerhz, *Am. J. Phys.* **38**, 429 (1970). Experiment 5, Laser Cavity Modes, is primarily based upon this article. It is recommended that this article be read before doing Experiment 5.

8. D. A. Robinson, *Am. J. Phys.* **43**, 652 (1975). Experiment 5 is partially based upon this short article. It is recommended that it be read before commencing Experiment 5.

INTRODUCTION

Laser, an acronym for *l*ight *a*mplification by *s*timulated *e*mission of *r*adiation, refers to any device for creating a population inversion and thus producing a beam of coherent radiation.

The basic components of a continuous laser are

1. An "active or amplifying material," which includes the atoms that interact with radiation.
2. A mechanism to create a greater population of electrons in an excited state than exists in a lower state, that is, population inversion.
3. A resonator or cavity that has a resonant frequency equal to the photon frequency.

Transition Probabilities and Population Inversion

Suppose there are n atoms in equilibrium with a radiation field. For simplicity, we consider here electron transitions between two quantum states—the upper state being labeled 2 and the lower state labeled 1. Suppose there are n_2 atoms instantaneously in the upper state and n_1 in the lower state. The populations n_1 and n_2 may change in three ways: stimulated absorption, spontaneous emission, and stimulated emission. The initial and final configuration for each transition are shown in Figure L.1. When many atoms, such as those in a discharge tube, undergo spontaneous emission, the emitted photons have random directions and phases, and the radiation is said to be **incoherent**. In the stimulated emission process the two photons have the same phase and direction, and a beam of such photons is said to be **coherent radiation**.

All three of the processes shown in Figure L.1 occur in a laser. Now, the transitions cause the populations of the two levels to change with time and we want to describe the rate of change of n_1, and the corresponding rate of change of energy in the field.

We first ask, "How does stimulated absorption cause n_1 to change"? Well, there is a certain probability per second of an electron in level 1 absorbing a photon of energy $hv = E_2 - E_1$ and making a transition to level 2. This transition probability is proportional to n_1 and the density of photons having frequency v. (The transition probability is usually expressed in terms of $\rho(v)$ instead of photon density, where $\rho(v)$ is the photon density per unit frequency $\times hv$. Units are $J/m^3 \cdot s^{-1}$.) The constant of proportionality is called the Einstein B coefficient and is labeled B_{12} (with units of $m^3 \cdot s^{-2}/J$):

$$\frac{dn_1}{dt} = -n_1 B_{12}\rho(v) \qquad \text{(number of stimulated upward transitions per second)} \qquad (1)$$

where the negative sign implies that stimulated absorption causes n_1 to decrease with time.

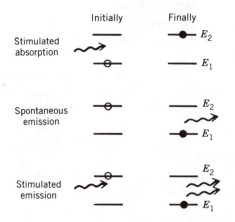

FIGURE L.1 Initial and final electron energies corresponding to stimulated absorption, spontaneous emission, and stimulated emission.

How does spontaneous emission cause n_1 to change? In this case the rate of change of n_1 is proportional to n_2. (If $n_2 = 0$, corresponding to no electrons in level 2, then spontaneous emission does not occur.) The constant of proportionality is the Einstein A coefficient, labeled A_{21} (in units of s^{-1}):

$$\frac{dn_1}{dt} = +n_2 A_{21} \quad \text{(number of spontaneous downward transitions per second)} \quad (2)$$

The rate of change of n_1 due to stimulated emission is proportional to the product $n_2 \times \rho(v)$. (If there are no electrons in level 2 to be stimulated or no photons to stimulate them, then stimulated emission does not occur.) The constant of proportionality is the Einstein B coefficient, labeled B_{21} (same units as B_{12}). Thus, we have

$$\frac{dn_1}{dt} = +n_2 B_{21} \rho(v) \quad \text{(number of stimulated downward transitions per second)} \quad (3)$$

From equations 1–3, the total rate of change of n_1 is given by

$$\frac{dn_1}{dt} = n_2 A_{21} + (n_2 B_{21} - n_1 B_{12})\rho(v) \quad \text{(number of transitions per second to and from level 1)}$$
$$(4)$$

Under thermal equilibrium conditions, n_1 and n_2 are related and the Einstein coefficients are also related. We now consider thermal equilibrium. If the atomic system is in thermal equilibrium with the radiation at a given temperature T, then the relative populations of any two energy levels, such as 1 and 2, are given by Boltzmann's equation:

$$\frac{n_2}{n_1} = \frac{e^{-E_2/kT}}{e^{-E_1/kT}} = e^{-hv/kT} \quad \text{(thermal equilibrium)} \quad (5)$$

where k is Boltzmann's constant. Note that $n_2 < n_1$.

Also under equilibrium conditions, the net number of downward transitions must be equal to the net number of upward transitions, namely,

$$n_2 A_{21} + n_2 B_{21} \rho(v) = n_1 B_{12} \rho(v) \quad \text{(thermal equilibrium)} \quad (6)$$

This is the **principle of detailed balance**, which states that for thermal equilibrium every process must be balanced by its exact opposite. By solving for $\rho(v)$, we obtain

$$\rho(v) = \frac{n_2 A_{21}}{n_1 B_{12} - n_2 B_{21}} \quad (\text{J/m}^3 \cdot \text{s}^{-1}) \quad (7)$$

By using equation 5 we obtain

$$\rho(v) = \frac{A_{21}}{B_{21}} \frac{1}{(B_{12}/B_{21}) e^{hv/kT} - 1} \quad (\text{J/m}^3 \cdot \text{s}^{-1}) \quad (8)$$

Planck's formula for emission of blackbody radiation by a body at temperature T is

$$\rho(v)\,dv = \frac{8\pi v^2}{c^3} \frac{hv}{e^{hv/kT} - 1}\,dv \quad (\text{J/m}^3) \quad (9)$$

For equation 8 to agree with Planck's formula, the following equations must hold:

$$B_{12} = B_{21} \qquad (\text{m}^3 \cdot \text{s}^{-2}/\text{J}) \tag{10}$$

$$\frac{A_{21}}{B_{21}} = \frac{8\pi h v^3}{c^3} \qquad (\text{J}/\text{m}^3 \cdot \text{s}^{-1}) \tag{11}$$

Historical Remark

In 1917, Einstein derived equations 10 and 11, but he did not know how to calculate the A and B coefficients. The calculation of these coefficients is a quantum mechanical, time-dependent perturbation theory calculation, and quantum mechanics was not developed until 1926.

The 1932 Nobel prize in Physics was awarded to Werner Heisenberg, Germany

For the creation of quantum mechanics, the application of which has, inter alia, led to the discovery of the allotropic forms of hydrogen.

The 1933 Nobel prize in Physics was awarded jointly to Erwin Schrödinger, Germany and Paul A. M. Dirac, Great Britain

For the discovery of new productive forms of atomic theory

Heisenberg developed "matrix mechanics," Schrödinger developed "wave mechanics," and Dirac developed relativistic quantum mechanics.

After the development of quantum mechanics the Einstein coefficients were calculated and B_{21}, in the electric dipole approximation, is

$$B_{21} = \frac{2\pi^2}{\varepsilon_0 h^2} \left| \int \psi_2^* (e\mathbf{r})\psi_1 \, dV \right|^2 \qquad (\text{m}^3 \cdot \text{s}^{-2}/\text{J}) \tag{12}$$

where ψ_1 and ψ_2 are the eigenfunctions of levels 1 and 2, $e\mathbf{r}$ is the electric dipole moment operator, and ε_0 is the permittivity of the vacuum.

Using equations 10 and 11 to express equation 4 in terms of B_{21}, we have

$$\frac{dn_1}{dt} = n_2 \frac{8\pi h v^3}{c^3} B_{21} + (n_2 - n_1)B_{21}\rho(v) \quad \text{(number of transitions per sec to and from level 1)} \tag{13}$$

Now, how does the laser provide *l*ight *a*mplification by *s*timulated *e*mission of *r*adiation? Well, we first ignore the spontaneous emission term in equation 13 because it gives rise to incoherent radiation and we are interested in coherent radiation, that is, "laser light." Then equation 13 becomes

$$\frac{dn_1}{dt} = (n_2 - n_1)B_{21}\rho(v) \tag{14}$$

Consider radiation passing through an optical medium that has atoms in levels E_1 and E_2. From equation 14 the rate of stimulated downward transitions will exceed that of upward transitions if $n_2 > n_1$, and, hence, the coherent radiation is amplified (an increase in the number of coherent photons).

The condition required for "lasing," namely, $n_2 > n_1$, is contrary to the thermal equilibrium distribution given by equation 5. The condition $n_2 > n_1$ is called **population inversion**.

The integral in equation 12 is the **matrix element** of the electric dipole moment operator between levels 1 and 2. If a matrix element is zero, the transitions between the levels are called **forbidden transitions**. The term forbidden means that transitions between the levels do not take place as a result of the interaction of the electric dipole moment of the atom with the radiation field. Forbidden transitions may occur by means of mechanisms other than electric dipole radiation. Such transitions occur at a much slower rate than the permitted (electric dipole) transitions.

A **metastable state** is a state from which all transitions to lower energy states are forbidden. A metastable state has a long lifetime, $\sim 10^{-3}$ s, and atoms making upward transitions due to collisions or radiation absorption tend to accumulate in metastable states. An ordinary state has permitted transitions to lower-energy states; hence, it has a short lifetime, $\sim 10^{-8}$ s.

In Figure L.2, the quantum states of helium are labeled in standard spectroscopic notation for an atom obeying LS coupling: $n^{2S+1}L_J$, where n is the principal quantum number and S, L, and J are the spin, orbital, and total angular momentum quantum numbers, respectively. The quantum states of neon are labeled by specifying the electron configuration ($2p^6$, $2p^5\,3s$, etc.), because these states cannot be described by LS or JJ coupling. (An intermediate coupling scheme must be used.)

Suppose 1 and 2 in equation 12 refer to atomic states describable by LS coupling, for example, helium. Then the subscripts 1 and 2 on ψ_1 and ψ_2 imply $L_1 S_1 J_1$ and $L_2 S_2 J_2$. Now, the general expression for an atomic jump between two such levels associated with electric dipole emission or absorption of a photon has been calculated (its just integral calculus—evaluation of the integral in equation 12) and they are stated as the **electric dipole selection rules**:

1. J may remain unchanged or it may increase or decrease by unity, but not by any larger amount. That is, either $J_1 = J_2$ or $J_1 - J_2 = -1$ or $J_1 - J_2 = +1$. Also a jump from a state with $J = 0$ to another state with $J = 0$ cannot occur.
2. L may remain unchanged or it may increase or decrease by unity, but not by any larger amount. That is, either $L_1 = L_2$ or $L_1 - L_2 = -1$ or $L_1 - L_2 = +1$.
3. S may not change. That is, $S_1 = S_2$.

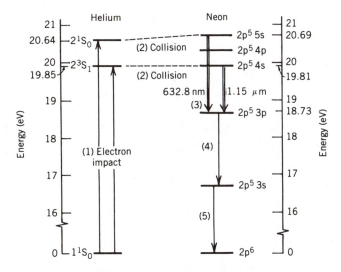

FIGURE L.2 Partial energy level diagram for helium and neon, where the five steps in the lasing process are numbered. Laser transitions are indicated with double lines.

In symbols, these results are

$$\Delta J = 0 \text{ or } \pm 1 \quad (\text{and not } 0 \rightarrow 0)$$
$$\Delta L = 0 \text{ or } \pm 1$$
$$\Delta S = 0$$

(15)

EXERCISE 1

Consider the three states of helium shown in Figure L.2. For each pair of levels determine ΔJ, ΔL, and ΔS, and use these results to ascertain if electric dipole transitions are allowed or forbidden. Specify whether each excited state of helium is a metastable state or not.

Active or Amplifying Medium: Helium and Neon Atoms

Helium and neon atoms constitute an active medium that interacts with the radiation field. The five steps in the lasing process, which are numbered in Figure L.2, are the following:

1. Electron impact excites helium atoms to the 2^1S_0 and 2^3S_1 states.
2. Resonant transfer of energy of excitation occurs in He–Ne collisions:

$$He(2^3S_1) + Ne(2p^6) \rightarrow He(1^1S_0) + Ne(2p^5 4s)$$
$$He(2^1S_0) + Ne(2p^6) \rightarrow He(1^1S_0) + Ne(2p^5 5s)$$

(16)

3. Stimulated emission occurs between the $2p^5 5s$–$2p^5 3p$ states and between the $2p^5 4s$–$2p^5 3p$ states. (These laser transitions are indicated with double lines.) The former transition produces 632.8 nm visible light and the latter 1.15 μm infrared radiation. Population inversion between each pair of states is maintained in the following way. The lifetimes of the three states are

$$\tau(2p^5 4s) = 0.96 \times 10^{-7} \text{ s}$$
$$\tau(2p^5 5s) = 1.1 \times 10^{-7} \text{ s}$$
$$\tau(2p^5 3p) = 1.85 \times 10^{-8} \text{ s}$$

(17)

Since $\tau(2p^5 3p)$ is less than $\tau(2p^5 4s)$ and $\tau(2p^5 5s)$, population inversion is maintained. The $2p^5 5s$ and $2p^5 4s$ states are not metastable. Photon or resonant trapping is the mechanism that increases their lifetimes. Photon trapping is the reabsorption of photons, in this case causing a $2p^5 3p \rightarrow 2p^5 5s$ transition at 632.8 nm or a $2p^5 3p \rightarrow 2p^5 4s$ transition at 1.15 μm. Each reabsorption process increases the effective lifetime of the state. In this instance the effective lifetime of both the $2p^5 4s$ and the $2p^5 5s$ states is increased by reabsorption of photons.

4. Electrons are removed from the $2p^5 3p$ state by (fast) spontaneous emission to the $2p^5 3s$ state.
5. Neon returns to the ground state by collisions with electrons and by collision with the container walls. Steps 1–5 are continuously repeated when the laser is operating.

Laser Cavity Modes

The schematic diagram of a helium–neon laser is shown in Figure L.3. The gas is confined in a glass or quartz tube typically 50 cm long and of 0.5 cm inside diameter. The tube is

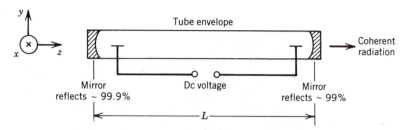

FIGURE L.3 Schematic diagram of a helium–neon laser.

terminated by two highly reflective circular mirrors with spherical curvature, which cause radiation to travel back and forth, and permit a fraction of the radiation to emerge as output.

The radiation that reflects back and forth between the mirrors interferes according to the superposition principle to produce standing waves. As you will learn in the discussion that follows, for the special case of plane mirrors the standing electromagnetic waves are analogous to standing waves on a string clamped at both ends. The possible wavelengths are

$$\lambda_m = \frac{2L}{m} \qquad m = 1, 2, 3, \ldots \qquad \text{(m)} \qquad (18)$$

where L is the mirror separation or the string length. The standing waves of a string are sketched in Figure L.4 for $m = 1, 2, 3$. The eigenfrequencies are given by

$$v_m = \frac{v}{\lambda_m} = \frac{v}{2L} m \qquad m = 1, 2, 3, \ldots \qquad \text{(Hz)} \qquad (19)$$

where for the string v is the wave velocity on the string and for the laser v is the velocity of light in the helium–neon gas. For the laser, these standing waves are called **longitudinal** or **axial modes**, and m is called the axial mode number or simply the mode number. Several axial modes may exist simultaneously in the laser cavity.

For every axial mode, there is a sequence of electromagnetic field configurations, called **transverse electromagnetic** (TEM$_{p\ell m}$) **modes**, some of which are shown in Figure L.5, where

1. p gives the number of nodal lines that lie in the xy plane and pass through the origin,
2. ℓ gives the number of nodal circles that lie in the xy plane, centered on the origin or axis of the cavity, and
3. m gives the eigenfrequencies (from equation 19) of the axial modes for given p and ℓ.

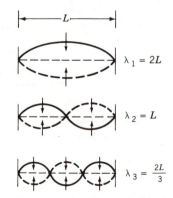

FIGURE L.4 Three standing waves on a string clamped at both ends. Dashed lines indicate the string position half a period later. Arrows show string motion.

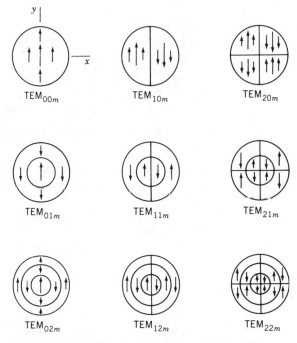

FIGURE L.5 Linearly polarized electric field configurations of $TEM_{p\ell m}$ modes for plane, circular mirrors, and arbitrary m. The circles and lines show where the electric field is zero. These figures are in the xy plane, the axes of which are shown in FIGURE L.3.

Figure L.5 is a diagram of the linearly polarized electric field configurations of $TEM_{p\ell m}$ modes for plane, circular mirrors, and arbitrary m. The arrows show the direction of the electric field, and the strength of the electric field is proportional to the length of the arrow. Depending on mirror alignment, two or more transverse modes may be simultaneously excited. Improvements in helium–neon laser construction have led to lasers that have only the TEM_{00m} mode excited, in which the electric field is uniform throughout the xy plane.

Spherical mirrors are easier to align than plane mirrors and they are typically used to form the resonant cavity. The electromagnetic fields in the cavity are calculated using numerical methods. For circular mirrors with a spherical curvature, the electric field for a particular direction of polarization is given by

$$E_{p\ell}(\rho, \phi, z) = E_0 \left(\frac{\rho \sqrt{2}}{w} \right)^{\ell} L_p^{\ell} \left(\frac{2\rho^2}{w^2} \right) \exp\left[\frac{-(\rho^2 \cos \ell \phi)}{w^2} \right] \sin kz \qquad \text{(V/m)} \qquad (20)$$

where ρ and ϕ are polar coordinates in the xy plane and z is parallel to the axis of the cavity. L_p^{ℓ} denotes a Laguerre polynomial and w denotes the spot size (for which the field has dropped to $1/e$ of its maximum value) of the TEM_{00m} mode at the mirror and is given by

$$w = \left(\frac{\lambda R}{\pi} \right)^{1/2} \left(\frac{L}{2R - L} \right)^{1/4} \qquad \text{(m)} \qquad (21)$$

where λ is the wavelength, L is the mirror separation, and R is the radius of curvature of the mirrors (assumed the same for both mirrors).

The eigenfrequencies of the standing waves are given by

$$v_{p\ell m} = \frac{c}{2L}\left[m + \frac{(2p+\ell+1)}{\pi}\arccos\left(1-\frac{L}{R}\right)\right] \quad \text{(Hz)} \qquad (22)$$

where it is assumed that the index of refraction of the helium–neon gas is one. For mirrors having different radii of curvature the arccos term becomes

$$\arccos\sqrt{\left(1-\frac{L}{R_1}\right)\left(1-\frac{L}{R_2}\right)} \qquad (23)$$

Note that if the mirrors are plane mirrors, then $R_1 = R_2 = \infty$ and the eigenfrequencies depend only on m; that is, from equation 22,

$$v_m = \frac{c}{2L}m \qquad m = 1, 2, 3, \ldots \quad \text{(Hz)} \qquad (24)$$

The wavelengths of the axial modes are then given by

$$\lambda_m = \frac{c}{v_m} = \frac{2L}{m} \quad \text{(m)} \qquad (25)$$

EXERCISE 2

For $\lambda_m = 632.8$ nm and $L = 50$ cm, show that the mode number m of a cavity with plane mirrors is 1.58×10^6. Show that the frequency spacing $v_m - v_{m-1}$ between two consecutive axial modes for such a cavity is 3×10^8 Hz.

For the same transverse modes, the frequency spacing between consecutive axial modes for spherical mirrors is, from equation 22,

$$v_{p\ell m} - v_{p\ell(m-1)} = \frac{c}{2L} \quad \text{(Hz)} \qquad (26)$$

which is the same as for plane mirrors.

The $2p^5\,5s$–$2p^5\,3p$ transition produces light centered at a frequency $v_0 = c/\lambda_0 = 4.738 \times 10^{14}$ Hz with a spectral width Δv, as shown in Figure L.6. The value of Δv is approximately 2×10^9 Hz.

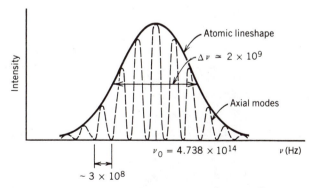

FIGURE L.6 Atomic lineshape of the $2p^5\,5s \rightarrow 2p^5\,3p$ neon transition. Note the axial modes that lie within the atomic lineshape.

EXERCISE 3

For $L = 50$ cm and plane cavity mirrors, show that the number of eigenfrequencies v_m that lie within the spectral width $\Delta v = 2 \times 10^9$ Hz is ~ 7. If $L = 7.5$ cm, then show that there is only one eigenfrequency within this spectral width.

The spectral width or width of the atomic lineshape is determined by the natural linewidth of the 632.8-nm spectral line and mechanisms that further broaden the line, such as pressure and Doppler broadening.

EXERCISE 4

The natural linewidth of the 632.8-nm line may be estimated by knowing the lifetimes of the $2p^5 5s$ and $2p^5 3p$ states, equation 17, and using the Heisenberg uncertainty principle. What is the estimated natural linewidth?

As the radiation is reflected back and forth between the mirrors of the cavity there are mechanisms that reduce the intensity. The primary loss mechanism is the mirror, which transmits 1 percent of the incident radiation (see Figure L.3). A secondary loss mechanism is misaligned mirrors, which cause light to pass out the sides of the cavity after multiple reflections. After one complete round-trip the losses decrease the intensity according to

$$I = I_0 e^{-2\gamma} \quad \text{(losses)} \qquad \text{(W/m}^2\text{)} \tag{27}$$

where γ is the (dimensionless) loss coefficient.

As a result of spontaneous and stimulated emission, light is generated within the laser. Coherent radiation intensity in the cavity is amplified by stimulated emission. After one complete round-trip the gain due to stimulated emission increases the intensity according to

$$I = I_0 e^{2\alpha L} \quad \text{(gain)} \qquad \text{(W/m}^2\text{)} \tag{28}$$

where L is the length of the cavity and α is the frequency-dependent gain coefficient per unit length.

When equations 27 and 28 are combined, the intensity after one complete round-trip is given by

$$I = I_0 e^{2(\alpha L - \gamma)} \qquad \text{(W/m}^2\text{)} \tag{29}$$

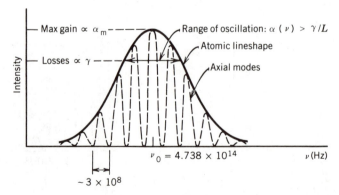

FIGURE L.7 The laser operates in the frequency interval where the gain exceeds the losses, that is, where $\alpha(v) > \gamma/L$.

If the losses exceed the gain ($\gamma > \alpha L$), then the standing waves inside the cavity die out. The standing waves may be sustained only if the gain is sufficient to compensate for the losses. The gain exceeds the losses, and, hence, the laser will operate, in the frequency interval in which $\alpha(v)$ is above γ/L:

$$\alpha(v) > \frac{\gamma}{L} \qquad (1/m) \tag{30}$$

Intensity, with losses and maximum gain shown, is sketched versus frequency in Figure L.7. The maximum gain coefficient is denoted by α_m. Three cavity modes are arbitrarily shown within the range of oscillation.

4. HELIUM, NEON, AND HELIUM−NEON LASER SPECTRA

APPARATUS

Spectrometer

Reflection grating ($\sim$30,000 reflecting surfaces per inch)

Helium and neon discharge tubes

Discharge tube power supply

Helium−neon 0.5-mW laser

Mercury vapor lamp

Prism

Safety goggles (To be worn when working with lasers. Some suppliers are Fish-Schurman, Glendale Optical, Korad/Hadron, Spectra Optics, and Ultra-Violet Products.)

OBJECTIVES

To use a reflection grating spectrometer to measure the wavelength of several spectral lines of both helium and neon.

To simultaneously observe the spectral line of the helium−neon laser and several spectral lines of neon.

To understand the physics involved in a helium−neon laser.

KEY CONCEPTS

Spectrometer alignment

Grating calibration

Emission spectral lines

REFERENCES

See the references listed in the Introduction to Laser Physics.

INTRODUCTION

See "Introduction to Laser Physics."

Reflection Grating and Spectrometer

Wavelength measurements are among the most accurate measurements in physics. The tool of measurement in this experiment is the spectrometer, principally a device for the accurate measurement of the angle of deflection of light (whether reflection, refraction, or diffraction). Figure 4.1 is a photograph of the spectrometer. Figure 4.2 shows the spectrometer viewed from above with the telescope in two different positions.

The reflection grating is an aluminized glass surface with precisely cut, parallel grooves. The number of reflecting surfaces per inch ($1/d$) is approximately 30 000 (see Figure 4.3).

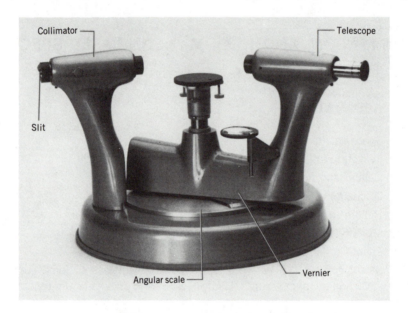

FIGURE 4.1 Optical spectrometer.

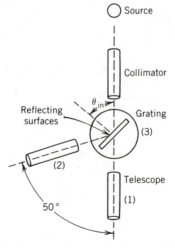

FIGURE 4.2 Top view of an optical spectrometer. Steps to align the spectrometer are numbered.

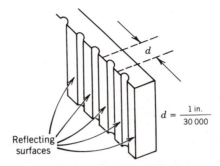

FIGURE 4.3 Diagram of a reflection grating.

NEVER TOUCH OR RUB THE SURFACE OF THE GRATING: TO DO SO DESTROYS THE GRATING, AND GRATINGS ARE EXPENSIVE TO REPLACE.

Each surface acts as a source of spherical waves (Huygen's principle), as shown in Figure 4.4. If the path difference from source to detector (the eye) of rays 1 and 2 in Figure 4.5 is

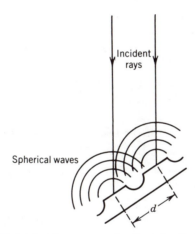

FIGURE 4.4 Incident rays and reflected spherical waves.

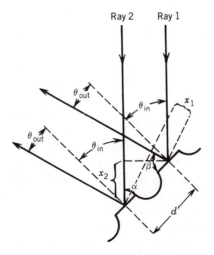

FIGURE 4.5 Diagram showing incident and reflected rays, and the geometry to determine the path difference.

$m\lambda$, where $m = 0, 1, 2, \ldots$, then constructive interference occurs. Note in Figure 4.5 that ray 2 travels a distance x_2 farther than ray 1 to reach the grating, where

$$x_2 = d \cos \alpha = d \cos\left(\frac{\pi}{2} - \theta_{in}\right) = d \sin \theta_{in} \quad \text{(m)} \quad (1)$$

Whereas, ray 1 travels a distance x_1 farther than ray 2 from the grating to reach the detector, where

$$x_1 = d \cos \beta = d \cos\left(\frac{\pi}{2} - \theta_{out}\right) = d \sin \theta_{out} \quad \text{(m)} \quad (2)$$

The path difference is $x_2 - x_1$ and for constructive interference

$$x_2 - x_1 = d(\sin \theta_{in} - \sin \theta_{out}) = m\lambda \quad \text{(m)} \quad (3)$$

Note that $\theta_{in} = \theta_{out}$ implies $m = 0$, which is the central maximum intensity.

EXPERIMENT

Alignment of Spectrometer

First move the mercury lamp near the slit and adjust it vertically. Then carry out the following optical adjustments:

1. Adjust the eyepiece of the telescope to focus on the crosshairs.
2. Focus the telescope at infinity. Why do we want the telescope focused at infinity?
3. Adjust the collimator lens for a sharp image of the slit viewed through the telescope.

The spectrometer will be aligned such that the angle θ_{in} is 65° (arbitrarily). The steps to do this are numbered as follows and are also shown in Figure 4.2.

1. With the grating removed align the telescope crosshairs on the nonmovable edge of the collimator slit.
2. Rotate the telescope 50° and read its angular position.
3. Place the grating on the table and rotate the table until the nonmovable edge of the slit is centered on the crosshairs. You are viewing the $m = 0$ spectrum in this position. The $m = 1$ spectral lines are to the left and the $m = -1$ spectral lines are to the right.

Calibration of Grating

The grating you are using is a replica; therefore, it has approximately the number of lines per inch of the original grating. Hence, we must use known wavelengths to calibrate the grating. It is suggested that the mercury arc lamp be used as the source of known wavelengths. Wavelengths of the spectral lines of mercury may be found in the *CRC Handbook of Chemistry and Physics*.

Since experimental physics has become extremely competitive in terms of funds, it is essential that you develop the ability to perform the measurement of interest while minimizing the required resources and time. Design an experiment, using the available resources, to calibrate the grating which will require the least time of the experimentalist.

EXERCISE 1

Does a single measurement to calibrate the grating risk accuracy? Does it make an undetected systematic error likely?

Design an experiment, using the available resources, which will yield the "best" value of the number of reflecting surfaces per unit length.

EXERCISE 2

Compare and contrast the expected results from the two designs. Do you expect greater accuracy from one design?

Calibrate the grating using one or both designs.

Helium and Neon Spectra

Replace the mercury lamp with the discharge tube containing the low-pressure helium gas. THE VOLTAGE SOURCE FOR THE DISCHARGE TUBE IS A FEW THOUSAND VOLTS. BE CAREFUL! **DO NOT TOUCH ANY EXPOSED WIRES THAT CONNECT TO THE POWER SUPPLY.** Determine the wavelength of four or five spectral lines. Compare your measured values with values listed in the *CRC Handbook of Chemistry and Physics*.

Replace the helium discharge tube with the neon tube and repeat the above measurements.

Helium–Neon–Laser

CAUTION

The helium–neon laser line should be observed with great caution to avoid any possibility of eye damage. **IT IS STRONGLY RECOMMENDED THAT SAFETY GOGGLES BE WORN WHEN WORKING WITH LASERS.**

Use a prism to reflect the laser as shown in Figure 4.6. Do NOT replace the prism with

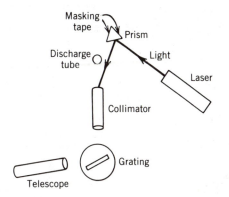

FIGURE 4.6 Arrangement of apparatus to observe neon lines and the laser line simultaneously.

a mirror. Attach tape to the other two sides of the prism to reduce the scattering of laser light throughout the laboratory.

Leave the neon tube so that its spectrum can be observed through the spectrometer. Calculate the angular position of the first-order, 632.8-nm helium–neon laser line. Rotate the telescope to this angular position. CLOSE THE COLLIMATOR SLIT. Arrange the helium–neon laser so that the laser and neon light fall on the CLOSED SLIT. This arrangement can be accomplished as shown in Figure 4.6 where some of the laser light is reflected from the prism to the CLOSED SLIT. OPEN THE SLIT JUST ENOUGH to observe the laser and neon spectra.

EXERCISE 3

Does the laser line coincide with any of the neon lines? Explain why it should or should not coincide.

5. LASER CAVITY MODES

APPARATUS

Helium–neon 0.5-mW laser
Spectral analyzer
Photodiode detector circuit (see Figure 5.4a)
2 convex lenses
    or (if a spectral analyzer is not available)
Helium–neon 0.5-mW laser
Oscilloscope
High-frequency oscillator, such as a General Radio 180- to 600-MHz oscillator
Photodiode detector circuit (see Figure 5.5a)
2 convex lenses

OBJECTIVES

To qualitatively observe the spatial distribution of laser light and, hence, to identify $TEM_{p\ell m}$ modes corresponding to different values of p and/or ℓ.

To measure the frequency differences of the various $TEM_{p\ell m}$ modes and to determine the effective linewidth $\Delta \nu$ of the $2p^5 5s \rightarrow 2p^5 3p$ transition of neon.

KEY CONCEPTS

$TEM_{p\ell m}$ modes	Signal mixing
Square-law detector	Intermediate frequency (IF) signal
Photodiode response time	Effective linewidth

REFERENCES

See the references listed in Introduction to Laser Physics.

INTRODUCTION

See "Introduction to Laser Physics."

EXPERIMENT

Transverse Modes

Use the two lenses, as shown in Figure 5.1 to expand the laser beam. How is the separation of the lenses related to their focal lengths? Observe the spatial distribution of light on a screen.

FIGURE 5.1 Laser beam expansion.

EXERCISE 1

Based on your observation, explain which transverse mode(s) are excited in the laser cavity. (Different transverse modes correspond to different values of p and/or ℓ. See Laser Cavity Modes in "Introduction to Laser Physics.")

The effect of mirror alignment on modes can be observed if a laser with adjustable mirrors is available. If possible vary the alignment of the mirrors while observing the expanded laser beam on a screen. This causes different modes to become active. It is possible to have two or more transverse modes active at the same time. Sketch each observed pattern in your notebook, and do Exercise 1.

TEM$_{p\ell m}$ Modes

A photodiode is a semiconductor diode used as a radiation detector. In operation the p-n junction is reversed biased and the junction region is illuminated with photons of energy $h\nu$ larger than the band gap of the semiconductor. Photon absorption creates electrons (●) and holes (○) on both sides of the junction, as shown in Figure 5.2. The created electrons and holes are swept across the junction by the applied electric field $\mathbf{E}_{applied}$, thereby increasing the reverse saturation current. Photodiode current density versus bias voltage is sketched in Figure 5.3 for three different incident radiation intensities. The reverse saturation current density increases with radiation intensity.

Photodiodes are square-law detectors; that is, the diode current density J is proportional to the square of the electric field. To illustrate this, let us consider two electromagnetic

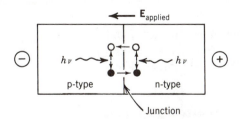

FIGURE 5.2 Schematic diagram of a photodiode with an applied field and incident photons that create electrons (●) and holes (○).

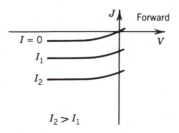

FIGURE 5.3 Photodiode current density versus bias
voltage for three radiation intensities.

waves having the same polarization, different frequencies, and different amplitudes. The
total electric field is

$$E = E_1 \cos \omega_1 t + E_2 \cos \omega_2 t \qquad \text{(V/m)} \tag{1}$$

and, after some algebra, the squared electric field may be written as follows:

$$E^2 = E_1^2 \cos^2 \omega_1 t + E_2^2 \cos^2 \omega_2 t + E_1 E_2 \cos(\omega_1 + \omega_2)t + E_1 E_2 \cos(\omega_1 - \omega_2)t \qquad \text{(V/m)}^2 \tag{2}$$

The high-frequency response of diode photodetectors is limited by the equivalent RC time
constant of the detector. The response time is typically 1 ns, or the high-frequency response
is about 10^9 Hz. The diode detects the average of the components in equation 2 whose
frequency exceeds 10^9 Hz. Assuming that ω_1 and ω_2 are greater than $2\pi \times 10^9$ rad/s and that
the difference frequency, $\omega_1 - \omega_2$, is less than $2\pi \times 10^9$ rad/s, then the current density J is

$$J \propto \frac{E_1^2}{2} + \frac{E_2^2}{2} + E_1 E_2 \cos(\omega_1 - \omega_2)t \qquad (\text{A/m}^2) \propto (\text{V/m})^2 \tag{3}$$

where the average of the linear cosine term, $\cos(\omega_1 + \omega_2)t$, is zero. If three electromagnetic
waves of different frequencies ω_1, ω_2, and ω_3 are incident on the diode, then

$$J \propto E_1 E_2 \cos(\omega_1 - \omega_2)t + E_1 E_3 \cos(\omega_1 - \omega_3)t + E_2 E_3 \cos(\omega_2 - \omega_3)t \qquad (\text{A/m}^2) \propto (\text{V/m})^2 \tag{4}$$

where the constant current terms, $E_1^2/2$, and so on, were dropped.

The eigenfrequencies of the laser radiation are given by equation 22 in the "Introduction
to Laser Physics." Calculate the frequency difference between two adjacent axial modes
TEM_{00m} and $\text{TEM}_{00(m+1)}$. You will need the manufacturer's values of the mirror separation
L and radii of curvature, R_1 and R_2. Also calculate the frequency difference between
TEM_{00m} and TEM_{01m} modes, and between the TEM_{01m} and $\text{TEM}_{00(m+1)}$ modes.

EXERCISE 2

Which of the above frequency differences could be detected by a diode having a 10^{-9}-s
response time? For a diode having such a response time determine the largest value of n
such that the diode could detect the frequency difference between the TEM_{00m} and
$\text{TEM}_{00(m+n)}$ modes.

If a spectral analyzer is available, wire the circuit shown in Figure 5.4a. Figure 5.4b is a
block diagram of the apparatus. Mount the circuit in a minibox using shielded leads. Place
the photosensitive surface of the diode opposite a hole drilled in the minibox.

A spectral analyzer measures the strength of each frequency of the input signal. The
procedure is to scan the frequency range of the analyzer while observing its output. When

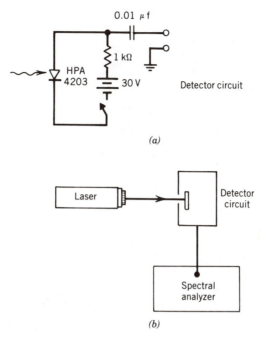

(a)

(b)

FIGURE 5.4 (a) Photodiode detector circuit used with a
spectral analyzer. (b) Block diagram.

the dial frequency on the analyzer coincides with a frequency component of the input signal,
a signal is displayed. Thus, a spectral analyzer would detect a signal when tuned to
$(\omega_1 - \omega_2)$ for a current density given by equation 3.

Scan the frequency of the analyzer and determine the frequencies of the input signal.

EXERCISE 3

Compare each observed frequency with its theoretical frequency, and identify the two
$TEM_{p\ell m}$ modes.

EXERCISE 4

Taking the effective linewidth of the atomic lineshape to be the range of oscillation shown
in Figure L.7 of the "Introduction to Laser Physics," determine the appropriate effective
linewidth from your measurements.

If a spectral analyzer is not available, then wire the circuit shown in Figure 5.5a. A block
diagram is shown in Figure 5.5b. Mount the circuit in a minibox using shielded leads and
place the photosensitive surface of the diode opposite a hole drilled in the box.

The photodiode produces signals from the laser at the difference frequencies, which are
then mixed in an uhf diode with a signal from a local oscillator. The intermediate frequency
(IF) output signal of the 1N23B is at the difference frequency of the two mixed signals; that
is, the 1N23B is also a square-law detector and the current density J is given by equation
3 where, in this case, ω_1, say, is the local oscillator frequency and ω_2 is the difference
frequency of two $TEM_{p\ell m}$ modes. The IF output signal, as observed on the oscilloscope, is
then set to 0 Hz by tuning the local oscillator frequency to the difference frequency of two
$TEM_{p\ell m}$ modes, that is, $\omega_1 - \omega_2$.

Answer Exercises 3 and 4.

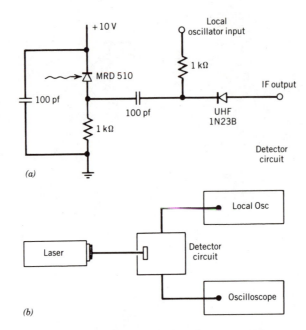

FIGURE 5.5 (*a*) Detector circuit that does not require a spectral analyzer. (*b*) Block diagram.

Microcomputers

6. INTRODUCTION TO COMPUTER-ASSISTED EXPERIMENTATION

APPARATUS

Microcomputer

Analog-to-digital/digital-to-analog conversion (ADC/DAC) card

Hall sample

Conditioning circuit for the Hall sample, see Figure 6.5

Calibrated electromagnet

1-μCi ^{137}Cs source

Geiger-Mueller tube and a scaler/timer or a scintillation counter and a single-channel analyzer
Conditioning circuit for scaler/timer, see Figure 6.11 or 6.13, or a conditioning circuit for a
    single-channel analyzer, see Figure 6.13
Conditioning circuit to measure time intervals, see Figure 6.11 or 6.13.

Figure 6.8 is a photograph of the conditioning circuit for the Hall sample. The circuit is easy to construct and is convenient for bench testing. The other circuits used in this experiment may be similarly constructed.

SOFTWARE

The publisher of this text has provided your instructor with software for this experiment. The software is dependent on both the ADC/DAC card and the computer. The software that is provided is written for both the Apple II and the IBM PC. The Apple II software is written for two cards: the 8-bit Mountain card and the 12-bit Sunset-Vernier card. Software for the IBM PC is also written for two cards: the 8-bit ML-16 Multi-Lab card and the 12-bit IBM DACA card. Instructions for each of these cards are at the end of the experiment.

For each card, the software displays the following menu on the monitor.

1. ADC-DAC Check.
2. Voltage Measurements.
3. Hall Effect.
4. Nuclear Counting.
5. Poisson Intervals.
6. Screen Test.

Program 1 is a brief introduction to ADC and DAC and involves connecting an analog input to the ADC and measuring the analog output of the DAC. Programs 2 to 5 are used with experiments. Program 6 is, as is indicated by its title, a test program.

OBJECTIVES

To bench test conditioning circuits and, hence, verify that the circuits are suitable for interfacing experimental apparatus to the computer via an ADC/DAC card.

To connect a variable dc voltage to the computer and use the provided software, Program 2, to do a calibration. The computer will then function as a voltmeter.

To interface the Hall sample to the computer via a conditioning circuit. To use the computer, Program 3, to measure the Hall voltage as a function of the Hall current (for constant magnetic field) and as a function of the magnetic field (for constant Hall current). To use the computer to control the experiment (the Hall current).

To interface a scaler/timer or a single-channel analyzer to the computer via a conditioning circuit. To use the computer, Program 4, to count the number of radionuclide that decay during a 3-s time interval and to analyze the Poisson distribution of counts. To also use the computer, Program 5, to measure the time interval between successive decays and to analyze the (time) interval distribution of times.

To understand the general approach of interfacing experimental apparatus to a microcomputer via an ADC/DAC card.

To learn how to create programs suitable for voltage, counting, and timing measurements.

To understand the use of a microcomputer to control an experiment, to collect the data, to analyze the data, and to provide a hard copy of the results.

KEY CONCEPTS

Bit	Analog range
Byte	Resolution
Analog signal	Hall voltage
Digital signal	Hall current
Analog-to-digital conversion (ADC)	Poisson distribution
Digital-to-analog conversion (DAC)	Interval distribution
Conditioning circuit	

REFERENCES

1. K. Ratzlaff, *Introduction to Computer-Assisted Experimentation*, Wiley, New York, 1987. This is a fairly general introduction to the subject matter.

2. C. Spencer, P. Seligmann, and D. Briotta, *Computers in Physics* **1**(1), 59 (Nov./Dec. 1987). The characteristics and specifications of a general-purpose interface for physics experiments are presented. The results of the development of such an interface are given. Also 50 references are listed, most of which are papers describing the use of computers in experimental physics.

3. D. Briotta, P. Seligmann, P. Smith, and C. Spencer, *Am. J. Phys.* **55**(10), 891 (1987). The use of microcomputers in undergraduate physics labs is discussed with examples given.

4. M. De Jong, *Apple II Applications*, Sams, Indianapolis, IN, 1983. Interfacing experimental apparatus to the Apple II microcomputer via the game port and an analog-to-digital converter card is discussed.

5. P. Horowitz and W. Hill, *The Art of Electronics*, Cambridge University Press, New York, 1980. The following circuit components that are used in this experiment are discussed in this book: transistors, monostable multivibrators, and operational amplifiers.

6. R. Evans, *The Atomic Nucleus*, reprint ed. of 14th (1972) printing, Krieger, Melbourne, FL, 1982. The interval distribution is discussed in general in Chapter 28.

7. *The TTL Data Book for Design Engineers*, 2d ed., Texas Instruments, Dallas, 1981. Specifications for the 74121 integrated circuit are given.

8. P. Bligh, J. Johnson, and J. Ward, *Phys. Educ.* **20**, 246 (1985). This article discusses interfacing the Hall sample to a microcomputer.

9. S. C. Gates with J. Becker, *Laboratory Automation Using the* IBM PC, Prentice-Hall, Englewood Cliffs, NJ, 1989.

10. T. Frederiksen, *Intuitive IC Electronics, A Sophisticated Primer for Engineers and Technicians*, McGraw-Hill, New York, 1982.

11. Appendix F: Writing Programs. Writing programs in BASIC and in assembly language are discussed.

12. Appendix G: A Software Tutorial for the IBM PC.

INTRODUCTION

Many experiments in this book have an optional section labeled Computer-Assisted Experimentation, which involves interfacing a microcomputer to the experimental apparatus and programming the computer to control or monitor the experiment. This experiment is a prerequisite for such optional sections.

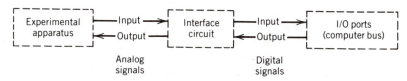

FIGURE 6.1 Block diagram of the computer-interface apparatus system.

Figure 6.1 is a simplified schematic diagram of the computer-interface apparatus system. This introduction is a general discussion of this system.

The signal output of the experimental apparatus is an analog signal, that is, it is continuous over a range of values. The analog signal could be sinusoidal, a pulse, or a decaying exponential, for example. A digital signal is composed of discrete values. Digital electronics and microcomputers deal with digital signals that have one of two values, commonly labeled on or off, 1 or 0. For computer analysis, the analog signal of the apparatus must be converted to digital form. The various integrated circuits (ICs) inside the computer interpret a voltage between about 3 and 5 V as a *bi*nary digi*t* (bit) value of 1, and a voltage between 0 and 1 V is interpreted as a bit value of 0. Simply stated, the purpose of the interface circuit is to convert the analog signal of the apparatus to a binary signal that is suitable for the computer.

Before discussing three methods of interfacing apparatus to the computer, it is worthwhile to discuss bits and bytes. A bit value is 0 or 1, and an 8-bit computer fetches and writes data and instructions 8 bits at a time. An 8-bit number ranges from 00000000 to 11111111 in the binary number system, or from 0 to 255 in the decimal number system. A group of 8 bits is called a byte. The memory capacity of a computer is usually given in kilobytes (Kbytes or just K), and 1 kilobyte is 1024 memory locations. A memory location, in terms of hardware, is 8 flip-flops. A flip-flop is a device that stores one of two states: on or off, 1 or 0.

The microcomputers presently available have a central processing unit (CPU) which is either an 8-bit, 16-bit, or 32-bit microprocessor with an 8- or 16-bit memory bus. The instructions of such microprocessors can manipulate 8-bit, 16-bit, or 32-bit data, and the data and instructions are fetched and written to memory 8 or 16 bits at a time. Inside the CPU, which is the brain of the computer, the numbers to be manipulated are held in registers. A register is just a set of flip-flops, whose contents are read or written simultaneously as a single group. The registers of a 16-bit microprocessor, for example, are sets of 16 flip-flops per register. A 16-bit register can hold a 16-bit binary number, that is, any one of the 65 536 binary numbers between 0000000000000000 and 1111111111111111.

Three ways to interface an experiment to a microcomputer are to

1. Use the game input/output (I/O) port.
2. Install an analog-to-digital conversion (ADC) card in one of the expansion card slots of the microcomputer.
3. Interconnect the electronic instrumentation used in the experiment with the microcomputer via the general-purpose interface bus (GPIB), where the GPIB is defined by the Institute of Electronic and Electrical Engineers (IEEE). The electronic instrumentation must be compatible with GPIB.

The ADC card method of interfacing is recommended for the following reasons. Many microcomputers have provision for a paddle or joystick, which is simply a variable resistor that connects to the game port. (The variable resistor could be a thermistor or a photoresistor, for example, whose resistance varies with temperature and light intensity, respectively.) The game port supplies 5 V to the variable resistor, and the resistor controls the current according to Ohm's law. The continuous resistance is the analog variable. An advantage of game port interfacing is simplicity; that is, the ADC circuitry is relatively simple. There are limitations to the game port: (1) For most computers the resolution is limited to 8 bits or, as we will see, 1 part out of 256; (2) one is limited to resistance as the analog variable. For these reasons interfacing via the game port is not the preferred approach.

The third method of interfacing requires electronic instruments (electrometers, oscilloscopes, counters, etc.) that have microcomputer-based processing capabilities. Such instruments communicate with a microcomputer through the GPIB. This method may be the best way for the microcomputer to run the experiment, collect and analyze the data, and present the experimentalist with a labeled graph of the finished results complete with error analysis. On the other hand, the electronic instrumentation required for this method is expensive and this method of interfacing is more complex than the other two methods. Hence, interfacing via the GPIB is not recommended. In the near future probably all instrumentation will be GPIB compatible, and for this reason a summary of the GPIB is given in Appendix C.

We will now discuss the ADC card in some detail. Usually an ADC card will also contain the circuitry for digital-to-analog conversion (DAC). The input to the DAC portion of the card is a digital voltage from the computer, and the DAC output is an analog voltage that may be applied to the experimental apparatus to control the experiment. The ADC input is an analog voltage from the experiment, via the interface circuit, and the output is a digital voltage to the computer. Two specifications of an ADC/DAC card are worth noting:

1. The analog range is the difference between the maximum and minimum ADC inputs and DAC outputs. It may be either a voltage or a current range. The voltage range is often 5 or 10 V; for example, a 10-V range could be from -5 to $+5$ V or from 0 to $+10$ V.
2. The second specification is resolution. 8-bit, 12-bit, and 16-bit cards are available. A 12-bit card, for example, means that the ADC circuit converts an analog voltage to a 12-bit number. This implies that a 12-bit card with a 10-V analog range from 0 to 10 V, say, would convert 0- and 10-V analog signals to binary numbers 000000000000 and 111111111111, respectively, or, in the decimal number system, 0 and 4095. That is, there are 4096 12-bit numbers, so that an analog voltage range of 10 V is divided into 4096 steps giving a resolution of 10 V/4096 = 2.441 mV. What this means is that an input analog voltage is converted to a 12-bit number by the ADC and two consecutive 12-bit numbers, for example, 000000000000 and 000000000001, correspond to an analog voltage difference of 2.441 mV. The resolution of 8- and 16-bit cards are 39.06 and 0.1526 mV, respectively, where a 10-V analog range was assumed. The 8-bit card has the least resolution and is the least expensive, whereas the 16-bit card has the most resolution and is the most expensive. The 12-bit card may be a reasonable compromise in terms of resolution and expense.

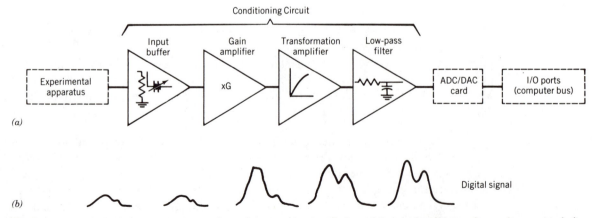

FIGURE 6.2 (a) Block diagram of a general conditioning circuit. (b) An arbitrary signal from the apparatus and the resulting signal following each stage of the circuit are shown.

In many experiments, the analog signal from the experimental apparatus is not a suitable input for the ADC for various reasons. To obtain a suitable input for the ADC, a signal conditioning circuit may be required. A general signal conditioning circuit is shown in Figure 6.2a. Figure 6.2b shows an arbitrary signal coming from the apparatus and the resulting signal following each stage of the circuit. (Note that the interface circuit shown in Figure 6.1 is the conditioning circuit plus the ADC/DAC card.) There are four possible stages of signal conditioning:

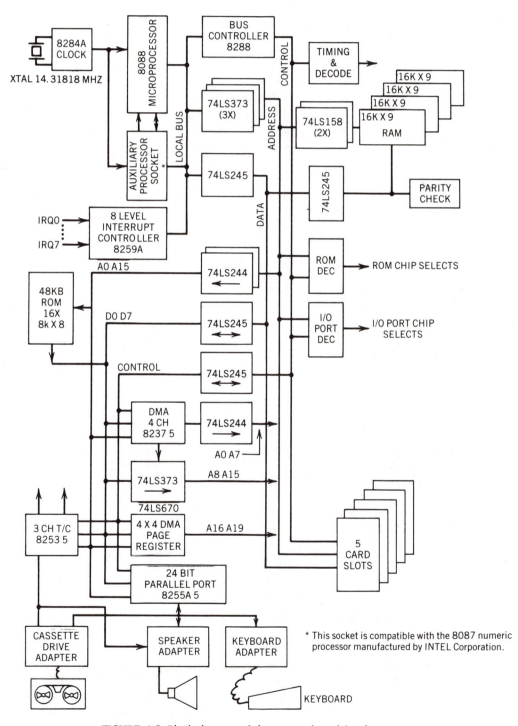

FIGURE 6.3 Block diagram of the system board for the IBM PC.

Stage 1. A gain 1, high-impedance buffer. Since it has a high input impedance, the buffer draws very little current from the output of the experiment (the output device of the experiment is often called a transducer) and it can provide whatever current is required for the next stage. The high-impedance buffer prevents "loading" the transducer; that is, it prevents the output of more current than the transducer can accurately provide, thereby preventing signal distortion.

Stage 2. An amplifier having fixed gain. Maximum precision requires that the range of the analog signal, which goes into the ADC, equal the analog range of the ADC. Thus, the fixed gain may have to be greater than or less than one, depending on the analog signal from the transducer and the analog range of the ADC. For example, suppose the transducer signal ranges from 0 to 0.1 V and the ADC analog is 0–5 V, then the desired gain is 50.

Stage 3. The transformation amplifier transforms the input signal. For example, the amplifier shown is a logarithmic amplifier that provides more gain for the low-level signals than for the high-level ones. Hence, good precision can be obtained at low input levels without the high-level signals exceeding the analog range of the ADC.

Stage 4. Filtering or removing frequencies in the analog signal that have no useful information. A low-pass filter removes the high-frequency signals and passes the low-frequency signals. A high-pass filter removes the low-frequency signals. A band-pass filter removes frequencies above and below the desired frequency band.

One or more of these four stages may be required, depending on the analog signal output of the transducer and the analog range of the ADC.

Figure 6.3 is a detailed block diagram of the system board for the IBM PC. The ADC/DAC card fits into one of the 5 card slots on the board. Note that the 5 card slots connect to the system board bus, composed of the data bus, address bus, and control bus.

Reference 2 reports that a wide range of physics experiments that use computers for data collection can be carried out with three types of measurements: (1) voltage, (2) the time between events, (3) the counts received during selectable time intervals. These three types of measurements will be carried out in this experiment.

EXPERIMENTS

Determine the analog range for your ADC/DAC card from the manufacturer's specifications. NEVER APPLY A VOLTAGE THAT IS OUTSIDE OF THIS RANGE. To apply such a voltage could damage the card or, worse, the computer.

The experiments are the following:

   I. Computer-assisted voltage measurements
      A. Variable dc voltage.
      B. Hall effect voltage
  II. Computer-assisted counting measurements
 III. Computer-assisted timing measurements

Computer-Assisted Voltage Measurements and Data Analysis

Variable Dc Voltage

The purpose of this exercise is to use the computer to measure dc voltages, that is, the computer will function as a voltmeter. The variable dc voltage is shown connected to the analog input of the ADC/DAC card in Figure 6.4. Note that no conditioning circuit is used. If maximum precision were required, we would insert a dc amplifier with a gain of 1.67

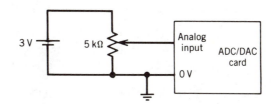

FIGURE 6.4 Variable dc voltage connected to the analog input. After calibration the computer will function as a voltmeter.

(assuming the ADC/DAC card has an analog range of 0–5 V) between the apparatus and the analog input of the ADC/DAC card.

Connect the circuit shown in Figure 6.4. The ADC/DAC card will convert the analog voltage input to a binary number. The software, Program 2, will do the following:

1. For a given input voltage, the software converts the corresponding binary number to an ordinary decimal number, labeled N, and displays it on the screen.
2. The software has the provision for inputing two known voltages (0 and 2 V, measured with a voltmeter). The software then determines the voltage V for each value of N, that is, it calibrates the apparatus by determining the scale factor, where $V =$ scale factor $\times N$. The software will then display N and V on the screen.

EXERCISE 1

Look up the analog range and the number of bits for your ADC/DAC card and determine the expected range of decimal numbers to be displayed on the screen as the input voltage is varied from 0 to $+3$ V.

Connect a voltmeter across the analog input of the ADC shown in Figure 6.4. Run the software and calibrate the apparatus. Once you have calibrated the apparatus vary the 5-kΩ potentiometer and observe the change in voltage displayed on the screen. The computer is now functioning as a voltmeter.

Hall Effect Voltage

The Hall effect is examined in detail in Experiment 17. Here we focus on interfacing the Hall sample to a computer to measure the so-called Hall voltage. We summarize some of the theory that is presented in Experiment 17.

Figure 17.1 of Experiment 17 shows a Hall sample of dimensions L, w, and t. The sample is in an external magnetic field **B** and it is connected to a potential difference V_0, which produces a current I. The Hall voltage V_H is the voltage between surfaces 3 and 4 of the sample, $V_H = V_3 - V_4$. From equation 16 of Experiment 17, V_H is given by

$$V_H = \frac{R_H I B_z}{t} \quad \text{(V)} \tag{1}$$

where R_H is the assumed constant Hall coefficient, t is the sample thickness, I is the current, and B_z is the applied field assumed in the z direction. Note in equation 1 that if I or B_z is held constant then V_H is a linear function of the other variable.

In this experiment you use the computer to calibrate the Hall sample, that is, to determine V_H as a function of B_z for fixed I. The computer will then function as a "tesla meter"; it can be used to measure unknown magnetic fields.

Circuit to Interface the Hall Sample to the Analog Input of the ADC

As you will learn, the conditioning circuit depends on the analog range and the maximum Hall voltage, $V_{H,max}$. The range of the Hall voltage is 0 to $V_{H,max}$, where for a semiconductor Hall sample $V_{H,max}$ may be a few tenths of a volt for maximum I and B_z.

Important Reminder: If the analog range of the ADC/DAC card is either 0 to $+K$ or $-K$ to $+K$, where K is typically 5 or 10 V, then for maximum precision the conditioning circuit must convert the 0 to $V_{H,max}$ input to an output of either 0 to $+K$ or $-K$ to $+K$.

We first consider an analog range of $-K$ to $+K$. Figure 6.5 shows the experimental apparatus, the conditioning circuits, and the card. Note that the analog output of the card connects to a conditioning circuit that provides the current I for the Hall sample. The Hall voltage, $V_H = V_3 - V_4$, is applied to a conditioning circuit whose output is the analog input for the card. The four 356N op-amps in Figure 6.5 are numbered 1 through 4, where 1 and 2 are the op-amps that interface the Hall sample to the analog input of the ADC. The pin assignment (top view) for the 356N is shown in Figure 6.6.

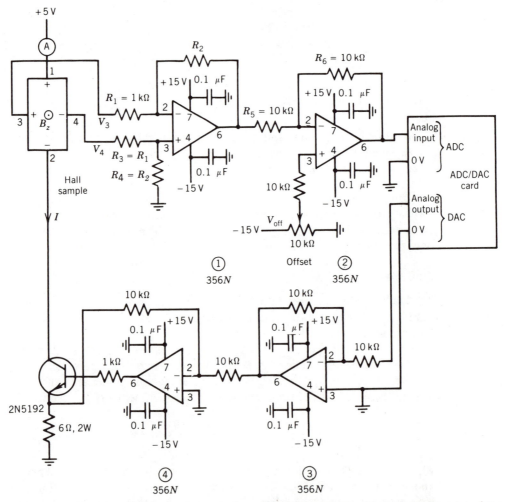

FIGURE 6.5 The output of the apparatus is the voltage difference $V_3 - V_4$, and op-amps 1 and 2 form the conditioning circuit for ADC analog input. The DAC analog output is the input to the conditioning circuit composed of op-amps 3 and 4 and the 2N 5192 transistor. This circuit controls the Hall current, and, hence, controls the experiment.

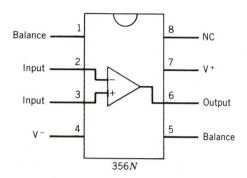

FIGURE 6.6 Top view of the pin assignments of the 356N op-amp.

Op-amp 1 is configured as a differential amplifier. (Op-amps, in general, and differential amplifiers, in particular, are discussed in reference 5.) The output of the differential amplifier $V_{out,1}$, at pin 6, is related to the voltage difference at the input $V_3 - V_4$, and the resistances R_1 and R_2:

$$V_{out,1} = -\frac{R_2}{R_1}(V_3 - V_4) = -\frac{R_2}{R_1}V_H = -\frac{R_2}{R_1}\frac{R_H IB_z}{t} \quad (V) \qquad (2)$$

The ratio R_2/R_1 is the gain G of the op-amp. Note that $V_{out,1}$ is G times the negative of the input. Pins 2 and 3 are the inverting (labeled $-$) and noninverting (labeled $+$) input pins of the op-amp. (Each pair of equal resistors that connects to op-amp 1 should be matched as closely as possible.)

Op-amp 2 has two inputs. One input is $V_{out,1}$, the output voltage of op-amp 1, and the other is V_{off}, the offset voltage. The output of op-amp 2, $V_{out,2}$, is related to the inputs and the resistances R_5 and R_6:

$$V_{out,2} = -\frac{R_6}{R_5}V_{out,1} + \frac{R_6 + R_5}{R_5}V_{off} = -V_{out,1} + 2V_{off} \quad (V) \qquad (3)$$

where R_6/R_5 and $(R_6 + R_5)/R_5$ are the gains. The negative sign occurs because the input $V_{out,1}$ is on the inverting pin, pin 2. From equation 2, the voltage $V_{out,1}$ ranges from 0 to $(-R_2/R_1)V_{H,max}$. With V_{off} set to 0 V, the output of op-amp 2 will range from 0 to $+(R_2/R_1)V_{H,max}$ as the input to op-amp 1 ranges from 0 to $+V_{H,max}$. The desired output of op-amp 2 is $-K$ to $+K$. By the appropriate choice of values for R_2 and V_{off}, the desired output can be obtained. We set the minimum value of $V_{out,2}$ to $-K$:

$$V_{out,2,min} = 0 + 2V_{off} = -K \qquad (V) \qquad (4)$$

and the maximum value to $+K$:

$$V_{out,2,max} = \frac{R_2}{R_1}V_{H,max} + 2V_{off} = +K \qquad (V) \qquad (5)$$

These two equations can be solved for R_2 and V_{off} in terms of the other known variables.

EXERCISE 2

Suppose $V_{H,max}$ is 100 mV and the analog range is $-K$ to $+K$, where $K = 5$ V. Determine V_{off} and R_2, and, hence, determine the gain of op-amp 1.

Remark: If the analog range is -10 V to $+10$ V, then equations 4 and 5 yield: $V_{off} = -5$ V and $R_2 = 200$ kΩ. Now $R_2 = 200$ kΩ and $V_{H,max} = 100$ mV implies $V_{out,1,max}$ is -20 V, which is not possible because the power supply limits the output to slightly more (positive) than -15 V. In this case choose R_2 such that $V_{out,1,max}$ is -14 V, and then use equations 4 and 5 to determine V_{off} and R_6 such that the output of op-amp 2 ranges from -10 V to $+10$ V as the input to op-amp 1 varies from 0 to 100 mV.

Some comments about this circuit. The 356N is a JFET input op-amp with an input impedance of 10^{12} Ω; hence, it is not likely to overload the transducer. Op-amp 1 acts as a buffer with gain, and op-amp 2 acts as a transformation amplifier in that it provides a constant dc level offset. The 0.1-μF capacitors couple any ac signals that may exist on the dc power supplies to ground.

If the analog range of the card is 0 to $+K$, then op-amp 2 is not necessary. However, it may be retained as a zero-adjust; that is, with the input to op-amp 1 set to zero, the offset-voltage can be adjusted such that the output of op-amp 2 is 0 V. If op-amp 2 is not retained, then a 20-kΩ potentiometer should be connected to the balance pins (pins 1 and 5) of op-amp 1, in order to set the output of op-amp 1 to 0 V with the input at 0 V. Also if op-amp 2 is not retained, then the input leads to op-amp 1 must be reversed so that the output of op-amp 1 is positive.

EXERCISE 3

Assume $V_{H,max} = 100$ mV and the analog range is 0 to $+5$ V. Determine R_2, and, hence, the gain of op-amp 1.

Circuit to Interface the Analog Output of the DAC to the Hall Sample

The conditioning circuit in Figure 6.5 that consists of op-amps 3 and 4 and the 2N 5192 transistor controls the Hall current I. Typically the current output of the DAC is about 10 mA, and the maximum current of a semiconductor Hall sample is about 500 mA. Hence, the input to the conditioning circuit is 0 to $+K$ volts at about 10 mA (we will not use negative voltages from the card in this circuit) and the output must deliver a selectable current ranging from 0 to 500 mA. Where do we start in designing such a circuit? A reasonable place to start is to select a device that can deliver the required current; that is, we start with the 2N 5192 npn transistor, which is a 4-A power transistor. This transistor can provide from 0 to 4 A. The collector current of a transistor is controlled by the bias of the emitter–base junction. As this bias voltage is varied the collector current will vary and, in the case of the 2N 5192, the collector current can be varied from 0 to 4 A. For an npn transistor, the collector current increases as the base voltage becomes more positive relative to the emitter. Op-amp 3 is an inverting buffer with unity gain. Op-amp 4 is an inverting amplifier with unity gain. As the analog output varies from 0 to $+K$ volts, the base of the 2N 5192 varies from 0 to $+K$ volts, and the collector current increases as the base voltage increases. Specifically, the circuit in Figure 6.5 will deliver from 0 to 400 mA as the analog output varies from 0 to $+3$ V.

Historical Remark

The 1956 Nobel prize was awarded jointly to
    William Shockley, the United States, John Bardeen, the United States, and Walter Houser Brattain, the United States
    For their researches on semiconductors and their discovery of the transistor effect.

John Bardeen is the only person awarded two Nobel prizes in the same field. The 1972 Nobel prize was awarded jointly to

John Bardeen, the United States, Leon N. Cooper, the United States, and Robert J. Schrieffer, the United States

For their jointly developed theory of superconductivity, usually called the BCS-theory.

Bench Testing the Circuits

Both circuits should be bench tested before they are connected to the computer. The circuit that provides the Hall sample current I can be bench tested without a magnetic field and the analog output of the card can be replaced with the circuit shown in Figure 6.7*a*. As part of the test, measure the current I as a function of the voltage input to op-amp 3. Is I a linear function of the input voltage?

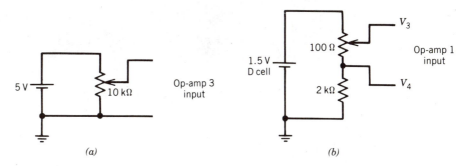

(a)   *(b)*

FIGURE 6.7 Inputs for bench testing the conditioning circuits in FIGURE 6.5.

Remark About Circuit Grounds: The 6-Ω, 2-W resistor should have its own lead that connects directly to ground; otherwise, oscillations may occur.

Before testing the conditioning circuit for the Hall voltage, mount the resistor R_2, using the resistance calculated in either Exercise 2 or 3, in the circuit. Then connect the output of the circuit (pin 6 of op-amp 1 or pin 6 of op-amp 2, depending on the analog range of your card) to a digital voltmeter and connect the circuit in Figure 6.7*b* to the input. As part of the test, measure the voltage output as a function of the input. Is it linear? (In general, it is good practice to measure the ac ripple of a dc source, and it is recommended that you use an oscilloscope to measure the ripple at the output of the circuit.)

The final value for R_2 may be determined once the maximum Hall voltage is known. Connect a digital voltmeter to leads 3 and 4 (see Figure 6.5) of the Hall sample. Use the circuit in Figure 6.7*a* as the input to op-amp 3. Place the Hall sample in the maximum allowed field (see the manufacturer's specifications), set the current I to $0.75 \times$ maximum current (again, see the manufacturer's specifications), and then rotate the Hall sample until the voltage observed on the digital voltmeter is a maximum. Leave the sample in this orientation. The Hall voltage is now a maximum. Determine R_2 and, if necessary, the offset voltage of op-amp 2.

Figure 6.8 is a photograph of the conditioning circuit for the Hall sample.

Running the Software

If you have one of the previously mentioned four ADC/DAC cards, then read the instructions for that card at the end of the experiment.

Connect the circuit in Figure 6.5 and run Program 3. It will carry out these steps:

1. ADC calibration. The ADC converts the analog input voltage V to a binary number, the software converts the binary number to a decimal number N, and $V = $ scale factor $\times N$.

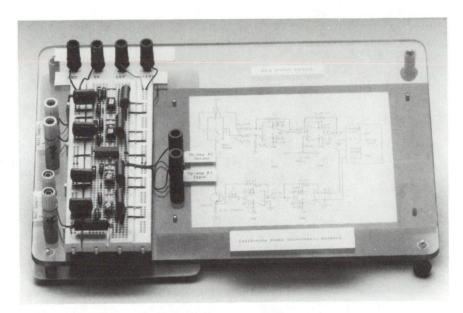

FIGURE 6.8 Photograph of the conditioning circuit for the Hall sample.

By using a voltmeter to set the analog input voltage to two known values (0 and 2 V were arbitrarily chosen), the software will calculate the scale factor and, hence, calibrate the ADC.

2. DAC calibration. The DAC output controls the applied current I. The DAC converts a binary number to an analog output voltage V_0, the software converts the binary number to a decimal number N, and $I =$ scale factor $\times N$ or $I =$ scale factor $\times V_0$. The scale factor is determined by inputing two currents, zero and the maximum permissible current, where the maximum permissible current is taken as $0.75 \times$ maximum current. The DAC output voltage and, hence, the applied current I are controlled by the left and right arrows, $\leftarrow$ and $\rightarrow$, on the keyboard. For a constant magnetic field, the software will plot the analog voltage output of the DAC versus I. The software limits the current to the maximum permissible value.

3. The "working current" for the Hall sample is about $0.75 \times$ maximum current. The current is set to this value. With a constant current the ADC analog input voltage V is proportional to the field B_z (see equation 1). Set the magnetic field to a known value and then enter that value in teslas into the computer. Repeat for several values of magnetic field; then the software will plot V versus B_z and provide the best fitting curve in the least-squares sense. The computer–Hall probe system is now functioning as a *tesla meter*. If magnets of unknown field strength are available use the tesla meter to measure them.

Computer-Assisted Counting and Data Analysis

The pulses to be counted and analyzed are produced by the decay of a ^{137}Cs radioactive source. Figure 18.1a of Experiment 18 shows the nuclear transitions and emitted particles in the decay of ^{137}Cs. The radioactive decay may be observed with either a scintillation counter, using a single-channel analyzer (SCA) to isolate the 0.663-MeV line, or a Geiger–Mueller (GM) tube with a scaler/timer. (The scintillation counter and single-channel analyzer are discussed in Appendix B.) The *decay time* and *transit time* of a scintillation counter determine the dead time or the time the counter is disabled, during which it will not count a second particle. The dead time of a scintillation counter is about 10^{-6} s. A count every 10^{-6} s means a count rate of 10^6 counts/s. If the count rate is kept below 10^4 counts/s, then dead time errors will be less than 1 percent.

A GM tube has a dead time of about 2×10^{-4} s, which implies a count rate of 5000 counts/s. To have dead time errors that are less than 1 percent, the count rate should be below 50 counts/s.

In addition to the dead time, the speed of the software may limit the maximum number of counts observed per second. The speed of the software is determined by what we ask the computer to do; for example, doing calculations, printing values, and so on, decreases the speed of the software. The software on the disk provided to your instructor limits the count rate to about 90 counts/s. For the SCA and the GM tube, position the radioactive source such that the count rates are about 80 and 50 counts/s, respectively.

In general, the manufacturers of scaler/timers do not provide a connector for output pulses. Rather it is necessary to open the chassis and wire up this connection. It is recommended that the instructor make this connection.

The output pulse depends on where the pulse is taken from the scaler/timer electronics. The pulse shown in Figure 6.9a is available from the input of the electronics, just after an isolation capacitor, and the pulse in Figure 6.9b is available from the output of a NAND gate. The place in the electronics where the pulse shown in Figure 6.9b is available will likely be indicated in the manufacturer's circuit diagram by a small square pulse.

Figure 6.10a is a sketch of the pulses from the PMT, using the ^{137}Cs source, with the

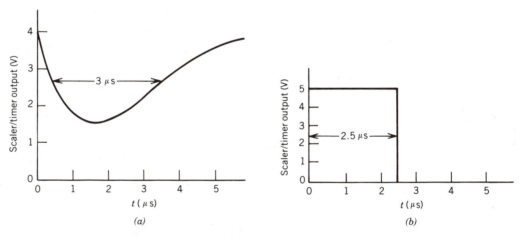

FIGURE 6.9 The output pulse from the scaler/timer depends on where it is taken from the electronics. Two possible output pulses are shown.

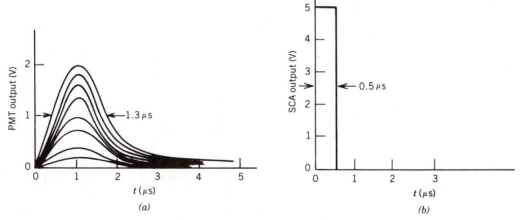

FIGURE 6.10 Output pulses from (a) the photomultiplier tube, PMT, and (b) the single-channel analyzer, SCA.

cathode-to-anode voltage set to 1000 V. The band of pulses with a 2-V amplitude corresponds to all of the energy of the 0.663-MeV gamma rays being deposited in the scintillator. These pulses are called the **total energy pulses**, and the amplitude is very dependent on the cathode-to-anode voltage. The pulses of smaller amplitude correspond to lesser energy of the gamma rays being deposited in the scintillator. Rather than attempting to count the nearly continuous range of pulse amplitudes coming from the PMT, we will use the SCA to isolate the *total energy* pulses.

Figure 6.10*b* shows the pulse from the lower-level-discriminator (LLD) output of the SCA. The LLD output provides a 5-V, 0.5 μs-wide pulse whenever the input pulse crosses the lower level threshold. The lower-level threshold is set by a linear 10-turn potentiometer. Hence, to count 2-V total energy pulses we set the potentiometer to a value just below 2.00, for example, 1.90.

Circuit to Interface the Apparatus to the Input of the ADC/DAC Card

The pulses shown in Figures 6.9*b* and 6.10*b* use the same conditioning circuit. The pulse in Figure 6.9*a* requires a different conditioning circuit, and this circuit is discussed first. This pulse is negative-going, with a full-width at half-maximum depth of 3 μs, and an offset voltage of about +1.5 V. The amplitude of the pulse does depend on the high-voltage setting of the GM tube. The conditioning circuit shown in Figure 6.11*a* consists of (1) a 2N 2222 transistor configured as a common-emitter amplifier, (2) a 74121 IC, which is a monostable multivibrator. We discuss these two parts of the conditioning circuit separately. Common-emitter amplifiers and monostable multivibrators are discussed in reference 5. Specifications of the 74121 are given in reference 7.

The purpose of the common-emitter amplifier is to provide an output pulse that is inverted and amplified, relative to the input pulse. The 2N 2222 is an npn transistor with a typical beta (h_{FE}), which is the current gain (ratio of collector to base current), of 150. The quiescent or dc voltages of the emitter and base are 0 and about +0.8 V, respectively. The quiescent base current is 4.2 V/56 kΩ or 75 μA, and a beta of 150 implies an 11.3-mA collector current. Such a collector current would produce a voltage drop of 52.9 V across

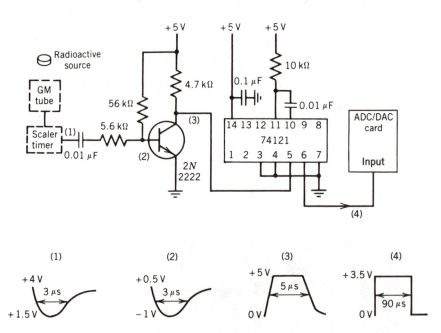

FIGURE 6.11 (*a*) Conditioning circuit for a scaler/timer output pulse shown in FIGURE 6.9*a*. (*b*) Pulses at the four labeled points in the circuit.

the 4.7-kΩ resistor, or a collector voltage of -47.9 V. This cannot occur, and the result is what is called saturation: The collector goes as close to ground as it can (about 0.3 V in this instance). The quiescent (saturation) current is then 4.7 V/4.7 kΩ or 1 mA. Thus, the voltage drop across the 4.7-kΩ resistor is about 5 V, and, hence, the collector current is about 1 mA.

Figure 6.11*b* shows the pulses at four points, which are labeled in Figure 6.11*a*. When the pulse on the base, shown in Figure 6.11*b*2, drives the base to a voltage of about 0.5 V or less, then the transistor ceases to conduct. Hence, the collector voltage becomes 5 V. Figure 6.11*b*3 shows the pulse on the collector, which is clipped off at 5 V because the supply is 5 V.

Additional Comments

The 5-V supply was chosen for the 2*N* 2222 because the 74121 IC requires 5 V, and this voltage is available from the ADC/DAC card. The 0.01-μF capacitor blocks the dc offset of the scaler/timer. The 4.7- and 56-kΩ resistors establish the quiescent collector and base currents at about 1 mA and 70 μA, respectively. The 5.6 kΩ gives rise to the pulse on the base lead as shown in Figure 6.11*b*2. The impedance seen by the scaler/timer is the following. The input signal sees, in series, 0.01 μF, 5.6 kΩ, and the impedance looking into the base (impedance to ground from the base). The base impedance is 56 kΩ (the internal impedance of the 5-V power supply is assumed to be negligible in comparison) in parallel with the internal emitter-base impedance, the product of the transistor beta and the emitter resistance, r_E, where

$$r_E = \frac{25}{I_c} \quad (\Omega) \qquad (6)$$

I_c is the collector current in milliamperes and, in this case, I_c is less than or equal to 1 mA. Hence, r_E is greater than or equal to 25 Ω. (Equation 6 is derived in reference 10, pages 53 and 66.) Thus, the magnitude of the input impedance is

$$Z_{in} = \sqrt{(5600 + 150 \times r_E)^2 + (1/\omega C)^2} \geqq 9398 \quad (\Omega) \qquad (7)$$

where a frequency corresponding to a 3-μs period was assumed. A large value for Z_{in} means the signal from the scaler/timer is not distorted by the amplifier. The above value is not particularly large.

The 5-μs full-width at half-maximum height shown in Figure 6.11*b*3 is too narrow and must be expanded. The reason for this is that a loop in the software reads the digital value in the ADC/DAC card every 50 μs; hence, the pulse width should be greater than 50 μs so that all of the pulses are counted. Note in Figure 6.11*b*4 that the output pulse of the 74121 has a 90-μs width.

The 74121 IC generates positive-going or negative-going square pulses of selectable width. It is triggered by the input pulse, and the width of the output pulse is determined by the values of the resistance and capacitance connected to pins 10 and 11. The output pulse is taken from pin 1 or pin 6, where the pulses from pins 1 and 6 are negative-going and positive-going, respectively. The 10-kΩ resistor and 0.01-μF capacitor shown connected to the 74121 in Figure 6.11*a* give rise to the 90-μs positive-going pulse shown in Figure 6.11*b*4.

We continue with a brief summary of the operation of the 74121. The function table for the 74121 is shown in Figure 6.12*a*, and the pin assignment (top view) is shown in Figure 6.12*b*. The 74121 has three input pins, numbered 3, 4, and 5, and two output pins, numbered 1 and 6. In Figure 6.11*a*, pins 3 and 4 are tied low (grounded) and pin 5 is the input pin. Pin 6 is the output pin. In the function table L or H means that the pin is tied low (0 V) or high ($\simeq 5$ V). X implies the pin can be high or low, that is, it does not matter if the pin is high or low. The arrows, $\downarrow$ and $\uparrow$, imply high-to-low and low-to-high transitions,

Function Table

Inputs			Outputs	
$A1$	$A2$	B	Q	$\bar{Q}$
L	X	H	L	H
X	L	H	L	H
X	X	L	L	H
H	H	X	L	H
H	↓	H	⊓	⊔
↓	H	H	⊓	⊔
↓	↓	H	⊓	⊔
L	X	↑	⊓	⊔
X	L	↑	⊓	⊔

(a)

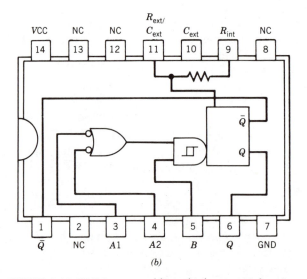

(b)

FIGURE 6.12 74121 monostable multivibrator: (a) function table, and (b) top view of the pin assignment. The bottom two entries in the function table, indicated by horizontal arrows, are both appropriate for the configuration of the 74121 in FIGURE 6.11a. NC means no internal connection.

respectively. The bottom two entries in the function table are appropriate for both pins 3 and 4 tied low, and then a low-to-high transition on pin 5 produces the output pulses shown in the table, where the pulse width is determined by the external resistance and capacitance connected to pins 10 and 11. In summary, the 74121, as configured in Figure 6.11a, has a positive-going output pulse with a 90-μs width.

The circuit that interfaces the SCA to the ADC/DAC card is shown in Figure 6.13a. Two pulses are shown in Figure 6.13b. In this instance, the pulse from the SCA drives the 74121 directly, and the 74121 is configured identically to that shown in Figure 6.11a.

The circuit shown in Figure 6.13a is also appropriate for the scaler/timer provided that the output pulse is that shown in Figure 6.9b.

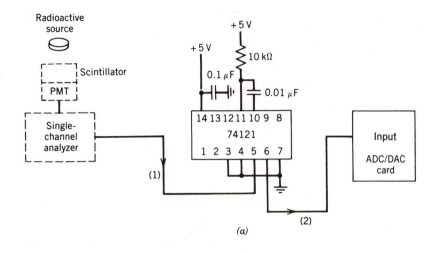

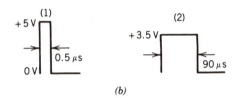

FIGURE 6.13 (a) Conditioning circuit for both the SCA and the scaler/timer, provided the scaler/timer output is that shown in FIGURE 6.9b. (b) Pulses at the labeled points in the circuit.

Bench Testing the Circuit

If you are using the GM tube and the scaler/timer, the following procedure is recommended. Disconnect the scaler/timer from the conditioning circuit and connect it directly to the oscilloscope. You should observe a pulse similar to that in Figure 6.9a or b. Measure the pulse characteristics, that is, height and width. Reconnect the scaler/timer output to the conditioning circuit, but DO NOT CONNECT THE CONDITIONING CIRCUIT TO THE COMPUTER. Use the oscilloscope to observe the pulses at the points specified in Figure 6.11 or 6.13. ARE THE 74128 OUTPUT PULSES SUITABLE FOR YOUR ADC/DAC CARD? If not, the conditioning circuit will have to be modified. Once you have suitable pulses, connect the circuit to the input of the ADC/DAC card as specified in the Instructions for ADC/DAC Cards, Program 4.

If you are using the PMT and SCA, then do the following. Using a BNC "tee" connector, connect the output of the PMT to both the SCA input and one input of a dual-trace oscilloscope. Set the lower level threshold voltage of the SCA to 1.90 V, that is, set the 10-turn potentiometer to 1.90 turns. With this setting, each SCA input pulse that exceeds 1.90 V will result in a 5-V square wave output pulse.

Connect the output of the SCA to the other oscilloscope input, and observe the signals as you vary the cathode-to-anode voltage, being careful not to exceed the high-voltage limit of the PMT. Then set the cathode-to-anode voltage such that the total energy pulse amplitude is 2 V. Disconnect the oscilloscope, connect the PMT directly to the SCA, and then connect the SCA output to the conditioning circuit as shown in Figure 6.13a. **DO NOT CONNECT THE CONDITIONING CIRCUIT TO THE COMPUTER.** Use the oscilloscope to observe the pulses at the points specified in Figure 6.13. **ARE THE 74121 OUTPUT PULSES SUITABLE FOR YOUR ADC/DAC CARD?** If not, the conditioning circuit will have to be modified. Once you have obtained suitable pulses, connect the circuit to the analog input of the ADC/DAC card.

Statistics of Radioactive Decay

Before considering what we want the computer to do, a brief review of the statistics of radioactive decay is in order. The decay of a radioactive nucleus is a random event and we cannot predict when any particular nucleus will decay. If we have a sample containing a very large number of radioactive nuclei, then the idealized probability that x nuclei will decay during a specified time interval is given by the Poisson distribution

$$P(x) = \frac{(\bar{x})^x e^{-\bar{x}}}{x!} \tag{8}$$

where $\bar{x}$ is the average number of decays in the time interval. For small $\bar{x}$, $P(x)$ versus x is a skewed distribution. As $\bar{x}$ increases, the Poisson distribution approaches the symmetrical Gauss distribution. In Figure 6.14a a Poisson distribution is sketched for small values of $\bar{x}$. In Figure 6.14b a theoretical distribution is sketched for $\bar{x} = 100$, and a standard deviation $s = 10$. (Recall from the Introduction that the standard deviation is defined as s for a finite number of observations.) $N(x)$ is the number of times that x occurs, that is, $N(x) = nP(x)$, where n is the total number of observations and $P(x)$ is the Poisson distribution. Note that Figure 6.14b resembles the symmetrical Gauss distribution discussed in the Introduction.

For any frequency distribution the standard deviation is defined as the square root of the average value of the square of the individual deviations from the mean $(x - \bar{x})$, for a large number of observations. Thus,

$$\sigma^2 \equiv \sum_{x=0}^{\infty} (x - \bar{x})^2 P(x) \qquad \text{(units of } x^2) \tag{9}$$

For the Poisson distribution,

$$\sigma^2 = \sum_{x=0}^{\infty} (x - \bar{x})^2 \frac{(\bar{x})^x}{x!} e^{-\bar{x}} = \bar{x} \tag{10}$$

Thus, the standard deviation for the Poisson distribution is equal to $\sqrt{\bar{x}}$. (The calculations showing the last equality in equation 10 are carried out in reference 2, under the Error Analysis section of the Introduction.)

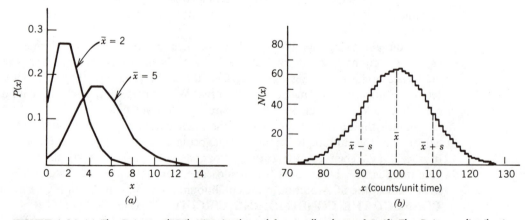

FIGURE 6.14 (a) The Poisson distribution is skewed for small values of $\bar{x}$. (b) The Poisson distribution approaches the symmetrical Gauss distribution as $\bar{x}$ increases.

Running the Software

Program 4 will carry out the following steps in the acquisition and analysis of the data.

1. The computer will count the number of pulses x that occurs in a 3-s time interval, and it will repeat this measurement a number of times n, where $24 < n < 1000$. The software does limit the number of counts in a 3-s time interval to 280. The software will display on the screen the number of the last 3-s time interval and the corresponding number of counts.

2. For a selected value of n the software calculates the mean $\bar{x}$, the standard deviation s, and the standard deviation of the mean s_m.

3. The software will plot a histogram: $N(x)$ (number of times x occurs) versus x. It will also mark $\bar{x}$, $\bar{x} - s$, and $\bar{x} + s$ on the histogram with vertical lines. Both the number of counts within and outside one standard deviation of the mean will be displayed on the histogram.

4. A theoretical Gauss distribution will be superimposed on the histogram. The theoretical distribution is given by equation 5 in the Introduction, with the exception that σ is replaced with s.

5. The software will calculate chi-square, χ^2, and display the value on the screen. (Chi-square is discussed in the Introduction.)

6. You may repeat the experiment for different values of n.

7. On the Apple II computer, you may toggle between the present and previous histogram by striking the X key.

Also, the software includes a test option. The test program allows you to select $\bar{x}$ and n. It will then generate random pulses that follow the Gauss distribution and carry out the analysis specified in steps 3 to 5. RUN THE TEST PROGRAM BEFORE RUNNING THE EXPERIMENT.

Computer-Assisted Time Measurements and Data Analysis

The time measurements will be the time intervals between successive pulses from the scaler/timer or the SCA. The statistical distribution of such time intervals is discussed first.

Statistics of Random Time Intervals

It was previously indicated that for a sample containing a very large number of radioactive nuclei, the number of nuclei x that decay per given time interval has a Poisson distribution. Suppose the time interval between successive decays is measured n times. There will, of course, be a variation in the size of the time intervals, and we want to know the corresponding distribution. This distribution is called the (time) interval distribution and it is derived from the Poisson distribution. The interval distribution describes the distribution in size of the time intervals between successive events in any random process in which the average rate has a constant value of ξ events per unit time. Referring to $\bar{x}$, the average number of decays in the time interval 0 to t, in equation 8, we write ξ in terms of $\bar{x}$ and t:

$$\xi \equiv \frac{\bar{x}}{t} \qquad \text{(average number of decays per unit of time)} \qquad (11)$$

(The value of ξ was determined in the previous measurements, and it should have been about 50 counts/s.) In terms of ξ and t, equation 8 becomes

$$P(x, t) = \frac{(\xi t)^x}{x!} e^{-\xi t} \qquad (12)$$

where, as before, $P(x, t)$ is the probability that x nuclei decay in the time interval 0 to t. We will use equation 12 to obtain the interval distribution for successive decays.

The probability that x nuclei decay in the infinitesimal time interval 0 to dt is given by

$$P(x, dt) = \frac{(\xi\, dt)^x}{x!}\, e^{-\xi\, dt} \tag{13}$$

Upon expanding the exponent in a Taylor series, equation 13 becomes

$$P(x, dt) = \frac{(\xi\, dt)^x}{x!}\left(1 - \xi\, dt + \frac{(\xi\, dt)^2}{2!} + \cdots\right) \tag{14}$$

For $x = 1$ we ignore very small terms involving dt raised to powers of 2 or more; hence, the probability of a single nucleus decaying in the time interval dt is

$$P(1, dt) \simeq \xi\, dt \tag{15}$$

The probability of observing x nuclei decaying, where $x > 1$, in the time interval dt involves dt raised to powers of 2 or more, which is negligibly small:

$$P(x, dt) \simeq 0 \qquad (x > 1) \tag{16}$$

The probability of observing no decay in the time interval 0 to t is

$$P(0, t) = e^{-\xi t} \tag{17}$$

Starting from time 0, the probability that the decay will occur in the interval t to $t + dt$ is given by the probability that 0 decays have occurred up to the time t multiplied by the probability that one decay occurs between t and $t + dt$:

$$P(1, t + dt) = P(0, t) \times P(1, dt) \simeq e^{-\xi t}\xi\, dt \tag{18}$$

where the starting point may be just after the previous decay. Equation 18 is the interval distribution. Note that equation 18 is a differential probability, often written as $dP(t)$.

The standard deviation for the interval distribution is obtained by replacing the sum in equation 9 with an integral, x with t, $\bar{x}$ with $1/\xi$, and $P(x)$ with $dP(t)$. Then equation 9 becomes

$$\sigma^2 = \int_0^\infty \left(t - \frac{1}{\xi}\right)^2 dP(t) \qquad (\text{s}^2) \tag{19}$$

where the differential probability is given by equation 18. Hence,

$$\sigma^2 = \int_0^\infty \left(t - \frac{1}{\xi}\right)^2 \xi\, e^{-\xi t}\, dt = \frac{1}{\xi^2} \qquad (\text{s}^2) \tag{20}$$

Thus, the standard deviation is equal to the average time interval $1/\xi$.

You will use the computer to measure the time interval between successive decays n times. A histogram of the number of observed time intervals versus time may be drawn by selecting a short but finite length Δt. The probable number of measured time intervals between t and $t + \Delta t$ is given by

$$n\, e^{-\xi t}\xi\, \Delta t \tag{21}$$

Note from equation 21 that the number of measured time intervals between t and $t + \Delta t$ decreases exponentially with the length of the time interval.

Circuit to Interface the Apparatus to the Input of the ADC/DAC Card

The circuit used in the counting experiment, Figure 6.11a or 6.13a, may be used in the timing experiment. In this case the computer determines the time interval between the successive 90-μs wide pulses.

Running the Software

Program 5 allows you to select n, the number of time intervals to be measured. The computer will measure and store the n time intervals, plot a histogram of the number of measured time intervals between t and $t + \Delta t$, plot a theoretical distribution (equation 21), and provide you with a video display of the results.

A test option is included in the software. RUN THE TEST PROGRAM BEFORE RUNNING THE EXPERIMENT.

Instructions for ADC/DAC Cards

8-Bit Mountain Card

These instructions are for an Apple II that has the 8-bit Mountain card. There are no internal adjustments to be made. The input and output analog ranges are -5 to $+5$ V. (The disk sets GS speed to normal.)

For the default slot number, place the board in slot 5. Attach the ADC input to pins 16 and 19 of the ADC connector, and the DAC output to pins 15 and 19 of the DAC connector. For the ADC this respresents channel 1, address 49361 or hex C0D1; for the DAC this represents channel 2, address 49362 or hex C0D2.

Place the disk in the disk drive and boot it; that is, turn on the machine or type PR # 6.

Choose the Mountain card from those displayed on the screen. Enter the slot number; default is slot 5.

Run Program 1 with a 1.5-V battery connected to the ADC input and a voltmeter connected to the DAC output to verify the operation of the input and output. That is, the analog voltage of the battery is converted to a binary number, and the program converts the binary number to a decimal number, which is displayed on the screen. The DAC output voltage is displayed on the voltmeter.

Programs 2 and 3 involve ADC and DAC conversions at the same address that is given in the preceding discussion.

For Programs 4 and 5 the $+5$ V for the conditioning circuit comes from DAC channel 2 (which will be set to $+5$ V by the program).

Because of the lack of counter-timers on the Mountain card, Programs 4 and 5 involve assembly-language routines, which are described in Appendix F. The pulses go in via the ADC input; they should be 3 to 5 V high and at least 50 μs long. The count rate should be around 50 counts per second.

12-Bit Sunset-Vernier Card

These instructions are for an Apple II that has the 12-bit Sunset-Vernier Advanced Interfacing Board (AIB). The analog range of the ADC input is set to -5 to $+5$ V with internal DIP switches. The analog range of the DAC output is 0 to 10 V. Line B7 on PIA # 2 should be connected internally to User pin 8 on the I/O connector. (The disk sets GS speed to normal.)

For the default slot number, place the card in slot 3. Attach the ADC input to channel 1, and the DAC output to the only DAC terminal.

Place the disk in the disk drive and boot it; that is, turn on the machine, or type PR # 6.

From the video display choose Sunset-Vernier AIB. Enter the slot number of the inserted card; default is slot 3.

Run Program 1 with a 1.5-V battery connected to the ADC input and a voltmeter connected to the DAC output to verify operation of the input and output. That is, the analog voltage of the battery is converted to a binary number by the card, and the program converts the binary number to a decimal number, which is displayed on the screen. The DAC output voltage is displayed on the voltmeter.

Programs 2 and 3 involve A/D and D/A conversions at the same addresses as mentioned in the discussion above.

For Programs 4 and 5 the +5 V for the conditioning circuit comes from the +5 V pin.

Because of the lack of counter-timers on the sunset card, Programs 4 and 5 involve assembly-language routines, which are described in Appendix F. The pulses go in via the USR 8 line, and they should be 3 to 5 V high and, at least, 50 μs long. The count rate should be around 50 or 80 counts per second.

8-Bit ML-16 Multi-Lab Card

These instructions are for the IBM PC, which has the ML-16 Multi-Lab general-purpose I/O (input/output) card, with a large 50-pin connector. Some adjustments have been already made internally:

1. The base address is set to hex 280 (&H280).
2. The 14-MHz video oscillator is divided by 16, and the resulting 1.14-μs pulses are fed to counter 0 clock.
3. Counter 0 output is fed to counter 1 clock; hence, they are cascaded for timing purposes.

The ADC analog range will be software-selected at -5 to $+5$ V, and the DAC range is -10 to $+10$ V. Counter 2 will be used for counting nuclear pulses. Each counter can count down from 65535 or less. Gates, clocks, and outputs should be 0 or 5 V.

Place the disk DOS Master (number 1) in disk drive a, which is the upper or the left-most disk. Turn on the computer; or, if it is already on, press ctrl-alt-del (all three at once). When it asks for the date, press the enter key (↵). After some time, the prompt appears: A⟩ or such.

Type: basica. Using capitals versus lowercase does not matter, but spaces often do. The BASIC prompt appears: ok.

Replace the DOS Master disk in its envelope, and insert the PC disk into the computer. Type: run "menu". Four choices of ADC/DAC cards will be displayed on the screen. Choose the ML-16 card.

For Programs 1, 2, and 3, we use A/D channel 0 and D/A channel 0. This means you put voltages in, up to 5 V, on pins 1 and 17, and you get voltages out from pins 19 and 21. Each channel reads 0 to 255 units internally (8 bits).

For Programs 4 and 5, the +5 V for the conditioning circuit comes from pin 50.

For Program 4, we use counter 1 out to stop the counting cycle via digital input 7 (DI 7); this means that you connect pin 26 to pin 39. We send pulses (no more than 5 V) to counter 2 clock and digital common; here, you connect the incoming pulse to pin 30 and the corresponding ground to pin 31. The count rate should be about 50 or 80 counts per second.

For Program 5, we use counter 2 out to communicate with the program via DI 7; this means that you connect pin 29 to pin 39. (An assembly language routine uses DI 7 information to control the timer.) Connect the incoming pulse as described for Program 4 above. The count rate should be around 50 or 80 counts per second.

12-Bit IBM DACA Card

These instructions are for an IBM PC that has the 12-bit IBM DACA general-purpose I/O (input/output) card, with a large 88-terminal distribution panel. Some adjustments have already been made internally:

1. The base address is set to hex 2E2 (&H2E2).
2. The ADC and DAC ranges are set to -5 to $+5$ V and -10 to $+10$ V, respectively.
3. The 14.32-MHz video oscillator is divided by 14 and the 0.98-μs pulses are fed to counter 0 clock.
4. Counter 0 output is fed to counter 1 clock; hence, they are cascaded for timing purposes.

Counter 2 is independent and is used in the discussion that follows for counting nuclear pulses. Each counter can count down from 65535 or less. Clocks and outputs should be 0 or 5 V.

Place the disk DOS Master (number 1) in disk drive a, which is the upper or left-most disk. Turn on the computer; or, if it is already on, press ctrl-alt-del (all three at once). When it asks for the date, press the enter key ($\dashv$). After some time, the system prompt appears: A$\rangle$ or such.

Type: basica. Using capitals versus lowercase letters does not matter, but spaces often do. The BASIC prompt should appear: ok.

Replace the DOS Master in its envelope, and insert the IBM PC disk instead. Type: run "menu". The names of four cards will be displayed on the screen. Choose the IBM-DACA.

For Programs 1, 2, and 3, connect the input voltage to A/D 0+ and 0−, with A/D 0− strapped to A GND. The DAC output is taken from D/A 0 and A GND (analog common). Each channel reads 0 to 4095 units internally (12 bits).

For Programs 4 and 5, the +5V for the conditioning circuit comes from a 1k resistor attached to the +10V terminal.

For Program 4, we use counter 1 out to stop the counting cycle via binary input 7 (BI 7); this means that you connect DELAY OUT to BI 7. We send the pulses (no more than 5 V) to counter 2 clock and digital common; hence, you connect the incoming pulses to COUNT IN and D GND. The count rate should be about 50 or 80 counts per second.

For Program 5, we use counter 2 out to communicate with the program; this means you connect COUNT OUT to BI 7. (An assembly language routine, which is described in Appendix F, uses BI 7 information to control the timer.) Connect the incoming pulse as described in Program 4. The count rate should be around 50 or 80 counts per second.

Printing Instructions for the Apple II

When a graph is displayed, press ctrl-P to stop the program. Then print the graph by using one of the following:

Some printer interfaces have built-in graphics capability (Grappler, Silentype). If so, then follow the simple instructions for your particular setup. Type: RUN to resume the program.

For Imagewriter II: turn on the printer (but do *not* type: PR # 1). Type: BLOAD PRIM for the first graph. Then type: CALL 6656,1 for a small print, or CALL 6656,2 for a large one. It takes less than a minute. Type: RUN to resume the program.

For the Epson-type printer: turn on the printer (but do *not* type: PR # 1). Type: BLOAD PREP for the first graph. Then type: CALL 6656. It takes about a minute. Type: RUN to resume the program.

Because of the wide variety of different printers and interfaces, PREP might not work as above. If your system supplies an extra line feed with each line, so that there are blank bars across the picture, then type: POKE 6762,146, which will suppress one line feed. If it does not print at all, try running it after typing PR # 1. If it prints miscellaneous symbols, try turning the printer off and on again.

PRIM and PREP default to the graphics page last plotted on. To select page 1, type: POKE 230,32. To select page 2, type: POKE 230,64.

Printing Instructions for the IBM PC

Turn on the printer.

After booting the system disk and before invoking basica, type: graphics, to install a graphics-printing routine in machine memory. Then, when you want to print a graph, press shift-PrtSc; it takes about 2 minutes. Afterward, press Enter to resume the program.

If your setup does not have the option just described, then, when the graph is on the screen and the program pauses, press ctrl-P. A basic program will be run that takes about 7 minutes to print the screen.

FUNDAMENTAL EXPERIMENTS

Thermodynamics of Photons and Electrons

7. ELECTRON PHYSICS: THERMIONIC EMISSION AND CHARGE-TO-MASS RATIO

Historical Note

The 1928 Nobel prize in Physics was awarded to: Sir Owen Willans Richardson, Great Britain
For his work on the thermionic phenomenon and especially for the discovery of the law named after him

APPARATUS [Optional Equipment in Brackets]

High vacuum system (e.g., rough and diffusion pumps), electron beam tube kit (available from Central Scientific Co.)
[Spot-welding apparatus]
5-in. length of 0.005-in.-diam tungsten filiament
piece of mica $2 \times \frac{1}{4}$ in.
Air-core solenoid (with inner dimensions at least large enough to accommodate electron tube)

POWER SUPPLIES OR BATTERIES

0- to 10-V, 0- to 3-A dc filament supply
0- to 200-V dc supply
0- to 3-A dc solenoid supply

METERS

Gaussmeter
0- to 3-A dc ammeters (2)
0- to 200-V dc voltmeter
0- to 10-V dc voltmeter
Electrometer

OBJECTIVES

To construct a vacuum thermionic diode and obtain and interpret its I-V characteristics for several different temperatures.

To obtain, by Hull's magnetron method, the value of e/m for the electron.

To understand the relationship between the free electron model for metals, quantum statistics, and the temperature behavior of the thermionic current as predicted by the Richardson–Dushman relation.

To understand the limitations that space charge imposes on thermionic current as predicted by Child's law.

KEY CONCEPTS

Thermionic emission
Image charge
Momentum space
Work function
Temperature-limited/space-charge-
  limited regimes

Space charge
Free electron
wave function
Fermi–Dirac distribution
Schottky emission

REFERENCES

1. J. S. Blakemore, *Solid State Physics*, 2d ed., Cambridge University Press, New York, 1985. Chapter 3.3 contains a discussion of the free electron gas and a derivation of the Richardson–Dushman relation for the temperature-limited thermionic current.

2. C. Kittel and H. Kroemer, *Thermal Physics*, 2d ed., Freeman, San Francisco, 1980. Chapter 5 discusses chemical potential; Chapter 6 discusses the Fermi–Dirac distribution.

3. C. Kittel, *Introduction to Solid State Physics*, 6th ed., Wiley, New York, 1986. The free electron model for metals is discussed in Chapter 6. Chapter 19 contains a concise, elegant derivation of the Richardson–Dushman relation using the concept of chemical potential developed in reference 2.

4. G. P. Harnwell and J. J. Livingood, *Experimental Atomic Physics*, McGraw-Hill, New York, 1933. Child's law for space-charge-limited emission is derived in Chapter 6; Chapter 4 discusses Hull's magnetron method for measuring e/m for electrons.

5. C. Herring and M. H. Nichols, *Rev. Mod. Phys.* **21**, 185 (1949). A review of theory and data pertaining to thermionic emission. Contains values of the work functions for several metals.

6. W. E. Forsythe and A. G. Worthing, *Astrophys. J.* **61**, 146 (1925). A useful collection of data concerning the electrical and thermal properties of tungsten filaments.

7. I. Langmuir and K. Blodgett, *Phys. Rev.* **22**, 347 (1923). Discusses Child's law.

8. A. W. Hull, *Phys. Rev.* **18**, 31 (1921). A description of the measurement of e/m for electrons by the magnetron method. Very thorough.

INTRODUCTION

Although the electron tube has been superseded by the transistor and the integrated circuit for many applications, thermionic emission remains a convenient and popular source of electrons for the operation of, for example, cathode-ray tubes and magnetrons. As background for this experiment, we discuss the physics of thermionic emission by calculating the thermionic flux of electrons from a hot filament in the absence of accelerating fields and space charges, after which we consider the effects the latter have on the current in a cylindrical diode with an axial filament.

The Thermionic Current

No Accelerating Fields

In the absence of accelerating fields due to an applied potential, an electron in the vicinity of the metal–vacuum interface depicted in Figure 7.1a is confined to the metal by a potential well of depth V_w, which represents the minimum amount of kinetic energy that an electron located at $z < 0$ must have to cross the barrier. As a refinement to the $eV(z)$ graphed in Figure 7.1a, we add the image potential

$$U_i = eV_i = \frac{-e^2}{16\pi\epsilon_0 z} \quad \text{(J)} \tag{1}$$

to represent the work required to draw the electron away from its positive image to a location $z > 0$, as shown in Figure 7.1b. To calculate the current density J_0 due to electrons with enough kinetic energy to cross this interface, consider the electrons contained within the dashed cylinder in Figure 7.1a. If the number of electrons per unit volume in the metal with velocities having z components in an interval dv_z about v_z is $N(v_z)\, dv_z$, the number in this interval hitting the barrier on the area dA in a time dt is just the number contained in the length $v_z\, dt$ of the cylinder, which is just $N(v_z)\, dv_z\, (v_z\, dt\, dA)$. If we assume that a fraction

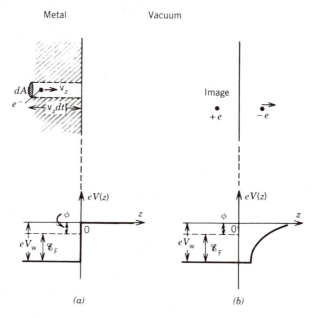

FIGURE 7.1 (a) Potential well for an electron near a vacuum–metal interface. (b) Effect of including the image potential.

r_e of these electrons are reflected from the barrier back into the metal, then the contribution of the electrons with v_z in this range to the current density is

$$dJ_0 = ev_z N(v_z)(1 - r_e)\, dv_z \qquad (A/m^2) \tag{2}$$

Note that only electrons with $v_z \geq \sqrt{2eV_w/m}$ have enough kinetic energy to cross the barrier and contribute to the current. J_0 is now calculated by integrating the expression for dJ_0:

$$J_0 = e(1 - r_e) \int_{\sqrt{\frac{2eV_w}{m}}}^{\infty} v_z N(v_z)\, dv_z \qquad (A/m^2) \tag{3}$$

where we have assumed that r_e varies slowly enough with v_z in this range so that it can be taken out of the integral. Performing this integral requires an explicit expression for $N(v_z)$ appropriate to electrons inside the metal at temperature T, which is now obtained using the free electron theory of metals.

Free Electron States and the Velocity Distribution $N(v_z)$

The wave functions and energies of the valence electrons in a metal can, to a first approximation, be calculated as if they were free, noninteracting particles confined to an infinite three-dimensional potential well. The wave functions describing the electronic states are solutions to the free electron (constant potential) Schrödinger equation and may be taken as

$$\Psi_{\mathbf{k}}(\mathbf{r}) = C\, e^{j(\mathbf{k} \cdot \mathbf{r})} \qquad (m^{-3/2}) \tag{4}$$

where C is a normalization constant and each electron state corresponds to a different value of the wave vector $\mathbf{k}$. If we consider, for simplicity, that the metal sample in which the electrons are confined is in the shape of a cube of side L with edges along the coordinate axes, then only discrete values of $\mathbf{k}$ correspond to allowed electronic states. These allowed wave vectors are found by imposing periodic boundary conditions on Ψ, that is, by demanding that $\Psi_{\mathbf{k}}(x + L, y, z) = \Psi_{\mathbf{k}}(x, y, z)$, with similar conditions for the y and the z dependence. The values of $\mathbf{k}$ for which these conditions are satisfied are specified by

$$k_x = \frac{2\pi}{L} n_x \qquad k_y = \frac{2\pi}{L} n_y \qquad k_z = \frac{2\pi}{L} n_z \qquad (rad/m) \tag{5}$$

where n_x, n_y, and n_z are integers. Using the linear momentum operator $\mathbf{p} = -i\hbar\nabla$, it is easy to show that each state specified by equations 4 and 5 can be assigned a discrete value of the momentum:

$$p_x = \frac{h}{L} n_x \qquad p_y = \frac{h}{L} n_y \qquad p_z = \frac{h}{L} n_z \qquad (kg \cdot m \cdot s^{-1}) \tag{6}$$

Thus, each state can be associated with a point in *momentum space*, as depicted by Figure 7.2a. The volume per state in momentum space is the volume of the indicated cube, h^3/τ, where $\tau = L^3$ is the *real space* volume of the metal. The number of electron states dS contained in the differential cube $dp_x\, dp_y\, dp_z$ in momentum space is thus,

$$dS = \frac{2\tau}{h^3} dp_x\, dp_y\, dp_z \tag{7}$$

where the factor of 2 arises because an electron in any of the states described above can have two possible spin orientations.

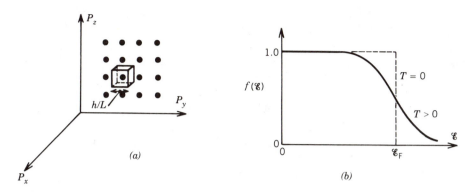

FIGURE 7.2 (a) Electron states in momentum space. (b) The Fermi–Dirac distribution.

For some absolute temperature T, the probability that any of the states described above will contain an electron is given in terms of the energy $\mathscr{E} = p^2/2m$ of the state by the Fermi–Dirac distribution:

$$f(\mathscr{E}) = \frac{1}{e^{(\mathscr{E} - \mathscr{E}_F)/k_B T} + 1} \tag{8}$$

where k_B is the Boltzmann constant and $\mathscr{E}_F$ is the Fermi energy for the metal, which is the energy of the highest occupied state at $T = 0$ K, as indicated by the plot of $f(\mathscr{E})$ in Figure 7.2b. Note that at temperatures $T > 0$, states of energy higher than $\mathscr{E}_F$ may be occupied, so that some electrons will be found above the dashed level drawn in Figure 7.1a. The only electrons that can escape the barrier to contribute to the current, however, are those with energies $\mathscr{E}$ such that $\mathscr{E} - \mathscr{E}_F > \phi$, where $\phi = eV_w - \mathscr{E}_F$ is called the **work function** of the metal. A convenient approximation to $f(\mathscr{E})$ for this range of energies can be obtained by noting (see Exercise 1 below) that for $\mathscr{E} - \mathscr{E}_F \gg k_B T$, the exponential term in equation 8 is quite large, so that

$$f(\mathscr{E}) \cong e^{(\mathscr{E}_F - \mathscr{E})/k_B T} \tag{9}$$

EXERCISE 1

The work function of tungsten is measured to be 4.5 eV (reference 5). Verify that, for a tungsten filament operating at 2500 K, this approximation to $f(\mathscr{E})$, for energies in the range of interest for thermionic emission, is quite good.

To obtain $N(v_z)$ for the calculation of the thermionic current density, we note that $N(\mathbf{p})\, dp_x\, dp_y\, dp_z$, the number of electrons per unit volume of sample having momenta in a differential volume $dp_x\, dp_y\, dp_z$ of momentum space, is calculated as the product of the corresponding number of momentum states per unit volume of sample (dS/τ from equation 7), and the occupation probability $f(\mathscr{E})$, if we express $\mathscr{E}$ in terms of p:

$$N(\mathbf{p})\, dp_x\, dp_y\, dp_z = \frac{2}{h^3} \exp\left\{ \left[\mathscr{E}_F - \frac{1}{2m}(p_x^2 + p_y^2 + p_z^2) \right] \Big/ k_B T \right\} dp_x\, dp_y\, dp_z \qquad (\mathrm{m}^{-3}) \quad (10)$$

which, expressed in terms of the electron velocity $\mathbf{v} = \mathbf{p}/m$, reads

$$N(v_x, v_y, v_z)\, dv_x\, dv_y\, dv_z = \frac{2m^3}{h^3} \exp\left\{ \left[\mathscr{E}_F - \frac{m}{2}(v_x^2 + v_y^2 + v_z^2) \right] \Big/ k_B T \right\} dv_x\, dv_y\, dv_z \qquad (\mathrm{m}^{-3})$$

$$\tag{11}$$

The expression for $N(v_z)$ is then obtained by integrating over all possible values of v_x and v_y, since any value of velocity parallel to the interface is allowed:

$$N(v_z)\, dv_z = \left(\frac{2m^3}{h^3} \int_{-\infty}^{\infty} \int_{-\infty}^{\infty} \exp\left\{\left[\mathscr{E}_F - \frac{m}{2}(v_x^2 + v_y^2 + v_z^2)\right]\middle/ k_B T\right\} dv_x\, dv_y\right) dv_z$$

$$= \frac{4\pi m^2}{h^3} k_B T \exp\left(\frac{\mathscr{E}_F - \frac{1}{2}mv_z^2}{k_B T}\right) dv_z \tag{12}$$

Finally, $N(v_z)$ is substituted back into equation 3 for J_0 to yield

$$J_0 = \frac{4\pi m e k_B^2}{h^3}(1 - r_e)T^2\, e^{-\phi/k_B T} \qquad (\text{A/m}^2) \tag{13}$$

which is the Richardson–Dushman expression for the thermionic current density in the absence of accelerating fields and space charges.

Note that the work function discussed above has several dependences that make it difficult to report with great precision. There is a slight dependence on temperature because of thermal expansion of the metal. Also, emissions from different crystallographic planes are characterized by slightly different values of ϕ; here we are concerned with a polycrystalline average appropriate to a tungsten wire. Finally, surface treatment can have a significant effect on the thermionic emission properties; cesiated tungsten, for example, has a work function less than one-third that of the pure metal.

Effect of an Accelerating Field

In most diode configurations, a potential difference is maintained by a battery between the thermionic emitter (cathode) and the collector or plate (anode). If all of the electrons that can be thermionically emitted are swept away from the cathode (no space-charge limitations) then application of an accelerating potential increases the emission current slightly by effectively lowering the energy barrier shown in Figure 7.1*b*. This is known as **field-aided** or **Schottky emission**.

To see how this works in the context of our simple model, consider the effect on $V(z)$ of the application of an accelerating field **E** near the cathode. As shown in Figure 7.3, the contribution of the accelerating field to the potential energy of the electron is added to the image charge contribution to produce a potential energy just outside the metal of the form

$$U(z) = eV(z) = \frac{-e^2}{16\pi\epsilon_0 z} - eE_z z \qquad (\text{J}) \tag{14}$$

This $U(z)$ has a maximum at

$$z_{\text{max}} = \sqrt{\frac{e}{16\pi\epsilon_0 E_z}} \qquad (\text{m}) \tag{15}$$

at which it has the value

$$U(z_{\text{max}}) = -\sqrt{\frac{e^3}{4\pi\epsilon_0}}\sqrt{E_z} \qquad (\text{J}) \tag{16}$$

The effective energy barrier, ϕ_{eff}, is thus lower than the work function ϕ by an amount $|U(z_{\text{max}})|$, so that

$$\phi_{\text{eff}} = \phi - \sqrt{\frac{e^3}{4\pi\epsilon_0}}\sqrt{E_z} \qquad (\text{J}) \tag{17}$$

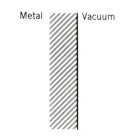

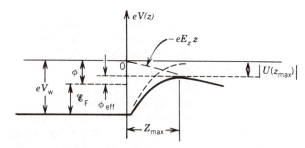

FIGURE 7.3 Potential energy of an electron near a vacuum–metal interface, including the effect of an accelerating field.

If this expression for ϕ_{eff} is substituted for ϕ in the expression for J_0 of equation 13, the field-aided thermionic current density becomes

$$J = J_0 \exp\left(\frac{\gamma\sqrt{E_z}}{k_{\text{B}}T}\right) \qquad (\text{A/m}^2) \tag{18}$$

with $\gamma \equiv (3/4\pi\epsilon_0)^{1/2}$.

EXERCISE 2

For the cylindrical diode shown schematically in Figure 7.4, calculate the magnitude of the electric field **E** near the surface of the cathode (neglecting space-charge effects, to be discussed below). Assume an inside diameter of 0.75 in. for the anode, a diameter of

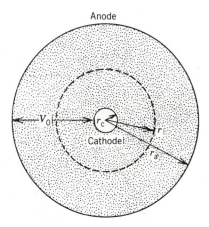

FIGURE 7.4 Cylindrical diode (top view).

0.005 in. for the cathode filament, and a potential difference between anode and cathode of 100 V. Use this value in conjunction with the expression of equation 18 to estimate, for a filament temperature of 2000 K, the ratio J/J_0.

The Effect of Space Charge

The above discussion of thermionic current density is based on the assumption that the flow of electrons from the filament is limited only by its temperature, which controls the number of electrons with sufficient energy to cross the metal–vacuum interface. This is a good assumption if the accelerating voltage is sufficiently large, in which case the diode is said to be operating in the saturation, or temperature-limited regime of the I-V characteristic, and the current density is given by equation 18. For lower voltages, in the space-charge-limited regime of the I-V characteristic, the current is limited not by the thermal availability of electrons, but by the retarding electric field due to the negative space charge that accumulates between the electrodes.

To determine the behavior of the I-V characteristic in the space-charge-limited regime for a diode with cylindrical geometry as depicted in Figure 7.4, we write Poisson's equation in cylindrical coordinates for the potential $V(r)$ in the region between the cathode and anode:

$$\frac{\partial^2 V}{\partial r^2} + \frac{1}{r}\frac{\partial V}{\partial r} = \frac{e}{\epsilon_0} n(r) \qquad (\text{V/m}^2) \tag{19}$$

where $n(r)$ is the number density of electrons, assumed to depend only on r. We can express $n(r)$ in terms of $V(r)$ by recognizing the current density J as

$$J = nev \qquad (\text{A/m}^2) \tag{20}$$

where v is the electron speed, given by conservation of mechanical energy as

$$v = \sqrt{\frac{2eV}{m}} \qquad (\text{m/s}) \tag{21}$$

if $V = 0$ at the cathode and thermal energies are neglected. If the length of the filament is l, then the current I crossing the dashed boundary in Figure 7.4 is $I = (2\pi rl)J$. Combining this with equations 20 and 21 yields $n(r)$ in terms of $V(r)$:

$$n(r) = \frac{I}{2\pi le}\left(\frac{2e}{m}\right)^{-1/2} r^{-1}\, V^{-1/2} \qquad (\text{m}^{-3}) \tag{22}$$

Substituting this expression for $n(r)$ back into equation 19 gives us a second-order nonlinear differential equation for $V(r)$:

$$\frac{\partial^2 V}{\partial r^2} + \frac{1}{r}\frac{\partial V}{\partial r} = \frac{I}{2\pi l\epsilon_0}\left(\frac{2e}{m}\right)^{-1/2} r^{-1}\, V^{-1/2} \qquad (\text{V/m}^2) \tag{23}$$

A solution of the form

$$V = Ar^\alpha \qquad (\text{V}) \tag{24}$$

satisfies this equation for particular values of A and α. The boundary conditions that both V and v vanish at the cathode are also satisfied in the limit of small cathode radius.

EXERCISE 3

Show by direct substitution that the proposed solution $V(r)$ of equation 24 does satisfy equation 23 for V provided that $\alpha = \frac{2}{3}$ and

$$A^{3/2} = \frac{9I}{8\pi l\epsilon_0}\left(\frac{2e}{m}\right)^{-1/2}$$

If the voltage between anode and cathode is V_0, then I as a function of V_0 is now obtained by writing $V(r)$ as in equation 24 (with A and α as determined in Exercise 3) and demanding that $V = V_0$ at $r = r_0$. The result is known as Child's law:

$$I = \frac{8\pi\epsilon_0}{9r_a}l\sqrt{\frac{2e}{m}}V_0^{3/2} \quad \text{(A)} \tag{25}$$

A correction factor, as discussed in reference 7, should be added to the right side of equation 25 to account for the nonzero cathode radius r_c. This factor is a function of r_c/r_a and is close to unity for the diode in our experiment. It should also be noted that, although we have not shown it here, the proportionality between I and $V_0^{3/2}$ is independent of the geometrical configuration of the electrodes.

Child's law should describe the I-V characteristic of the diode in the voltage range for which the current is space-charge limited; for higher voltages the current is temperature-limited and is determined by the expression of equation 18.

Magnetron Method for the Measurement of e/m

If, in addition to the accelerating electric field associated with the potential V_0, a magnetic field **B** is applied parallel to the filament of the diode, the trajectory of the electrons is curved, as indicated in Figure 7.5a. As B is increased for some constant V_0, the curvature becomes sharper until, at some critical $B = B_0$, the electrons never reach the anode at $r = r_a$ and the current abruptly drops to zero. The trajectory of the electrons for $B = B_0$ is as shown in Figure 7.5(b). If B_0 is measured experimentally, then the charge-to-mass ratio e/m can be deduced from the relationship between B_0, V_0, and r_a, which we now obtain.

Although the trajectory can be derived by writing $\mathbf{F} = m\mathbf{a}$ with the Lorentz force $\mathbf{F} = -e(\mathbf{E} + \mathbf{v} \times \mathbf{B})$ and then solving the resulting differential equation, it is possible to obtain the required relationship using elementary considerations of the energy and angular momentum of the emitted electrons, without solving for their detailed motion.

For this calculation, we use cylindrical coordinates (r, θ, z) with the z axis along the

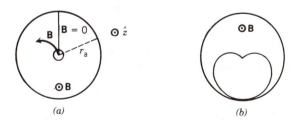

(a)   (b)

FIGURE 7.5 (a) Trajectories of electrons for zero magnetic field (straight line) and for fields less than the critical field (curved path). (b) Trajectory of electrons for the critical magnetic field.

filament. If the cathode is at ground ($V = 0$) and an electron is emitted at $t = 0$ with a velocity $\mathbf{v} = v_{r0}\hat{r} + v_{\theta0}\hat{\theta} + v_{z0}\hat{z}$, then conservation of energy reads

$$\tfrac{1}{2}m(v_{r0}^2 + v_{\theta0}^2) = \tfrac{1}{2}m(v_r^2 + v_\theta^2) - eV \qquad \text{(J)} \qquad (26)$$

where v_{z0} disappears from both sides because $F_z = 0$. Additionally, the angular momentum of the electrons will change because they are subjected to a magnetic torque $\mathbf{T}$. If the initial angular momentum of the emitted electrons is $L_0 = mv_{\theta0}r\hat{z}$, then $\mathbf{L}$ at any time t is just $\mathbf{L} = \mathbf{L}_0 + \int_0^t \mathbf{T}\,dt$. The torque $\mathbf{T}$ exerted on the electron by the magnetic field is just $\mathbf{r} \times \mathbf{F} = Berv_r\hat{z}$. Thus, for the angular momentum at time t we have

$$|\mathbf{L}| = mv_\theta r = mv_{\theta0}r_c + \int_0^t Berv_r\,dt \qquad \text{(kg} \cdot \text{m}^2/\text{s)} \qquad (27)$$

Substituting dr for $v_r\,dt$ and integrating from r_c to r in equation 27 gives the magnitude of the angular momentum at any position as

$$mv_\theta r = mv_{\theta0}r_c + \tfrac{1}{2}Be(r^2 - r_c^2) \qquad \text{(kg} \cdot \text{m}^2/\text{s)} \qquad (28)$$

At the critical field B_0, electrons arrive at $r = r_a$ (where $V = V_0$) with zero radial velocity v_r. Rewriting equations 26 and 28 with $r = r_a$, $V = V_0$, $B = B_0$, and $v_r = 0$ yields

$$\tfrac{1}{2}m(v_{r0}^2 + v_{\theta0}^2) = \tfrac{1}{2}mv_\theta^2 - eV_0 \qquad \text{(J)} \qquad (26')$$

$$mv_\theta r_a = mv_{\theta0}r_c + \tfrac{1}{2}B_0e(r_a^2 - r_c^2) \qquad \text{(kg} \cdot \text{m}^2/\text{s)} \qquad (28')$$

The tangential speed v_θ at $r = r_a$ may be eliminated between the above equations to yield a relationship between B_0 and V_0:

$$V_0 = B_0^2\left(\frac{e}{8m}\right)r_a^2\left(1 - \frac{r_c^2}{r_a^2}\right)^2 + \left(\frac{B_0r_cv_{\theta0}}{2} - \frac{mv_{\theta0}^2}{2e}\right)\left(1 - \frac{r_c^2}{r_a^2}\right) - \frac{mv_{r0}^2}{2e} \qquad \text{(V)} \qquad (29)$$

from which e/m may be deduced if everything else is measured. If the ratio $r_c/r_a \ll 1$, as for our experiment, then this relation simplifies to

$$V_0 = B_0^2\left(\frac{e}{8m}\right)r_a^2 + B_0\frac{r_cv_{\theta0}}{2} - \frac{m}{2e}(v_{r0}^2 + v_{\theta0}^2) \qquad \text{(V)} \qquad (30)$$

If v_{r0} and $v_{\theta0}$ can be considered small for most of the emitted electrons, then this simplifies further to

$$V_0 = \frac{B_0^2r_a^2}{8}\left(\frac{e}{m}\right) \qquad \text{(V)} \qquad (31)$$

from which e/m can easily be calculated by measuring B_0, the cutoff field for the current that flows in response to the applied voltage V_0.

EXPERIMENT

Although many variations and options are possible, the diode should be assembled approximately as shown in Figure 7.6 and evacuated. Be sure to measure the filament diameter and length as well as the length and inner diameter of the anode. The filament should be aligned coaxially with the cylindrical anode. The mica insulator, if cut as shown

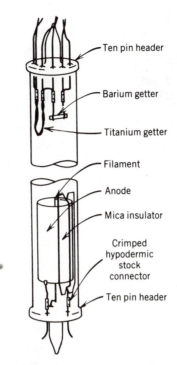

FIGURE 7.6 Diode assembly.

and compressed slightly, will minimize sagging of the hot tungsten filament. If it is desired to seal off the tube so that continuous evacuation is not necessary, installation and firing of the getters after sealing will improve vacuum quality. In any case, the filament and getters (if installed) should be heated to a dull red and outgassed during evacuation.

Thermionic Emission

The circuit for the measurement of the thermionic emission current is shown schematically in Figure 7.7. For several values of the filament current, measure I against V_p throughout

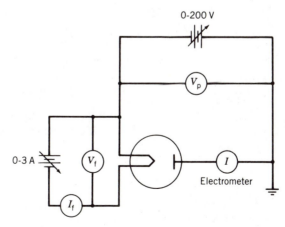

FIGURE 7.7 Circuit for the measurement of the thermionic emission current.

the range of available voltages. Measurements of I_f and V_f should also be recorded for each set of data for use in filament temperature determination.

Note that the electrometer is actually a voltmeter in parallel with a high resistance appropriate to the scale being used, so that the voltage drop across the input may or may not be negligible, depending on the scale multiplier setting. You will need to correct your values of V_p for this drop to obtain V_0 or choose your scale so that this correction is negligible.

On a single set of axes, plot $\log I$ versus $\log V_0$ for each filament current. Identify the space-charge-limited and temperature-limited regime of your curves.

EXERCISE 4

What effect do you expect the voltage gradient along the filament to have on the appearance of your plot? What about the effect of the contact potential difference between the anode and cathode?

EXERCISE 5

From a least-squares fit of the data in the space-charge-limited regime, determine the exponent β in the law $I \propto V^\beta$. Does your result agree, to within experimental uncertainty, with the prediction of Child's law?

To analyze the data in the saturation regime, it is necessary to associate a temperature with each value of the filament current. To do this:

(a) Plot the data from Table 7.1 so that the temperature of the filament can be determined if its resistivity ρ is known.

(b) From the measured V_f corresponding to each filament current I_f, calculate the total resistance R of the series combination of leads plus filament. Plot R versus I_f and extrapolate to $I_f = 0$ to determine the room temperature value of R.

(c) From the value of R at room temperature, the dimensions of the filament and the room temperature resistivity of tungsten, determine the resistance of the leads alone, which is assumed to remain constant as the filament is heated. Recall that the resistance of a wire is $\rho L/A$, where L is its length and A its cross-sectional area.

(d) From the value of R at each current and the lead resistance, determine the filament resistance for each current I_f. From this determine the resistivity of the filament and, from the plot in (a), the filament temperature for each I_f.

For each filament temperature for which I versus V_0 data were taken, determine the saturation current density J_0 from equation 18 and the measured I. Produce a Richardson plot of $\log(J_0/T^2)$ versus $1/T$. From a least-squares estimate of the slope of this line, determine the work function of tungsten.

EXERCISE 6

How does your value of ϕ compare with the value reported in reference 5 of 4.5 eV? What might explain any discrepancies?

TABLE 7.1 RESISTIVITY OF TUNGSTEN VERSUS TEMPERATURE

Temperature, T (K)	Resistivity, ρ ($\mu\Omega/cm$)
300	5.64
400	8.06
500	10.74
600	13.54
700	16.46
800	19.47
900	22.58
1000	25.70
1100	28.85
1200	32.02
1300	35.24
1400	38.52
1500	41.85
1600	45.22
1700	48.63
1800	52.08
1900	55.57
2000	59.10
2100	62.65
2200	66.25
2300	69.90
2400	73.55
2500	77.25
2600	81.0
2700	84.7
2800	88.5
2900	92.3
3000	96.2
3100	100.0
3200	103.8
3300	107.8
3400	111.7
3500	115.7
3655	121.8

Source: W. E. Forsythe and A. G. Worthing, "The Properties of Tungsten and the Characteristics of Tungsten Lamps", *Astrophysical Journal* **61**, 146 (1925).

Charge-to-Mass Ratio

Use the gaussmeter to measure and plot the magnetic field B at the center of the air-core solenoid as a function of the solenoid current i. Place the diode so that the center of the filament is approximately coincident with the center of the solenoid and align it so that the filament is along the solenoid axis.

For several combinations of filament current I_f and potential difference V_0, measure and plot the I versus i curves for the tube. In each case derive the cutoff field B_0 from your measurements along with the corresponding value of e/m.

EXERCISE 7

You can use equation 31 to find e/m for any combination of B_0 and V_0 for which you can show that it is a good approximation to equation 30. Justify the use of the simpler equation by estimating the magnitude of the last two terms in equation 30. *Hint*: It can be shown from the above discussion of the thermal distribution of speeds that for the emitted electrons

$$(v_{r0})_{rms} = \sqrt{\frac{2k_B T}{m}}$$

while

$$(v_{\theta 0})_{rms} = \sqrt{\frac{k_B T}{m}}$$

EXERCISE 8

Describe, at least qualitatively, the effect (if any) of each of the following on the appearance of the I versus i data and on the experimental value of e/m. Can you think of a way in which each effect can be minimized or eliminated?

(a) The existence of a voltage gradient along the filament.
(b) The magnetic field of the current that heats the filament.
(c) The contact potential difference between the anode and cathode.
(d) Misalignment of the diode with respect to the solenoid axis.
(e) Poor vacuum in the tube.
(f) The "end effects" in the electric field experienced by the electrons near the ends of the filament where **E** has a z component.

8. BLACKBODY RADIATION

Historical Note

The 1911 Nobel prize in Physics was awarded to Wilhelm Wien, Germany
   For his discoveries regarding the laws governing the radiation of heat.

The 1918 Nobel prize in Physics was awarded to Max Planck, Germany
   In recognition of the services he rendered to the advancement of Physics by his discovery of energy quanta.

APPARATUS

Tungsten strip lamp (e.g., General Electric Model 18A/T10/6V)

Optical pyrometer

Variable dc power supply (12 V, 200 W)

Shunt resistance

Oscillator with audio power amplifier (200 W) or parts listed in Figure 8.8 along with variable dual 12-V dc, 200-W power supply

Monochromator (with range extending to about 1100 nm)

Photodetector (photomultiplier tube is recommended, cooled if possible. The operation of this device is discussed in Appendix B.)

Current-to-voltage converter (labeled "current converter" in Figures 8.8 and 8.9)

Integrator

Lock-in amplifier

Zero-voltage switch CA 3059

Two digital voltmeters (DVMs)

Large converging lens

OBJECTIVES

To investigate the temperature dependence of both the total radiated power and the radiation at a single wavelength.

To understand and use the technique of modulation spectroscopy to look for deviations in the character of the radiation from the predictions of Planck's law.

To understand the characteristics of the radiation emitted by bodies at elevated temperatures as described by Planck's law.

KEY CONCEPTS

Blackbody	Spectral distribution
Planck's law	Wien displacement law
Emissivity	Modulation (or derivative) spectroscopy
Optical pyrometry	

REFERENCES

1. R. Eisberg and R. Resnick, *Quantum Physics of Atoms, Molecules, Solids, Nuclei, and Particles*, 2d ed., Wiley, New York, 1985. Chapter 1 briefly discusses Planck's radiation law and its role in thermometry (the optical pyrometer). Chapter 11 discusses the Bose distribution and the blackbody spectrum that characterizes the photon gas.

2. F. Reif, *Fundamentals of Statistical and Thermal Physics*, McGraw-Hill, New York, 1965. Chapter 9 contains a careful discussion of the fundamental concepts (boundary conditions, detailed balance, etc.) involved in the discussion of blackbody radiation.

3. T. R. Harrison, *Radiation Pyrometry and its Underlying Principles of Radiant Heat Transfer*, Wiley, New York, 1960. This was the first comprehensive treatment of the theory of radiation thermometry and its practice.

4. D. P. DeWitt and G. D. Nutter (Eds.), *Theory and Practice of Radiation Thermometry*, Wiley, New York, 1988.

5. H. J. Kostkowski and R. D. Lee, *Theory and Methods of Optical Pyrometry*, National Bureau of Standards Monograph 41, 1962.

6. D. H. Jaecks and R. DuBois, Am. J. Phys. **40**, 1179 (1972). Describes the experimental determination of hc/k_B.

7. C. N. Manikopoulos and J. F. Aquirre, Am. J. Phys. **45**, 576 (1977). Describes a technique for obtaining the IR spectrum of a hot cavity.

8. J. Dusek, R. J. Kearny, and G. Baldini, Am. J. Phys. **48**, 232 (1980). Describes the derivative (modulation) spectroscopy technique for examining the Planck formula.

9. J. C. DeVos, *Physica* **20**, 690 (1954). Provides data on the spectral emissivity of tungsten ribbon as a function of temperature and wavelength.

INTRODUCTION

Theoretical and experimental investigations of the radiation emitted by bodies at elevated temperature formed one of the cornerstones of the quantum theory of light, which was developed in the early 1900s. Aside from its obvious historical and practical importance, the character of this radiation, called *blackbody* radiation in the ideal case, is of current importance in the study of astrophysics and cosmology, since radiation from stars and the cosmic background radiation are characterized in terms of blackbody spectra. The study of blackbody radiation is also relevant to microscopic processes, and thus arises in connection with atomic physics and quantum electronics. In this introduction we give a brief discussion of the nature of this radiation.

Figure 8.1 depicts a rectangular cavity of dimensions L_x, L_y, and L_z with a small hole of area A in one side. The hole, with regard to its radiative properties, is the best approximation to a blackbody, or perfect absorber, since photons entering the hole from without do not reemerge. The radiation emitted from the hole as the walls of the cavity are maintained at some absolute temperature T is blackbody radiation; its intensity and spectral distribution depend only on T. To derive this distribution, we consider the electromagnetic field produced inside the cavity as a result of emission and absorption of radiant energy by the heated atoms of the walls.

The electric field **E** associated with the radiation in the cavity satisfies the wave equation

$$\nabla^2 \mathbf{E} = \frac{1}{c^2} \frac{\partial^2 \mathbf{E}}{\partial t^2} \qquad (\text{V/m}^3) \qquad (1)$$

so that **E** may be considered to be a superposition of plane wave solutions of the form

$$\mathbf{E_k} = \mathbf{A_k}\, e^{j[\mathbf{k} \cdot \mathbf{r} - \omega t]} \qquad (\text{V/m}) \qquad (2)$$

where **k** is the wave vector with magnitude $k = \omega/c$, and $A_\mathbf{k}$ depends on **k** but is independent of position. If periodic boundary conditions are imposed on **E** so that

$$\mathbf{E_k}(x + L_x, y, z) = \mathbf{E_k}(x, y, z)$$

$$\mathbf{E_k}(x, y + L_y, z) = \mathbf{E_k}(x, y, z) \qquad (\text{V/m}) \qquad (3)$$

$$\mathbf{E_k}(x, y, z + L_z) = \mathbf{E_k}(x, y, z)$$

then the components of **k** in equation 2 take on only discrete values corresponding to the

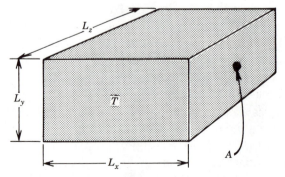

FIGURE 8.1 Rectangular cavity with a small hole as an approximation to a blackbody.

integers m_x, m_y, m_z:

$$k_x = \frac{2\pi}{L_x} m_x \qquad k_y = \frac{2\pi}{L_y} m_y \qquad k_z = \frac{2\pi}{L_z} m_z \qquad \text{(rad/m)} \qquad (4)$$

In the quantum (photon) picture of this radiation field, photons of energy $\hbar\omega = \hbar ck$ and with one of two possible transverse polarizations are associated with each possible $\mathbf{k}$ given by equation 4; if the number of photons in a state with wave vector $\mathbf{k}$ and polarization α is $n_{\mathbf{k},\alpha}$, then the radiant energy at angular frequency ω is $\hbar\omega n_{\mathbf{k},\alpha}$.

The spectral distribution of the energy within the cavity can be calculated if we know the equilibrium value of $n_{\mathbf{k},\alpha}$ for any temperature T. If we consider each state $(\mathbf{k}, \alpha)$ independently, the relative probability that this state will contain n photons is given by the Boltzmann factor

$$P_n = e^{-\beta \mathscr{E}_n} \qquad (5)$$

where $\mathscr{E}_n = n\hbar ck$ is the energy in the state with wave vector $\mathbf{k}$ and where $\beta \equiv 1/k_{\mathbf{B}}T$. The average value of $n_{\mathbf{k},\alpha}$ is thus

$$\bar{n}_{\mathbf{k},\alpha} = \frac{\sum_{n=0}^{\infty} n P_n}{\sum_{n=0}^{\infty} P_n} = \frac{\sum_{n=0}^{\infty} n\, e^{-\beta n\hbar ck}}{\sum_{n=0}^{\infty} e^{-\beta n\hbar ck}} \qquad (6)$$

Both of the series involved in this expression can be easily summed; the result for $\bar{n}_{\mathbf{k},\alpha}$ is

$$\bar{n}_{\mathbf{k},\alpha} = \frac{1}{e^{\beta\hbar\omega} - 1} \qquad (7)$$

EXERCISE 1

Show that equation 7 follows from equation 6 by summing the series explicitly. Note that the numerator can be expressed as

$$-(\hbar ck)^{-1} \frac{\partial}{\partial \beta} \sum_{n=0}^{\infty} e^{-\beta n\hbar ck}$$

To obtain $u(\omega; T)\, d\omega$, the energy density in an interval $d\omega$ about ω, we write

$$u(\omega; T)\, d\omega = \bar{n}_{\mathbf{k},\alpha}(\hbar\omega)[g(\omega)\, d\omega] \qquad \text{(J/m}^3\text{)} \qquad (8)$$

where $g(\omega)\, d\omega$ is the number of states $(\mathbf{k}, \alpha)$ per unit volume within an interval $d\omega$ about ω. The expression for $g(\omega)$ is obtained from an examination of the allowed values of $\mathbf{k}$ given by equation 4. The number of allowed values of k_x contained in the one-dimensional interval dk_x is just $(L_x/2\pi)\, dk_x$, with similar expressions for the intervals dk_y and dk_z. The number of photon states contained in the infinitesimal "volume" of k space $d^3\mathbf{k} = dk_x\, dk_y\, dk_z$ is thus

$$2\left(\frac{L_x}{2\pi} dk_x\right)\left(\frac{L_y}{2\pi} dk_y\right)\left(\frac{L_z}{2\pi} dk_z\right) = 2\left(\frac{V}{8\pi^3}\right) d^3\mathbf{k}$$

where $V = L_x L_y L_z$ is the volume of the cavity and the extra factor of 2 arises from the two possibilities for the polarization. To translate this into the number of states per unit angular frequency ω, we note that since $\omega = ck$, ω depends only on the magnitude of $\mathbf{k}$, and the intervals dk and $d\omega$ are related by $dk = d\omega/c$. Thus, the states contained in the interval $d\omega$

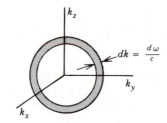

FIGURE 8.2 Spherical shell in k space.

are just those contained in the spherical shell in k space of thickness $d\omega/c$, as illustrated in Figure 8.2. The volume of such a shell is given by

$$4\pi k^2 \, dk = 4\pi \left(\frac{\omega}{c}\right)^2 \frac{d\omega}{c}$$

The number of states in the interval $d\omega$ is now found by multiplying

[number of states per unit volume of k space]

$\times$ [volume of k space corresponding to the interval ω]

$$= 2\left(\frac{V}{8\pi^3}\right)\left(4\pi\left(\frac{\omega}{c}\right)^2\left(\frac{d\omega}{c}\right)\right)$$

The result for the number of states per unit volume of cavity in the interval $d\omega$ is thus

$$g(\omega) \, d\omega = \frac{\omega^2}{\pi^2 c^3} \, d\omega \qquad (\text{m}^{-3}) \tag{9}$$

The energy density $u(\omega; T)$ is now given by substituting the expressions in equations 7 and 9 into equation 8:

$$u(\omega; T) \, d\omega = \frac{\hbar}{\pi^2 c^3} \frac{\omega^3}{e^{\beta\hbar\omega} - 1} \, d\omega \qquad (\text{J/m}^3) \tag{10}$$

From here it is a matter of geometry to calculate $I_{\text{BB}}(\omega; T) \, d\omega$, the power per unit area (intensity) emitted in the frequency interval $d\omega$ by the hole in the cavity wall. From the above discussion it is evident that the photons in the cavity have wave vectors that are distributed isotropically, that is, they approach the hole uniformly from all directions with speed c. If all of the photons approached the hole of area A from an angle θ with respect to the normal to the wall, as depicted in Figure 8.3, then the energy contained in the dashed

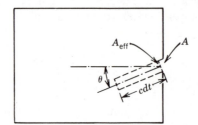

FIGURE 8.3 Geometry for the calculation of the flux of photons emerging from a hole.

volume of length $c\,dt$ and cross-sectional area $A_{\text{eff}} = A\cos\theta$ would emerge from the hole in a time dt. In this instance the power emitted in $d\omega$ would be $u(\omega; T)(cA_{\text{eff}})\,d\omega$, but because all directions of approach are equally probable, A_{eff} must be replaced with $\bar{A}_{\text{eff}}$, its average over all possible directions of approach:

$$\bar{A}_{\text{eff}} = \frac{1}{2\pi} \int_{\theta=0}^{\theta=\pi/2} \int_{\phi=0}^{\phi=2\pi} (A\cos\theta)\,d\Omega = \frac{A}{2} \qquad (\text{m}^2) \tag{11}$$

In addition, a factor of $\frac{1}{2}$ is needed to account for the fact that one half of the photons in the cavity have velocities that are, at any instant, away from the hole. Using equations 10 and 11, the power emitted per unit area in $d\omega$ becomes

$$I_{\text{BB}}(\omega; T)\,d\omega = \frac{\hbar}{4\pi^2 c^2} \frac{\omega^3}{e^{\beta\hbar\omega} - 1}\,d\omega \qquad (\text{W/m}^2) \tag{12}$$

Since in most experimental situations it is the wavelength rather than the angular frequency of the radiation that is measured directly, a more useful result is the expression for $I_{\text{BB}}(\lambda; T)\,d\lambda$, which is the power emitted per unit area in a wavelength interval $d\lambda$ about λ by an ideal blackbody. This is obtained from equation 12 by equating the power emitted in a frequency range $d\omega$ with that emitted in the corresponding wavelength range $d\lambda$:

$$I_{\text{BB}}(\lambda; T)\,d\lambda = I_{\text{BB}}(\omega; T)\,d\omega \qquad (\text{W/m}^2) \tag{13}$$

This gives $I_{\text{BB}}(\lambda; T)\,d\lambda$ as

$$I_{\text{BB}}(\lambda; T)\,d\lambda = \frac{2\pi hc^2}{\lambda^5} \frac{d\lambda}{e^{\beta hc/\lambda} - 1} \qquad (\text{W/m}^2) \tag{14}$$

which is Planck's law for blackbody radiation. The function $I_{\text{BB}}(\lambda; T)$ is plotted versus λ in Figure 8.4a for several temperatures.

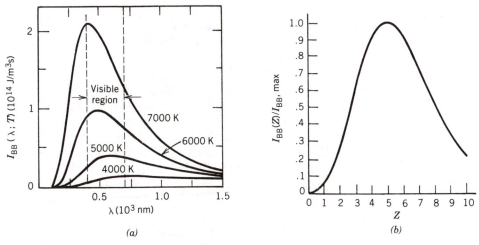

FIGURE 8.4 (a) Blackbody intensity versus wavelength for several temperatures. (b) Normalized blackbody intensity versus the dimensionless parameter z.

EXERCISE 2

Verify equation 14 by noting that the differential intervals $d\omega$ and $d\lambda$ are related by the relation $d\omega = (-2\pi c/\lambda^2)\,d\lambda$.

Two important features of the radiation from an ideal blackbody at temperature T can be deduced readily from the expression for $I_{BB}(\lambda; T)$: the total power radiated and the wavelength at which $I_{BB}(\lambda; T)$ is a maximum.

To calculate the total power radiated per unit area, one only has to integrate the intensity given by equation 12 over all angular frequencies from 0 to ∞. The result for the total power P is

$$P = \sigma T^4 \qquad (\text{W/m}^2) \tag{15}$$

where $\sigma = 2\pi^5 k_B^4 / 15c^2 h^3$ is called the Stefan–Boltzmann constant and has the numerical value 5.67×10^{-8} W/m²K⁴.

The wavelengths at the peaks of the curves in Figure 8.4a are given in terms of T by considering the dimensionless form of equation 14. Letting $z = \beta hc/\lambda$, the expression for I_{BB} becomes

$$I_{BB}(z)\, d\lambda = \frac{2\pi}{\beta^5 h^4 c^3} \frac{z^5}{e^z - 1}\, d\lambda \qquad (\text{W/m}^2) \tag{16}$$

The function $I_{BB}(z)$, plotted in normalized form in Figure 8.4b, has a maximum value at $z = 4.96511$. From the definition of z, it can be seen that the relationship between the peak wavelength λ_{max} and T is

$$\lambda_{max} T = 0.2014 \frac{hc}{k_B} \qquad (\text{m} \cdot \text{K}) \tag{17}$$

This is known as the Wien displacement law. Thus, the peak wavelength decreases with increasing T, as illustrated by Figure 8.4a.

EXERCISE 3

The frequency at which the function $I_{BB}(\omega; T)$ (given by equation 12) attains its peak does not correspond to the wavelength at which the function $I_{BB}(\lambda; T)$ (given by equation 14) peaks. How can this be so?

To apply the above theoretical results to the measurements you will make in this set of experiments, it is necessary to know the relationship between the radiation emitted by the ideal perfectly absorbing cavity hole discussed above and that emitted by an object (a tungsten ribbon in our experiments) at elevated temperatures. It can be argued (see reference 2) that the emitted spectral intensity characterizing a perfectly absorbing object is just $I_{BB}(\lambda; T)$, which is derived above for the cavity. If, as is almost always true, the object in question is not perfectly absorbing, the spectral intensity, $I(\lambda; T)$, of the radiation it emits is given as a fraction of I_{BB}:

$$I(\lambda; T)\, d\lambda = \epsilon(\lambda; T) I_{BB}(\lambda; T)\, d\lambda \qquad (\text{W/m}^2) \tag{18}$$

The factor $\epsilon(\lambda; T)$ is called the **normal spectral emissivity** (or simply the **emissivity**) that characterizes the radiation emitted normal to the surface of the hot object; it is generally a function of both wavelength and temperature and has a value between 0 (for a perfect reflector) and 1 (for a perfect absorber). The emissivity is a quantity that must be empirically determined for each object. Such a determination has been made by DeVos (reference 9) for tungsten ribbon over the range of wavelengths and temperatures relevant to this set of experiments. These results for $\epsilon(\lambda; T)$, shown graphically in Figure 8.5, were

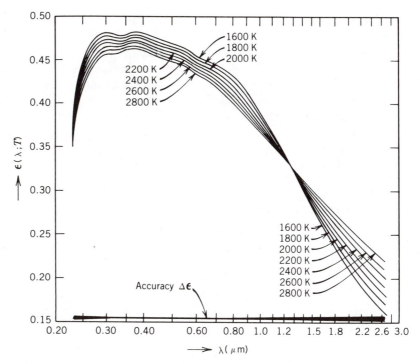

FIGURE 8.5 Normal spectral emissivity for tungsten as a function of wavelength and temperature.

obtained by a direct method in which the brightness of a cavity with tungsten walls is compared directly to the brightness of the walls themselves; the definition of equation 18 was then used to obtain ϵ for a particular temperature and wavelength. Note that $\epsilon(\lambda; T)$ for constant T exhibits a wide variation over the range of wavelengths involved, whereas for any particular wavelength, ϵ varies slowly and almost linearly with temperature.

EXERCISE 4

Discuss the restrictions on the behavior of the spectral emissivity of an object so that equation 15 for the power radiated per unit area from an ideal blackbody may be rewritten for this object as

$$P = \epsilon_T \sigma T^4 \qquad (\text{W/m}^2) \tag{15'}$$

The fraction ϵ_T defined in equation 15' is referred to as the **total emissivity**. In addition, if the ambient temperature is not 0 K, but rather some temperature T_0, then the object also absorbs some energy from the environment; hence, the net rate at which energy is being radiantly transferred from the object is

$$\frac{dU_R}{dt} = \epsilon_T \sigma A(T^4 - T_0^4) \qquad (\text{W}) \tag{15''}$$

where A is the surface area of the emitter.

EXPERIMENTS

In this set of experiments you will investigate the radiation from an incandescent tungsten ribbon. Although measuring $I(\lambda)$ for several temperatures would be the most straightforward way to accomplish this, difficulties arise because of absorption by air in the wavelength region of interest as well as from the exigencies of infrared spectroscopy. Such experiments, however, have been described in the literature (e.g., reference 7). We describe here three experiments for which neither of these difficulties is a consideration: investigation of the temperature dependence of the total radiated power, determination of hc/k_B, and investigation of the spectral intensity by modulation spectroscopy (an advanced experiment).

Temperature Determination

All of the experiments listed above will require some means of independently determining the temperature of a lamp filament. There are two convenient techniques that may be used separately or in combination.

1. *Resistance Measurement.* Independent measurements of the resistivity (ρ) of tungsten versus temperature, such as that given in Table 7.1 of Experiment 7, Electron Physics, may be used to determine the temperature of the filament from its measured resistance. The resistance R of the filament is given by $R = \rho L/a$, where L and a are its length and cross-sectional area, respectively. From the tabulated value of ρ and the measured value of R at room temperature, the value of the constant (L/a) may be deduced. A curve of $R(T)$ versus T may thus be constructed from the resistivity data; this is then used to deduce T from the measured R.

2. *Optical Pyrometry.* The determination of the temperature with an optical pyrometer is accomplished by visually matching the brightnesses, over a narrow range of wavelengths transmitted by a red filter, of the calibrated pyrometer lamp with that of the source whose temperature is being measured. The matching is done in one of two ways, as depicted somewhat schematically by Figure 8.6. In the pyrometer pictured in Figure 8.6a the match

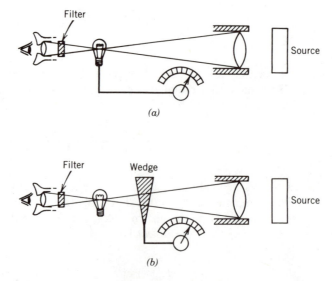

FIGURE 8.6 (a) Pyrometer in which the lamp current is varied to match the source brightness. (b) Pyrometer in which the apparent source brightness is varied to match the lamp.

is accomplished by adjusting the lamp current until the filament disappears into the image of the source. The matching may also be accomplished by keeping the lamp current constant and varying the apparent source brightness with a wedge of absorbing glass (Figure 8.6b). In either case the pyrometer has been calibrated in such a way that, when the brightnesses are matched, a scale gives the temperature of a blackbody which, at these wavelengths, has a brightness equal to that of the source. In our experiments, the source will be a tungsten filament with $\epsilon < 1$, so that it is necessary to correct the temperature reading of the pyrometer to take account of the emissivity. Since the brightness of a blackbody at temperature T_P given by the pyrometer is equal to the brightness of the tungsten filament at the "true" temperature T_t, we have, if the comparison is made over a narrow range of wavelengths centered around λ_0,

$$I_{BB}(\lambda_0; T_P) = I(\lambda_0; T_t) \qquad \left(\frac{W}{m^2\,m}\right) \tag{19}$$

Using equations 14 and 18 for I_{BB} and I and assuming that $\beta hc/\lambda_0 \gg 1$, we have, for the relationship between T_t and T_P,

$$\frac{1}{T_t} = \frac{1}{T_P} + \frac{\lambda_0 k_B}{hc} \ln[\epsilon(\lambda_0; T)] \qquad (K^{-1}) \tag{20}$$

EXERCISE 5

Show that $\beta hc/\lambda_0 \gg 1$ at 2500 K for $\lambda_0 = 653$ nm. Verify the above relationship between T_t and T_P.

Measurements

Temperature Dependence of the Radiated Power. If the rate at which the heated filament loses energy by conduction and convection is assumed to be linear in the temperature difference $T - T_0$, then the total rate at which energy is lost is given by

$$\frac{dU}{dt} = K(T - T_0) + \epsilon_T \sigma A(T^4 - T_0^4) \qquad (W) \tag{21}$$

where K is a constant. In steady state this is equal to the electrical power supplied to the filament, so that

$$I^2 R = K(T - T_0) + \epsilon_T \sigma A(T^4 - T_0^4) \qquad (W) \tag{22}$$

At sufficiently high temperatures the first term becomes negligibly small, and above 1000 K, T_0^4 is less than 1 percent of T^4.

Construct the circuit of Figure 8.7. The shunt resistance R should be chosen with due consideration to power dissipation requirements and to the maximum voltage output of the supply. Measure the voltages across the bulb and the shunt for a wide range of current. For each current, calculate both the electrical power supplied to the bulb and the temperature T determined by either of the methods suggested above. Plot $\log(I^2 R)$ against $\log T$.

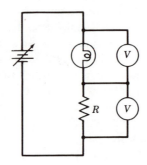

FIGURE 8.7 Circuit for investigating the temperature dependence of the radiated power.

EXERCISE 6

From this plot, determine the exponent n in the expression for the radiated power: $P = \epsilon_T \sigma A T^n$. According to your data, for what temperature range is this expression valid?

Determination of hc/k_B. The setup for this experiment is shown in Figure 8.8. Equations 14 and 18 predict that, for visible wavelengths and for temperatures attainable in a tungsten filament, the spectral intensity of the radiation from the source will be

$$I(\lambda; T)\, d\lambda \cong (2\pi hc^2)\epsilon(\lambda; T)\lambda^{-5}\, e^{-hc/\lambda k_B T}\, d\lambda \qquad (\text{W/m}^2) \qquad (23)$$

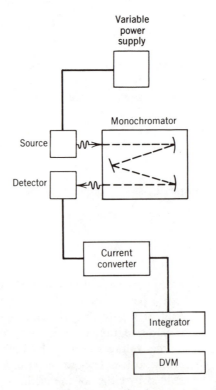

FIGURE 8.8 Configuration for the determination of hc/k_B.

The signal $S(\lambda; T)$ from the detector can thus be expressed as a proportionality:

$$S(\lambda; T) \propto r(\lambda)\epsilon(\lambda; T)\lambda^{-5} e^{-hc/\lambda k_B T} \tag{24}$$

where $r(\lambda)$ is the wavelength-dependent overall efficiency of the system. Note that $r(\lambda)$ takes into account such factors as aperture, efficiency and resolution ($\Delta\lambda$) of the monochromator, the gain of the detector, the transmission of the bulb envelope, and the area of the source. Taking the natural logarithm of both sides of equation 24 yields, to within a constant that depends on detector calibration,

$$\ln S = \text{constant} + \ln[r(\lambda)\epsilon(\lambda; T)\lambda^{-5}] - \frac{hc}{\lambda k_B} T^{-1} \tag{25}$$

Thus, if the natural logarithm of the detector signal S is plotted versus $1/T$ for some constant wavelength, the result should be a straight line with slope $-hc/\lambda k_B$ if the temperature dependence of $\ln \epsilon$ can be neglected.

With the apparatus configured as shown, focus an image of the center of the filament onto the entrance slit of the monochromator and mask the slit so that light from only a small region around the hottest point on the filament can be detected. Choose four wavelengths in the visible and, for each one, measure S versus T, the temperature of the center of the filament. If the optical pyrometer is used to measure T, align its axis perpendicular to the tungsten surface. For each wavelength, plot $\ln S$ versus $1/T$.

EXERCISE 7

From the data, determine the best value of hc/k_B. Determine from the emissivity data in Figure 8.5 whether or not we were justified in neglecting the temperature dependence of $\ln \epsilon$.

Modulation Spectroscopy of the Spectral Intensity. In this experiment you will look for possible deviations of the spectral intensity from the form predicted above by theory. If the theory is correct, we expect the signal S from a detector configured as in the previous section to be

$$S(\lambda; T) = Cr(\lambda)\epsilon(\lambda; T)I_{BB}(\lambda; T) \tag{26}$$

where C is a constant and I_{BB} is given by equation 14. If we suppose that the actual form of S, denoted S_{exp}, deviates from this and can be written as

$$S_{exp}(\lambda; T) = Cr(\lambda)\epsilon(\lambda; T)I_{BB}(\lambda; T)(1 + \delta) \tag{27}$$

then we can measure the temperature derivative of δ with the use of the apparatus in Figure 8.9. This is accomplished by noting that, from the definition of δ given by equation 27, we have, approximately,

$$\frac{(S'/S)_{exp}}{(I'/I)_{BB}} - \frac{\epsilon'/\epsilon}{(I'/I)_{BB}} \cong 1 + \frac{\delta'}{(I'/I)_{BB}} \tag{28}$$

where the prime denotes a derivative with respect to T. The ratio $(S'/S)_{exp}$ will be measured experimentally, $(I'/I)_{BB}$ is calculated theoretically, and ϵ'/ϵ can be deduced from Figure 8.5. Equation 28 thus permits an indirect experimental measurement of $\Delta \equiv 1 + \delta'/(I'/I)_{BB}$ as a function of wavelength for fixed temperature. A result of $\Delta = 1$ (i.e., $\delta' = 0$) would correspond to perfect agreement between experiment and theory.

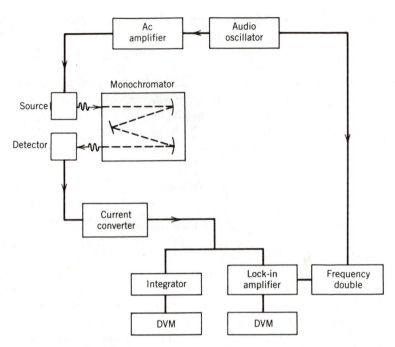

FIGURE 8.9 Experimental configuration for the modulation spectroscopy measurement.

EXERCISE 8

What approximations were made in writing equation 28? Is it important to know $r(\lambda)$ for any of these experiments?

The quantities S'_{exp} and S_{exp} are taken from the readings of the two DVMs shown in Figure 8.9. For any particular wavelength, the DVM on the output of the integrator reads S_{exp}. The amplifier and audio oscillator provide a small (typically less than 0.5 percent) temperature modulation of the source at audio frequencies. The modulation signal is fed through a frequency doubler to the lock-in amplifier, which outputs a dc signal to a DVM that is proportional to S'_{exp}. (For a discussion of the operation of the lock-in, see Appendix A.) Possible circuits for the source modulator and the frequency doubler are provided in Figures 8.10a and 8.10b, respectively.

EXERCISE 9

Explain the necessity of the frequency doubler in this detection scheme. *Hint*: Sketch the temperature versus time curve for the filament while it is being supplied with a modulated voltage and compare this to the audio oscillator output versus time.

It must be noted that the output of the lock-in is not S'_{exp} directly, but rather a quantity proportional to it, namely, $S'_{exp} \Delta T / \sqrt{2}$, where ΔT is the extent of the temperature modulation. Since ΔT is not directly measurable, it is necessary to determine the constant of proportionality between the lock-in output and S'_{exp} indirectly by assuming that theory and experiment coincide for one wavelength; for this wavelength the constant is calculated from the measured DVM outputs assuming $\delta' = 0$ in equation 28.

Choose an operating point for the bulb and measure the temperature of the center of the filament with the optical pyrometer positioned with its axis normal to the tungsten surface.

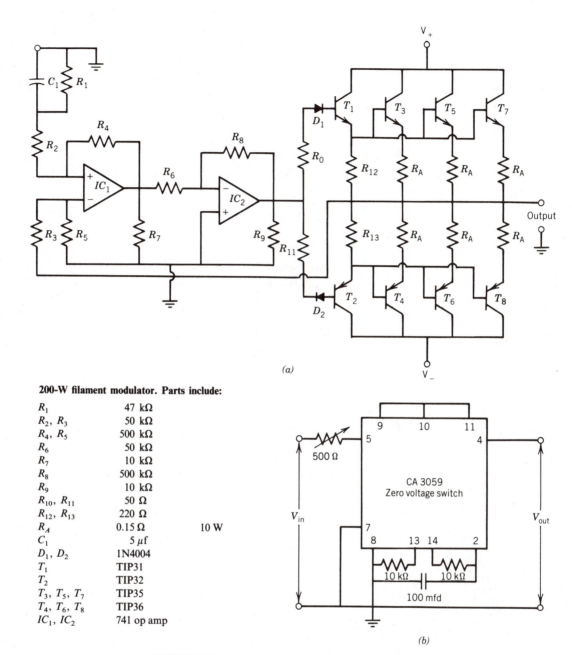

200-W filament modulator. Parts include:

R_1	47 kΩ	
R_2, R_3	50 kΩ	
R_4, R_5	500 kΩ	
R_6	50 kΩ	
R_7	10 kΩ	
R_8	500 kΩ	
R_9	10 kΩ	
R_{10}, R_{11}	50 Ω	
R_{12}, R_{13}	220 Ω	
R_A	0.15 Ω	10 W
C_1	5 μf	
D_1, D_2	1N4004	
T_1	TIP31	
T_2	TIP32	
T_3, T_5, T_7	TIP35	
T_4, T_6, T_8	TIP36	
IC_1, IC_2	741 op amp	

FIGURE 8.10 (a) Source modulator. (b) Frequency doubler.

Focus an image of the center of the filament on the monochromator entrance slit. Tabulate the readings on both DVMs against the wavelength setting of the monochromator. Compute and plot the experimental value of Δ versus λ.

EXERCISE 10

What can you conclude about the comparison between theory and experiment in the wavelength region you investigated? Would a 5 percent uncertainty in the values of ϵ'/ϵ account for the observed deviations of the points on the Δ versus λ curve from the theoretically predicted line $\Delta = 1$?

Waves and Particles

9. PHOTOELECTRIC QUANTUM YIELD, h/e: PHOTOELECTRON SPECTROSCOPY

Historical Note

The 1921 Nobel prize in Physics was awarded to Albert Einstein, Germany

For his services to Theoretical Physics, and especially for his discovery of the law of the photoelectric effect

Note that Einstein was not specifically cited for his work in relativity, although the special theory was published in 1905 and the last major paper on the general theory was published in 1915. Pais (reference 10) suggests two reasons for this: (1) The Nobel Academy believed the experimental results were not adequately clear. (2) The Nobel Academy lacked members who could competently evaluate the content of relativity theory.

APPARATUS

Phototube

Monochromatic light source (Hg lamp and interference filters or a grating or a prism)

0- to 50-V dc power supply

Current-voltage converter circuit (see Figure 9.8)

Electrometer

Photodiode of known quantum yield in the visible spectrum

OBJECTIVES

To obtain data for the stopping potential V_{sp} as a function of photon frequency ν, and to analyze the data to determine a value for h/e and a value for $\phi_a + E_G/2$ (if the photocathode is an intrinsic semiconductor) or a value for ϕ_a (if the photocathode is a metal).

To determine the quantum yield $Y(h\nu)$ as a function of the photon frequency ν, and to compare this data with the theoretical quantum yield.

To understand the physical basis for the contact potential between a semiconductor and a metal, and between two different metals.

KEY CONCEPTS

Photons

Stopping potential

Spectral response curves

Quantum yield or efficiency

Fermi–Dirac distribution

Contact potential

Work function

Photoemission current

Valence band

Conduction band

Band-gap energy

Electron affinity

Fermi energy

I-V-characteristic curve

Photoemission threshold

Reverse current

Dark current

REFERENCES

1. R. Eisberg and R. Resnick, *Quantum Physics of Atoms, Molecules, Solids, Nuclei, and Particles*, 2d ed., Wiley, New York, 1985. The photoelectric effect is discussed on pages 27–34. Contact potential difference is discussed on pages 407–409. Energy band theory of solids is discussed in Chapter 13.

2. C. Kittel, *Introduction to Solid State Physics*, 6th ed., Wiley, New York, 1986. The free electron model for metals, band theory of semiconductors, and the Fermi energy are treated thoroughly.

3. R. Engstrom, *Photomultiplier Handbook*, RCA Corporation, Lancaster, PA, 1980. This handbook includes information on the design, construction, and theory of operation of photomultiplier tubes. The phototube used in this experiment is a single-stage photomultiplier tube. The photomultiplier tube is discussed in Appendix B of this book.

4. W. Spicer, *Phys. Rev.* **112**, 114 (1958). A model is presented for photoemission from semiconductors, which includes a calculation of the quantum yield for a semiconductor photocathode.

5. R. Powell, *Am. J. Phys.* **46**, 1046 (1978). The quantum yield is derived for a phototube having a metal photocathode. Experimental results are also presented.

6. E. Snyder, *The Physics Teacher*, **98** (Feb. 1985). A method to correct for the reverse photocurrent is given.

7. A. Sommer, *Photoemissive Materials*, Wiley, New York, 1968. This book emphasizes the physical and chemical properties of photoemissive materials.

8. R. Mielke, *J. Undergraduate Res. Phys.* **1** (1) (Apr. 1982). A method is given for the direct measurement of the stopping potential in the photoelectric effect.

9. A. Dekker, *Solid State Physics*, Prentice-Hall, Englewood Cliffs, NJ, 1957. On page 231 it is shown that for two sets of electronic energy levels in equilibrium the Fermi energies must be the same.

10. A. Pais, *Subtle is the Lord, The Science and the Life of Albert Einstein*, Oxford University Press, New York, 1982.

INTRODUCTION

In 1905 Einstein postulated that the electromagnetic field was quantized, and he was then able to explain photoemission of electrons from solids. The circuit shown in Figure 9.1*a* may be used to measure the current or number of electrons ejected per second as a function of light intensity. The circuit in Figure 9.1*b* may be used to measure the maximum kinetic energy K_{max} of an ejected electron as a function of light frequency v. Conservation of energy applied to the ejected electron in Figure 9.1*b* yields

$$K_{max} = eV'_{sp} \qquad (eV) \qquad (1)$$

where V'_{sp} is the stopping potential, the potential when the photocurrent goes to zero. The work done on the electron is proportional to the potential V'_{sp} which is not the same as the

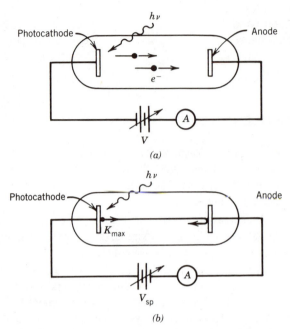

FIGURE 9.1 (a) Circuit to measure the photoelectric current as a function of the light intensity. (b) Circuit to measure the stopping potential as a function of the light frequency.

applied potential V_{sp} because of the contact potential, which will be discussed later. Conservation of energy applied to the photon and electron yields

$$eV'_{sp} = K_{max} = h\nu - \phi \qquad \text{(eV)} \qquad (2)$$

where ϕ is the minimum energy required to remove an electron from the photocathode.

We now recognize that the photoelectric effect is a part of a subfield of modern solid-state physics called photoelectron or photoemission spectroscopy. In 1958 Spicer (reference 4) proposed a model for photoemission of electrons from a solid as a three-step process:

1. A photon is absorbed and an electron makes a transition from an occupied state to one of the higher-energy empty states.
2. The excited electron then moves through the solid, possibly suffering numerous collisions.
3. An excited electron that reaches the solid–vacuum interface with sufficient energy can escape over the potential barrier.

Steps 1–3 are numbered in Figure 9.2. For those electrons that escape, their total number per absorbed photon, the quantum yield $Y(h\nu)$, and their distribution in energy $N(E, h\nu)$ can be measured. $Y(h\nu)$ and the energy distribution curves (EDC) are of interest because they are found to contain information about the detailed electronic energy band structure of the solid.

In reference 4 Spicer reports on photoemission studies of alkali antimonide semiconductors, for example, Cs_3Sb, K_2CsSb, and $Na_2KSb:Cs$, where the notation :Cs implies trace amounts of Cs. Such semiconductors and alkali metals are common photocathode materials. Typical spectral response curves are shown in Figure 9.3 for some semiconductors. The

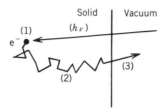

FIGURE 9.2 Model for the photoemission of electrons from a solid.

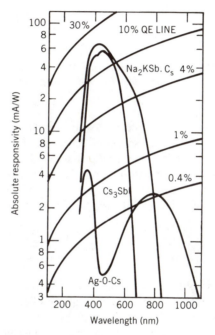

FIGURE 9.3 Spectral response curves for some semi-conductors.

decrease in response at short wavelengths is determined by the transmission characteristics of the glass window to the photocathode.

EXERCISE 1

Use the spectral response curve given in Figure 9.3 for Cs_3Sb to calculate its quantum yield (in units of the number of emitted electrons per incident photon) at a wavelength of 400 nm. In Figure 9.3, QE is the quantum efficiency which is the same as quantum yield.

Since both alkali antimonide and alkali metals are common photocathode materials, the three steps of the photoemission process are numbered on both the simplified semiconductor energy diagram in Figure 9.4a and the simplified energy diagram of a metal in Figure 9.4b. At 0 K the valence band and lower bands of the semiconductor are filled and the conduction band is empty. The valence band and conduction band are separated by a band-gap energy E_G; that is, there are no states between the bands. The electron affinity E_A is the energy required to remove an electron from the bottom of the conduction band to the vacuum level

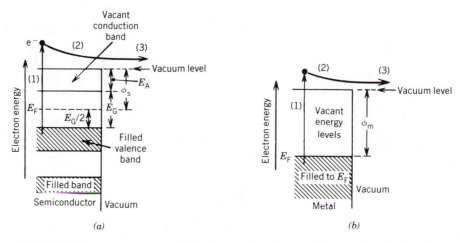

FIGURE 9.4 Three-step photoemission process for (a) an intrinsic semiconductor at 0 K and (b) a free electron model of a metal at 0 K.

(minimum energy of a free electron). Thus, the minimum energy required to produce photoemission from a semiconductor is $E_G + E_A$. For Cs_3Sb, $E_G = 1.60$ eV and $E_A = 0.45$ eV. The Fermi energy E_F is shown in Figure 9.4a for an intrinsic semiconductor, that is, a semiconductor without impurity atoms. The work function for both a semiconductor and a metal is defined to be the energy difference between the vacuum level and the Fermi level; hence, for an intrinsic semiconductor,

$$\phi_s = E_A + \frac{E_G}{2} \qquad (eV) \qquad (3)$$

Note that ϕ_s is not the minimum energy required to produce photoemission.

At 0 K the energy states of the metal are filled to the Fermi energy (see Figure 9.4b). The metal has unoccupied states above E_F at 0 K. ϕ_m is the work function of the metal and, unlike the semiconductor, it is the minimum energy required to produce photoemission. For potassium, $\phi_m = 2.24$ eV.

EXERCISE 2

The probability that an electron has an energy E is given by the Fermi–Dirac distribution function $f(E)$:

$$f(E) = \frac{1}{\exp[(E - E_F)/kT] + 1}$$

where E_F is the Fermi energy. (a) Assuming Cs_3Sb is an intrinsic semiconductor, calculate the probability that an energy level at the bottom of the conduction band is occupied at room temperature. (b) Calculate the probability that an energy level of a metal located 0.1 eV above E_F is occupied at room temperature.

Contact Potential and Stopping Potential

For the circuit shown in Figure 9.1b the electron traveling from the cathode to the anode has to overcome a retarding potential. Because of the contact potential difference between the cathode and anode, the retarding potential is not just the applied potential V_{sp}. To see

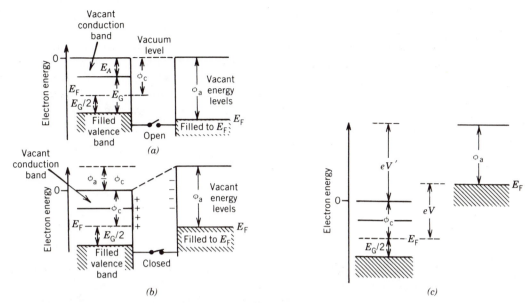

FIGURE 9.5 Electron energy states of an intrinsic semiconductor photocathode and a metal anode when the photocathode and anode (a) are not in electrical contact, (b) are electrically shorted together, and (c) are connected by an applied voltage V.

this consider Figure 9.5a, which shows electron energy states of the photocathode (arbitrarily assumed to be an intrinsic semiconductor) on the left and energy states of the anode (assumed to be a metal) on the right. The energy of an electron outside the cathode or anode is zero, and the two solids have a common vacuum level. The work function of the anode ϕ_a is assumed to be greater than the work function of the cathode ϕ_c, where for an intrinsic semiconductor ϕ_c is $E_A + E_G/2$. Note in Figure 9.5a that electrons near the top of the valence band of the cathode have more energy than electrons near the Fermi energy of the anode. Some of the more energetic electrons of the cathode will transfer to the anode when the two solids come in electrical contact.

In Figure 9.5b the solids are in electrical contact. A certain number of electrons have passed through the wire going from the cathode to the anode. Consequently, the anode becomes negatively charged and the cathode positively charged. The two solids are shown in Figure 9.5b after they have come to equilibrium, where the Fermi energies are the same (see reference 9). There is then a contact potential difference CPD between the cathode and anode given by

$$\text{CPD} = -\frac{\phi_a - \phi_c}{e} \quad \text{(V)} \tag{4}$$

where for $\phi_a > \phi_c$ the CPD is negative corresponding to a positive change in the electric potential energy of a photoelectron moving from the cathode to the anode. You should convince yourself that if the photocathode is a metal, then ϕ_c in equation 4 is the work function of that metal.

EXERCISE 3

If the anode is tungsten with a work function of 4.56 eV and the photocathode is Cs_3Sb, then determine the CPD.

In Figures 9.5b and c the vacuum level of the cathode is arbitrarily assigned the value zero.

If the cathode and anode are connected in a circuit with an applied voltage V across them, then the energy spacing between the Fermi energies will be eV as shown in Figure 9.5c. Figure 9.5c is for a grounded cathode with the anode at a negative potential, which corresponds to the circuit in Figure 9.1b with the cathode grounded and the applied voltage equal to V. As an electron moves from the cathode to the anode, the work done on it is eV' and from Figure 9.5c

$$eV' + \phi_c = eV + \phi_a \qquad (\text{eV}) \qquad (5)$$

or

$$V' = V + \frac{\phi_a - \phi_c}{e} = V - \text{CPD} \qquad (\text{V}) \qquad (6)$$

Note that the voltage V' differs from the applied voltage V by minus the CPD.

We now relate the photon energy $h\nu$ to the applied stopping potential V_{sp} for which the photocurrent is zero. From equation 6 the stopping potential seen by the electron traveling from the cathode to the anode is given by

$$V'_{sp} = V_{sp} + \frac{\phi_a - \phi_c}{e} \qquad (\text{V}) \qquad (7)$$

Multiplying equation 7 by e and using equation 2 to eliminate V'_{sp} we have

$$eV_{sp} + (\phi_a - \phi_c) = h\nu - E_A - E_G \qquad (\text{eV}) \qquad (8)$$

where for the semiconductor photocathode the minimum energy ϕ to remove an electron from the valence band to the vacuum level is $E_G + E_A$. Solving for V_{sp}, we obtain the stopping potential as a function of the light frequency

$$V_{sp} = \frac{h}{e}\nu - \frac{\phi_a + E_G/2}{e} \qquad (\text{V}) \qquad (9)$$

where we used equation 3, $\phi_c = E_A + E_G/2$. The equation analogous to equation 9 for a metal photocathode is

$$V_{sp} = \frac{h}{e}\nu - \frac{\phi_a}{e} \qquad (\text{V}) \qquad (10)$$

EXERCISE 4

Obtain equation 10 by using arguments similar to those used to obtain equation 9. Reference 1 discusses the CPD between two metals on pages 407 to 409.

Typical I-V-characteristic curves for a phototube having a semiconductor cathode and a metal anode are shown in Figure 9.6 for two different frequencies and intensities. Curves for a cathode and an anode made of different metals are similar to those in Figure 9.6 with the stopping potential given by equation 10. The dash lines show the phototube behavior when a reverse current exists. A reverse current is present when small quantities of the photo-cathode material, for example, Cs_3Sb or potassium, evaporate and deposit on the anode

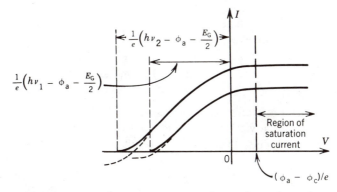

FIGURE 9.6 *I-V* characteristic curves for a phototube with a semiconductor photocathode and a metal anode.

surface. When light strikes the anode, photoemission occurs and a reverse current (anode to cathode electron flow) exists. This effect can be reduced by masking the tube so that no direct light strikes the anode, but some scattered light inevitably strikes it.

The reverse current means the observed photoelectric current is the sum of the electron emission from the cathode and from the anode. The stopping potential in both equations 9 and 10 is the potential to stop the electron current from the cathode to the anode; hence, the reverse current prevents us from measuring V_{sp} directly. Reference 6 gives a method for correcting the photocurrent and obtaining the "true" value for V_{sp}.

When the potential seen by the photoelectron is zero or positive, that is, $V' \geq 0$, then ideally all of the photoelectrons reach the anode and the resultant constant current is the saturation current (see Figure 9.6). The onset of the saturation current occurs when $V' = 0$ or from equation 6 when $V = CPD = (\phi_a - \phi_c)/e$ (see Figure 9.6).

Quantum Yield $Y(h\nu)$

Spicer (reference 4) examines the photoemission from a slab of photocathode material of thickness dx which is located a distance x from the emitting surface (see Figure 9.7). His expression for the photoemission current from the slab $di(x, h\nu)$ is

$$di(x, h\nu) = \alpha_p(h\nu)I(x, h\nu)P(x, h\nu)\, dx \qquad \text{(number of emitted electrons per second)} \qquad (11)$$

where $\alpha_p(h\nu)$ is the absorption coefficient (1/distance) for excitation into states above the vacuum level, $I(x, h\nu)$ is the light intensity at x, and $P(x, h\nu)$ is the probability of an electron excited at x by a photon of energy $h\nu$ escaping from the material. $I(x, h\nu)$ is given by

$$I(x, h\nu) = I'e^{-\alpha_t(h\nu)x} \qquad \text{(number of photons per second)} \qquad (12)$$

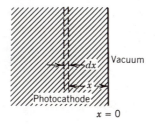

FIGURE 9.7 A slab of photocathode material of thickness dx and distance x from vacuum.

where I' is the intensity of the transmitted light at $x = 0$ in photons per second and $\alpha_t(hv)$ is the total absorption coefficient. Spicer points out that the simplest form for $P(x, hv)$, which will fit the experimental data, is

$$P(x, hv) = G(hv) e^{-\beta x} \qquad (13)$$

where β is a constant and $G(hv)$ is an, as yet, undefined function of hv.

The quantum yield $Y(hv)$ is obtained by dividing equation 11 by I' and integrating over an assumed infinite thickness. The result is

$$Y(hv) - \frac{1}{I'} \int_0^\infty di(x, hv) = \frac{\alpha_p(hv)}{\alpha_t(hv) + \beta} G(hv) \qquad \text{(number of emitted electrons per photon)}$$

$$(14)$$

Spicer found that near the photoemission threshold the quantum yield has the form

$$Y(hv) = G[hv - (E_G - E_A)]^{3/2} \qquad \text{(number of emitted electrons per photon)} \qquad (15)$$

over a spectral range of about 1 eV, where G is a constant over this range and E_G and E_A are the band gap and electron affinity of the semiconductor.

Powell (reference 5) obtains an expression for the quantum yield of a metal photocathode near the photoemission threshold. Powell finds

$$Y(hv) \propto (hv - \phi_c)^2 \qquad \text{(number of emitted electrons/photon)} \qquad (16)$$

for photon energies that are less than 3 eV, where ϕ_c is the work function of an alkali–metal cathode.

Table 9.1 is a partial listing of vacuum phototubes, where the S number is a spectral response designation. The different tube types have different physical dimensions and/or different electro-optic characteristics. Ag–O–Cs and Ag–O–Rb photoemissive materials are discussed in reference 7.

TABLE 9.1 VACUUM PHOTOTUBES

Type	S number	Photocathode Composition	Window
1P39	S-4	Cs_3Sb	Lime glass
1P42	S-9	[a]	(Lime glass)
917	S-1	Ag–O–Cs	Lime glass
919	S-1	Ag–O–Cs	Lime glass
922	S-1	Ag–O–Cs	Lime glass
926	S-3	Ag–O–Rb	Lime glass
929	S-4	Cs_3Sb	Lime glass
934	S-4	Cs_3Sb	Lime glass
935	S-5	Cs_3Sb	Corning 9741
2022	S-1	Ag–O–Cs	Lime glass
6570	S-1	Ag–O–Cs	Lime glass

[a]Obsolete; formerly similar to S-11, which is Cs_3Sb.

EXPERIMENTS

Each measurement suggested under the objectives of this experiment requires (1) a photo-tube, (2) a monochromatic light source, and (3) a circuit to measure the output of the phototube. In addition, the quantum yield experiment requires a method of calibrating the power output of the monochromatic light source. The two experiments are discussed separately below.

I–V Characteristic Curve

Reference 8 discusses a circuit that uses an operational amplifier to measure the stopping potential directly. The circuit suggested uses an operational amplifier as shown in Figure 9.8a. Figure 9.8b shows the 356N operational amplifier and the electrical connections. A top view of the pin assignments for the 356N is shown in Figure 6.6, Experiment 6.

The 25-kΩ potentiometer shown in Figure 9.8b serves the purpose of balancing the operational amplifier such that the output voltage V_2 is zero when the "dark current" provides the input. To balance the op-amp, wrap the phototube with a dark cloth and then adjust the potentiometer until V_2 reads zero.

Note in Figure 9.8a that the polarity of the photocathode can be switched by reversing the 3-V leads, that is, by connecting the positive terminal to ground and the negative terminal to the ungrounded side of the potentiometer.

The relation between V_2 and the anode current I can be easily obtained knowing the general property of op-amps, namely, that the (+) and (−) input potentials are nearly

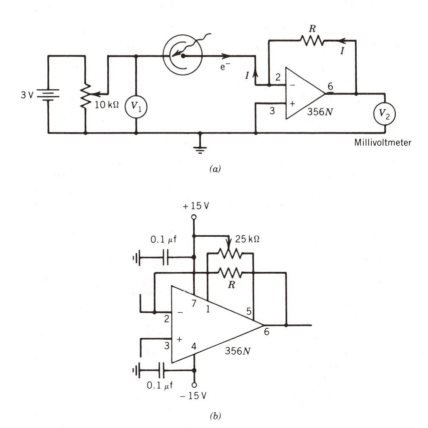

FIGURE 9.8 (a) Current–voltage converter circuit. (b) 356N op-amp connections.

equal, $V_+ - V_- \simeq 0$. Since we have grounded V_+, we must have $V_- \simeq 0$, so the phototube anode is effectively grounded. The result is that V_1 measures the potential difference across the cathode and anode, and a current the same magnitude as the anode current is drawn through the feedback resistor R, that is, $I = V_2/R$. R should be about $1\,M\Omega$ to get a convenient reading of V_2.

I versus V curve. Using a single filter measure V_2 as a function of V_1 from -3 to $+3\,V$. To obtain a detailed plot of I (or just V_2) versus V_1 it is suggested that data be taken in approximately 0.2-V increments. Plot your data.

Remark: The current I or voltage V_2 is proportional to the light intensity, and the intensity is dependent on the lamp-phototube separation and the probability per unit time of an atomic transition occurring between two energy levels of mercury, that is, the intensity will vary with the wavelength of the emitted spectral line. Do not alter the lamp-phototube separation while you are recording the data for any one spectral line. It is suggested that this separation be adjusted such that the maximum voltage V_2 is about $200\,mV$.

h/e

Stopping potential. Measure the stopping potential V_{sp} as a function of the photon frequency. Plot your data and do a least-squares fit to determine the equation of the line. Look up the accepted values of h and e, and calculate the ratio h/e to four significant figures.

Obtain the manufacturer's specifications for your phototube to determine the composition of the cathode and anode. Then look up the work function of the anode ϕ_a and the band gap E_G for the cathode if it is a semiconductor.

EXERCISE 5

Does the accepted value for ϕ_a (or $\phi_a + E_G/2$ if the cathode is a semiconductor) fall within the error range of your experimental value?

Quantum Yield

To determine the quantum yield of the phototube it is necessary to calibrate the power output of the monochromatic light source. The suggested method to do the calibration is to use a photodiode detector of known quantum yield in the energy range of the light source, that is, the manufacturer provides a spectral response curve over the visible part of the spectrum.

The quantum yield of the phototube is defined to be

$$Y \equiv \frac{\text{number of electrons ejected per second}}{I'} \qquad \text{(number of emitted electrons per photon)} \tag{17}$$

where I' is the number of photons/second transmitted at the photocathode surface. I' is related to the number of photons incident per second on the window of the phototube I_0 (see Figure 9.9):

$$I' = T(1 - R)I_0 \qquad \text{(number of photons transmitted per second)} \tag{18}$$

where T is the transmission coefficient of the glass window and R is the reflection coefficient of the photocathode. The index of refraction of the glass is about 1.5 and, therefore,

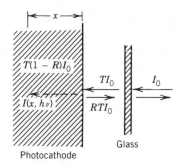

FIGURE 9.9 $I(x, h\nu)$ is the intensity of light of frequency ν after penetrating the photocathode a distance x.

$T \simeq 0.92$ in the visible part of the spectrum. For a potassium photocathode $R \simeq 0.70$ and is nearly constant in the visible part of the spectrum. For a Cs_3Sb photocathode $R \simeq 0.28$ over the visible part of the spectrum. The Ag–O–Cs photocathode has a reflection coefficient of about 0.17 over the visible part of the spectrum; however, it does depend on the concentration of silver.

Substituting equation 18 into 17 we obtain

$$Y = \frac{\text{electrons ejected per second}}{T(1-R)I_0} \qquad \text{(number of emitted electrons per photon)} \quad (19)$$

The phototube will be operated in the saturation current region, and we assume all of the ejected electrons are collected by the anode; hence, I is proportional to the number of electrons ejected per second, where I is the phototube current. The number of incident photons per second is proportional to i/κ, where i is the current of the standard detector and κ is the spectral response (number of electrons/photon) of the detector. Thus, the quantum yield becomes

$$Y = \frac{I\kappa}{T(1-R)i} \qquad \text{(number of emitted electrons per photon)} \quad (20)$$

Set the anode in Figure 9.10 at a large positive voltage (do not exceed the maximum

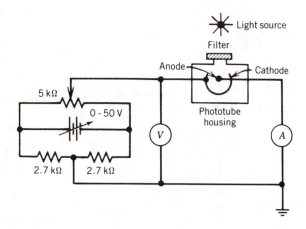

FIGURE 9.10 Circuit to determine the quantum yield of a phototube as a function of light frequency.

value specified by the manufacturer) and measure I as a function of the light frequency v. With the voltage across the standard detector set at the manufacturer's recommended value, measure the current i at the same light frequencies, then calculate Y. Plot Y^n versus hv, where n is either $\frac{1}{2}$ (metal photocathode) or $\frac{2}{3}$ (semiconductor photocathode), and do a least-squares fit.

EXERCISE 6

From your graph, what is the value of ϕ_c(or $E_G + E_A$)? Compare ϕ_c(or $E_G + E_A$) with the accepted value.

COMPUTER-ASSISTED EXPERIMENTATION (OPTIONAL)

Prerequisite

Experiment 6, Introduction to Computer-Assisted Experimentation.

Experiment

Circuits for interfacing the phototube to the computer via an ADC/DAC card having an analog range of 0 to $+K$ V or $-K$ to $+K$ V are discussed in the paragraphs that follow. You should read the discussion that is appropriate for the analog range of your card.

ADC/DAC Analog Range of 0 to $+K$ V

The conditioning circuit is shown in Figure 9.11. The DAC output controls the experiment via op-amp 3. The 25-kΩ potentiometer shown in Figure 9.8b is not needed for op-amp 3. The output of this op-amp is given by

$$V_{out,\,3} = -\frac{R_3}{10k} V_{in} + \frac{R_3 + 10k}{10k} V_{off} \quad (V) \tag{21}$$

where V_{in} is the DAC output, V_{off} is the offset voltage, and the desired range $V_{out,\,3}$ is -3 to $+3$ V.

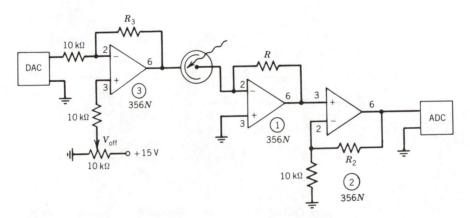

FIGURE 9.11 Conditioning circuits for an ADC/DAC analog range of 0 to $+K$ V.

EXERCISE 7

Assume that $K = 5$ V. Show that an input ranging from 0 to 5 V results in an output ranging from $+3$ to -3 V, provided that $R_3 = 12$ kΩ and $V_{off} = +1.36$ V.

The circuit to interface the output of the phototube to the ADC input is also shown in Figure 9.11. Note that op-amp 1 in Figure 9.11 is the op-amp shown in Figure 9.8a. The output of op-amp 2 is given by

$$V_{out,\,2} = \left(1 + \frac{R_2}{10k}\right) V_{out,\,1} \qquad (V) \qquad (22)$$

where $V_{out,\,1}$ is the output of op-amp 1, which ranges from 0 to an adjusted value of about 200 mV.

EXERCISE 8

Assuming that $K = 5$ V and the maximum output of op-amp 1 is 200 mV, show that the maximum output of op-amp 2 is 5 V provided that $R_2 = 240$ kΩ.

Bench Testing   Knowing the analog range of your ADC/DAC card, select and install R_2. Replace the DAC output in Figure 9.11 with a 0 to K-V power supply, and connect the output of op-amp 2 to a voltmeter in place of the ADC. Set V_{off} to the appropriate value. With the dark current as the input to op-amp 1, set the outputs of op-amp 1 and 2 to zero in the order that they are numbered by adjusting the 25-kΩ potentiometer for each op-amp. (See Figure 9.8b.) By using the 546-nm spectral line, say, with a lamp-phototube separation such that the maximum output of op-amp 1 is 200 mV, vary the input of op-amp 3 from 0 to K V and observe the output of op-amp 2. If the circuit is functioning properly, connect it to the computer as is shown in Figure 9.11.

Measurements and solftware are suggested at the end of the experiment.

ADC/DAC ANALOG RANGE OF $-K$ TO $+K$ V

In this case the circuit to interface the apparatus to the computer is shown in Figure 9.12. The DAC output can provide a maximum current of about 1 mA and, hence, it connects directly to the phototube.

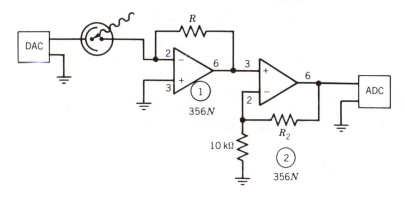

FIGURE 9.12 Conditioning circuit for an ADC/DAC analog range of $-K$ to $+K$ V.

EXERCISE 9

Refer to Figure 9.8. If $R = 1$ MΩ and the maximum output of the op-amp is 200 mV, what is the maximum current I?

The output of op-amp 2 is given by

$$V_{\text{out, 2}} = \left(1 + \frac{R_2}{10\text{k}}\right) V_{\text{out, 1}} \qquad (\text{V}) \tag{23}$$

where $V_{\text{out, 1}}$ is the output of op-amp 1.

EXERCISE 10

Assuming that $K = 5$ V and op-amp 1 output ranges from 0 to 200 mV, show that the output of op-amp 2 ranges from 0 to $+5$ V provided that R_2 is 240 kΩ. (Note only one half of the analog range is utilized. This is because of the need to adjust the output of op-amp 2 to zero when the input to op-amp 1 is the dark current.)

Bench Testing   Install the appropriate resistance R_2. In the circuit shown in Figure 9.12, replace the DAC output with the 3-V power supply and 10-kΩ potentiometer shown in Figure 9.8a; replace the ADC input with a voltmeter. With the dark current as the input to op-amp 1, set the outputs of 1 and 2 to zero in the order that they are numbered by adjusting the 25-kΩ potentiometer for each op-amp. The potentiometer is shown in Figure 9.8b. By using the 546-nm spectral line, say, with the lamp-phototube separation such that the maximum output of op-amp 1 is 200 mV, vary the voltage to the phototube and observe the output of 2. If the circuit functions properly, connect it to the computer as is shown in Figure 9.12.

MEASUREMENTS AND SOFTWARE

Some possible things to do are the following.

1. Write software that increments the DAC output, reads the ADC input for each DAC output, and stores each pair of values.
2. Run the software for each spectral line, and instruct the computer to plot an I-V characteristic curve and to determine the stopping potential.
3. Write software that plots stopping potential versus frequency, analyzes the data, and prints a hard copy.

---

10. DIFFRACTION OF X RAYS AND MICROWAVES BY PERIODIC STRUCTURES: BRAGG SPECTROSCOPY

Historical Note

The 1914 Nobel prize in Physics was awarded to Max Von Laue, Germany
For his discovery of the diffraction of X-rays by crystals.

The 1915 Nobel prize in Physics was awarded jointly to Sir William Henry Bragg, and his son, Sir William Lawrence Bragg, both of Great Britain
For their serivces in the analysis of crystal structures by means of X-rays.

APPARATUS

X-ray diffractometer system (e.g., Tel-X-Ometer crystallography system, available from PASCO Scientific) which includes:
   Ratemeter or scaler
   Geiger–Mueller (G–M) tube
   Motor drive
   X-ray tube with copper anode
   Scatter shield
NaCl and KCl in powder and single-crystal forms
Ni foil
Microwave Bragg diffraction apparatus (available from Sargent-Welch Scientific Co.) which includes:
   Goniometer
   Cubic lattice of Al spheres in a foam plastic matrix
   Microwave transmitter and receiver

OBJECTIVES

To obtain and analyze single-''crystal'' data from the microwave diffraction apparatus as an analog to the Bragg scattering of x rays.

To obtain and analyze x-ray Bragg reflections from single alkali–halide crystals of known orientation.

To obtain and analyze x-ray reflections from a powder sample.

To become acquainted with some aspects of crystal symmetry, particularly as applied to the cubic lattices.

To understand the physical basis of the Bragg reflections from crystal planes as predicted by the Bragg condition and a consideration of the structure factor.

KEY CONCEPTS

Bragg diffraction (scattering)   Structure factor
Bravais lattice   Form factor
Primitive cell   Bremsstrahlung
Basis   Characteristic radiation
Miller indices   Absorption edge
Fourier analysis   Powder (Debye–Scherrer) method
Reciprocal lattice vector

REFERENCES

1. C. Kittel, *Introduction to Solid State Physics*, 6th ed., Wiley, New York, 1985. Chapters 2 and 3 provide a concise, readable introduction to crystal geometry, the reciprocal lattice, and Bragg diffraction.

2. R. Eisberg and R. Resnick, *Quantum Physics of Atoms, Molecules, Solids, Nuclei and Particles*, 2d ed., Wiley, New York, 1985. Section 9 contains a discussion of the production and nomenclature of atomic x-ray emission lines.

3. B. E. Warren, *X-ray Diffraction*, Addison-Wesley, Reading, MA, 1969. Discusses crystal symmetry and diffraction techniques.

4. P. J. Barry and A. D. Brothers, *Am. J. Phys.* **54**, 186 (1986). Describes a technique for interfacing a diffractometer system with a microcomputer for an automated data acquisition system.

5. *International Tables for X-Ray Crystallography*, Vol. 3, Reidel, Boston, 1962. Contains pertinent numerical data such as wavelengths of characteristic x-ray emission lines and absorption edges and atomic form factors.

6. L. V. Azaroff and M. J. Buerger, *The Powder Method in X-Ray Crystallography*, McGraw-Hill, New York, 1958.

7. H. F. Meiners (ed.), *Physics Demonstration Experiments*, Vol. 2, Ronald Press, New York, 1970. The microwave scattering experiment discussed below is based on the arrangement described in Chapter 33, Section 7.15, which features apparatus developed by H. F. Meiners.

INTRODUCTION

When electromagnetic radiation is incident upon a periodic array of scattering centers as in Figure 10.1*a*, there are certain discrete directions for the incident ray that result in strong reflections; this is because of constructive interference of the radiation scattered from each of the centers. The directions for which these strong reflections occur are related through the Bragg law, described in the discussion that follows, to the geometry of the arrangement. Measurements of the angular positions and intensities of these Bragg reflections can be used to deduce the arrangement and spacings of the scatterers. If the scatterers are the atoms or molecules of a crystal of unknown geometry and the radiation is a monochromatic x-ray beam of known wavelength, measurements of the angular distribution and intensities of the reflected beams can be used to determine the crystallographic symmetry and interatomic spacings. Conversely, a crystal with a known structure may be used to spectrally analyze an x-ray beam or as an x-ray monochromator.

X-ray scattering, sometimes referred to as Bragg diffraction, is widely used in research laboratories around the world. One recent estimate puts the number of x-ray diffraction

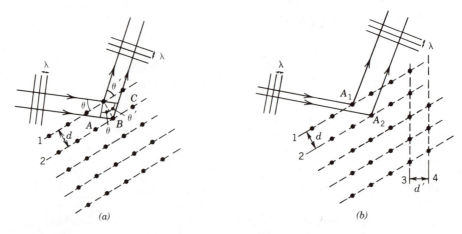

FIGURE 10.1 (*a*) Periodic array of scatterers. (*b*) Periodic array with atoms "misaligned."

users worldwide at about 25 000, one-third of which are in the United States. In addition to basic crystallographic structure analysis, diffraction techniques are also currently useful in such applications as the qualitative and quantitative analysis of material composition, the analysis of stress/strain conditions within a given polycrystalline material, and the study of phase transitions at elevated temperatures. X-ray scattering is also currently being used in the study of macromolecular systems of interest to molecular biologists. Researchers have, for example, studied diffraction patterns from magnetically oriented solutions of macromolecular assemblies that yield subcellular structural information; "movies" of proteins in motion have been produced using high-intensity nanosecond pulses of x rays; the three-dimensional internal structure of complex organic molecules is also currently being probed with x rays.

The analysis of the x-ray diffraction patterns to be observed in this experiment is based on the Bragg law of equation 2, which follows. We derive this law first from a simplified, heuristic, two-dimensional viewpoint, following which is a more general (and somewhat more complex) three-dimensional treatment.

The Bragg Law in Two Dimensions

The Bragg law may be derived in a simple way by considering the reflection (or scattering) of x rays from the planes of atoms indicated in Figure 10.1a. If the x rays are treated classically as monochromatic electromagnetic waves of wavelength λ, then the reflections from successive planes of atoms will interfere constructively if the total difference δ in optical path lengths for waves reflected from planes 1 and 2 is an integral number of wavelengths. If the spacing between the indicated planes is d and the incident beam makes an angle θ with these planes, then the path length difference is given by

$$\delta = d(\sin \theta + \sin \theta') \qquad (\text{m}) \qquad (1)$$

If we assume that $\theta = \theta'$, as is the case for specular reflection of visible radiation from dielectric or metallic surfaces, then the condition for constructive interference is

$$2d \sin \theta = n\lambda \qquad (\text{m}) \qquad (2)$$

where n is a positive integer. Thus, we expect reflections from this family of planes whenever $\theta = \theta'$ and the Bragg law of equation 2 is satisfied.

Note that although we assumed $\theta = \theta'$, this condition emerges as a natural consequence of the requirement that, for a Bragg reflection, atoms within a given plane, as well as atoms in different planes, must scatter constructively. It should also be noted that the Bragg law is a consequence only of the spatial periodicity of the scatterers in a direction perpendicular to the reflecting planes and does not depend on the alignment of the atoms of plane 1 with those of plane 2.

EXERCISE 1

Show that if the Bragg law is satisfied for the situation in Figure 10.1b, the radiation scattered from atoms A_1 and A_2 will constructively interfere even though they are not "aligned" in a direction perpendicular to planes 1 and 2 as in Figure 10.1a.

Three-Dimensional Description: Bravais Lattices and Miller Indices

The crystalline *lattices* discussed above contain several families of parallel planes in addition to the ones pictured in Figure 10.1a, each with its own orientation and spacing (e.g., those parallel to planes 3 and 4 in Figure 10.1b), which have the potential to produce Bragg reflections if equation 2 is satisfied. Additionally, if these lattices are considered as representing an arrangement of scatterers that exhibits periodicity in each of three dimensions, the enumeration of all the possible reflections from every family of planes is a formidable task that requires an understanding of the geometry of crystals in three dimensions.

The structure of a crystalline arrangement of atoms or molecules is described by specifying a basic repetitive unit of the lattice, the unit cell. The fourteen fundamental types of three-dimensional crystal lattices, the so-called Bravais lattices, are divided into seven crystal systems according to the geometry of the unit cell; these are listed in Table 10.1. Figure 10.2a shows the cell geometry and location of the lattice points for each cell. The choice of unit cell for each type of lattice is not unique; the cells shown here are the conventional ones rather than the "primitive" cells of minimum volume. Each cell is described conveniently in terms of a set of axes and three translation vectors $\mathbf{a}_1$, $\mathbf{a}_2$, and $\mathbf{a}_3$, as pictured in Figure 10.2b. The restrictions on the angles α, β, and γ are given in the third column of Table 10.1. The relationship between the translation vectors, the unit cell, and the structure of the crystal is illustrated by Figure 10.2c. As can be seen here, a cell can be translated into any other cell in the lattice by a displacement of the form $l\mathbf{a}_1 + m\mathbf{a}_2 + n\mathbf{a}_3$, where l, m, and n are integers. To complete the specification of the crystal structure it is necessary to specify a basis, that is, a group of atoms or molecules to be associated with each point of the lattice.

Of particular interest in this experiment is the cubic system with its three lattice types shown in the top row of Figure 10.2: simple cubic (sc), body-centered cubic (bcc), and face-centered cubic (fcc). The unit cell for all three structures is a cube, but the location of the lattice points within the cube differs for each structure. The NaCl crystal, an important example of a crystal structure with an fcc lattice, can be described by choosing a basis consisting of one Na^+ ion and one Cl^- ion. If an Na^+ ion is considered to be located at each fcc lattice point, then a Cl^- ion is found displaced by a vector $\frac{1}{2}\mathbf{a}_1 + \frac{1}{2}\mathbf{a}_2 + \frac{1}{2}\mathbf{a}_3$ with

TABLE 10.1 THE SEVEN CRYSTAL SYSTEMS

System	Number of Lattices	Restrictions on Conventional Cell Axes and Angles
Triclinic	1	$a_1 \neq a_2 \neq a_3$ $\alpha \neq \beta \neq \gamma$
Monoclinic	2	$a_1 \neq a_2 \neq a_3$ $\alpha = \gamma = 90° \neq \beta$
Orthorhombic	4	$a_1 \neq a_2 \neq a_3$ $\alpha = \beta = \gamma = 90°$
Tetragonal	2	$a_1 = a_2 \neq a_3$ $\alpha = \beta = \gamma = 90°$
Cubic	3	$a_1 = a_2 = a_3$ $\alpha = \beta = \gamma = 90°$
Trigonal	1	$a_1 = a_2 = a_3$ $\alpha = \beta = \gamma < 120°, \neq 90°$
Hexagonal	1	$a_1 = a_2 \neq a_3$ $\alpha = \beta = 90°$ $\gamma = 120°$

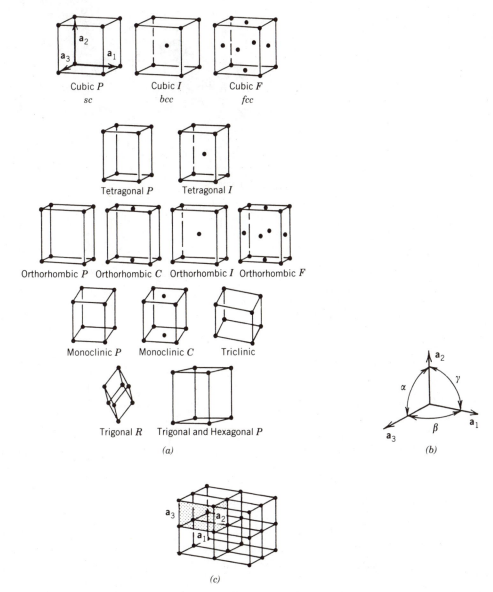

FIGURE 10.2 (a) Cell geometry for the Bravais lattices. (b) Translation vectors for the unit cell. (c) Relationship between the unit cell, the translation vectors, and the crystal structure.

respect to each Na^+, where $|\mathbf{a}_1| = |\mathbf{a}_2| = |\mathbf{a}_3| = a$, the length of the side of the unit cube. This arrangement is shown in Figure 10.3.

The standard way to specify the three-dimensional orientation of the planes of scatterers associated with each possible Bragg reflection is by use of Miller indices. The Miller indices specifying a plane such as the one depicted in Figure 10.4a is a set of three numbers (hkl) determined from the intercepts of the plane on the three crystal axes. To find the indices for a plane: (1) find the intercepts along each axis, expressed in units of the translation vector parallel to that axis; (2) take the reciprocal of each of these three numbers; (3) multiply each of the three numbers by the smallest integer necessary to clear the fractions. For the plane of Figure 10.4a, the intercepts are expressed in ordered triplet form as (322), from which the reciprocals are ($\frac{1}{3}, \frac{1}{2}, \frac{1}{2}$). Clearing fractions gives (233) as the Miller indices of this family of planes.

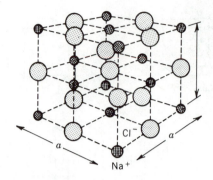

FIGURE 10.3 The structure of an NaCl crystal.

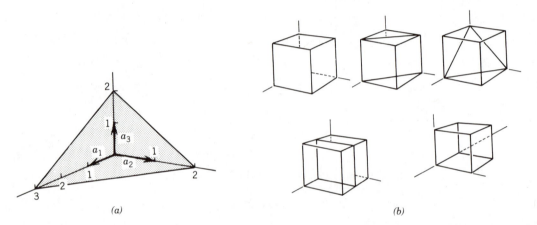

FIGURE 10.4 (a) Crystal plane shown with intercepts on axes. (b) Various planes within cubic unit cell.

EXERCISE 2

Determine the Miller indices for each of the planes shown with the cubic unit cell in Figure 10.4b. Note that for negative digits in a set of indices, minus signs are conventionally written as dashes above the digit, for example, $(1, \bar{1}, 0)$.

Bragg Diffraction in Three Dimensions; Reciprocal Lattice Vectors

To predict the directions in which x rays will be Bragg reflected from a crystal of given geometry, consider the situation of Figure 10.5, in which the x-ray source emits a plane wave with wave vector $\mathbf{k}$ $(k = 2\pi/\lambda)$ of the form

$$\mathbf{E}_i = \mathbf{E}_{0i} \, e^{j(\mathbf{k}\cdot\mathbf{r} - \omega t)} \qquad (\text{N/C}) \tag{3}$$

which is incident upon the volume element of crystal dV located at $\mathbf{r}$ with respect to an origin O. The electrons within this element respond to the wave by scattering radiation which, at the detector, has an amplitude proportional to

$$E_i \, e^{j[\mathbf{k}'\cdot(\mathbf{r}_\mathrm{D} - \mathbf{r})]} \, n(\mathbf{r}) \, dV \qquad (\text{N/C}) \tag{4}$$

where $n(\mathbf{r})$ is the electron density in dV and $\mathbf{k}'$ is the wave vector of the scattered wave.

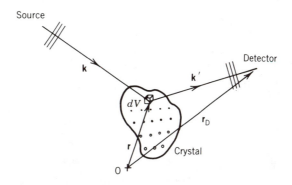

FIGURE 10.5 X-ray plane wave scattered from a volume element of a crystal.

Ignoring both the time dependence and the constant phase $\mathbf{k}' \cdot \mathbf{r}_D$, the behavior of the amplitude of the scattered wave E_s at the detector is given by

$$E_s \propto n(\mathbf{r}) \, e^{j[(\mathbf{k} - \mathbf{k}') \cdot \mathbf{r}]} \, dV \qquad (5)$$

The amplitude of the radiation at the detector due to scattering by the entire crystal is proportional to a quantity obtained by performing a volume integral of equation 5 over the electron distribution of the crystal. The result is the scattering amplitude F:

$$F = \int dV \, n(\mathbf{r}) \, e^{-j[\Delta \mathbf{k} \cdot \mathbf{r}]} \qquad (6)$$

where $\Delta \mathbf{k} \equiv \mathbf{k}' - \mathbf{k}$ is called the scattering vector. To obtain from equation 6 the values of Δk for which there will be strong Bragg reflections, it is necessary to Fourier analyze the electron density function $n(\mathbf{r})$. This is done by writing $n(\mathbf{r})$ as a sum:

$$n(\mathbf{r}) = \sum_{\mathbf{G}} n_{\mathbf{G}} \, e^{j(\mathbf{G} \cdot \mathbf{r})} \qquad (\mathrm{m}^{-3}) \qquad (7)$$

where the $n_{\mathbf{G}}$ values are possibly complex and the summation ranges over all possible reciprocal lattice vectors $\mathbf{G}$, which we now define as

$$\mathbf{G}(hkl) = h\mathbf{b}_1 + k\mathbf{b}_2 + l\mathbf{b}_3 \qquad (\mathrm{m}^{-1}) \qquad (8)$$

Each combination of integers (hkl) specifies a different $\mathbf{G}$. The vectors $\mathbf{b}_1$, $\mathbf{b}_2$, and $\mathbf{b}_3$ constitute a basis for the *reciprocal lattice* and are defined in terms of the crystal translation vectors

$$\mathbf{b}_1 = 2\pi \frac{\mathbf{a}_2 \times \mathbf{a}_3}{\mathbf{a}_1 \cdot \mathbf{a}_2 \times \mathbf{a}_3} \qquad \mathbf{b}_2 = 2\pi \frac{\mathbf{a}_3 \times \mathbf{a}_1}{\mathbf{a}_1 \cdot \mathbf{a}_2 \times \mathbf{a}_3} \qquad \mathbf{b}_3 = 2\pi \frac{\mathbf{a}_1 \times \mathbf{a}_2}{\mathbf{a}_1 \cdot \mathbf{a}_2 \times \mathbf{a}_3} \qquad (\mathrm{m}^{-1}) \qquad (9)$$

EXERCISE 3

Calculate the reciprocal lattice vectors $\mathbf{G}$ for a simple cubic lattice of side a. Verify that the expansion of equation 7 for the function $n(\mathbf{r})$, which repeats in each unit cell of Figure 10.2c, is simply the three-dimensional Fourier series representation of this function.

EXERCISE 4

Show that, for the vectors $\mathbf{b}_1$, $\mathbf{b}_2$, and $\mathbf{b}_3$ defined above, $a_i \cdot b_j = 2\pi \delta_{ij}$. Use this orthogonality relationship to show that each reciprocal lattice vector $\mathbf{G}(hkl)$ is perpendicular to the set of planes with Miller indices (hkl).

EXERCISE 5

Use the result of Exercise 4 to show that the spacing between planes (hkl) is given by $d(hkl) = 2\pi/|G(hkl)|$. Calculate $d(hkl)$ for a cubic lattice of side a.

We can invert the sum in equation 7 by the standard procedures of Fourier analysis to obtain an expression for $n_\mathbf{G}$, the Fourier coefficient corresponding to $\mathbf{G}$:

$$n_\mathbf{G} = V_C^{-1} \int\limits_{\text{cell}} dV\, n(\mathbf{r})\, e^{-j(\mathbf{G} \cdot \mathbf{r})} \qquad (\text{m}^{-3}) \qquad (10)$$

where V_C is the volume of the unit cell over which the integral is to be done.

Using the expansion of equation 7 for $n(\mathbf{r})$ in the expression for the scattering amplitude F of equation 6 gives us a useful form for F:

$$F = \sum_\mathbf{G} \int\limits_{\text{crystal}} dV\, n_\mathbf{G}\, e^{j[(\mathbf{G} - \Delta\mathbf{k}) \cdot \mathbf{r}]} \qquad (11)$$

Each term in the above sum contains an exponential which, if its argument is nonzero, oscillates in such a way that integrating over the crystal volume gives zero for that term. Thus, the only way F can be nonzero, so that a reflection can occur, is if the scattering vector $\Delta\mathbf{k}$ happens to equal one of the $\mathbf{G}$ values, making the exponential equal to unity for one term of the sum. Hence, a necessary condition on $\Delta\mathbf{k}$ ($=\mathbf{k}' - \mathbf{k}$) for the occurrence of a Bragg reflection is

$$\Delta\mathbf{k} = \mathbf{G} \qquad (\text{m}^{-1}) \qquad (12)$$

where $\mathbf{G}$ is any of the reciprocal lattice vectors defined by equation 8. If we limit our

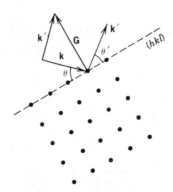

FIGURE 10.6 Initial wave vector, final wave vector, and the scattering vector for a Bragg reflection.

consideration to elastic (coherent) scattering in which the wavelength is unaltered, we have the additional restriction

$$\mathbf{k}' = \mathbf{k} \qquad (\text{m}^{-1}) \tag{13}$$

An example of a Bragg reflection satisfying the conditions of equations 12 and 13 is illustrated in Figure 10.6, where $\mathbf{G}$, the scattering vector for this reflection, is drawn perpendicular to the reflecting plane (hkl).

EXERCISE 6

Verify with the aid of Figure 10.6 that the two conditions for Bragg reflection cited above imply $\theta = \theta'$.

It is straightforward to show that the conditions expressed by equations 12 and 13 for Bragg reflections imply the two-dimensional form of the specular reflection law of equation 2. Rewriting equation 12 as

$$\mathbf{k}' = \mathbf{k} + \mathbf{G} \qquad (\text{m}^{-1}) \tag{14}$$

and taking the squared magnitude of both sides gives

$$k'^2 = k^2 + G^2 + 2\mathbf{k} \cdot \mathbf{G} \qquad (\text{m}^{-2}) \tag{15}$$

Applying the condition of equation 13 and writing (with reference to Figure 10.6) $\mathbf{k} \cdot \mathbf{G}$ as $-kG \sin \theta$ gives

$$2k \sin \theta = G \qquad (\text{m}^{-1}) \tag{16}$$

Recall that $\mathbf{G}$ can be any reciprocal lattice vector that is (according to the result of Exercise 4) perpendicular to the plane (hkl) doing the reflecting, so that $\mathbf{G} = n\mathbf{G}(hkl)$, where n is any integer. Using the result of Exercise 5, we have

$$\frac{2\pi}{d(hkl)} = G(hkl) = \frac{G}{n} \qquad (\text{m}^{-1}) \tag{17}$$

for the relationship between $d(hkl)$ and G. Using this to eliminate G in equation 16 and inserting the definition of the wave vector $\mathbf{k}$ yields the Bragg law

$$2\,d(hkl) \sin \theta = n\lambda \qquad (\text{m}) \tag{18}$$

as in equation 2 above.

The preceding discussion of the space and reciprocal lattices and their relationship to the scattering amplitude not only yields a more general interpretation of the Bragg law, as derived in a simple way from Figure 10.1, but also gives us information with regard to the expected intensities of the reflections from different sets of planes (hkl). In particular, the conditions cited above for Bragg reflections are necessary but not sufficient, and some of the reflections permitted by equation 18 will be absent because of the possibility of destructive interference between waves scattered by atoms in the same cell. The discussion of this feature of Bragg reflection requires a further examination of the scattering amplitude of equation 11 and a brief discussion of form and structure factors.

Structure Factor and Form Factor

If equation 12 is satisfied so that for some $\mathbf{G}$, $\Delta\mathbf{k} = \mathbf{G}$, then the exponential appearing in the expression for the scattering amplitude F in equation 11 is just unity, and F (denoted by $F_{\mathbf{G}}$ for this particular $\mathbf{G}$) is then given by

$$F_{\mathbf{G}} = \int n_{\mathbf{G}}\, dV \tag{19}$$

where the integration is performed over the entire volume of the crystal. Substituting for $n_{\mathbf{G}}$, the Fourier coefficient of $n(\mathbf{r})$ given by equation 10, gives an expression for the scattering amplitude in terms of the electron distribution within a single cell:

$$F_{\mathbf{G}} = N\left[\int_{\text{cell}} dV\, n(\mathbf{r})\, e^{-j(\mathbf{G}\cdot\mathbf{r})}\right] \tag{20}$$

where N is the number of cells in the crystal and the expression within the square brackets is called the structure factor $S_{\mathbf{G}}$. The electron density function $n(\mathbf{r})$ is often most conveniently broken up into chunks associated with each of the atoms contained within the cell, which is the region of integration in equation 20. If this is done, then this integral can be expressed as a simple sum:

$$F_{\mathbf{G}} = NS_{\mathbf{G}} = N\left[\sum_i f_i\, e^{-j(\mathbf{G}\cdot\mathbf{r}_i)}\right] \tag{21}$$

in which the f_i are known as form factors for the atoms and the sum is over all atoms in the cell located at positions $\mathbf{r}_i$; each f_i is effectively a portion of the integral in equation 20 over the charge distribution associated with the ith atom. The value of $F_{\mathbf{G}}$, for any $G(hkl)$, determines whether there will be a reflection corresponding to the atomic plane (hkl) and, hence, to $\Delta\mathbf{k} = \mathbf{G}(hkl)$.

EXERCISE 7

Using the conventional cubic cell, calculate $\mathbf{G}(hkl)$ from the definitions of equations 8 and 9. Evaluate $S_{\mathbf{G}}$ for the fcc lattice of single atoms, assuming identical form factors f_i for all atoms. Note that each conventional cell in the fcc lattice contains four atoms, so that $S_{\mathbf{G}}$ will contain four terms. You should show that if h, k, and l are either all even or all odd, then the structure factor for $\mathbf{G}(hkl)$ will be nonzero, but that for other combinations of the indices, $S_{\mathbf{G}}$ will be zero. Reflections from a plane (hkl) that are permitted by the Bragg law but for which the structure factor corresponding to $\mathbf{G}(hkl)$ is zero will not be detected.

EXERCISE 8

KCl has the structure shown in Figure 10.3. The lattice is fcc with a basis consisting of one K atom and one Cl atom; one pair of atoms is associated with each lattice site. Because K^+ and Cl^- have the same number of electrons, their form factors f are nearly equal and so they appear as identical ions to an x-ray beam. Calculate the structure factors $S_{\mathbf{G}}$ for this crystal, keeping in mind that each unit cell now contains 8 ions. What additional restrictions are placed on planes (hkl) that can produce Bragg reflections over and above those for a general monatomic fcc lattice considered in Exercise 7? Can you give a physical explanation for the additional reflections that are now absent?

Production of X Rays

In this set of experiments Bragg diffraction of both x rays and microwaves from periodic structures will be studied. The production of microwaves is discussed in connection with Experiment 2; here we give a brief account of the production the x rays we will use.

Figure 10.7*a* is a schematic representation of an x-ray tube. Electrons are emitted thermionically from the heated cathode, which is maintained at a large potential difference V with respect to the anode target. As these electrons impact the anode, x rays are emitted with a spectral intensity distribution similar to the one shown in Figure 10.7*b*. This spectrum exhibits two main features:

(1) There is a broad continuous spectrum of radiation, referred to as **bremsstrahlung**, caused by the sudden deceleration of the electrons as they impact the anode. (The word is derived from the German words *brems* (braking) and *strahlung* (radiation).) This radiation extends spectrally out to long wavelengths (low photon energies) with decreasing intensity and down to a minimum wavelength $\lambda_0 = hc/Ve$, which is the wavelength of a photon that carries away all the kinetic energy of an electron incident on the anode. From the point of view of x-ray diffraction, this component of the tube emission is often considered as background.

(2) Superimposed on the *bremsstrahlung* continuum is the nearly monochromatic set of

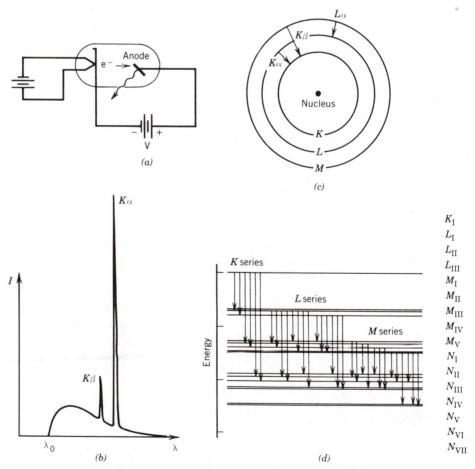

FIGURE 10.7 (*a*) Production of x rays. (*b*) Typical spectral intensity distribution of the output of an x-ray tube. (*c*) Atomic transitions associated with the production of characteristic radiation. (*d*) Energy-level diagram for a vacancy (hole) and the allowed x-ray transitions.

x-ray lines that reflect the atomic structure of the atoms of the anode. The mechanism for the production of this **characteristic radiation** is suggested by Figure 10.7c. A high-energy electron impacts the anode and knocks out an inner shell electron from an anode atom. An x-ray photon is emitted when the vacancy thus created is filled by means of a downward transition made by an electron in one of the higher energy shells. The process can be represented on an energy diagram like that of Figure 10.7d, which represents the energy of an atom with a vacancy in a particular shell along with transitions allowed by the selection rules. The nomenclature for the various lines is derived from the initial and final states of the transition, as is suggested by Figure 10.7c. Note that the spin–orbit interaction, along with other relativistic effects, creates a splitting of the energy levels of the various shells according to the quantum number j, which indexes the total angular momentum. The line K_α, for example, is really a multiplet consisting of two lines (K_{α_1}, K_{α_2}), which are seldom resolved in x-ray diffraction work.

Because of its monochromatic nature, the characteristic radiation described above is quite useful for x-ray diffraction. In this experiment you will use an anode made of Cu, for which the important emission lines are K_α ($\lambda = 0.154178$ nm) and K_β ($\lambda = 0.139217$ nm). The wavelength given for the K_α radiation is a weighted average for the doublet $K_{\alpha_{1,2}}$. The K_β radiation from Cu is about six times weaker than the K_α so that diffraction patterns can be made easier to interpret if the K_β is selectively filtered out. This can be done conveniently for Cu K radiation by means of a foil made of Ni, which has an absorption *edge* (due to photoelectric absorption) at $\lambda = 0.148802$ nm, so that wavelengths shorter than this are selectively absorbed. A Ni foil with a "thickness" of 19 mg/cm, for example, when used as a filter for Cu K, will produce a beam with a K_α component that is 500 times more intense than the K_β.

EXPERIMENT

Microwave Bragg Diffraction

This part of the experiment is an investigation of scattering of microwaves with $\lambda \cong 3$ cm from a cubic lattice of aluminum spheres as an analog to the diffraction of x rays from a crystalline substance. The principles of operation of the reflex klystron and the diode detector are discussed in connection with Experiment 2.

(a) *Wavelength Measurement.* After an appropriate warm-up period and with the klystron transmitter and diode receiver configured as in Figure 10.8a, adjust the repeller voltage on the klystron for a peak reading on the receiver meter. With the units arranged as in Figure 10.8b in front of a sheet of aluminum (about 30×30 cm), move the receiver along the meter stick and observe maxima (antinodes) and minima (nodes) in the standing-wave pattern produced by reflection from the sheet. Obtain the best value for the separation between nodes by measuring the distance through which the receiver travels in producing some large number of these peaks and dips in the signal. Repeat this measurement a few times.

EXERCISE 9

How is the distance between nodes obtained above related to the wavelength of the microwaves? Report the best value of λ, along with an estimate of the uncertainty.

If the apparatus for Experiment 2 is available, you can make a precision measurement of λ with the wavemeter as an alternative to the technique described above.

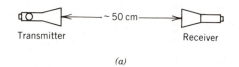

Transmitter   ~50 cm   Receiver

(a)

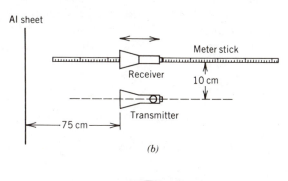

Al sheet

Meter stick

Receiver

10 cm

75 cm

Transmitter

(b)

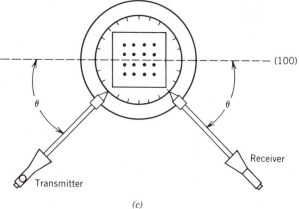

(100)

θ   θ

Transmitter   Receiver

(c)

FIGURE 10.8 (*a*) Klystron transmitter and diode receiver.
(*b*) Configuration for wavelength measurement.
(*c*) Configuration for scattering measurement.

(b) *Bragg Reflection from a "Crystal."* Align the layers of Styrofoam with the Al balls so that a cubic arrangement of scatterers is formed. Peak the transmitter output as in (a) and arrange the transmitter, receiver and crystal on the goniometer as in Figure 10.8*c* so that reflections off the (100) plane of scatterers can be measured. Record readings from the receiver as θ is varied in 1° increments. Note that both receiver and transmitter positions must be changed between readings to keep the angles of reflection and incidence equal. Make an intensity versus θ plot.

EXERCISE 10

From your data, calculate the spacing between (100) planes in the crystal. Compare this with a direct measurement of the side of the unit cube.

EXERCISE 11

As stated in connection with Exercise 4, the reciprocal lattice vectors $G(hkl)$ are perpendicular to the planes (*hkl*). Use this fact to calculate the angle between the (100) planes and those with indices (110) and (210). Also, use the expression given in Exercise 5 for $d(hkl)$ to calculate the spacing between these two additional sets of planes in terms of the (100) spacings.

Use the angles calculated above to rotate the crystal so that reflection data can be taken for the (100) and (210) planes. Collect, plot, and analyze this data to deduce experimental values for the interplanar spacings. Compare these to the spacings calculated from the result of Exercise 11 by using the directly measured (100) spacings.

X-Ray Bragg Diffraction

Note: As with all ionizing radiation, caution should be exercised to avoid unnecessary bodily exposure. A scattering shield (standard equipment on the Tel-X-Ometer) should be in place when the x-ray tube is in operation.

(a) *Single Crystal*. Mount the NaCl single crystal in the crystal mount of the diffractometer arrangement so that the (100) face is parallel to the back of mount, as shown schematically in Figure 10.9. Use an accelerating voltage of 30 kV for the x-ray tube and a current consistent with its power rating to produce a collimated beam of x rays that strikes the crystal face as shown. The diffractometer should be in $\theta-2\theta$ mode, that is, the G–M detector arm should move through an angle 2θ whenever the crystal holder turns through θ so that the angles of incidence and reflection remain equal as the angle θ is scanned.

For values of θ within the range of motion of the diffractometer, record the detected intensity versus θ. Plot the data.

Repeat the scan with the Ni foil in the incident x-ray beam.

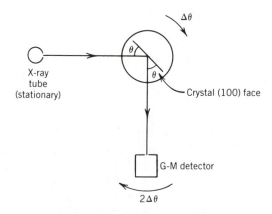

FIGURE 10.9 Arrangement for measurement of single-crystal diffraction.

EXERCISE 12

Identify each observed peak with respect to the value of n (the order) in the Bragg condition of equation 18 and with respect to the wavelength of the radiation responsible for it. Tabulate the angular positions of the peaks and, from the known wavelengths involved, calculate values for $d(100)$ in NaCl. You should keep in mind the restrictions that the discussion of Exercises 7 and 8 place on the reflections. For NaCl the edge of the unit cube of Figure 10.3 has an accepted length of $a = 0.563$ nm. How does this compare with the best value derived from your data?

Repeat the measurements and analysis for the (100) face of a KCl single crystal, for which the accepted value of a is 0.629 nm. Account for the differences in the KCl and NaCl patterns.

(b) *Powder (Debye–Scherrer) Method.* In a powder sample composed of a large number of small crystallites with random orientation, each set of planes (hkl) exhibits all possible orientations with respect to the incident x-ray beam, as depicted in Figure 10.10a. The Bragg condition of equation 18 for each set of indices (hkl) is thus always satisfied for some small fraction of the crystallites. This means that, for each set of planes associated with a nonzero structure factor, we expect a specular reflection at the Bragg angle with respect to these planes, that is, at an angle 2θ with respect to the direction of the incident beam. The locus of all such directions is a set of cones with half-angles 2θ and with the incident beam direction as a common axis, as shown in Figure 10.10b.

If a cylindrical band of photographic film is positioned with its diameter coincident with the incident beam, the Bragg angles may be determined by measuring the angular positions of the exposed rings. In this arrangement, known as a **powder camera**, the sample should be a powder formed into a cylinder perpendicular to the plane of the film so that all reflections can be recorded photographically.

If a powder camera is unavailable, the angular positions of the Bragg reflections may be determined by scanning the G–M tube position while keeping the sample and x-ray tube stationary, as suggested by Figure 10.10a. The sample can be prepared by grinding the substance to a fine powder with a ceramic mortar and pestle and allowing it to stand until enough moisture has been absorbed from the air so that it can be conveniently packed onto a glass microscope slide.

Obtain the angular positions of the powder peaks for NaCl and KCl as discussed above. For each peak, tabulate the Bragg angle θ, $d(hkl)$, and the quantity $m^2 d^2(hkl)$, where m^2 takes on all integer values up to 20.

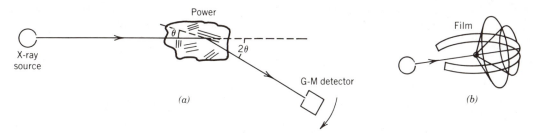

FIGURE 10.10 (a) Diffraction from a powder sample. (b) Locus of diffraction angles in a powder camera.

EXERCISE 13

From your tabulation and the relation between $d(hkl)$ and the cubic cell side a (see Exercise 5), identify, for each peak, the entry in your tabulation which corresponds to the common value of a. Assign Miller indices (hkl) to each peak reflection.

EXERCISE 14

Determine the best value of a for NaCl and KCl. How do these compare with the accepted values given above? You may wish to weight your average, keeping in mind that smaller values of the measured angles 2θ are likely to contain larger relative errors.

COMPUTER-ASSISTED EXPERIMENTATION (OPTIONAL)

Prerequisite

Experiment 6, Introduction to Computer-Assisted Experimentation.

Experiment

A complete computer interface for an x-ray Bragg spectrometer should consist of (a) a means of acquiring a count rate from a detector, (b) a means of recording the angular position of the detector arm, (c) a mechanism by which the computer can increment the angular position of the detector arm and/or crystal mount between readings (if possible for a particular system), and (d) software instructions to plot the data so that peak positions may be obtained. We discuss the details for implementing (a) and (b) below.

(a) The conditioning circuit shown in Figure 6.11a is suitable for counting pulses input from a G–M tube to a scaler/timer. Figure 6.11b shows the output of each stage of the conditioning circuit, which finally presents, for each input pulse, a digital pulse to the analog input of the ADC. The text of Experiment 6 contains a thorough discussion of the details of this circuit.

(b) One possibility for recording the angular position of the detector arm is discussed in reference 4 for the Tel-X-Ometer unit in particular. A miniature 10-turn, 10-kΩ potentiometer is configured as a voltage divider as indicated in Figure 10.11 and the center lead is connected to the input of the ADC card, so that the angular position of its shaft is digitized. The potentiometer is mounted in a bracket attached to the thumbwheel on the detector arm; its shaft is then mechanically coupled, by means of a small section of rubber tubing, to the end of a metal shaft that is installed coaxially with the thumbwheel. As the detector arm is taken through a Bragg angle range of 100°, the thumbwheel and potentiometer shafts complete about nine revolutions. For maximum angular resolution, the dc voltages V_1 and V_2 should be adjusted so that taking the detector through the angular range of interest will cause the input of the ADC to traverse the entire allowable range.

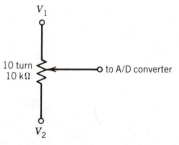

FIGURE 10.11 Voltage divider for digitizing the angular position of the detector.

EXERCISE 15

Determine, for your A/D card, optimal values of V_1 and V_2 if data are to be taken over the range $10° < \theta < 110°$. From the number of bits in your card's digital output, determine the angular resolution of the system for this range of Bragg angles. How does this compare with the uncertainty in the angle readings taken from the scale of the spectrometer?

Measurements and Software

Software needs to convert each potentiometer voltage to an angle, acquire the number of pulses received for each position during some fixed counting period, store the data, and plot the results.

Calibration should be achieved at the beginning of each run by setting the detector at the end points of the range. Background counts need to be acquired and used to adjust the data. The maximum allowable count rate will be determined by the G–M tube and by the software. This last consideration is discussed further in the text of Experiment 6, in the section entitled Computer-Assisted Counting and Data Analysis.

Atomic and Molecular Physics

---

11. FRANCK–HERTZ EXPERIMENT: ELECTRON SPECTROSCOPY

Historical Note

The 1925 Nobel prize in Physics was awarded jointly to James Franck, Germany and Gustav Hertz, Germany

For their discovery of the laws governing the impact of an electron upon an atom

Franck and Hertz performed the experiment in 1914, 12 years before the development of quantum mechanics, and it provided striking evidence that atomic energy states are quantized.

APPARATUS [Optional Apparatus in brackets]

Franck–Hertz tube
Electric oven
Variac
Thermocouple and temperature potentiometer
Electrometer
Circuit to provide dc voltages for the Franck–Hertz tube, see Figure 11.5
6.3-V ac filament supply
Oscilloscope [with an available sawtooth voltage]

[Oscilloscope camera]
[Microcomputer, ADC/DAC card, conditioning circuits (see Figures 11.5, 11.8, and 11.9)]
[xt chart recorder]

OBJECTIVES

To verify the quantization of atomic electron energy states of mercury atoms by observing the maxima and minima of an electron current passing through a gas of mercury atoms.

To understand how ordinary and metastable atomic electron energy states of mercury affect the transmission of electrons.

To understand how temperature affects the number density of mercury atoms, the mean free path of a transmission electron, and the kinetic energy of a transmission electron.

KEY CONCEPTS

Elastic and inelastic scattering   Metastable states
LS coupling   Mean free path
Ordinary states   Contact potential difference
                                         Forbidden transitions
                                         Allowed transitions

REFERENCES

1. R. Eisberg and R. Resnick, *Quantum Physics of Atoms, Molecules, Solids, Nuclei, and Particles*, 2d ed., Wiley, New York, 1985. Contact potential is discussed on pages 407–408 and the Franck–Hertz experiment is briefly discussed on pages 107–110.

2. D. Halliday and R. Resnick, *Physics*, Part I, 3d ed., Wiley, New York, 1978. Mean free path is discussed on pages 522–524.

3. K. Rossberg, *A First Course in Analytical Mechanics*, Wiley, New York, 1983. Collision of two particles is discussed on pages 157–160.

4. L. Loeb, *Atomic Structure*, Wiley, New York, 1938. An older book which you may find in the library. A good discussion of the Franck–Hertz experiment appears on pages 249–254. The energy levels of mercury are discussed on pages 256–264.

5. E. Leybold, Manufacturer's description of the Franck–Hertz tube. This is a discussion of the Franck–Hertz tube, which should be available from your instructor.

6. R. Eisberg, *Fundamentals of Modern Physics*, Wiley, New York, 1961. LS coupling is discussed on pages 428–441.

INTRODUCTION

The Franck–Hertz experiment, first performed in 1914, verifies that the atomic electron energy states are quantized by observing maxima and minima in transmission of electrons through mercury vapor. The variation in electron current is caused by inelastic electron scattering that excites the atomic electrons of mercury. The Franck–Hertz tube, which contains a drop of mercury, is shown in Figure 11.1 along with electrical connections. The tube requires the following operating voltages

Filament, f: 6.3 V ac. Operation at a lower voltage will increase the lifetime of the tube.

Space-charge grid voltage across grid G_1 and cathode k: $V_s = 2.7$ V dc. This voltage determines the space charge about the cathode and, thus, the emission current. See the

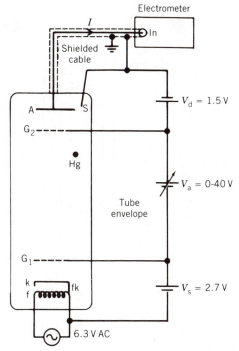

FIGURE 11.1 Schematic diagram of the Franck–Hertz
tube with electrical connections.

discussion of space charge and emission current in Experiment 7, Electron Physics. Some
tubes do not have grid G_1. For such tubes the voltage V_s is omitted and the negative
terminal of V_a connects directly to fk.

Accelerating voltage across grid G_2 and cathode k: $V_a = 0$–40 V dc.

Decelerating voltage across anode A and grid G_2: $V_d = 1.5$ V dc. Only those electrons that
arrive at G_2 with an energy greater than eV_d will reach the anode A.

The electrodes of the Franck–Hertz tube are coaxial, cylindrical electrodes as shown in
Figure 11.2. The tubes that do not have grid G_1 have noncylindrical electrodes.

Energy Levels of Mercury

A mercury atom has 80 electrons. For an atom in the ground state the K, L, M, and N
shells of mercury are filled and the O and P shells have the following electrons:
0 shell: $5s^2$, $5p^6$, $5d^{10}$
P shell: $6s^2$

Energy levels of mercury, which are relevant to this experiment, are shown in Figure 11.3.
(See reference 4 for a discussion of the energy levels of mercury.) The energy levels are
labeled with two notations:

$n\ell$, where n is the principal quantum number and ℓ is the orbital angular momentum quantum
number, designated by $s(\ell = 0)$ and $p(\ell = 1)$.

$^{2S+1}L_J$, where S, L, and J are the total spin quantum number, total orbital angular momentum
quantum number, and total angular momentum quantum number (see reference 6).

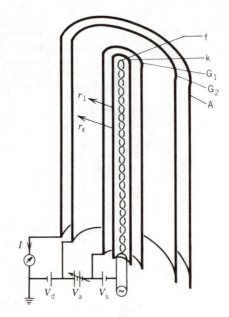

FIGURE 11.2 The electrodes of the Franck–Hertz are coaxial, cylindrical electrodes.

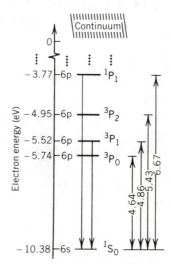

FIGURE 11.3 Energy levels of mercury that are relevant to this experiment. Energy level separation in electron volts is indicated on the right.

The $^1P_1$ and $^3P_1$ are ordinary states, having lifetimes of about 10^{-8} s before decaying to the $^1S_0$ ground state by photon emission. The $^3P_2$ and $^3P_0$ are metastable states, having lifetimes of about 10^{-3} s or 10^5 times as long as an ordinary state. (See the discussion of metastable states in the "Introduction to Laser Physics.") Hence, the probability per second of an electron making a transition from either the $^3P_2$ or $^3P_0$ state to the $^1S_0$ ground state by photon emission is 10^5 times smaller than the transition from either the $^3P_1$ or $^1P_1$ to $^1S_0$. Thus, the transitions from $^3P_2$ and $^3P_0$ to $^1S_0$ are *forbidden* transitions, while the transitions from $^1P_1$ and $^3P_1$ to $^1S_0$ are *allowed* transitions. The allowed transitions for photon emission are indicated by the two arrows on the left in Figure 11.3, and the four arrows on the right indicate energy spacing in units of electron volts.

Direct excitation of $^3P_0$, $^3P_1$, $^3P_2$, and $^1P_1$ from $^1S_0$ by electron impact is essentially equally probable.

Atomic Excitation by Inelastic Electron Scattering

An electron traveling from the cathode k toward the anode A has a mean free path $\bar{\ell}$ (see reference 2) given by

$$\bar{\ell} = \frac{1}{\sqrt{2}\pi n R_0^2} \quad \text{(m)} \tag{1}$$

where $R_0 \simeq 1.5 \times 10^{-10}$ m is the radius of a mercury atom and n is the number of atoms per unit volume. At the end of one mean free path the electron has gained a kinetic energy K from the electric field E:

$$K = eE\bar{\ell} \quad \text{(J)} \tag{2}$$

where e is the electron charge and E is the electric field established by the accelerating voltage V_a. If $\bar{\ell}$ is long, then K will be large.

The number density n is very sensitive to the tube temperature; therefore ℓ and, hence, K are very temperature sensitive.

EXERCISE 1

What is the mean free path $\bar{\ell}$ of an electron in a Franck–Hertz tube heated to 373 K? 423 K? 473 K? Assume the gas of mercury atoms behaves as an ideal gas. A table of vapor pressure of mercury and temperature may be found in the *CRC Handbook of Chemistry and Physics*.

When an electron of kinetic energy K approaches a mercury atom with $K < 4.6$ eV, the energy difference between the first excited state and the ground state, then the collision is elastic. In an elastic collision the electron loses some kinetic energy determined by the laws of conservation of momentum and kinetic energy.

EXERCISE 2

The loss of kinetic energy by an electron when it collides elastically with a mercury atom is greatest when the collision is head-on. For an elastic head-on collision with the mercury atom, assumed initially at rest, show that the change in electron energy is given by

$$\Delta K = \frac{4mM}{(m + M)^2} K_0 \quad \text{(J)} \tag{3}$$

where m and M are the masses of an electron and a mercury atom and K_0 is the initial electron energy. (See reference 3 for a discussion of a head-on, two-particle collision.) What is the fractional loss of kinetic energy by the electron for such a collision?

From your answer to Exercise 2 it should be clear that the loss of electron energy due to a single elastic collision is negligibly small. The probability of an inelastic collision occurring is large when the electron's energy equals the energy difference between an excited state and the ground state of the mercury atom, that is, 4.6, 4.9, 5.4, and 6.7 eV (see Figure 11.3). At some radial distance $r_1 \pm \Delta r$ from the cylindrical cathode k, the kinetic energy K of the electron will equal 4.6 eV and the first inelastic collision occurs. An arbitrary r_1 is shown in Figure 11.2. The inelastic collisions occurring in the cylindrical shell of radius $r_1 \pm \Delta r$ populates the $^3P_0$ metastable state of the mercury atoms in the shell. The electrons that later enter the shell will collide elastically with the mercury atoms that are in the $^3P_0$ state; hence, these electrons pass through the shell with negligible energy loss. At some larger radius $r_\varepsilon \pm \Delta r$ (shown in Figure 11.2) these same electrons will have a kinetic energy of $\varepsilon = 4.9$ eV and they collide inelastically with mercury atoms in the ground state. The $^3P_1$ state decays to the $^1S_0$ state after about 10^{-8} s by photon emission, and then the atom is ready for another inelastic collision; that is, the atoms in this shell continuously convert electron kinetic energy to radiant energy. If the accelerating voltage V_a is high enough, this process may be repeated in cylindrical shells of radii $r_{2\varepsilon} \pm \Delta r$, $r_{3\varepsilon} \pm \Delta r$, ... Note that r_ε, $r_{2\varepsilon}$, and so on are the distances of travel required for the electron to gain a certain energy.

What effect do these inelastic collisions have on the current measured by the electrometer in Figure 11.1? The decelerating voltage V_d is 1.5 V and as V_a is increased from 0 V a current is first observed when V_a exceeds 1.5 V and the observed current will increase as V_a increases until $V_a = 4.9$ V. When $V_a = 4.9$ V the electrons lose energy from inelastic collisions; hence, they no longer have enough energy to overcome the 1.5-V decelerating voltage and the observed current decreases. (With $V_a = 4.9$ V, r_ε is approximately the distance from the cathode k to the grid G_2.) As V_a is increased from 4.9 V the current will again increase until $V_a = 9.8$ V, which corresponds to a second cylindrical shell a distance of $r_{2\varepsilon} \pm \Delta r$ from the cathode where inelastic collisions which populate the $^3P_1$ state occur. Thus, when $V_a = n \times 4.9$ V, $n = 1, 2, 3, \ldots$, there is a decrease in current. A curve of expected current I versus accelerating voltage V_a is sketched in Figure 11.4. Each peak represents the onset of inelastic collisions that populate the $^3P_1$ state. The first peak does not occur at 4.9 V because of the contact potential difference between cathode and anode. Contact potential difference is discussed in reference 1.

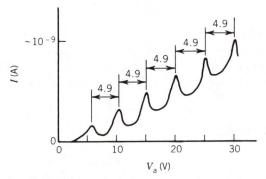

FIGURE 11.4 Electrometer current as a function of the accelerating voltage V_a.

EXERCISE 3

Determine the accelerating voltage V_a as a function of the radius r measured from the cathode. Then show that the ratio of the radii of the first two cylindrical shells is given by

$$\frac{r_{2\varepsilon}}{r_\varepsilon} = e^{2\pi\epsilon_0 4.9/\lambda}$$

where $\epsilon_0 = 8.85 \times 10^{-12}$ F/m, λ is the magnitude of the charge per unit axial length of the cathode, and the numerical value 4.9 has units of volts. Assume that there is no space charge surrounding the cathode and that the length of the cylindrical electrodes is much greater than the diameter; that is, assume an infinite length.

EXPERIMENT

Read reference 5, the manufacturer's description of the Franck–Hertz tube. Note the precautions specified in this description.

Circuit to Provide Dc Voltages for the Franck–Hertz Tube

The electrical circuit to provide the dc voltages for the Franck–Hertz tube is shown in Figure 11.5. The point labeled S in 11.5 connects to the shield, G_1 connects to grid 1, G_2 connects to grid 2, and fk connects to the filament–cathode. Compare these connections with the equivalent connections shown in Figure 11.1. In Figure 11.5 the 1.5-V D cell provides the decelerating voltage V_d, the accelerating voltage V_a is the voltage between the emitter and collector of the SK 3025 transistor, $V_a = V_{G_2} - V_{G_1}$, and the 2.7-V Zener diode provides the space-charge voltage V_s. If a tube does not have grid G_1 then the 2.7-V Zener and the lead labeled G_1 are omitted. Note that the ac filament supply for the tube is not included in Figure 11.5.

In terms of the accelerating voltage the desired output of the circuit is $V_a = 0$ to $+40$ V. How does the circuit provide this voltage? Well, the collector voltage, and, hence, the accelerating voltage, is controlled by the collector current; the collector current is controlled by the bias voltage of the emitter–base junction; and the bias voltage is controlled by the output of op-amp 2. The output of op-amp 2 is given by

$$V_{out, 2} = -\frac{R_2}{R_1} V_{out, 1} = +\frac{R_2}{R_1} V_{in} \quad (V) \tag{4}$$

where the output of op-amp 1 is the negative of its input, V_{in}. Op-amps are discussed in reference 5, Experiment 6.

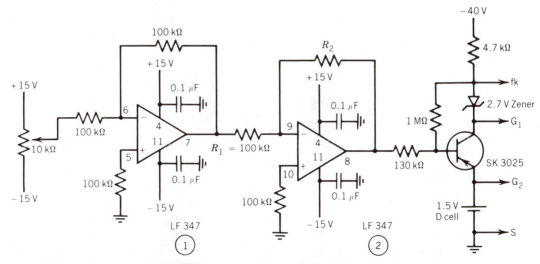

FIGURE 11.5 Circuit to provide the dc voltages for the Franck–Hertz tube. This circuit is also used in computer-assisted experimentation with the replacement of the dc voltage input with the analog output of the DAC, where the analog range of the ADC/DAC card is $-K$ to $+K$ volts.

An op-amp 2 output of -6.5 V provides a "forward" bias to the emitter–base junction and the resulting collector current is large, creating a voltage drop across the 4.7-kΩ resistor of about 39 V. In this case the accelerating V_a is approximately zero. An op-amp 2 output of $+6.5$ V provides a "reverse" bias to the emitter–base junction and the collector current is then zero; hence, the collector voltage is -40 V and V_a is $+41$ V. As the output of op-amp 2 ranges from -6.5 to $+6.5$ V the accelerating voltage ranges from 0 to 41 V.

The input to op-amp 1 is the voltage on the "slider" of the 10-kΩ potentiometer. We (arbitrarily) use an input of -5 to $+5$ V, and then select R_2 such that op-amp 2 has the desired output. (The $+15$ and -15 V shown connected to the 10-kΩ potentiometer may be obtained from the op-amp power supplies.)

EXERCISE 4

For an input V_{in} ranging from -5 to $+5$ V what value of the resistance R_2 is required?

A top view of the pin assignment of the LF 347 op-amp is shown in Figure 11.6. Note that the package is actually four op-amps. Each op-amp is a JFET input op-amp with an input impedance of $10^{12}\,\Omega$. JFET op-amps are discussed in reference 5, Experiment 6.

It is suggested that you bench test the circuit before connecting it to the Franck–Hertz tube. To do so, measure the voltages at the points labeled G_2, G_1, and fk as the input voltage to op-amp 1 is varied from -5 to $+5$ V.

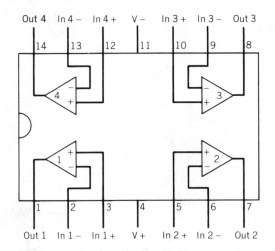

FIGURE 11.6 Top view of the pin assignments of the LF 347 op-amp. The package is actually 4 op-amps.

Electrometer

Electrometers are extremely sensitive electronic instruments designed to measure both $(+)$ and $(-)$ currents and voltages. A function switch permits the selection of current or voltage mode. A range selector switch allows voltage measurements from about 1 μV to 1 V and current measurements from about 1 pA to 1 mA. The output of the electrometer ranges from 0 to plus or minus a few volts for an input ranging from 0 to full scale.

The output current of the Franck–Hertz tube ranges from 0 to about -10^{-9} A as the accelerating voltage varies from 0 to $+30$ V (voltage of G_2 relative to fk). Two instruments for measuring the output of the Franck–Hertz tube are briefly described below.

The manufacturer lists the specifications of the Hewlett-Packard Model 425A dc microvolt ammeter as the following:

Voltmeter input impedance	$10^6\,\Omega \pm 3\%$
Voltage range	10^{-6} V full scale to 1 V full scale
Ammeter range	10^{-12} A full scale to 3×10^{-3} A full scale
Output	0 to ± 1 V for 0 to full-scale input; output polarity is the same as the input
Ac rejection (at amplifier output)	At least 1 db at 1.0 Hz, 50 db at 50 Hz

The specifications of the Keithley Model 610C electrometer, as listed by the manufacturer, are:

Voltmeter input impedance	$10^{14}\,\Omega$
Voltage range	10^{-3} V full scale to 10^2 V full scale
Ammeter range	10^{-14} A full scale to 0.3 A full scale
Output	0 to ± 3 V for 0 to full-scale input; output polarity is opposite input polarity
Frequency response	Dc to 100 Hz

The ac rejection at the amplifier output of the Hewlett-Packard instrument implies that it is a dc instrument, and the frequency response of the Keithley instrument indicates it is a dc or quasi-dc measuring device.

A circuit to bench test the electrometer is shown in Figure 11.7. Select a value of R appropriate for an electrometer input of 10^{-3} V full scale, say, then measure the electrometer output as a function of the input. Reverse the polarity of the 1.5-V D cell and repeat the measurements. To test a different input range of the electrometer change R to a value appropriate for that range.

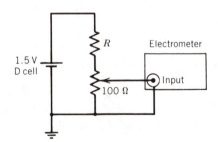

FIGURE 11.7 Circuit to bench test the electrometer.

Connecting the Circuit to the Franck–Hertz Tube

Turn off all voltages and then connect the circuit and filament power supply to the Franck–Hertz tube. NEVER APPLY VOLTAGES TO THE TUBE UNLESS IT IS IN THE OVEN AT THE DESIRED TEMPERATURE. To do otherwise could burn out the tube, and it is expensive. After connecting the circuits to the tube, then:

1. Attach the thermocouple to the tube (approximately centered on the anode), place the metal shield around the tube and connect it to ground (this will reduce ac pickup), and then insert the tube into the oven as deeply as possible.

2. Plug the oven into the variac and experiment with the oven temperature as a function of voltage. You want to determine the variac setting(s) required to bring the tube–oven system to the desired temperature as quickly as possible. The voltages to the tube may be applied after a stable operating temperature is reached.

With the anode connected to the electrometer as shown in Figure 11.1 vary the input voltage to op-amp 1 from -5 to $+5$ V, recording appropriate currents and voltages in your notebook. Also measure the temperature periodically and adjust the oven voltage as required to maintain a reasonably constant temperature.

Analyze the data, interpreting it in terms of the theory discussed in the Introduction.

EXERCISE 5

The cathode at temperature T emits electrons with a distribution of speeds. The average kinetic energy $\bar{K}$ of the electrons is

$$\bar{K} = 2kT \quad \text{(J)}$$

where k is the Boltzmann constant and T is the absolute temperature. Assuming that T is 2500 K, what effect will $\bar{K}$ have on your measured excitation potentials? What effect will the distribution of speeds of the electrons have on the sharpness of the peaks?

EXERCISE 6

What effect would contact potential have on peak spacing? What peak positions would be affected by contact potential?

Optional Experimental Method

The experimental method described below requires two additional pieces of equipment, namely, an oscilloscope with an available sawtooth voltage and an *xt* recorder. If this equipment is available you may want to omit the previous measurements and carry out the experiment below.

A problem with the preceding experiment is the temperature variation that occurs while recording data. If the data could be recorded in an appropriately short time interval then the temperature variation would be small. We now discuss such a method of recording the data.

Consider the circuit in Figure 11.8. The 10-kΩ potentiometer is connected across the horizontal sweep voltage of the oscilloscope, the *scope sawtooth*. The slider on the potentiometer may be adjusted such that the slider voltage is a 10-V sawtooth, that is, the input to op-amp 1 is a sawtooth ranging from 0 to $+10$ V. The desired output of op-amp 2 is -6.5 to $+6.5$ V. How do we obtain this output? Well, the output of op-amp 2 is

$$V_{\text{out, 2}} = -\frac{R_2}{R_1} V_{\text{out, 1}} + \frac{R_1 + R_2}{R_1} V_{\text{off}} \quad \text{(V)} \tag{5}$$

where V_{off} is the adjustable offset voltage and $V_{\text{out, 1}}$ is the output of op-amp 1. Op-amp 1 is configured to have a gain of -1; hence, its output will be a sawtooth voltage ranging from 0 to -10 V.

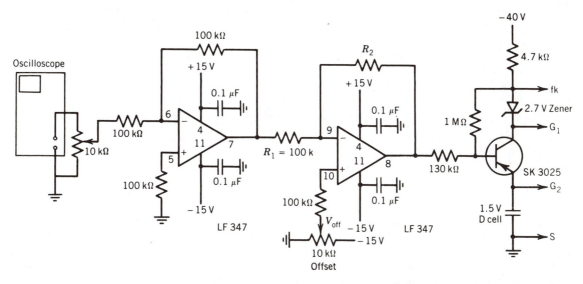

FIGURE 11.8 Circuit with a 0 to +10 V sawtooth input that provides the dc voltages for the Franck–Hertz tube. This circuit is also used in computer-assisted experimentation with the replacement of the sawtooth input with the analog output of the DAC, where the range of the ADC/DAC card is 0 to +K volts.

EXERCISE 7

When the input to op-amp 1 ranges between 0 and +10 V the desired output of op-amp 2 is −6.5 and +6.5 V, respectively. Use equation 5 to show that $V_{off} = -2.83$ V and $R_2 = 1.3R_1$.

If the above values of V_{off} and R_2 are used, the circuit in Figure 11.8 will provide an accelerating sawtooth voltage ranging from 0 to +40 V as the input sawtooth ranges from 0 to +10 V. The period of the sawtooth is determined by the time per centimeter setting on the horizontal oscilloscope sweep.

Bench test the circuit before connecting it to the Franck–Hertz tube. The test should include observing the collector voltage on the oscilloscope as the time per centimeter setting is changed. If the circuit is functioning properly turn the voltages off and connect the circuit to the Franck–Hertz tube.

Connect the output of the electrometer to the oscilloscope vertical input. Bring the Franck–Hertz tube to a temperature of about 175 °C and then apply voltages. If a Polaroid camera is available photograph the oscilloscope screen, which should resemble Figure 11.4.

Disconnect the electrometer from the oscilloscope and connect it to an xt chart recorder. Record the data on the chart recorder and then analyze it.

When running the experiment the time per centimeter setting of the scope will be determined by the response time of the electrometer. For the H-P 425A instrument the oscilloscope sweep must be about 1 s/cm or longer, and for the Keithley 610C it may be about 10 ms/cm or longer.

EXERCISE 8

Suppose you wanted to carry out this experiment such that the data recorded correspond to the $^1S_0 \rightarrow {}^1P_1$ transition, rather than the $^1S_0 \rightarrow {}^3P_1$ transition. What tube temperature would be required to do this? To answer this question calculate the ratio of pressure to temperature required to observe current maxima separated by 6.7 eV, corresponding to the $^1S_0 \rightarrow {}^1P_0$

transition, and then use a mercury vapor pressure table to determine the values for pressure and temperature that satisfy the ratio. You may assume the mercury gas is an ideal gas and that the accelerating voltage is 30 V. Also, to calculate the electric field E you will need to know (or approximately measure) the distance from the cathode to grid 2.

OPTIONAL: COMPUTER-ASSISTED EXPERIMENTATION

Prerequisite

Experiment 6, Introduction to Computer-Assisted Experimentation.

Experiment

This experiment will utilize two conditioning circuits, one that couples the output of the experimental apparatus to the ADC input and another that couples the output of the DAC to the apparatus, providing the accelerating voltage. The circuit that couples the apparatus to the ADC will be discussed first. Also, this experiment involves voltage measurement and you may want to review such measurements as described under Hall Effect Voltage, Experiment 6, Introduction to Computer-Assisted Experimentation.

In Experiment 6 it was pointed out that the conditioning circuit depends on the analog range of the ADC/DAC card and the maximum voltage output of the experimental apparatus, in this case specified as $V_{E, max}$. The circuits shown in Figures 11.9a and b are for analog ranges of $-K$ to $+K$ and 0 to $+K$, respectively.

Consider the circuit for an analog range of $-K$ to $+K$ volts, Figure 11.9a. The output of the op-amp is given by

$$V_{out} = -\frac{R_2}{R_1} V_E + \frac{R_1 + R_2}{R_1} V_{off} \quad (V) \tag{6}$$

where V_E is the electrometer output.

EXERCISE 9

Assuming $V_{E, max} = -1$ V and $K = 5$ V, show that $R_2 = 10R_1$ and $V_{off} = -0.455$ V.

The circuit shown in Figure 11.9b is for an ADC/DAC card with an analog range of 0 to $+K$ volts. The output of the op-amp is

$$V_{out} = -\frac{R_2}{R_1} V_E \quad (V) \tag{7}$$

where R_2 is chosen such that the range of the output is 0 to $+K$ volts.

It is good laboratory procedure to bench test a circuit before using it in an experiment. To do so, replace the electrometer in Figure 11.9 with the circuit shown in Figure 11.10, and connect the output of the op-amp to a voltmeter.

The analog output of the DAC can be used to control dc voltages for the Franck–Hertz tube. For an analog range of $-K$ to $+K$ volts the circuit in Figure 11.5 may be used, where the 10-kΩ potentiometer system is replaced with the analog output of the DAC. For an analog range of 0 to $+K$ volts the circuit in Figure 11.8 is adequate, where the oscilloscope–potentiometer system is replaced with the analog output of the DAC. In either circuit it is likely that you will have to change the value of R_2.

Interface the Franck–Hertz tube to the computer using the two circuits that are

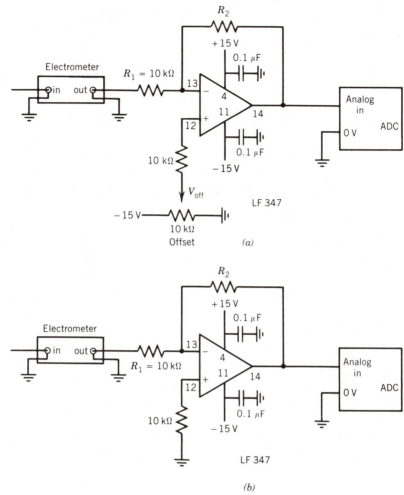

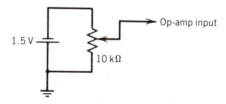

FIGURE 11.9 Conditioning circuits for an ADC/DAC card with analog range (a) from $-K$ to $+K$ volts, and (b) from 0 to $+K$ volts.

FIGURE 11.10 Input for bench testing either circuit in FIGURE 11.9.

appropriate for the analog range of your ADC/DAC card. Some possible things you can do are the following:

1. Write software that increments the accelerating voltage, reads the ADC input voltage for each accelerating voltage, and stores each pair of values. Also write software that plots and analyzes the data.

2. Instruct the computer to print a hard copy of the data and analysis.

12. ISOTOPE STRUCTURE AND FINE STRUCTURE OF HYDROGEN: OPTICAL SPECTROSCOPY

Historical Note

The 1955 Nobel prize in Physics was divided equally between Willis Eugene Lamb, the United States

For his discoveries concerning the fine structure of the hydrogen spectrum

and Polykarp Kusch, the United States

For his precision determination of the magnetic moment of the electron.

The theory that predicts the fine structure of the hydrogen spectrum and the magnetic moment of the electron with high precision also won a Nobel prize.

The 1965 Nobel prize in Physics was awarded jointly to Sin-Itiro Tomonaga, Japan, Julian Schwinger, and Richard P. Feynman, the United States

For their fundamental work in quantum electrodynamics, with deep-ploughing consequences for the physics of elementary particles.

APPARATUS [Optional Apparatus in brackets]

Spectrograph
Hydrogen–deuterium discharge tube
Mercury lamp
Traveling microscope
[Iron arc]

A spectrograph with a resolving power of at least 50 000 is needed to resolve the fine structure of hydrogen and a resolving power of at least 4000 is needed to resolve the isotope structure of hydrogen. Appendix D describes a spectrograph that has the required resolution.

OBJECTIVES

To record both the fine structure and the isotope structure spectra of hydrogen and to analyze the data.

To become familiar with a spectograph and to understand two of its important characteristics: dispersion and resolution.

To understand the fine structure and isotope structure of hydrogen

KEY CONCEPTS

Fine structure   Wave number
Isotope structure   Spectroscopic notation
Spin–orbit interaction   Spectral line broadening
Lamb shift   Dispersion
Reduced mass   Resolution

REFERENCES

1. W. Hindmarsh, *Atomic Spectra*, Pergamon, Elmsford, NY, 1967. Part I is the theory of atomic spectra, which includes the isotope structure of hydrogen on page 10 and an excellent description of the fine structure of hydrogen in Chapter 4. Part II is 17 articles pertaining to atomic spectra reprinted from several journals. Article 15 is Bethe's paper explaining the so-called Lamb shift.

2. R. Eisberg and R. Resnick, *Quantum Physics of Atoms, Molecules, Solids, Nuclei, and Particles*, 2d ed., Wiley, New York, 1985. On pages 284–288, the contribution of the spin–orbit interaction to the fine structure of hydrogen is discussed and the other two interactions that contribute to the fine structure of hydrogen, namely, the relativistic variation of electron mass with velocity and the Lamb shift, are briefly discussed.

3. F. Richtmyer, E. Kennard, and T. Lauritsen, *Introduction to Modern Physics*, 5th ed., McGraw-Hill, New York, 1955. The discussion of the fine structure of hydrogen is on pages 258–266.

4. S. Pollack and E. Wong, *Am. J. Phys.* **39**, 1386 (1971). This paper reports experimental results on both the isotope structure and the fine structure of hydrogen using a 1.5-m, Ebert-mount spectrograph. (A 2-m, Ebert-mount spectrograph, which was designed by Professor Wong, is described in Appendix D of this text.)

5. W. Lamb, Jr., and R. Retherford, *Phys. Rev.* **72**, 241 (1947). This paper reports on the measurement of fine structure of hydrogen using radio-frequency methods. (Lamb received the Nobel prize for this work in 1955.)

6. H. Bethe, *Phys. Rev.* **72**, 339 (1947). This is a theoretical paper that explains the measurements of Lamb and Retherford, reference 5.

INTRODUCTION

The history of atomic spectroscopy is the story of the measurement of spectra and the development of a theory that explains the observed spectra. In 1947 Lamb and Retherford (see reference 5) measured the fine-structure splitting of two energy levels of hydrogen. (Fine structure is a phrase used to describe the splitting of spectral lines into several distinct components. Fine structure is found in all atomic spectra.) Their results could not be explained by relativistic quantum mechanics. That same year Bethe (reference 6), who treated the atomic electron of hydrogen nonrelativistically, used quantum electrodynamics to predict Lamb and Retherford's observation to an accuracy of 2 percent. In 1949 Tomonaga, Schwinger, and Feynman developed independently a relativistically covariant formulation of quantum electrodynamics, which predicts the observation with more precision than Bethe's theory.

The interactions that produce the fine structure of hydrogen are (1) the spin–orbit interaction, (2) relativistic variation of electron mass with velocity, and (3) interaction between the electron and the vacuum radiation field (Lamb shift).

Isotope structure is a phrase used to describe the difference in frequency between the corresponding lines of two isotopes of the same chemical element. This frequency shift in the spectral lines of one isotope relative to another is due to the mass difference of the two nuclei. The isotope structure of hydrogen is predicted by the Bohr theory.

Isotope Structure

Often in general physics the mass of the nucleus is assumed to be infinitely large compared with the mass of the orbiting electron, so that the nucleus remains at rest. However, for accurate spectroscopic measurements the finite mass of the nucleus must be taken into account.

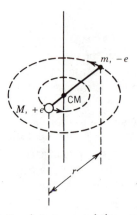

FIGURE 12.1 Revolution around the common center of mass.

For a real hydrogen atom, the nucleus of finite mass M and the single electron of mass m revolve around a common center of mass as shown in Figure 12.1. By transforming the kinetic energy of the two-particle system from laboratory coordinates to the coordinates of the center of mass and relative coordinates, it is not difficult to show that the motion of the electron relative to the nucleus is as if the nucleus is fixed and the electron orbits with a reduced mass μ, where

$$\frac{1}{\mu} = \frac{1}{M} + \frac{1}{m} \quad (1/\text{kg}) \tag{1}$$

or

$$\mu = \frac{Mm}{M+m} \quad (\text{kg}) \tag{2}$$

This is shown in Figure 12.2.

The Bohr theory predicts that the frequency of a spectral line of hydrogen is given by

$$v = cR\left(\frac{1}{n_f^2} - \frac{1}{n_i^2}\right) \quad (\text{Hz}) \tag{3}$$

where n_f and n_i are the principal quantum numbers for the final and initial states, respectively, c is the velocity of light, and R is the Rydberg constant for an electron of

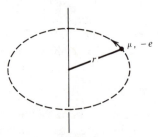

FIGURE 12.2 Motion of the electron of reduced mass μ relative to the nucleus.

reduced mass μ. R is defined to be

$$R \equiv \frac{\mu e^4}{8\varepsilon_0^2 c h^3} \qquad (1/\text{m}) \tag{4}$$

where e is the electronic charge, ε_0 is the permittivity of empty space, and h is Planck's constant. Note that R depends on μ and, hence, on M. The dependence of R on M gives rise to a difference in frequency $\Delta \nu \equiv \nu_2 - \nu_1$ between the corresponding lines of two isotopes of hydrogen with masses M_1 and M_2. Using equations 2–4, we obtain

$$\frac{\nu_2 - \nu_1}{\nu_1} = \frac{R_2 - R_1}{R_1}$$

$$= \frac{\mu_2 - \mu_1}{\mu_1}$$

$$= \frac{M_2}{M_1}\left(\frac{M_1 + m}{M_2 + m}\right) - 1 \tag{5}$$

EXERCISE 1

The Balmer series is the set of transitions having $n_f = 2$. For the first lines of the Balmer series ($n_i = 3$) determine the frequency difference $\Delta \nu$ between the lines for ordinary hydrogen and deuterium. Using your result for $\Delta \nu$ show that the wavelength difference $\Delta \lambda$ is 1.79 Å. Determine $\Delta \lambda$ for the lines of the Balmer series for $n_i = 6$. (Figure 1, reference 4, is an enlarged photograph of both the fine structure and isotope structure spectral lines.)

Fine Structure

The fine structure of the $n = 2$ and $n = 3$ energy levels of hydrogen is shown in Figure 12.3c. The goal of this section is to understand the origin of this fine structure.

We start with the following nonrelativistic, time-independent Schrödinger equation:

$$\mathbf{H}_0 \psi_{n\ell m} \equiv \left[\frac{\mathbf{p}^2}{2\mu} + V(r)\right]\psi_{n\ell m} = E_n^0 \psi_{n\ell m} \qquad (\text{J/m}^{3/2}) \tag{6}$$

where $\mathbf{p}$ is the momentum operator for the electron of reduced mass μ and $V(r) = -e^2/4\pi\varepsilon_0 r$. The eigenvalues of H_0 are

$$E_n^0 = -chR_H \frac{1}{n^2} \qquad n = 1, 2, 3, \ldots \qquad (\text{J}) \tag{7}$$

where R_H is the Rydberg constant for (ordinary) hydrogen. The eigenfunctions are given by

$$\psi_{n\ell m}(r, \theta, \phi) = R_{n\ell}(r)Y_\ell^m(\theta, \phi) \qquad (1/\text{m}^{3/2}) \tag{8}$$

where ℓ and m are the orbital angular momentum and magnetic quantum numbers, respectively. For a given value of n, $\ell = 0, 1, 2, \ldots, n - 1$ and for a given value of ℓ, $m = -\ell, -\ell + 1, \ldots, \ell - 1, \ell$. $R_{n\ell}(r)$ and $Y_\ell^m(\theta, \phi)$ are the radial wave function and spherical harmonic function, respectively. The first 14 eigenfunctions for the one-electron atom are given on page 243 of reference 2.

To communicate with spectroscopists one must be comfortable with the concept of wave number $\bar{\nu}$, which is defined as the reciprocal of the wavelength, $\bar{\nu} \equiv 1/\lambda$, and the units are

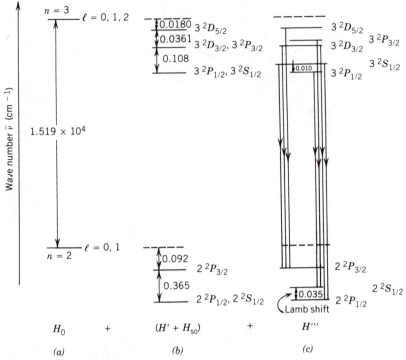

FIGURE 12.3 The splitting of the $n = 2$ and $n = 3$ levels of hydrogen by the various interactions. Part c shows the fine-structure energy levels and the transitions that produce the seven fine-structure spectral lines.

typically inverse centimeters, cm^{-1}. Since λ is the number of centimeters per wavelength, the interpretation of $\bar{v}$ is the number of wavelengths per centimeter. Wave number is energy divided by hc, where h is Planck's constant and c is the velocity of light. The eigenvalues of interest in this experiment are those for $n = 2$ and $n = 3$. The eigenvalues of H_0 are shown in Figure 12.3a.

EXERCISE 2

Show that the wave number spacing between the quantum states with $n = 2$ and $n = 3$ is 1.519×10^4 cm^{-1}.

The relativistic kinetic energy T of an electron of rest mass μ is

$$T = \sqrt{\mu^2 c^4 + p^2 c^2} - \mu c^2$$

$$= \mu c^2 \sqrt{1 + \frac{p^2 c^2}{\mu^2 c^4}} - \mu c^2 \qquad \text{(J)} \qquad (9)$$

Carrying out a binomial expansion of the square root and keeping only the first two terms in the expansion, we obtain

$$T \simeq \frac{p^2}{2\mu} - \frac{p^4}{8\mu^3 c^2} \qquad \text{(J)} \qquad (10)$$

The Hamiltonian H_0 includes the first term, and the second term, included as a perturbation, is

$$H' \equiv -\frac{p^4}{8\mu^3 c^2} \qquad \text{(J)} \qquad (11)$$

We will write H' in a more useful form. From equation 6 we have

$$\frac{p^2}{2\mu} = E_n^0 - V(r) \qquad \text{(J)} \qquad (12)$$

Squaring (12) and substituting for p^4 in (11) we obtain

$$H' = -\frac{[E_n^0 - V(r)]^2}{2\mu c^2} \qquad \text{(J)} \qquad (13)$$

From first-order perturbation theory the effect of H' is to possibly cause a shift of each energy E_n^0 and to remove some of the degeneracy. The additional energy $E'_{n\ell}$ of the nth energy level E_n^0 is calculated from first-order perturbation theory:

$$E'_{n\ell} = \int \int \int \psi^*_{n\ell m} H' \psi_{n\ell m} \, d\tau \qquad \text{(J)} \qquad (14)$$

It remains to be seen that the additional energy E' depends on both n and ℓ, but not m. Substituting (13) into (14) and evaluating the integrals yields

$$E'_{n\ell} = -\frac{1}{2\mu c^2}\left[E_n^{02} + \frac{2e^2 E_n^0}{4\pi\varepsilon_0 a_0}\frac{1}{n^2} + \frac{e^4}{(4\pi\varepsilon_0 a_0)^2}\frac{1}{(\ell + \frac{1}{2})n^3} \right] \qquad \text{(J)} \qquad (15)$$

where $a_0 \equiv 4\pi\varepsilon_0 h^2/4\pi^2\mu e^2 = 5.29 \times 10^{-11}$ m is called the Bohr radius. Note in equation 15 that the quantum mechanical averages of $1/r$ and $1/r^2$, called the expectation values, $\langle 1/r \rangle$ and $\langle 1/r^2 \rangle$, are $1/(a_0 n^2)$ and $1/[a_0^2 n^3(\ell + \frac{1}{2})]$, respectively.

EXERCISE 3

Show that equation 15 may be written as

$$E'_{n\ell} = -\frac{2h^2 R_H^2}{\mu n^3}\left(\frac{1}{\ell + \frac{1}{2}} - \frac{3}{4n} \right) \qquad \text{(J)} \qquad (16)$$

Note that $E'_{n\ell}$ does indeed depend on n and ℓ; hence, for a given n states of different ℓ will have different energies.

The corrections to the energy levels that result from a second-order perturbation are small and will be ignored.

The next perturbation we consider is the spin–orbit interaction. The spin–orbit interaction H_{so} (see equation 5, Experiment 16, and the discussion that follows the equation) is given by

$$H_{so} = \zeta(r)\mathbf{L} \cdot \mathbf{S} \qquad \text{(J)} \qquad (17)$$

where $\mathbf{L}$ and $\mathbf{S}$ are the orbital and spin angular momenta of the electron and $\zeta(r)$ is the function of r that multiplies $\mathbf{L} \cdot \mathbf{S}$.

EXERCISE 4

For hydrogen show that

$$\zeta(r) = \frac{ge^2}{16\pi\varepsilon_0\mu^2 r^3 c^2} \qquad (1/Js^2) \qquad (18)$$

where g is the electron spin g factor, $g = 2.00232$. (To answer this question see the discussion of H_{so} in Experiment 16.)

We write the spin–orbit interaction in terms of the total angular momentum $\mathbf{J}$, where $\mathbf{J}$ is defined to be

$$\mathbf{J} = \mathbf{L} + \mathbf{S} \qquad (Js) \qquad (19)$$

Forming the dot product of $\mathbf{J} \cdot \mathbf{J}$, we solve for $\mathbf{L} \cdot \mathbf{S}$ and obtain

$$\mathbf{L} \cdot \mathbf{S} = \tfrac{1}{2}(\mathbf{J}^2 - \mathbf{L}^2 - \mathbf{S}^2)$$

$$= \frac{\hbar^2}{2}[j(j+1) - \ell(\ell+1) - s(s+1)] \qquad (Js)^2 \qquad (20)$$

where the magnitudes of the vectors were used, that is, $\mathbf{J}^2 = \hbar^2 j(j+1)$, and so on. The spin quantum number for a single electron is $s = \tfrac{1}{2}$, and the possible values of the total angular momentum quantum j are $|\ell - s|, |\ell - s| + 1, \ldots, \ell + s - 1, \ell + s$; therefore, $j = |\ell \pm \tfrac{1}{2}|$ for a single electron.

Using equations 18 and 20, H_{so} becomes

$$H_{so} = \frac{ge^2}{16\pi\varepsilon_0\mu^2 c^2 r^3}\frac{\hbar^2}{2}[j(j+1) - \ell(\ell+1) - s(s+1)] \qquad (J) \qquad (21)$$

From first-order perturbation theory H_{so} changes the energy of the nth level E_n^0, by an amount

$$E_{n\ell}'' = \iiint \psi_{n\ell m}^* H_{so} \psi_{n\ell m}\, d\tau$$

$$= \frac{ge^2}{16\pi\varepsilon_0\mu^2 c^2}\frac{\hbar^2}{2}[j(j+1) - \ell(\ell+1) - \tfrac{3}{4}]\frac{1}{a_0^3 n^3 \ell(\ell+\tfrac{1}{2})(\ell+1)} \qquad (J) \qquad (22)$$

Note from equation 22 that the expectation value of $1/r^3$ equals $1/[a_0^3 n^3 \ell(\ell+\tfrac{1}{2})(\ell+1)]$. Since j and ℓ are related by $j = |\ell \pm \tfrac{1}{2}|$, E'' is arbitrarily shown to be dependent on ℓ rather than j.

EXERCISE 5

Express equation 22 in terms of R_H and show that

$$E_{n\ell}'' = \frac{h^2}{\mu}R_H^2 \frac{j(j+1) - \ell(\ell+1) - \tfrac{3}{4}}{n^3 \ell(\ell+\tfrac{1}{2})(\ell+1)} \qquad (J) \qquad (23)$$

where $g \simeq 2$ was used.

EXERCISE 6

Using equations 16 and 23 determine the sum $E' + E''$ in terms of the quantum numbers n and j for the case $j = |\ell + \frac{1}{2}|$ by substituting $\ell = j - \frac{1}{2}$, and, hence, show that

$$E'_{nj} + E''_{nj} = \frac{2h^2 R_H^2}{\mu n^2} \left(-\frac{1}{j} + \frac{3}{4n} + \frac{1}{j(2j+1)} \right) \qquad j = |\ell + \tfrac{1}{2}| \qquad (24)$$

(We now write E' and E'' with subscripts n and j, rather than n and ℓ.) For the case $j = |\ell - \frac{1}{2}|$ show that

$$E'_{nj} + E''_{nj} = \frac{2h_2 R_H^2}{\mu n^2} \left[-\frac{1}{j+1} + \frac{3}{4n} - \frac{1}{(2j+1)(j+1)} \right] \qquad j = |\ell - \tfrac{1}{2}| \qquad (25)$$

The quantum states in Figure 12.3b and c are labeled using spectroscopic notation $n^{2s+1}\ell_j$, where $s = \frac{1}{2}$, $\ell = 0(S)$, 1(P), 2(D), ..., $n - 1$, and $j = |\ell \pm \frac{1}{2}|$. See the discussion of spectroscopic notation in Experiment 16, Electron Spin Resonance.

EXERCISE 7

For $n = 2$, $\ell = 0$, 1 and for $n = 3$, $\ell = 0$, 1, 2 use equations 24 and 25 to calculate $E'_{nj} + E''_{nj}$ and, hence, verify the energy levels shown in Figure 12.3b.

The theory that has been summarized up to this point, and which predicts the energy levels shown in Figure 12.3b, is a simplified version of the Dirac theory of relativistic quantum mechanics. In 1947 Lamb and Retherford obtained experimental results on the fine structure of hydrogen that could not be explained by the Dirac theory (see reference 6). Lamb and Retherford examined the structure of the $n = 2$ level of hydrogen by radio-frequency methods and found that the $2\,^2S_{1/2}$ level is shifted from the $2\,^2P_{1/2}$ level by $0.035\ \mathrm{cm}^{-1}$ as shown in Figure 12.3c. This energy shift is called the Lamb shift.

The theory that predicts the fine structure shown in Figure 12.3c is quantum electrodynamics (QED). The agreement between theory (QED) and experiment (usually referred to as the Lamb shift) is one of the outstanding achievements of twentieth century physics. It is the standard by which other agreements are measured.

The final interaction H''' needed to produce the energy levels in Figure 12.3c is the interaction of the atomic electron with the vacuum radiation field. This interaction is briefly discussed in reference 2, and in reference 1 a simplified calculation is outlined, showing the splitting of the levels in Figure 12.3c.

For one-electron atoms, transitions take place only between levels whose quantum numbers ℓ and j satisfy the selection rules

$$\Delta \ell = \pm 1$$
$$\Delta j = 0, \pm 1 \qquad (26)$$

(See reference 2, page 288.) These selection rules permit the seven transitions shown in Figure 12.3c. Seven distinct spectral lines are not observed because of spectral line broadening. (Broadening of spectral lines is discussed with the spectrograph in Appendix D.) Typically, the three transitions shown on the left of Figure 12.3c are broadened into a single line and the four transitions on the right in Figure 12.3c are broadened into one line, and these two broadened lines overlap as shown in Figure 12.4. The accepted separation of the two peaks is $\Delta \bar{v} = 0.32\ \mathrm{cm}^{-1}$.

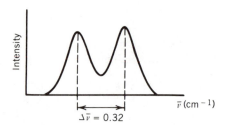

FIGURE 12.4 Broadening of the seven fine-structure lines gives rise to two broadened lines that overlap.

EXERCISE 8

Show that the wave-number difference of $\Delta\bar{\nu} = 0.32 \text{ cm}^{-1}$ corresponds to a wavelength difference of $\Delta\lambda = 0.14 \text{ Å}$. (See Figure 1, reference 4, which is an enlarged photograph of both the isotope structure and fine-structure lines of hydrogen.)

Spectrograph

The 2-m, Ebert-mount spectrograph is described in Appendix D. The dispersion is derived and the resolution is discussed in some detail. Also the adjustments for the Ebert-mount spectrograph are explained (other spectrographs require similar adjustments). Two methods of determining unknown wavelengths, the dispersion curve method and the iron arc method, are described. You should read Appendix D and answer the questions, even if your spectrograph has a different mounting.

EXPERIMENT

Set the angular position of the grating such that the H_α lines of the Balmer series fall in about the central region of the photographic plate. (You should also decide what order of the spectrum you want to photograph.) Using the 20-μm slit, expose the film for about one minute. Develop the film and, use one of the two methods to determine unknown wavelengths, then determine the wavelength difference $\Delta\lambda$ for the isotope structure. For both hydrogen and deuterium determine $\Delta\lambda$ for the fine structure. Compare your results with the accepted values.

EXERCISE 9

Use your data from the isotope structure to determine M_D/M_H, where M_D and M_H are the mass of deuterium and hydrogen, respectively.

EXERCISE 10

Look carefully at the photograph of the hydrogen and deuterium lines. Which lines have greater resolution? What is the source of broadening that causes one set of lines to have a greater linewidth?

13. MOLECULAR STRUCTURE AND PROPERTIES OF $^{14}N_2$: OPTICAL SPECTROSCOPY

APPARATUS [Optional Apparatus in Brackets]

Spectrograph
Nitrogen spectrum tube
Mercury lamp
Traveling microscope
[Iron arc]

A spectrograph with a resolving power of at least 10 000 is needed to resolve the rotational structure of the nitrogen molecule N_2. Appendix D describes a spectrograph that has the required resolution.

OBJECTIVES

To record and analyze the molecular spectrum of $^{14}N_2$, and to calculate
1. The average internuclear separation
2. The vibrational energies
3. Dissociation energy
4. The Morse potential
5. The molecular force constant

To understand the rotational, vibrational, and electronic structure of diatomic molecules.

To become familiar with a spectrograph and to understand two of its major characteristics, dispersion and resolution.

KEY CONCEPTS

Rotational structure
Vibrational structure
Electronic structure
Rigid and nonrigid rotor selection rules
Harmonic oscillator selection rules
Anharmonic oscillator selection rules
Morse potential
Vibrating rotor selection rules

Rotation–vibration band
Band origin
Band head
P, Q, and R branches
Vibrating–rotating molecule selection rules
Electronic selection rules

REFERENCES

1. R. Eisberg and R. Resnick, *Quantum Physics of Atoms, Molecules, Solids, Nuclei, and Particles*, 2d ed., Wiley, New York, 1985. Molecules are discussed in Chapter 12.
2. G. Herzberg, *Molecular Spectra and Molecular Structure, 1. Spectra of Diatomic Molecules*, Van Nostrand, New York, 1950. This is the basic text for diatomic molecules.

3. A. Gaydon and R. Pearse, *The Identification of Molecular Spectra*, Chapman & Hall, London 1963. The band structure of the N_2 molecule is summarized on pages 209–218. Photographic plate 3 includes the "2nd positive" bands of N_2, where 2nd positive signifies the $C^3\Pi_u \rightarrow B^3\Pi_g$ transition. This is the set of bands you are asked to study in this experiment.

INTRODUCTION

When atoms combine into a molecule, the inner electrons in each atom can be regarded as remaining with the nucleus of that atom, but the outer electrons come to belong to the molecule as a whole rather than to any individual nucleus. For the diatomic molecule N_2 there are $2P^6$ electrons that belong to the molecule as a whole. The Hamiltonian corresponding to the kinetic and Coulomb potential energies of the molecule N_2 is

$$H = -\sum_{i=1}^{6} \frac{\hbar^2}{2m} \nabla_i^2 - \sum_{k=1}^{2} \frac{\hbar^2}{2M} \nabla_k^2 + \sum_{\substack{i,j=1 \\ i \neq j}}^{6} \frac{e^2}{4\pi\varepsilon_0 r_{ij}} - \sum_{i=1}^{6} \sum_{k=1}^{2} \frac{Ze^2}{4\pi\varepsilon_0 r_{ik}} + \frac{Z^2 e^2}{4\pi\varepsilon_0 R} \quad \text{(J)} \qquad (1)$$

where m is the mass of the electron, M is the mass of the nucleus, the labels i and j signify the electrons and k signifies the nuclei, and R is the separation of the nuclei. Analytical solutions of the corresponding Schrödinger equation are not possible. To find the eigenfunctions and eigenvalues, which can be compared with spectra, approximations must be made.

The first approximation is to divide the energy into three types: (1) rotational kinetic energy of the nuclei about the center of mass of the molecule, (2) vibrational kinetic energy of the nuclei with respect to the center of mass of the molecule, and (3) electronic energy, for example, the energy of the $2P^6$ electrons for the N_2 molecule. Consistent with this approximation the molecular energy levels can be divided into the three types: (1) rotational energy levels, (2) vibrational energy levels, and (3) electronic energy levels. As you will see later, these three types of energy are not independent of each other.

The spectra emitted or absorbed by a molecule can be divided into three spectral ranges that correspond to the transitions between the three types of molecular energy levels: (1) rotation spectra (this spectra is in the far infrared), (2) vibration spectra (in the near infrared), and (3) electronic spectra (in the visible or UV). The general type of transition between molecular energy levels is one in which changes occur in the electonic state of the molecule as well as in the vibration and rotation states.

In what follows the discussion is restricted to diatomic molecules and for each energy type we discuss the quantized energy levels, selection rules that must be satisfied for transitions between these levels, and the emission spectral lines that result from transitions between these levels. We start our discussion with the rotational energy of a diatomic molecule.

Rotational Energy Levels

The initial model is a rigid rotor, which is two point masses connected by a massless rod and separated by a fixed distance R_0. This model is shown in Figure 13.1a. By transforming the kinetic energy of the two-particle system from laboratory coordinates to the coordinates of the center of mass and relative coordinates, it is not difficult to show that the motion of one nucleus relative to the other is as if one nucleus is fixed and the other orbits with a reduced mass $\mu = M_1 M_2/(M_1 + M_2)$. The one-particle system is shown in Figure 13.1b and it is dynamically equivalent to the two-particle system shown in Figure 13.1a. The classical rotational energy is

$$E_r = \frac{N^2}{2I} \quad \text{(J)} \qquad (2)$$

where N is the angular momentum of the system and $I = \mu R_0^2$ is the moment of inertia.

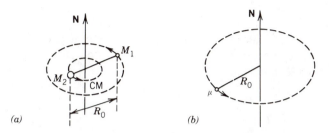

FIGURE 13.1 (a) Rigid rotor model of a diatomic molecule.
(b) Dynamically equivalent one-particle system.

The magnitude of the quantized angular momentum is $N = \sqrt{r(r+1)}\hbar$, where the rotational quantum number r has the values $0, 1, 2, \ldots$. The quantized energy for the system is given by

$$E_r = \frac{\hbar^2}{2I} r(r+1) \qquad \text{(rigid rotor)} \qquad \text{(J)} \tag{3}$$

Energy levels of the rigid rotor are shown in Figure 13.2a.

If the rigid rotor has an electric dipole moment, then absorption and emission of radiation occurs. To calculate the frequencies that are actually emitted or absorbed, it is necessary to know the selection rule for the quantum number r. This selection rule is obtained by evaluating the matrix elements of the electric dipole moment operator using the eigenfunctions of the rotational states (see reference 2, page 72). The selection rule is

$$\Delta r = \pm 1 \qquad \text{(rigid and nonrigid rotor)} \tag{4}$$

Transitions for $\Delta r = -1$ (emission spectral lines) are shown in Figure 13.2a, and Figure 13.2b shows the spectral lines, where v is the frequency.

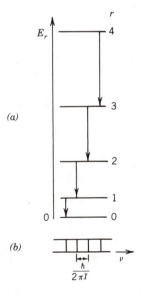

FIGURE 13.2 Rigid rotor (a) energy levels and (b) emission spectral lines.

EXERCISE 1

Figure 13.2*b* indicates that the frequency separation of the successive spectral lines is $\hbar/2\pi I$. Show this. Note that the equilibrium separation of the nuclei R_0 can be determined from the rotational spectrum.

EXERCISE 2

In Table 12-1, reference 1, R_0 for the diatomic nitrogen molecule N_2 is reported as 1.09 Å. Using this value for R_0, calculate the frequencies of the emitted radiation for the four transitions shown in Figure 13.2*a*. What is the frequency spacing $\Delta\nu$ between consecutive spectral lines?

A better model for representing the rotations of a diatomic molecule is a nonrigid rotor, that is, a rotating system consisting of two point masses which are connected by a massless spring having force constant k. The system is shown in Figure 13.3*a* and the dynamically equivalent system is shown in Figure 13.3*b*. The classical energy of the system is obtained by the following argument. Suppose the internuclear separation of the nonrigid rotor is R, where R_0 is the equilibrium length of the spring. The classical energy is the sum of the kinetic and potential energies, and assuming a harmonic potential energy, we have

$$E = \frac{N^2}{2\mu R^2} + \frac{1}{2}k(R - R_0)^2 \quad \text{(nonrigid rotor)} \quad \text{(J)} \tag{5}$$

where the magnitude of the angular momentum is $N = I\omega = \mu R^2 \omega$ and ω is the angular velocity. Applying Newton's second law to the reduced mass μ, we have

$$k(R - R_0) = \mu\omega^2 R \quad \text{(N)} \tag{6}$$

where $\omega^2 R$ is the centripetal acceleration of the reduced mass μ. Solving equation 6 for $R - R_0$ and substituting into equation 5, we find that

$$E = \frac{N^2}{2\mu R^2} + \frac{1}{2}k\frac{\mu^2\omega^4 R^2}{k^2} = \frac{N^2}{2\mu R^2} + \frac{N^4}{2\mu^2 R^6 k} \quad \text{(J)} \tag{7}$$

In quantum theory the magnitude of the angular momentum is $N = \sqrt{r(r + 1)}\hbar$. Therefore, the quantum-mechanical energy of the nonrigid rotor is

$$E_r = \frac{\hbar^2}{2\mu R^2}r(r + 1) - \frac{\hbar^4 r^2(r + 1)^2}{2\mu^2 R^6 k} \quad \text{(nonrigid rotor)} \quad \text{(J)} \tag{8}$$

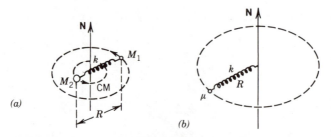

(a)

(b)

FIGURE 13.3 (*a*) Nonrigid rotor model of a diatomic molecule.
(*b*) Dynamically equivalent one-particle system.

where $r = 0, 1, 2, \ldots$, and the negative sign that results from a more rigorous calculation has been inserted. The selection rule, equation 4, is valid for both rigid and nonrigid rotors.

Vibrational Energy Levels

The simplest motion for the vibrations in a diatomic molecule is examined first; namely, each atom moves toward or away from the other in simple harmonic motion. The motion of the two masses can be reduced to the harmonic motion of a single mass of reduced mass μ about an equilibrium position. The equivalent systems are shown in Figure 13.4. The classical vibrational frequency of the system is

$$v_{\text{vib}} = \frac{1}{2\pi} \sqrt{\frac{k}{\mu}} \quad \text{(Hz)} \tag{9}$$

The quantum-mechanical energy for a one-dimensional harmonic oscillator is obtained by solving the time-independent Schrödinger equation, and the resulting eigenvalues are

$$E_v = h v_{\text{vib}}(v + \tfrac{1}{2}) \quad v = 0, 1, 2, \ldots \quad \text{(harmonic oscillator)} \quad \text{(J)} \tag{10}$$

where v is the vibrational quantum number. The harmonic oscillator potential energy and the vibrational energy levels are shown in Figure 13.5a. If the molecule has an electric dipole moment, then absorption and emission of radiation may occur. The selection rule for the quantum number v is obtained by evaluating the matrix elements of the electric dipole

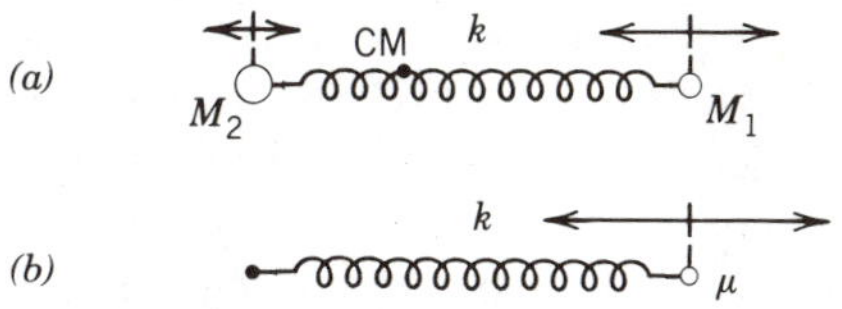

(a)

M_2 CM k M_1

(b)

k μ

FIGURE 13.4 (a) One-dimensional harmonic oscillator model of a diatomic molecule.
(b) Dynamically equivalent one-particle model.

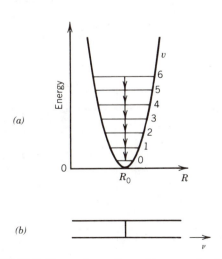

FIGURE 13.5 Harmonic oscillator (a) energy levels and
(b) spectral line.

moment operator using the eigenfunctions of the vibrational states (see reference 2, page 80). The selection rule is

$$\Delta v = \pm 1 \qquad \text{(harmonic oscillator)} \tag{11}$$

Transitions corresponding to emission ($\Delta v = -1$) are shown in Figure 13.5a, and the resulting single emission spectral line is shown in Figure 13.5b.

EXERCISE 3

In Table 12-1, reference 1, the wave number corresponding to the frequency v_{vib} is reported as 2360 cm^{-1} for the N_2 molecule, where wave number is the reciprocal of the wavelength: 2360 cm^{-1} = $1/\lambda$ = v_{vib}/c, where c is the velocity of light. Calculate the frequency of the single vibrational emission line and compare it with the frequencies calculated in Exercise 2.

EXERCISE 4

It was previously stated that the second term in the nonrigid rotor energy (see equation 8) is small. For the N_2 molecule calculate both constants that multiply the quantum numbers in equation 8, and show that the second term is indeed smaller than the first term. For the value of R use the value of R_0 given in Exercise 2, and the force constant k can be calculated from equation 9.

For the harmonic potential energy in Figure 13.5a, the potential energy and therefore the restoring force increase indefinitely with increasing distance from the equilibrium position. In an actual molecule the force is zero for large separation of the two atoms and therefore the potential energy is constant. Hence, the potential energy of the actual molecule has the form shown by the solid line curve in Figure 13.6. The broken line curve in Figure 13.6 is the harmonic potential energy. Note that the two potential curves overlap at the bottom; hence, the first few eigenvalues should be approximately the same for both potential energies.

Expanding the actual potential energy of the molecule about R_0 gives

$$V(R) = V(R_0) + \left.\frac{dV}{dR}\right|_{R_0} (R - R_0) + \left.\frac{1}{2}\frac{d^2V}{dR^2}\right|_{R_0} (R - R_0)^2$$
$$+ \left.\frac{1}{6}\frac{d^3V}{dR^3}\right|_{R_0} (R - R_0)^3 + \cdots \qquad \text{(J)} \tag{12}$$

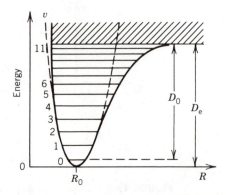

FIGURE 13.6 Energy levels of an anharmonic oscillator.

The first term on the right side is a constant that we may ignore, since we observe a change in the energy rather than the absolute energy. The second term is zero because the slope is zero at R_0. The third term is the harmonic potential energy and the higher terms are nonharmonic potential energy terms.

A point mass that has a potential energy curve of the form of the solid line curve in Figure 13.6 is called an anharmonic oscillator. The eigenvalues for an anharmonic oscillator are obtained by solving the time-independent Schrödinger equation with the potential given by equation 12. The eigenvalues are

$$E_v = hc\omega_e(v + \tfrac{1}{2}) - hc\omega_e x_e(v + \tfrac{1}{2})^2$$

$$+ hc\omega_e y_e(v + \tfrac{1}{2})^3 + \cdots \quad \text{(anharmonic oscillator)} \quad \text{(J)} \quad (13)$$

where $v = 0, 1, 2, \ldots$. The so-called vibrational constants (ω_e, x_e, y_e) are written in standard (internationally adopted) notation. ω_e is a wave number (typical units are cm^{-1}), c is the velocity of light ($\omega_e c = $ frequency), and x_e and y_e are dimensionless, where $y_e \ll x_e \ll 1$. Also, $\omega_e x_e$ is positive and $\omega_e y_e$ may be positive or negative. The energy levels E_v of the anharmonic oscillator are shown in Figure 13.6.

The energy D_0 shown in Figure 13.6 is called the **dissociation energy**. It represents the work that must be done to dissociate the molecule. There are no discrete vibrational levels above D_0, and, hence, there are a finite number of discrete vibrational levels. The number of discrete levels is arbitrarily shown to be eleven in Figure 13.6.

A potential energy that approximates the anharmonic potential energy is the so-called Morse potential. The Morse potential energy is given by

$$V(R - R_0) = D_e(1 - e^{-\beta(R - R_0)})^2 \quad \text{(J)} \quad (14)$$

where D_e, shown in Figure 13.6, is the sum of D_0 plus the vibrational energy E_0. From equation 14, note that the Morse potential approaches D_e as $R \to \infty$ and it equals zero when $R = R_0$, both of which are in agreement with the solid line curve shown in Figure 13.6. However, when $R \to 0$, the Morse potential does not approach infinity, as it must do for a correct potential energy. This failure of the Morse potential for small R is not serious, since the part of the potential for small R is of no practical importance. It is shown in reference 2, on page 100, that D_e and β are related to the vibrational constants by

$$D_e = \frac{\omega_e^2}{4\omega_e x_e} \quad \text{(J)} \qquad \beta = \sqrt{\frac{2\pi^2 c\mu}{D_e h}} \times \omega_e \quad \text{(1/m)} \quad (15)$$

These equations for D_e and β ignore the very small cubic and higher order terms in the anharmonic potential energy.

To consider the spectrum of the anharmonic oscillator we need the selection rules, which are determined by evaluating the matrix elements of the electric dipole moment operator between eigenfunctions of the anharmonic oscillator. The selection rule is

$$\Delta v = \pm 1, \pm 2, \pm 3, \ldots \quad \text{(anharmonic oscillator)} \quad (16)$$

The intensity of the spectral lines decreases rapidly as Δv increases.

Figure 13.7 shows five vibrational levels and transitions between these levels. The vibrational constants, and, hence, the dissociation energy, the Morse potential, and the force constant, can be determined from the vibrational spectrum.

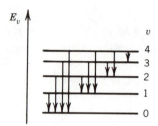

FIGURE 13.7 Five anharmonic oscillator energy levels and transitions.

EXERCISE 5

Consider the frequencies of two vibrational spectral lines: $h\nu_1 \equiv E_1 - E_0$ and $h\nu_2 \equiv E_2 - E_0$, where E_0, E_1, and E_2 are energy levels of the anharmonic oscillator, and ν_1 and ν_2 are empirically determined. Ignoring cubic and higher order terms in the anharmonic energy levels, show that the vibrational constants are given by $\omega_e = (3\nu_1 - \nu_2)/c$ and $x_e = (2\nu_1 - \nu_2)/(6\nu_1 - 2\nu_2)$. (**Remark**: For the first few energy levels the potential energy is approximately harmonic and, hence, $c\omega_e \simeq \nu_{\text{vib}} = (\sqrt{k/\mu})/2\pi$, and the force constant k can be determined.)

Vibration–Rotation Energy Levels

Thus far we have treated rotation and vibration of the molecule as occurring separately. Molecular spectra indicates that rotation and vibration occur simultaneously. The model in which rotation and vibration occurs simultaneously is called the vibrating rotor or the rotating vibrator.

If we neglect the interaction of the vibration and rotation, the energy of the vibrating rotor would be the sum of the vibrational energy of the anharmonic oscillator, equation 13, and the rotational energy of the nonrigid rotor, equation 8. However, in a more accurate treatment we must take into consideration the fact that during the vibration the internuclear distance R and, consequently, the moment of inertia $I = \mu R^2$ change.

EXERCISE 6

Using your results from Exercises 2 and 3, show that the period of vibration, $T_{\text{vib}} = 1/\nu_{\text{vib}}$, is very small compared with the period of rotation, $T_{\text{rot}} = 1/\nu_{\text{rot}}$, for the lower lying rotational states.

Since $T_{\text{vib}} \ll T_{\text{rot}}$ we use a mean value for R^2 and R^6, which occur in the energy of the nonrigid rotator, equation 8, namely, $\overline{I/R^2}$ and $\overline{1/R^6}$, where the mean value is over one vibration. The mean value of the internuclear distance depends on the vibrational quantum number v, and we express this dependence on v with a subscript: $\overline{I/R_v^2}$ and $\overline{I/R_v^6}$.

We obtain the energy of a nonrigid rotor in a given vibrational level having quantum number v by replacing R^2 and R^6 in equation 8 by their mean values:

$$E_r = \frac{\hbar^2}{2\mu}\left(\overline{\frac{1}{R_v^2}}\right)r(r+1) - \frac{\hbar^4}{2\mu^2 k}\left(\overline{\frac{1}{R_v^6}}\right)r^2(r+1)^2$$

$$\equiv B_v r(r+1) - D_v r^2(r+1)^2 \qquad \text{(nonrigid rotor in vibrational level } v) \qquad \text{(J)} \qquad \text{(17)}$$

where $r = 0, 1, 2, \ldots$. B_v and D_v are called rotational constants, and, as defined in equation 17, they have energy units. In some texts, for example, reference 2, they are defined such that the units are cm^{-1}. Note that $\bar{R}_v$ is the mean internuclear separation when the system is in the vibrational state with quantum number v.

The energy of the vibrating rotor is the sum of the rotational energy, equation 17, and the vibrational energy of the anharmonic oscillator, equation 13:

$$E_{rv} = B_v r(r + 1) - D_v r^2(r + 1)^2 + hc\omega_e(v + \tfrac{1}{2})$$

$$- hc\omega_e x_e(v + \tfrac{1}{2})^2 \qquad \text{(vibrating rotor)} \qquad \text{(J)} \qquad (18)$$

where $v = 0, 1, 2, \ldots$, $r = 0, 1, 2, \ldots$, and small, higher order terms have been ignored. An energy level diagram for the vibrating rotor is given in Figure 13.8. Note that each vibrational energy level is split into a series of rotational energy levels.

The selection rules for transitions between the energy levels of the vibrating rotor are

$$\Delta r = \pm 1 \qquad \Delta v = \pm 1, \pm 2, \pm 3, \ldots \qquad \text{(vibrating rotor)} \qquad (19)$$

Consider a particular vibrational transition from v' to v'', where $v' > v''$ (emission spectra). We denote the initial and final rotational quantum numbers by r' and r'', respectively, where $\Delta r = r' - r'' = \pm 1$. The frequencies of the resulting spectral lines (ignoring the rotational constant D_v and the vibrational constant $hc\omega_e x_e$) are

$$\nu = \frac{E'_{rv} - E''_{rv}}{h}$$

$$= \frac{1}{h}[B'_v r'(r' + 1) - B''_v r''(r'' + 1) + hc\omega_e v' - hc\omega_e v'']$$

$$= \frac{1}{h}[B'_v r'(r' + 1) - B''_v r''(r'' + 1)] + \nu_{\text{vib}}(v' - v'') \qquad \text{(Hz)} \qquad (20)$$

where $c\omega_e = \nu_{\text{vib}}$ was used, $v' - v'' = 1, 2, 3, \ldots$, and $\Delta r = r' - r'' = \pm 1$.

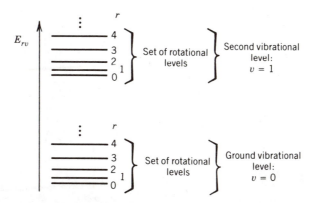

FIGURE 13.8 Energy levels of a vibrating rotor. Each vibrational level splits into a set of rotational levels.

Consider $\Delta r = +1$; then equation 20 may be written as

$$\nu_{\mathrm{R}} = \nu_{\mathrm{vib}}(v' - v'') + 2\frac{B_v'}{h} + \left(\frac{3B_v' - B_v''}{h}\right)r + \left(\frac{B_v' - B_v''}{h}\right)r^2 \qquad r = 0, 1, 2, \ldots \qquad \mathbf{(Hz)} \qquad \mathbf{(21)}$$

where r'' has been replaced with r; hence, r is the final rotational quantum number. Since r can take a whole series of values, this equation represents a series of spectral lines, which is called the **R branch**. For $\Delta r = -1$, equation 20 may be written

$$\nu_{\mathrm{P}} = \nu_{\mathrm{vib}}(v' - v'') - \left(\frac{B_v' + B_v''}{h}\right)r + \left(\frac{B_v' - B_v''}{h}\right)r^2 \qquad r = 1, 2, \ldots \qquad \mathbf{(Hz)} \qquad \mathbf{(22)}$$

where again r'' has been replaced with r. As before, r is the final rotational quantum number. The series of spectral lines represented by equation 22 is called the **P branch**. Figure 13.9a shows the vibrational levels with quantum numbers v' and v'', each with five rotational levels. Also the four transitions for both the R and P branches are shown in Figure 13.9a. Figure 13.9b shows the spectral lines. (We have assumed that $B_v' - B_v''$ is small and positive; hence, ν_{R} and ν_{P} increase and decrease, respectively, as r increases. Note the nonuniform frequency spacing of the spectral lines.) The spectral lines shown in Figure 13.9b are called a **rotation–vibration band** or simply a **band**. Note that the spectral line of frequency $\nu_{\mathrm{vib}}(v' - v'')$ is missing, since a transition with $\Delta r = 0$ is not allowed by the selection rule. This missing line is called the **band origin**.

Note from equations 21 and 22 that the spectrum of a vibrating rotor will be an R and P branch for each value of Δv; that is, $\Delta v = v' - v'' = 1, 2, \ldots$. Also a detailed calculation shows that the larger Δv is the larger is the difference between B_v' and B_v''. Figure 13.10 is the absorption spectrum of the diatomic molecule HCl for a single value of Δv; that is, the spectrum is a single P and a single R branch. The double-peak structure is because of the

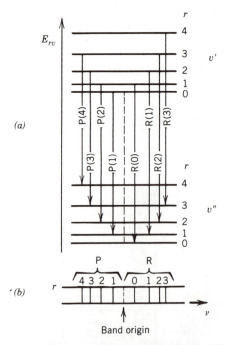

FIGURE 13.9 Vibrating rotor (a) energy levels with R and P branch transitions and (b) spectral lines, a band.

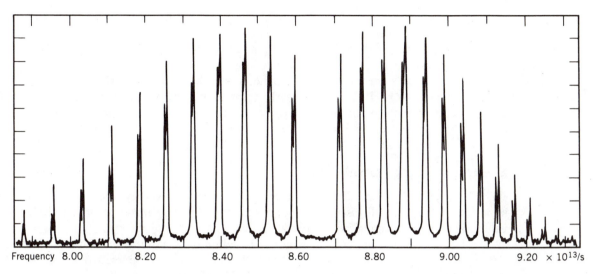

FIGURE 13.10 Infrared absorption spectra of the diatomic molecule HCl.

two isotopes of chlorine, ^{35}Cl (abundance of 75.5 percent) and ^{37}Cl (abundance of 24.5 percent).

EXERCISE 7

On Figure 13.10 label each spectral line with the notation used in Figure 13.9, that is, P(r) and R(r) for the P and R branches, respectively. Assuming $\Delta v = 1$, calculate v_{vib} and $\overline{R}_v$ using Figure 13.10. Use the R(0) transition to calculate $\overline{R}_v$. Compare your values with v_{vib} and R_0 given in Table 12-1, reference 1.

Often a **band head** is formed in a band. A band head is where the density of the spectral lines (number of spectral lines per unit frequency interval) suddenly falls to zero, while on the other side of the band head the density of lines decreases slowly. We now consider the conditions under which such a band head appears.

A band head may appear in the P or the R branch. A band head appears in the P branch if $B'_v - B''_v$ is positive. (Note from equation 17, where B_v is defined, that $B'_v - B''_v > 0$ implies that $\overline{R}'_v < \overline{R}''_v$, that is, the average nuclear separation in the vibrational state v' is less than that in the state v''.) In this case the term in equation 22 that is linear in r is negative and the term that is quadratic in r is positive; therefore, v_P decreases as r increases until it reaches a minimum value at certain value of r. For larger values of r, the positive quadratic term exceeds the negative linear term and v_P increases as r increases. This is shown qualitatively in Figure 13.11a, where the dashed spectral lines in the P branch are for v_P increasing as r increases. The band head is where the branch "turns back on itself."

A band head appears in the R branch if $B'_v - B''_v < 0$ (we assume $3B'_v - B''_v$ is positive and greater than $|B'_v - B''_v|$). In this case the term in equation 21 that is linear in r is positive

FIGURE 13.11 Spectral lines in (a) the P branch and (b) the R branch.

and the term that is quadratic in r is negative; hence, ν_R increases as r increases until it reaches a maximum value at a certain value of r. For larger values of r, ν_R decreases as r increases, creating a band head as shown qualitatively in Figure 13.11b, where the dashed spectral lines are for ν_R decreasing as r increases.

A band head does not form for every band; a band head does not occur in the spectrum shown in Figure 13.10. If B_v' is very nearly equal to B_v'', the band head may lie at such a great distance from the origin that it is not observed, since for the corresponding r value the intensity may have decreased to zero.

Electronic Energy Levels

We will not solve the Schrödinger equation for the electronic energy levels of the diatomic molecule; rather, we will discuss the approximations that are made to obtain the eigenvalues, and we will summarize the results.

The Born–Oppenheimer approximation is based on the great difference of masses of the electrons and nuclei in a molecule. When the nuclei move the electrons can almost instantaneously adjust to their new positions. Therefore, instead of trying to solve the Schrödinger equation for a collection of mobile electrons and nuclei, we regard the nuclei as frozen with a separation R and solve the Schrödinger equation for the electrons moving in the potential generated by the stationary nuclei. Different separations of the nuclei may then be taken and, hence, R becomes a parameter in the Schrödinger equation. The dependence of the energy of the molecule on R is called a molecular potential energy curve. The molecular potential energy curve does depend on the electronic state. The solid curve in Figure 13.6 shows the molecular potential energy curve $V(R)$ versus R for the electronic ground state.

When we make the Born–Oppenheimer approximation, equation 1 becomes

$$H = -\sum_{i=1}^{6} \frac{\hbar^2}{2m} \nabla_i^2 + \sum_{\substack{i,j=1 \\ i \neq j}}^{6} \frac{e^2}{4\pi\varepsilon_0 r_{ij}} - \sum_{i=1}^{6} \sum_{k=1}^{2} \frac{e^2}{4\pi\varepsilon_0 r_{ik}}$$

$$+ \frac{Z^2 e^2}{4\pi\varepsilon_0 R} \quad (R - a \text{ parameter}) \quad \text{(J)} \quad (23)$$

Equation 23 represents a six-body problem, since the six electrons are free to move and the two nuclei are fixed. There is no hope of finding analytical solutions to the corresponding Schrödinger equation, and we have to make another approximation. One approximation is called the linear combination of atomic orbitals (LCAO). In this approximation we assume the molecular orbitals (eigenfunctions) are linear combinations of the atomic orbitals (eigenfunctions) of the parent atoms. With these two approximations, Born–Oppenheimer and LCAO, solutions to the Schrödinger equation can be found. Before summarizing the electronic energy levels that result from solving the Schrödinger equation, we need to consider the total angular momentum of the diatomic molecule.

Molecules have four sources of angular momentum, and they must be coupled together properly:

1. **N**, angular momentum of the nuclei about the center of mass of the molecule, where $|\mathbf{N}| = [r(r+1)]^{1/2}\hbar$.
2. **L**, total orbital angular momentum of the electrons, where $|\mathbf{L}| = [\ell(\ell+1)]^{1/2}\hbar$.
3. **S**, total spin angular momentum of the electrons, where $|\mathbf{S}| = [s(s+1)]^{1/2}\hbar$.
4. **I**$_k$, nuclear spin angular momentum of the kth nucleus, where $k = 1, 2$ for the diatomic molecule.

Nuclear spin gives rise to the hyperfine interaction, which has a small effect on the energy levels; hence, we ignore nuclear spin in our discussion of the total angular momentum of a molecule.

Four basic types of coupling have been proposed by Hund, which are called cases a to d (see reference 2, page 219). Any molecule is approximately described by one of Hund's cases, and the N_2 molecule is best described by case b, which we now consider. In case b it is assumed that the spin–orbit interaction is zero or small. The total orbital angular momentum of the electrons $\mathbf{L}$ is coupled by the strong axial electrostatic field to the internuclear axis; hence, $\mathbf{L}$ precesses rapidly about this axis, as shown in Figure 13.12a. The magnitude of $|\mathbf{L}|$ along this axis is denoted by $|\mathbf{\Lambda}| = \Lambda\hbar$, where the quantum number Λ is $0, 1, 2, 3, \ldots, \ell$, and where $|\mathbf{L}| = \hbar[\ell(\ell+1)]^{1/2}$. (Note the similiarity between $|\mathbf{\Lambda}|$ and L_z of atomic physics, where $L_z = m\hbar$ and $|m| = 0, 1, 2, 3, \ldots, \ell$.)

Each value of Λ corresponds to a distinct electronic state of the molecule. With $\Lambda = 0, 1, 2, 3, \ldots$, the corresponding state of the molecule is designated a $\Sigma, \Pi, \Delta, \Phi, \ldots$ state. (Note the similarity of the quantum number Λ and the quantum number ℓ of atomic physics: For $\ell = 0, 1, 2, 3, \ldots$, the atomic states are designated as S, P, D, F, $\ldots$ states.)

The angular momenta $\mathbf{\Lambda}$ and $\mathbf{N}$ form a resultant angular momentum designated as $\mathbf{K}$, where the magnitude of $|\mathbf{K}| \equiv \hbar[\kappa(\kappa+1)]^{1/2}$. $\mathbf{K}$ is the total angular momentum of the molecule apart from spin, and therefore the quantum number κ must have integral values. Since $\mathbf{\Lambda}$ and $\mathbf{N}$ are perpendicular, the quantum number κ cannot be less than Λ. Therefore for a given Λ we have

$$\kappa = \Lambda, \Lambda+1, \Lambda+2, \Lambda+3, \ldots \tag{24}$$

If $\Lambda = 0$, a Σ state, the angular momentum $\mathbf{K}$ is identical with $\mathbf{N}$, and $\kappa = 0, 1, 2, \ldots$. Note that quantum states with $\kappa < \Lambda$ do not occur.

The curved arrow shown in Figure 13.12a indicates the rotation of the whole molecule about $\mathbf{K}$. In our previous discussions of the rotational energy of the molecule, we considered only the angular momentum $\mathbf{N}$ of the nuclei (see equations 7 and 8). We now obtain the rotational energy for the whole molecule, and we start with the classical rotational energy.

The classical kinetic energy of rotation of a rigid body is given by

$$E = \tfrac{1}{2}I_x\omega_x^2 + \tfrac{1}{2}I_y\omega_y^2 + \tfrac{1}{2}I_z\omega_z^2$$

$$= \frac{P_x^2}{2I_x} + \frac{P_y^2}{2I_y} + \frac{P_z^2}{2I_z} \quad \text{(3D rigid rotor)} \quad \text{(J)} \tag{25}$$

where x, y, z are the directions of the principal axes; I_x, ω_x, P_x are the moment of inertia, angular velocity, and angular momentum, respectively, about the x axis and similarly for

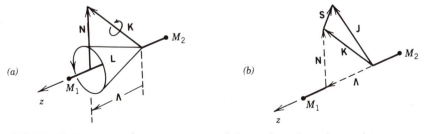

FIGURE 13.12 (a) Angular momenta $\mathbf{N}$ and $\mathbf{\Lambda}$ couple to form the resultant $\mathbf{K}$. (b) $\mathbf{S}$ and $\mathbf{K}$ couple to form $\mathbf{J}$, the total angular momentum of the molecule.

other axes. For the diatomic molecule $P_z^2 = \Lambda^2$, $P_x^2 + P_y^2 = N^2 = K^2 - \Lambda^2$, $I_x = I_y = \mu R^2$, and I_z is the moment of inertia of the electrons about the internuclear axis. Hence, equation 25 may be written

$$E = \frac{K^2 - \Lambda^2}{2\mu R^2} + \frac{\Lambda^2}{2I_z} \quad \text{(J)} \tag{26}$$

Using quantized angular momenta, $K^2 = \kappa(\kappa + 1)\hbar^2$ and $\Lambda^2 = \Lambda^2 \hbar^2$, we find that the quantized rotational energy is

$$E = \frac{\hbar^2}{2\mu R^2} \kappa(\kappa + 1) + \left(\frac{\hbar^2}{2I_z} - \frac{\hbar^2}{2\mu R^2}\right)\Lambda^2 \quad \text{(3D rigid rotor)} \quad \text{(J)} \tag{27}$$

Compare equations 27 and 3.

It was previously shown that the energy for the nonrigid rotor equals the energy of the rigid rotor plus the term $-\hbar^4 r^2(r + 1)^2/2\mu^2 R^6 k$ (see equation 8). We obtain the energy of the 3D nonrigid rotor in a similar way:

$$\begin{aligned} E_\kappa &= \frac{\hbar^2}{2\mu R^2} \kappa(\kappa + 1) + \left(\frac{\hbar^2}{2I_z} - \frac{\hbar^2}{2\mu R^2}\right)\Lambda^2 - \frac{\hbar^4}{2\mu^2 R^6 k}\kappa^2(\kappa + 1)^2 \\ &\equiv B_v \kappa(\kappa + 1) + (A - B_v)\Lambda^2 - D_v \kappa^2(\kappa + 1)^2 \end{aligned}$$

$$\text{(3D nonrigid rotor in vibrational state } v \text{ and electronic state } \Lambda) \quad \text{(J)} \tag{28}$$

where, as for the nonrigid rotor, the defined rotational constants, B_v and D_v, depend on the vibrational quantum number v (see equation 17). For given electronic and vibrational states both Λ and v are constants and, hence, the rotational energy depends on κ. You should compare equations 17 and 28. Note that the quantum number κ has replaced the quantum number r.

I_z is much smaller than μR^2 because the electron mass is much less than the nuclear mass; hence, A is much larger than B_v. I_z depends on the distance of the electrons from the z axis and, hence, it depends on the electronic state occupied by the electrons. Thus, for a given electronic state both Λ and A are constant. Therefore, for given electronic and vibrational states, the rotational energy levels of the molecule, equation 28, are the same as those given by equation 17 except that there is a shift of magnitude $(A - B_v)\Lambda^2$, which is constant for given electronic and vibrational states, except that according to equation 24 levels with $\kappa < \Lambda$ are absent. Henceforth, we will drop the term $(A - B_v)\Lambda^2$. As an example, a rotational energy level diagram for $\Lambda = 2$ is shown in Figure 13.13. The rotational levels that do not occur ($\kappa < \Lambda$) are indicated by dashed lines.

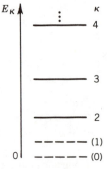

FIGURE 13.13 Rotational energy levels for $\Lambda = 2$. Levels indicated by dashed lines do not exist.

The energy $E_{\kappa v}$ for the vibrating–rotating molecule is the sum of equations 28 and 13:

$$E_{\kappa v} = B_v \kappa(\kappa + 1) - D_v \kappa^2(\kappa + 1)^2$$
$$+ hc\omega_e(v + \tfrac{1}{2}) - hc\omega_e x_e(v + \tfrac{1}{2})^2 \qquad \text{(vibrating–rotating molecule)} \qquad \text{(J)} \quad (29)$$

where $v = 0, 1, 2, \ldots$; $\kappa = \Lambda, \Lambda + 1, \Lambda + 2, \ldots$, and the small, higher order terms in equation 13 have been dropped. You should compare equations 18 and 29. The selection rules for electric dipole transitions between the energy levels given by equation 29 are

$$\Delta\kappa = 0, \pm 1 \qquad \Delta v = \pm 1, \pm 2, \pm 3, \ldots \qquad \text{(vibrating–rotating molecule)} \quad (30)$$

with the following restrictions: transitions between $\kappa = 0$ and $\kappa = 0$ levels are forbidden and transitions between $\Lambda = 0$ and $\Lambda = 0$ are permitted only if $\Delta\kappa \neq 0$. When we compare equation 30 with equation 19, we see that Δv has not changed, but $\Delta\kappa$ may be 0 or ± 1, where Δr could only be ± 1. The frequencies of the R and P branches are given by equations 21 and 22, provided that we replace r with κ and let κ take the values $\Lambda, \Lambda + 1, \Lambda + 2, \ldots$ and $\Lambda + 1, \Lambda + 2, \ldots$ for the R and P branches, respectively, where for both branches κ is the final rotational quantum number. We may now have an additional branch called, the **Q branch** corresponding to $\Delta\kappa = 0$. Consider $\Delta\kappa = \kappa' - \kappa'' = 0$ or $\kappa' = \kappa''$; then if we replace r' and r'' with κ' and κ'', equation 20 may be written

$$\nu_Q = \nu_{\text{vib}}(v' - v'') + \frac{B_v' - B_v''}{h}\kappa(\kappa + 1) \qquad \kappa = \Lambda, \Lambda + 1, \Lambda + 2, \ldots \qquad \text{(Hz)} \quad (31)$$

where κ'' has been replaced with κ. Note from equation 30 that if the electronic transition is from a state having $\Lambda = 0$ to another state having $\Lambda = 0$, then $\Delta\kappa \neq 0$; hence, the Q branch will not exist.

The total spin angular momentum **S** of the molecule remains to be considered. **S** and the angular momentum **K** couple to form the total angular momentum **J** of the molecule. This is shown in Figure 13.12b, where **S** points in an arbitrary direction. The magnitude of **J** is $[j(j + 1)]^{1/2}\hbar$, where according to the rules of coupling angular momenta, the quantum number j is given in terms of the quantum numbers κ and s by

$$j = |\kappa - s|, |\kappa - s + 1|, |\kappa - s + 2|, \ldots, \kappa + s - 1, \kappa + s \qquad (32)$$

where the magnitude of **S** is $\sqrt{s(s + 1)}\hbar$.

In general (except when $\kappa < s$), the interaction between **S** and **K** splits each level with a given κ into $2s + 1$ levels. For example, if $\kappa = 2$ and $s = 1$, then $j = 1, 2, 3$ and the $\kappa = 2$ level splits into three levels. Figure 13.14a shows the split rotational levels for $s = \tfrac{1}{2}$, and

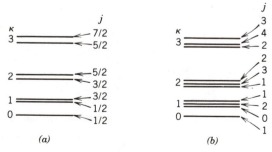

FIGURE 13.14 Splitting of rotational levels for (a) $s = \tfrac{1}{2}$ and (b) $s = 1$.

Figure 13.14b shows the splitting for $s = 1$. The splitting shown in Figure 13.14a and b is drawn to a much larger scale than the separation of levels with different κ. For the purposes of this experiment the small splitting of the rotational levels due to the interaction between **S** and **K** can be ignored.

The total energy E of the molecule is the sum of the electronic energy E_e, which is an eigenvalue of the Hamiltonian given by equation 23, and the energy $E_{\kappa v}$ given by equation 29:

$$E = E_e + E_{\kappa v} \qquad (\text{J}) \tag{33}$$

A partial level diagram of the N_2 molecule is shown in Figure 13.15. The units are those used in spectroscopy, namely, inverse centimeters (cm^{-1}) or wave number. (Wave number is discussed in Experiment 15.) The states shown in the diagram are Σ, Π, or Δ states; that is, $\Lambda = 0$, 1, or 2. The upper left numerical superscript on the state designation is $2s + 1$, where s is the electron spin quantum number, and it is called the multiplicity of the state. The ground state and the excited states on the right side of Figure 13.15 have $s = 0$ and, hence, they are called spin singlet states. The excited states on the left side have $s = 1$, and they are called spin triplet states. The letters X, A, B, C, D, E and a, w, x, y, z have no quantum significance; they are simply used to distinguish quantum states. X is frequently used for the ground state of the molecule. The Σ states have a plus or minus sign superscript, where the sign implies that the electronic wave function does not or does change sign under reflection in a plane perpendicular to the internuclear axis. The subscripts g and u (from the German *gerade* and *ungerade*) imply that the electronic wave function does or does not change sign under inversion.

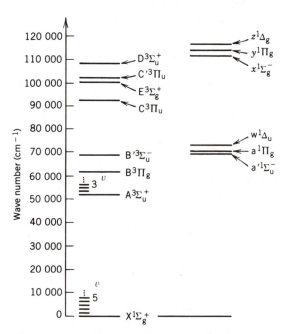

FIGURE 13.15 Electronic energy levels of N_2. Each electronic level splits into a series of vibrational levels, which is shown for the ground and first excited levels only. The ground state is a singlet state ($s = 0$) and the excited states to the left are triplet states ($s = 1$), while those to the right are singlet states.

Each electronic level splits into a series of vibrational levels. This splitting is shown in Figure 13.15 for the ground and first excited states only. Also, each vibrational level splits into a series of rotational levels (see Figure 13.8). It was indicated previously that the general type of transition between molecular energy levels is one in which changes occur in the electronic state of the molecule as well as in the vibration and rotation states. We now consider general transitions between molecular levels.

We label the electronic energy levels with the quantum numbers Λ and s; that is, we now write the electronic energy as $E_{\Lambda s}$ instead of E_e, and, hence, the total energy is $E_{\Lambda s} + E_{\kappa v}$. Consider transitions that result in the emission spectral lines of a single band. Let the initial and final states have quantum numbers $\Lambda' s' \kappa' v'$ and $\Lambda'' s'' \kappa'' v''$, respectively. The frequency of the emitted radiation is given by

$$v = \frac{1}{h}(E'_{\Lambda s} - E''_{\Lambda s} + E'_{\kappa v} - E''_{\kappa v}) \qquad \text{(Hz)} \qquad (34)$$

where $E_{\kappa v}$ is given by equation 29.

The selection rules for the emission and absorption of electric dipole radiation having frequency v, given by equation 34, are

Rotation–vibration selection rules:

$$\Delta\kappa = 0, \pm 1 \qquad \text{but } \kappa = 0 \nleftrightarrow \kappa = 0 \text{ and for } \Lambda = 0 \to \Lambda = 0, \Delta\kappa \neq 0$$

$$\Delta v = 0, \pm 1, \pm 2, \pm 3, \ldots$$

Electronic selection rules:

$$\Delta\Lambda = 0, \pm 1$$

$$\Delta s = 0 \qquad (35)$$

$$g \leftrightarrow u, g \nleftrightarrow g, u \nleftrightarrow u$$

$$\Sigma^+ \leftrightarrow \Sigma^+, \Sigma^- \leftrightarrow \Sigma^-, \Sigma^+ \nleftrightarrow \Sigma^-$$

where $\leftrightarrow$ and $\nleftrightarrow$ mean the transition is allowed and forbidden, respectively. Note that there is now no strict selection rule for the vibrational quantum number v. In principle, transitions between each vibrational level of the upper electronic state to each vibrational level of the lower electronic state may occur.

EXERCISE 8

By applying the electronic selection rules to the energy level diagram in Figure 13.15, determine all of the allowed electronic transitions. Which transition or transitions would have a forbidden Q branch?

EXERCISE 9

The transition $C^3\Pi_u \to B^3\Pi_g$ occurs very readily in a gas discharge tube. By using Figure 13.15, calculate an approximate frequency of the emitted radiation for this transition.

It was previously pointed out that the molecular potential energy $V(R)$ depends on the electronic state; that is, there is a distinct molecular potential energy for each electronic state. In Figure 13.16, $V(R)$ versus R is drawn for two electronic states. Each electronic state

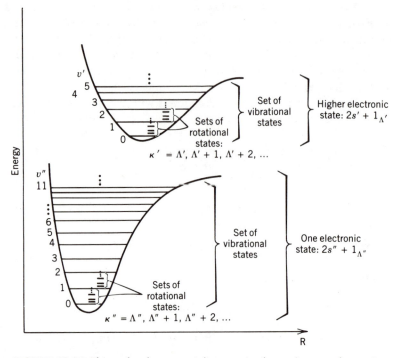

FIGURE 13.16 The molecular potential energy is shown for two electronic states. Each electronic state is split into a set of vibrational states, and each vibrational state is split into a set of rotational states.

has its own set of vibrational levels, and each vibrational level has its own set of rotational levels.

Consider Figure 13.9 again. If v' is a vibrational level in one electronic state and v'' is a vibrational level in a lower electronic state, then the frequency of the emitted radiation is in the visible or ultraviolet (UV). This case is a general molecular transition and the frequency is given by equation 34. Figure 13.9b shows a single band composed of two branches (a third branch, the Q branch, may also exist). Now a single band corresponds to a particular value of Δv, and since $\Delta v = v' - v'' = \pm 1, \pm 2, \pm 3, \ldots$ there can be many such bands. We now examine the system of bands that results from the transitions between two electronic states.

To get the general picture, we first ignore the rotational structure by setting $\kappa' = 0$ and $\kappa'' = 0$. When we define $h\nu_e \equiv E'_{\Lambda s} - E''_{\Lambda s}$ and use equation 29, then we may write equation 34 in terms of frequency as

$$\nu = \nu_e + [c\omega'_e(v' + \tfrac{1}{2}) - c\omega'_e x'_e(v' + \tfrac{1}{2})^2]$$
$$- [c\omega''_e(v'' + \tfrac{1}{2}) - c\omega''_e x''_e(v'' + \tfrac{1}{2})^2] \quad \text{(Hz)} \quad (36)$$

where the prime and double prime on the vibrational constants imply they depend on the electronic states, $^{2s'+1}\Lambda'$ and $^{2s''+1}\Lambda''$.

In terms of wave number $\bar{\nu}$, equation 36 becomes

$$\bar{\nu} = \bar{\nu}_e + [\omega'_e(v' + \tfrac{1}{2}) - \omega'_e x'_e(v' + \tfrac{1}{2})^2]$$
$$- [\omega''_e(v'' + \tfrac{1}{2}) - \omega''_e x''_e(v'' + \tfrac{1}{2})^2] \quad \text{(cm}^{-1}) \quad (37)$$

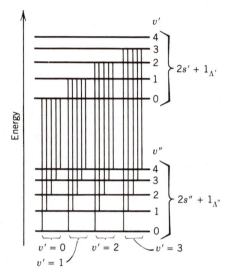

FIGURE 13.17 Two electronic states, each split into a set of vibrational states, are shown. If transitions between rotational levels were included, then the 20 transitions shown would give rise to 20 bands.

where $\Delta v = v' - v'' = 0, \pm 1, \pm 2, \dots$. Equation 37 is the wave number $\bar{v}$ of the $v'-v''$ band origin (see Figure 13.9b). Equation 37 represents all possible transitions between the different vibrational levels of the two participating electronic states.

Consider the transitions shown in Figure 13.17, where each transition having the same value of v' are grouped together. In each group, $v'' = 0, 1, 2, 3, 4$ where the v'-0 band has the shortest wavelength and the v'-4 band has the longest wavelength; that is, these v'' progressions extend from the first band with $v'' = 0$ toward longer wavelengths with increasing v''. The 20 transitions shown in Figure 13.17 correspond to 20 band origins. Note that if we included transitions between the rotational levels, then there would be 20 bands. Hence, the totality of transitions between two different electronic states of a molecule is a system of bands.

Consider the transition $v' = 0 \rightarrow v'' = 0$ (the 0-0 band). The wave number $\bar{v}_{00}$ of this band origin is

$$\bar{v}_{00} = \bar{v}_e + (\tfrac{1}{2}\omega'_e - \tfrac{1}{4}\omega'_e x'_e) - (\tfrac{1}{2}\omega''_e - \tfrac{1}{4}\omega''_e x''_e) \qquad (\text{cm}^{-1}) \tag{38}$$

Note that the terms in parentheses on the right of equation 38 represent the zero-point vibrational energies in the upper and lower states. (Identify this transition on the energy diagram in Figure 13.16.) The wave number $\bar{v}_{00}$ is called the **origin of the bands**. Figure 13.18 shows the spectrum for eight of the band origins. (The band origins are shown as dashed lines because they are missing from each band.) Note that wavelength increases to the right in this figure. The wavelength spacing between consecutive band origin lines in Figure 13.18

FIGURE 13.18 Spectrum for eight band origins.

assumes that $\omega_e' > \omega_e''$, that is, that the energy separation of consecutive vibrational levels in the upper electronic state is larger than the energy separation between the corresponding consecutive vibrational levels in the lower electronic level.

EXERCISE 10

Write out the equation for $\bar{v}_{v'0}$ and $\bar{v}_{0v''}$, where both v' and v'' have the values 1, 2, and 3, and then show that $\bar{v}_{03} < \bar{v}_{02} < \bar{v}_{01} < \bar{v}_{00} < \bar{v}_{10} < \bar{v}_{20} < \bar{v}_{30}$.

An analysis of the band origin spectrum gives the wave number of each band origin spectral line as an empirically determined constant. Knowing these wave numbers, we can calculate vibrational constants for both the upper electronic state (ω_e' and x_e') and the lower electronic state (ω_e'' and x_e''). Once we know the vibrational constants we can calculate the following:

1. The set of vibrational energies for both electronic states.
2. The dissociation energy, and, hence, the Morse potential, for both electronic states.
3. hv_e, the difference in electronic energy of the two states.
4. The force constant k of the molecule for both electronic states.

EXERCISE 11

Using equation 37 write out the equations for the following wave number differences: $\bar{v}_{00} - \bar{v}_{01}$ and $\bar{v}_{10} - \bar{v}_{12}$. Solve these two equations for ω_e'' and x_e'' in terms of the wave number differences. Do a similar calculation for the wave number differences $\bar{v}_{00} - \bar{v}_{10}$ and $\bar{v}_{10} - \bar{v}_{20}$, and then solve for ω_e' and x_e'.

Note that $\bar{v}_e$ can be calculated from equation 38. The dissociation energies and the Morse potential energies can be calculated from equations 14 and 15. Finally, the molecular potential energy of both electronic states is approximately harmonic for the first few vibrational levels, and, hence, $c\omega_e \simeq v_{vib} = (\sqrt{k/\mu})/2\pi$. This equation can be solved for k in terms of ω_e.

The bands of N_2 are summarized on pages 209–218 of reference 3. Plate 2 of reference 3 includes a spectrogram of N_2 with a label of "2nd positive." The label 2nd positive means the electronic transition is $C^3\Pi_u \rightarrow B^3\Pi_g$. The spectrogram shows many bands corresponding to different values of Δv.

EXERCISE 12

In doing Exercises 2 and 9 you calculated the frequency spacing Δv and the frequency v. Use these results to show that the wavelength spacing between consecutive spectral lines in a given band is approximately $\Delta \lambda = 0.45$ Å.

We now include the rotational structure in our discussion of bands. Previously, we specified the quantum numbers of the upper and lower states as $\Lambda's'\kappa'v'$ and $\Lambda''s''\kappa''v''$, respectively. Including the rotational energy levels, the frequency of the emitted radiation is given by

$$v = \frac{1}{h}(E'_{\Lambda s} - E''_{\Lambda s} + E'_{\kappa v} - E''_{\kappa v})$$

$$= v_e + [c\omega'_e(v' + \tfrac{1}{2}) - c\omega'_e x'_e(v' + \tfrac{1}{2})^2]$$

$$\quad - [c\omega''_e(v'' + \tfrac{1}{2}) - c\omega''_e x''_e(v'' + \tfrac{1}{2})^2]$$

$$\quad + \frac{1}{h}[B'_v \kappa'(\kappa' + 1) - D'_v \kappa'^2(\kappa' + 1)^2]$$

$$\quad - \frac{1}{h}[B''_v \kappa''(\kappa'' + 1) - D''_v \kappa''^2(\kappa'' + 1)^2] \qquad \text{(Hz)} \qquad (39)$$

For a given vibrational transition, v_e and the first two brackets are constants, and we defined the sum to be v_0. Using this definition and writing (39) in terms of wave number, we have

$$\bar{v} = \bar{v}_0 + [\bar{B}'_v \kappa'(\kappa' + 1) - \bar{D}'_v \kappa'^2(\kappa' + 1)^2]$$

$$\quad - [\bar{B}''_v \kappa''(\kappa'' + 1) - \bar{D}''_v \kappa''^2(\kappa'' + 1)^2] \qquad (\text{cm}^{-1}) \qquad (40)$$

where $\bar{v}_0$ is the wave number of the v'-v'' band origin and each barred letter is defined to be the unbarred letter divided by hc, for example, $\bar{B}'_v \equiv B'_v/hc$. For a constant value of $\bar{v}_0$, a given band origin, all of the transitions represented by equation 40 form a single band. The single band will include at least P and R branches, and it may include a Q branch. The wave number for each branch is

$$\bar{v}_P = \bar{v}_0 - (\bar{B}'_v + \bar{B}''_v)\kappa + (\bar{B}'_v - \bar{B}''_v)\kappa^2 \qquad \kappa = \Lambda'' + 1, \Lambda'' + 2, \ldots \quad (\text{cm}^{-1}) \quad (41)$$

$$\bar{v}_Q = \bar{v}_0 + (\bar{B}'_v - \bar{B}''_v)\kappa + (\bar{B}'_v - \bar{B}''_v)\kappa^2 \qquad \kappa = \Lambda'', \Lambda + 1, \ldots \quad (\text{cm}^{-1}) \quad (42)$$

$$\bar{v}_R = \bar{v}_0 + 2\bar{B}'_v + (3\bar{B}'_v - \bar{B}''_v)\kappa + (\bar{B}'_v - \bar{B}''_v)\kappa^2 \qquad \kappa = \Lambda', \Lambda' + 1, \ldots \quad (\text{cm}^{-1}) \quad (43)$$

where $\kappa'' = \kappa$; hence, κ is the rotational quantum number of the final rotational state. (Compare the above equations with equations 21, 22, and 31.) An energy level diagram for a single band with P, Q, and R branches is shown in Figure 13.19. The electronic transition is arbitrarily taken to be the $w^1\Delta_u \rightarrow a^1\Pi_g$ transition of the N_2 molecule (see Figure 13.15). Note that the head of the Q branch lies very close to $\bar{v}_0$. Thus, in a band system in which Q branches appear the spectral lines of the Q heads should be used to determine the vibrational constants.

To determine the rotational constants $\bar{B}'_v$ and $\bar{B}''_v$, we assume the lines of each branch have been assigned the correct value of κ and the wave number of each line determined. Using equations 41–43, the following differences in wave number are readily obtained:

$$\bar{v}_R(\kappa) - \bar{v}_Q(\kappa) = 2\bar{B}'_v(\kappa + 1) \qquad (\text{cm}^{-1}) \qquad (44)$$

$$\bar{v}_Q(\kappa + 1) - \bar{v}_P(\kappa + 1) = 2\bar{B}'_v(\kappa + 1) \qquad (\text{cm}^{-1}) \qquad (45)$$

$$\bar{v}_R(\kappa) - \bar{v}_Q(\kappa + 1) = 2\bar{B}''_v(\kappa + 1) \qquad (\text{cm}^{-1}) \qquad (46)$$

$$\bar{v}_Q(\kappa) - \bar{v}_P(\kappa + 1) = 2\bar{B}''_v(\kappa + 1) \qquad (\text{cm}^{-1}) \qquad (47)$$

You should examine Figure 13.19 and convince yourself that the wave number difference of both equations 44 and 45 equals the separation of two successive rotational levels in the upper set of rotational states, and that the wave number difference of both equations 46 and 47 equals the separation of two successive rotational levels in the lower set of states.

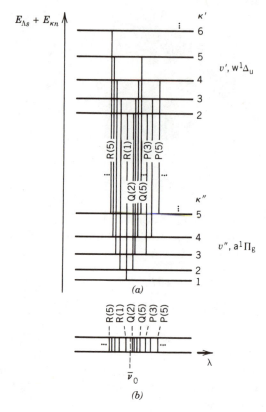

FIGURE 13.19 (a) $w^1\Delta_u$ and $a^1\Pi_g$ are two of the electronic states shown in FIGURE 13.15. v' and v'' specify single vibrational states of each electronic state, and each vibrational state is split into a set of rotational states as shown. (b) Spectrum of a single band with P, Q, and R branches.

Equations 44 and 45 can be used to calculate $\bar{B}_v'$ for several values of κ, and then the mean value and its error can be determined. In a similar way, equations 46 and 47 can be used to determine the mean value of $\bar{B}_v''$.

Spectrograph

The 2-m, Ebert-mount spectrograph is described in Appendix D. The dispersion is derived and the resolution is discussed in some detail. Also the adjustments for the Ebert-mount spectrograph are explained (other spectrographs required similar adjustments). Two methods of determining unknown wavelengths, the dispersion curve method and the iron arc method, are described. You should read Appendix D and answer the questions, even if your spectrograph has a different mounting.

EXPERIMENT

Set the angular position of the grating such that the wave number $\bar{\nu}_{00}$, the origin of the bands of the $C^3\Pi_u \rightarrow B^3\Pi_g$ transition, falls in about the central region of the photographic plate. (You should decide what order of the spectrum you want to photograph.) Using the

20-μm slit expose the film for about one minute. Develop the film and, using one of the two methods described in Appendix D to determine unknown wavelengths, determine the wavelengths of the spectral lines.

You should first identify which $v' \rightarrow v''$ transition corresponds to which band. You will find reference 3 helpful in this identification.

For both the $C^3\Pi_u$ state and the $B^3\Pi_g$ state determine the rotational constants, $\bar{B}'_v$ and $\bar{B}''_v$, and the vibrational constants, ω'_e, x'_e, ω''_e, and x''_e, and then calculate the following:

1. The average internuclear separation
2. The vibrational energies
3. Dissociation energy
4. The Morse potential
5. Force constant k

Also calculate $h v_e$, the difference in electronic energy of the two states. Draw the Morse potential for both states and indicate the vibrational energies.

14. ZEEMAN EFFECT: OPTICAL SPECTROSCOPY

Historical Note

The 1902 Nobel prize in Physics was awarded jointly to
   Hendrik Antoon Lorentz, the Netherlands, and Pieter Zeeman, the Netherlands
   In recognition of the extraordinary service they rendered by their researches into the influence of magnetism upon radiation phenomena.

APPARATUS [Optional Equipment in Brackets]

^{198}Hg, Na, or He discharge lamp (preferably a Geissler tube)
Laboratory electromagnet [with tapered pole pieces]
Polarizer
Two converging lenses
Ebert spectrograph with photographic plates **or** Fabry–Perot etalon (mounted in vacuum can)
Constant-deviation prism spectrometer
Photomultiplier
Electrometer
Chart recorder

OBJECTIVES

To observe the Zeeman components for one or more spectral lines, measure the splittings, and compare the results with the predictions of theory.

To understand the physical origin of the Zeeman effect.

To be able to predict, for the case of weak external field and *LS* coupling, the number, relative intensities, polarizations, and splittings of the various components of a Zeeman multiplet.

KEY CONCEPTS

Multiplet	Zeeman energy
LS coupling	Polarization
Magnetic moment	*g* factor
Bohr magneton	Hamiltonian
Selection rules	Dipole matrix element
Fabry–Perot spectroscopy	Free spectral range
Finesse	Chromatic resolution

REFERENCES

1. R. Eisberg and R. Resnick, *Quantum Physics of Atoms, Molecules, Solids, Nuclei and Particles*, 2d ed., Wiley, New York, 1985. *LS* coupling and Zeeman effect are discussed in Chapter 10. Magnetic moments, the spin–orbit interaction, and transition matrix elements are treated in Chapter 8.

2. M. Nayfeh and M. Brussel, *Electricity and Magnetism*, Wiley, New York, 1985. Section 8.8 treats magnetic moments and orientational potential energy; the dipole radiation field is discussed in Chapter 15.

3. C. Candler, *Atomic Spectra*, 2d ed., Van Nostrand, Princeton, NJ, 1964. Chapters 6 and 16 treat the spectral line intensity rules for the Zeeman effect.

4. E. U. Condon and G. H. Shortley, *The Theory of Atomic Spectra*, Cambridge University Press, Cambridge, England, 1959. Quantum mechanical discussion of the Landé *g* factor (Chapter V) and of the spectral intensities (Chapter XVI). A classic reference.

5. B. Cagnac and J.-C. Pebay-Peyroula, *Modern Atomic Physics: Fundamental Principles*, Wiley, New York, 1975. Chapter 8 contains a development of the classical theory of the Zeeman effect that accounts for the observed polarizations.

6. E. Hecht, *Optics*, 2d ed., Addison-Wesley, Reading, MA, 1987. Section 9.6 discusses multiple beam interference devices, the Fabry–Perot interferometer in particular. Constant-deviation prisms are discussed briefly in Section 5.5.

7. G. Hernandez, *Fabry–Perot Interferometers*, Cambridge University Press, Cambridge, England, 1986. A detailed reference.

8. W. A. Hilton, *Am. J. Phys.* **30**, 724 (1962). Describes the construction and use of an inexpensive Fabry–Perot etalon.

9. S. Pollack and E. Wong, *Am. J. Phys.* **39**, 1387 (1971). Discusses light sources for the observation of the Zeeman effect. Since the writing of this article, Na Geissler tubes have become commercially available.

10. C. Manka and K. Mittelstaedt, *Am. J. Phys.* **41**, 287 (1973). Discusses the observation of the Zeeman effect for the 4046- and 5461-Å Hg lines using an Ebert spectrograph.

11. *Nobel Lectures: Physics*, Vol. 1, Elsevier, Amsterdam, 1967. On page 33 is a narrative account by Zeeman describing his investigations.

INTRODUCTION

When an atom makes a transition from an initial state to a final state of lower energy, a photon is emitted with an energy equal to the difference in energy of these states: A single spectral line is observed that corresponds to this transition. If either of the two states involved in the transition has a magnetic moment, then the application of an external magnetic field **B** causes the spectral line to split into several closely spaced components, the extent of the splitting being proportional to B. In this experiment you will try to understand and observe this splitting, known as the Zeeman effect. In what follows, we first treat the relatively straightforward case of the Zeeman effect in "one-electron atoms" for large external magnetic fields; we then move to the more practical, but slightly more complex, case of small fields. Finally, we discuss the situation for multielectron atoms.

One-Electron Atoms

For many atoms, (e.g., Na) the optical spectrum is associated with transitions between states of a single electron outside a closed inner shell. This single *optically active* electron may have both an orbital and a spin magnetic moment. We can calculate the magnetic moment μ_ℓ associated with the orbital motion of this electron by use of a classical model in which it orbits in a circle of radius r with speed v, as in Figure 14.1a. The magnetic moment μ is defined for any charge distribution as

$$\mu \equiv \frac{1}{2} \int \mathbf{r} \times \mathbf{v}\, dq \qquad (\text{C} \cdot \text{m}^2/\text{s}) \qquad (1)$$

Evaluating this for an electron in a circular orbit, in terms of its angular momentum $\mathbf{L} \equiv \mathbf{r} \times m\mathbf{v}$, yields for μ_ℓ

$$\mu_\ell = -\frac{e}{2m} \mathbf{L} \qquad (\text{C} \cdot \text{m}^2/\text{s}) \qquad (2)$$

where the minus sign signifies that, for an electron, μ_ℓ is always opposed to **L**. This result is conventionally written in terms of g_ℓ, called the **orbital g factor**, and $\mu_0 \equiv e\hbar/2m$, called the **Bohr magneton**:

$$\mu_\ell = -\frac{g_\ell \mu_0}{\hbar} \mathbf{L} \qquad (\text{C} \cdot \text{m}^2/\text{s}) \qquad (3)$$

where $g_\ell = 1$. If an external magnetic field **B** is applied, then, according to classical mechanics, μ_ℓ undergoes Larmor precession about **B** with angular frequency $\omega = g_\ell \mu_0 B/\hbar$, thus altering the motion of the electron, as discussed in reference 5. The total energy of the electron now depends on the relative orientation of μ_ℓ and **B** because the presence of the field contributes a magnetic (Zeeman) energy

$$E_Z^\ell = -\mu_\ell \cdot \mathbf{B} \qquad (\text{J}) \qquad (4)$$

Different orientations of **L**, as depicted in Figure 14.1b, thus correspond to different values of E_Z^ℓ.

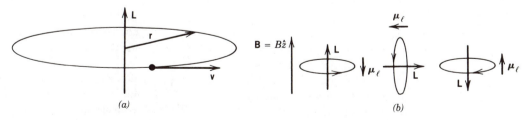

FIGURE 14.1 (a) Classical picture of an electron orbiting a nucleus. (b) Different orientations for the orbital angular momentum.

Since the electron also possesses a spin angular momentum **S**, there is an additional contribution to the magnetic moment of the atom, μ_s, given by analogy with equation 3 as

$$\mu_s = -\frac{g_s \mu_0}{\hbar} \mathbf{S} \qquad (\text{C} \cdot \text{m}^2/\text{s}) \tag{5}$$

with a corresponding contribution to the energy of

$$E_Z^s = -\mu_s \cdot \mathbf{B} \qquad (\text{J}) \tag{6}$$

The total magnetic (Zeeman) energy of the atom is thus

$$E_Z = -(\mu_\ell + \mu_s) \cdot \mathbf{B} \qquad (\text{J}) \tag{7}$$

It must be noted that in the expression for μ_s, the **spin g factor** g_s is approximately 2, rather than 1, as for g_ℓ, a result of relativistic quantum mechanics. Precise spectroscopic measurements yield a value of $g_s = 2.00232$, as predicted by the more refined theory of quantum electrodynamics.

In the quantum-mechanical description of the one-electron atom, the state of the electron is described approximately, in the *absence* of any external **B**, by the wave function $\psi_{n,\ell,m_\ell,m_s}(r, \theta, \phi)$, which is an eigenfunction of the approximate Hamiltonian

$$H_0 = \frac{p^2}{2m} + V(r) \qquad (\text{J}) \tag{8}$$

where V is the Coulomb potential energy of the electron, assumed spherically symmetric, and where we have ignored, for the moment, the spin–orbit interaction between the moving electron spin and the nucleus (discussed in Experiment 12 and briefly below). The four subscripts for ψ are the quantum numbers that describe the electron state. The positive integer n is the principal quantum number. The integer ℓ indexes the magnitude L of the orbital angular momentum **L** and can take on values from 0 through $n - 1$; L is then given as $[\ell(\ell + 1)]^{1/2}\hbar$. The integer m_ℓ ranges from $-\ell$ to $+\ell$ and indexes the z component (or the component along any fixed direction) of **L**:

$$L_z = m_\ell \hbar \qquad (\text{J} \cdot \text{s}) \tag{9}$$

This number then specifies the orientation of the orbit mentioned above in connection with Figure 14.1b. The electron has a spin quantum number s which is not included in the subscript because it is always understood to be $\frac{1}{2}$; the magnitude of the spin angular momentum **S** is then given as $S = [s(s + 1)]^{1/2}\hbar$ in analogy with L. The fourth quantum number m_s then gives the orientation of the spin (up or down) by specifying its z component:

$$S_z = m_s \hbar \qquad (\text{J} \cdot \text{s}) \tag{10}$$

where m_s is either $+\frac{1}{2}$ or $-\frac{1}{2}$. The energy of the state corresponding to ψ now depends on n and ℓ but, since there is no magnetic field, is independent of m_ℓ and m_s, which determine the orientation of the moments μ_ℓ and μ_s.

We now go the high-field limit by applying a magnetic field along $\hat{z}$ that is so strong that the spin–orbit correction to the Hamiltonian H_0 of equation 8 is negligible by comparison with the Zeeman interaction. The Hamiltonian for the system is then approximated as $H \simeq H_0 + H_Z$, where H_Z is the Zeeman interaction Hamiltonian represented as the operator equivalent of equation 7. The energy E is now calculated using the time-independent

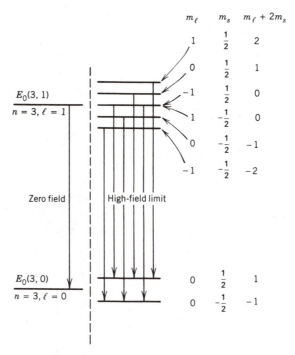

FIGURE 14.2 Effect of a very strong magnetic field on two atomic states.

Schrödinger equation, $H\psi = E\psi$, to give

$$E = E_0(n, \ell) + E_Z = E_0(n, \ell) + \mu_0 B(m_\ell + 2m_s) \quad \text{(J)} \quad (11)$$

where E_0 is the energy corresponding to H_0, the zero-field Hamiltonian, and the values $g_\ell = 1$ and $g_s = 2$ have been used. Figure 14.2 shows the effect of a very strong field on two states, one with $n = 3$, $\ell = 0$, and the other with $n = 3$, $\ell = 1$. Each of these levels, degenerate for $\mathbf{B} = 0$, splits in the presence of $\mathbf{B}$, as indicated to the right of the vertical dashed line; the extent of the splitting, since we have ignored the spin–orbit interaction, depends only on the sum $m_\ell + 2m_s$. Note that while only one transition (spectral line) is possible if $\mathbf{B} = 0$, the application of $\mathbf{B}$ produces a system of lines corresponding to the indicated transitions. For dipole radiation, only transitions obeying the selection rules $\Delta m_s = 0$, $\Delta m_\ell = 0$, ± 1 are observed (the "allowed" transitions). These **selection rules** are based on the evaluation of the dipole matrix elements between initial and final states, and are discussed in Chapters 8 and 10 of reference 1.

EXERCISE 1

Estimate the energy splittings, in electron-volts, of the spectral lines corresponding to the transitions shown in Figure 14.2, as predicted by equation 11 for the high-field limit. Assume a magnetic field of $B \sim 10^6$ G (i.e., 10^2 T). What additional information would you need to determine the separation in wavelength for these lines?

Although the above calculation for the Zeeman splittings in the high-field limit is straightforward conceptually, magnetic fields appropriate to this limit are beyond the capabilities of the electromagnets found in most undergraduate laboratories. In this experiment, the fields used are more appropriate to the weak-field limit, in which the effect of the spin–orbit interaction on the electronic states must be explicitly considered.

The spin–orbit interaction can be viewed as the interaction of the spin magnetic moment $\boldsymbol{\mu}_s$ of the electron with the internal magnetic field attributed to the motion of the positively charged nucleus as viewed in the rest frame of the orbiting electron. Since this internal field can be shown to be proportional to $\mathbf{L}$, the form of the interaction can be written as $H_{so} = \zeta(r)\mathbf{L}\cdot\mathbf{S}$, where $\zeta(r)$ is a function of the distance of the electron from the nucleus. The electron states in zero external field are now determined from a Hamiltonian $H' = H_0 + H_{so}$:

$$H' = \frac{p^2}{2m} + V(r) + \zeta(r)\mathbf{L}\cdot\mathbf{S} \qquad \text{(J)} \qquad (8')$$

The states ψ' are determined as eigenfunctions of H' with energies E', and the Zeeman energy E_Z, now assumed very small compared with the spin–orbit energy contribution E_{so}, is then calculated as a perturbation, using these states.

EXERCISE 2

The spin–orbit interaction may be written as $H_{so} = -\boldsymbol{\mu}_s \cdot \mathbf{B}_i$, where $\mathbf{B}_i$ represents the (spatially varying) internal magnetic field "seen" by the spinning electron due to its orbital motion with respect to the positively charged nucleus. Estimate the average magnitude of this internal field for an atom in which the average value of E_{so} is roughly 2×10^{-4} eV.

The states ψ' are different from the states ψ determined by H_0 above in that for the former, the angular momenta $\mathbf{L}$ and $\mathbf{S}$, although constant in *magnitude*, are no longer constant in *direction*: the $\mathbf{L}\cdot\mathbf{S}$ interaction causes them to precess about their sum, the *total* angular momentum $\mathbf{J} = \mathbf{L} + \mathbf{S}$, which is strictly constant. Thus, ℓ and s are still good quantum numbers, but m_ℓ and m_s are not; replacing these last two are j and m_j, which index the magnitude of $\mathbf{J}$ and its z component J_z, respectively. According to the quantum rules for the addition of angular momenta, j can assume any value from $|\ell - s|$ to $|\ell + s|$ in steps of 1, depending on the relative orientation of $\mathbf{L}$ and $\mathbf{S}$, as illustrated in Figure 14.3a. By analogy with these, the magnitude of $\mathbf{J}$ is given in terms of its index j as $J = [j(j+1)]^{1/2}\hbar$, while J_z is given in terms of m_j as

$$J_z = m_j\hbar \qquad \text{(J}\cdot\text{s)} \qquad (12)$$

Figure 14.3b shows the relationship between $\mathbf{J}$, $\mathbf{L}$, and $\mathbf{S}$ for an electron in the particular

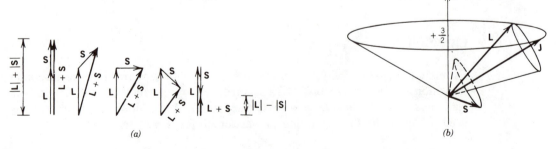

FIGURE 14.3 (a) Addition of angular momenta. (b) The relationship between $\mathbf{J}$, $\mathbf{L}$, and $\mathbf{S}$ for the case $\ell = 2$, $j = \frac{5}{2}$, $m_j = \frac{3}{2}$.

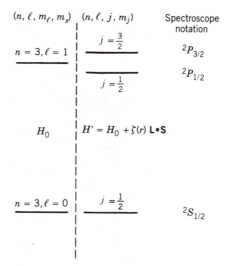

FIGURE 14.4 The effect of the spin–orbit interaction on $n = 3$ states.

case $\ell = 2$, $j = \frac{5}{2}$, $m_j = \frac{3}{2}$; note the vector **J** for this state is constant in time, but may lie anywhere in the cone for which $J_z = \frac{3}{2}\hbar$.

The states ψ' may thus be labeled as ψ'_{n,ℓ,j,m_j} and may be considered as arising from the states ψ_{n,ℓ,m_ℓ,m_s} when the **L** $\cdot$ **S** interaction is "turned on." This is depicted in Figure 14.4 for $n = 3$ states: The states (n, ℓ, m_ℓ, m_s), eigenfunctions of H_0, are drawn to the left of the vertical dashed line, while those of H' (n, ℓ, j, m_j) are to the right. Note that the $\ell = 1$ states are split into a *multiplet* of states, one for each value of j, by the spin–orbit interaction; the magnitude of the energy splitting is determined by the difference in the values of E_{so} calcuated for each state. States with different values of m_j, but the same value of j, have the same energy (are degenerate) in the absence of an external **B** field. The spectroscopic symbol to the right of each level is the conventional, compact notation used to specify the set of quantum numbers s, ℓ, j for each state in the form ${}^{2s+1}\ell_j$, where the capital letters correspond to different values of ℓ: $S = 0$, $P = 1$, $D = 2$, and so on.

EXERCISE 3

Verify that the coupling of ℓ and s for the states of H_0 in Figure 14.4 produces states with the j values indicated and that the spectroscopic notation is appropriate to each state of H'.

We can now calculate the Zeeman splittings of the levels characterized by the set of quantum numbers (n, ℓ, j, m_j) by evaluating E_Z of equation 7 for each of these states. This is a bit more difficult than for the high-field case, since μ_ℓ and μ_s, being proportional to **L** and **S**, respectively, are no longer constant in direction for these states. The calculation can be done by evaluating the matrix elements of the Zeeman Hamiltonian H_Z in operator form, but the vector model of the angular momenta discussed above provides an approach that is physically much more insightful.

Figure 14.5a shows how **L** and **S** (and, hence, μ_ℓ and μ_s) each precess about the constant vector **J** in the absence of an externally applied **B** field. The angular frequency of this precession is proportional to B_i, the strength of the internal field discussed in Exercise 2. If we now apply an external field **B** along the z direction that is much weaker than B_i, then the precession of μ_ℓ and μ_s about **J** is much more rapid than the precession of the total

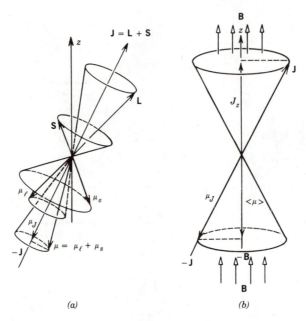

FIGURE 14.5 (a) **L** and **S** precessing about **J**. (b) **J** precessing about **B**.

moment $\mu = \mu_\ell + \mu_s$ about **B**. Since we wish to evaluate the time average of $E_Z = -\mu \cdot \mathbf{B}$, μ can be considered to be oriented, on the average, along $-\mathbf{J}$ as **J** precesses about **B** (shown in Figure 14.5b), even though, instantaneously, μ is never actually antiparallel to **J**. Thus, we need to calculate $\langle \mu_J \rangle$, the average component of μ along **J**; E_Z will then be given by

$$E_Z = -\langle \mu_J \rangle B \cos \theta_{JB} \qquad (\mathrm{J}) \qquad (13)$$

where θ_{JB} is the constant angle that **J** makes with **B**. This average of μ_J will be given by

$$\langle \mu_J \rangle = \mu_\ell \cos \theta_{LJ} + \mu_s \cos \theta_{SJ} \qquad (\mathrm{C \cdot m^2/s}) \qquad (14)$$

where θ_{LJ} and θ_{SJ} are the angles that **L** and **S** make with **J**. The cosines of these three angles can be written as

$$\cos \theta_{JB} = \frac{J_z}{J} \qquad \cos \theta_{LJ} = \frac{\mathbf{L} \cdot \mathbf{J}}{LJ} \qquad \cos \theta_{SJ} = \frac{\mathbf{S} \cdot \mathbf{J}}{SJ} \qquad (15)$$

These can be evaluated in terms of the quantum numbers for the state. The dot product $\mathbf{L} \cdot \mathbf{J}$ can be evaluated by expanding it as $\mathbf{L} \cdot \mathbf{J} = \mathbf{L} \cdot (\mathbf{L} + \mathbf{S}) = L^2 + \mathbf{L} \cdot \mathbf{S}$. For $\mathbf{L} \cdot \mathbf{S}$, note that $J^2 = (\mathbf{L} + \mathbf{S})^2 = L^2 + S^2 + 2\mathbf{L} \cdot \mathbf{S}$, so that $\mathbf{L} \cdot \mathbf{S}$ can be rewritten as $\frac{1}{2}(J^2 - L^2 - S^2)$. So we have $\mathbf{L} \cdot \mathbf{J} = \frac{1}{2}(L^2 + J^2 - S^2)$. A similar procedure for $\mathbf{S} \cdot \mathbf{J}$ yields $\mathbf{S} \cdot \mathbf{J} = \frac{1}{2}(S^2 + J^2 - L^2)$. Equation 14 now gives $\langle \mu_J \rangle$ in terms of μ_ℓ and μ_s, which are expressed in terms of L and S by using equations 3 and 5 with $g_\ell = 1$ and $g_s = 2$. The result for E_Z, from equation 13, is

$$E_Z = \frac{\mu_0}{2\hbar J}(3J^2 + S^2 - L^2)B\frac{J_z}{J} \qquad (\mathrm{J}) \qquad (16)$$

Substituting $m_j\hbar$ for J_z, as in equation 12, and the expressions for J, S, and L in terms of

j, s, and ℓ gives E_Z in the form

$$E_Z = \mu_0 B g_j m_j \qquad \text{(J)} \qquad (17)$$

where g_j, known as the **Landé g factor**, is given by

$$g_j = 1 + \frac{j(j+1) + s(s+1) - \ell(\ell+1)}{2j(j+1)} \qquad (18)$$

The factor g_j is computed for each level in a multiplet, and is a function j, ℓ, and s for each state; the Zeeman energy can then be calculated for each m_j using equation 17. The total energy for each state is thus $E = E'(n, \ell, j) + E_Z$. Note that for the one-electron atom being discussed here, $S = \frac{1}{2}$ for all states.

EXERCISE 4

Explain why, in the weak-field limit, the average, but not the instantaneous, magnetic moment vector is antiparallel to **J**.

EXERCISE 5

Verify the steps leading to equations 17 and 18.

An important example of a one-electron atom is Na, which has a single optically active electron ($n = 3$) outside closed spherically symmetric electron shells. The first two excited states of the electron along with the ground state are shown in Figure 14.6a. The closely spaced yellow D lines, observed for **B** = 0, are accounted for by transitions to the $^2S_{1/2}$ from the P states. If a weak external field is applied, each state is split, as indicated to the right, so that each D line has several components, separated in energy from the zero-field level by E_Z. The D line spectrum, as it would appear on the photographic plate of a grating spectrometer, is shown schematically in Figure 14.6b.

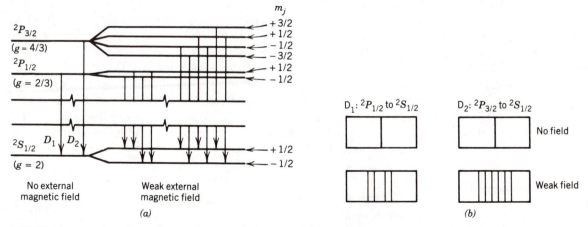

FIGURE 14.6 (a) The effect of a weak external magnetic field on three states of Na. (b) The effect of a weak magnetic field on the appearance of the Na D-line spectrum on a grating spectrogram.

EXERCISE 6

Verify the assignment of g factors indicated by Figure 14.6a to each of the levels. Determine, for each of the Zeeman-split D lines, the shift of the photon energies of the indicated transitions relative to the $B = 0$ values if an electromagnet provides a field of $5\,kG$. Summarize the results in a rough sketch that indicates schematically how the lines would appear on the photographic plate of a grating spectrograph.

Atoms with Several Optical Electrons

The calculation of the Zeeman patterns can be extended to atoms with more than one optically active electron outside closed shells. The one-electron Hamiltonian of equation 8' is not adequate to describe the interactions of these electrons with each other. The potential energy is no longer spherically symmetric. The system must be described by a many-electron Hamiltonian that includes the interactions between the optical electrons, and by wave functions that give information about the entire system of optical electrons. The calculation of the electronic states and the Zeeman patterns for these atoms is, in general, quite complex.

For some atoms of intermediate atomic number (for Hg, e.g.), the spin–orbit interaction is weak enough so that its effect may be considered after the stronger effects of the Coulomb interaction between the outer electrons have been taken into account. The effects of a weak external magnetic field are then considered as a perturbation, so that the treatment is quite analogous to that discussed above for a single optical electron. In this scheme, referred to as **Russell–Saunders** (or *LS*) coupling, the states of the outer electrons are built up by considering the effects of the various interactions in order of their importance. For the simplest case of two outer electrons, the following occurs:

1. These electrons are each assigned quantum numbers n_1, ℓ_1 and n_2, ℓ_2 as if they were individually in states appropriate to the Hamiltonian of equation 8.

2. The relative orientations of the spins are correlated to the spatial motions of the electrons in such a way that states of constant total spin $\mathbf{S}' = \mathbf{S}_1 + \mathbf{S}_2$ are formed, with corresponding quantum number s'. The possibilities for s' are determined from the angular momentum addition rules: s' can have any value from $s_1 + s_2$ down to $|s_1 - s_2|$ in steps of 1. Since $s = \frac{1}{2}$ for electrons, this means, for two electrons, that s' can be 0 or 1. States in which $s' = 0$ are called **singlet** states; $s' = 1$ states are **triplets**. Triplet states are usually lower in energy than singlets.

3. The orbital angular momenta next add to form a total $\mathbf{L}' = \mathbf{L}_1 + \mathbf{L}_2$, so that, in analogy with the spins, the magnitude L' is indexed by the quantum number ℓ' which takes on values from $\ell_1 + \ell_2$ down to $|\ell_1 - \ell_2|$ in steps of 1. The value of ℓ' for a state is designated spectroscopically, as for one electron, by S, P, D, and so on, for $\ell' = 0$, 1, 2, ... States with the lowest value of ℓ' usually have the largest energy.

4. The spin–orbit interaction couples $\mathbf{L}'$ and $\mathbf{S}'$ for a state to form a constant total angular momentum $\mathbf{J}' = \mathbf{L}' + \mathbf{S}'$ as for the single electron. The resulting states of the **multiplet** each have a value of j' that ranges from $\ell' + s'$ down to $|\ell' - s'|$. The spacing between these states of different j' depends on the strength of the spin–orbit interaction; the states with higher j' values have the higher energies.

5. Each state with a given j' corresponds a series of states corresponding to different orientations of $\mathbf{J}$, that is, to different values of $J'_z = m_{j'}\hbar$, where $m_{j'}$ ranges in steps of 1 from $-j'$ to $+j'$. In zero magnetic field, states with different $m_{j'}$ are degenerate; when a weak field is applied, the Zeeman energy shifts are given, as in the one-electron case, by equations 17 and 18 with s', ℓ', j', and $m_{j'}$ substituted for their unprimed counterparts:

$$E_z = \mu_0 B g_{j'} m_{j'} \qquad (J) \tag{17'}$$

$$g_{j'} = 1 + \frac{j'(j'+1) + s'(s'+1) - \ell'(\ell'+1)}{2j'(j'+1)} \tag{18'}$$

Each state built from an initial configuration $(n_1\ell_1, n_2\ell_2)$ is now characterized by the set of quantum numbers $(s', \ell', j', m_{j'})$. The notation $^{2s'+1}\ell_{j'}$ for each term in a multiplet is the same as for the one-electron case.

This coupling scheme is illustrated schematically in Figure 14.7, for the case of zero external field, for three configurations of the two outer electrons of Hg: the ground state $6s^2$ (i.e., $n_1 = n_2 = 6$, $\ell_1 = \ell_2 = 0$), 6s6p, and 6s7s. Note that for the $6s^2$ configuration, only $s' = 0$ is possible, since the Pauli exclusion principle does not permit the spins S_1 and S_2 to be parallel. The indicated transition $^3S_1 \rightarrow {}^3P_0$ produces the 4046-Å line, which yields a relatively simple Zeeman pattern upon application of an external field. Figure 14.8a shows the effect of an external field on these two states. Note that the $^3S_1$ state splits into three, while the $^3P_0$ state is unaffected, thus giving rise to three components in the Zeeman pattern, as indicated in Figure 14.8b. The notation employed in this schematic representation of the Zeeman pattern is explained below in connection with selection rules, intensities, and polarizations.

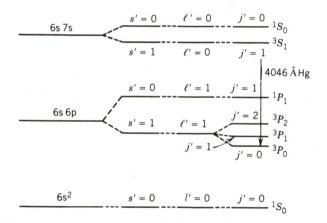

FIGURE 14.7 The *LS* coupling scheme for three states of the two outer electrons of Hg.

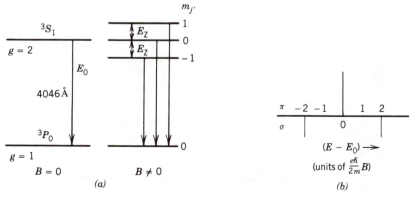

FIGURE 14.8 (*a*) Effect of a magnetic field on the states involved in the emission of the 4046-Å line in Hg. (*b*) The Zeeman pattern for the 4046-Å line in Hg.

EXERCISE 7

Show that the Zeeman components of the 4046 Å in Hg are separated in energy by $E_z = e\hbar B/m$.

Selection Rules, Intensities, and Polarizations

The formalism of quantum mechanics not only allows us to predict the wavelengths of the Zeeman components of a spectral line, but also can relate the appearance of the pattern to the nature of the states involved in the various transitions.

The intensity with which a given wavelength is emitted in an atomic transition is proportional to the probability with which that transition occurs. This can be calculated by computing the matrix element of the electric dipole moment $\mathbf{P} \equiv \Sigma_i e\mathbf{r}_i$, where $\mathbf{r}_i$ is the position of the ith optically active electron in the atom, by evaluating the expression

$$\mathbf{P}_{fi} = \int \psi_f^* \mathbf{P} \psi_i \, dV \qquad (\text{C} \cdot \text{m}) \tag{19}$$

where ψ_i and ψ_f are the initial and final states of the atom. The relative intensities of the various components of a Zeeman pattern are then each approximately proportional to P_{fi}^2. Calculated for the set of transitions corresponding to any pair of states ${}^{2s'+1}\ell'_{j'}$ (e.g., to the D_1 line shown in Figure 14.6), the relative intensities depend only on the initial and final values of j' and $m_{j'}$ for each individual component, independent of the kind of atom to which the LS coupling scheme has been applied.

Selection rules arise from the observation that certain of the dipole matrix elements are almost zero, so that transitions between certain states are *forbidden*, that is, make almost no contribution to the observed spectral pattern. If there is no external magnetic field, the "allowed" ($P_{fi} \neq 0$) transitions must, for the LS coupling scheme, satisfy the conditions

$$\Delta s' = 0$$
$$\Delta \ell' = 0, \pm 1 \tag{20}$$
$$\Delta j' = 0, \pm 1 \quad (\text{but } j' = 0 \to j' = 0 \text{ is forbidden})$$

for the changes in these quantum numbers during a transition. When an external magnetic field is applied, the number of observable transitions increases, but these are subject to the additional selection rule

$$\Delta m_{j'} = 0, \pm 1 \quad (\text{but } m_{j'} = 0 \to m_{j'} = 0 \text{ is forbidden if } \Delta j' = 0) \tag{21}$$

EXERCISE 8

Verify that all of the transitions indicated in Figure 14.6 are consistent with the selection rules stated above.

Evaluation of P_{fi}^2 for each pair of initial and final states that satisfy the selection rules gives expressions for the total relative radiated intensity I of each component of the Zeeman pattern, in arbitrary units (reference 3):

For $j' \to j' - 1$ transitions

$$m_{j'} \to m_{j'} \pm 1 \qquad I = \tfrac{1}{2}a(j' \mp m_{j'} - 1)(j' \mp m_{j'})$$
$$m_{j'} \to m_{j'} \qquad I = a(j' + m_{j'})(j' - m_{j'}) \tag{22}$$

For $j' \to j'$ transitions

$$m_{j'} \to m_{j'} \pm 1 \qquad I = b(j' \mp m_{j'})(j' \pm m_{j'} + 1)$$

$$m_{j'} \to m_{j'} \qquad I = bm_{j'}^2$$

where a and b are constants. The expressions for the $j' \to j' + 1$ transitions may be added by considering the reverse of the $j' \to j' - 1$ transitions and noting that forward and reverse intensities are the same ($P_{\text{fi}}^2 = P_{\text{if}}^2$).

Experimental observations of the Zeeman effect seldom measure the total intensities directly, since detection systems generally measure light emerging from the atom in a narrow range of directions. It is important to note that the radiation emitted by the transitions considered above is anisotropic: The intensity as well as the polarization of the light in each component (for each value of $\Delta m_{j'}$) depends on the direction of emission relative to **B** and on $\Delta m_{j'}$ for the associated transition. Figure 14.9 is a top view that illustrates the polarization of light emitted parallel to **B** (longitudinally) and perpendicular to **B** (transversely).

For the *transverse* observation, the emitted radiation is *linearly* polarized: for transitions in which $\Delta m_{j'} = 0$, called the π components, the electric field ϵ of the radiation is parallel to the applied field **B**; for the $\Delta m_{j'} = \pm 1$ transitions, or the σ components, ϵ is perpendicular to **B** and thus oscillates along a direction perpendicular to the plane of the paper, as indicated by the symbol in the figure.

For *longitudinal* observations, the σ components are *circularly* polarized, and the π components do not appear. Since all components of a Zeeman pattern are visible when viewed transversely, this is the experimental arrangement most often used. The relative intensities for the individual components in the transverse direction are given by

For $j' \to j' - 1$ transitions

$$m_{j'} \to m_{j'} \pm 1 \qquad I = a'(j' \mp m_{j'} - 1)(j' \mp m_{j'})$$

$$m_{j'} \to m_{j'} \qquad I = 4a'(j' + m_{j'})(j' - m_{j'}) \qquad (23)$$

For $j' \to j'$ transitions

$$m_{j'} \to m_{j'} \pm 1 \qquad I = b'(j' \pm m_{j'} + 1)(j' \mp m_{j'})$$

$$m_{j'} \to m_{j'} \qquad I = 4b'm_{j'}^2$$

where a' and b' are constants and the units for I are again arbitrary. For a given Zeeman pattern, the total intensity of the emitted radiation from all allowed transitions is isotropic and, on the average, has no net polarization.

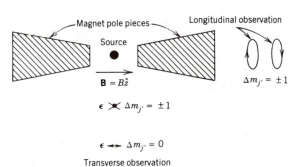

FIGURE 14.9 Polarization of the components of a Zeeman pattern.

Figure 14.8b illustrates the way in which the Zeeman pattern is conventionally represented for a transverse observation. The horizontal axis represents the transition energy, relative to that of the zero-field transition. The π and σ components are represented above and below this axis, respectively, by vertical lines with heights proportional to the intensities.

The observed polarization states are sometimes explained by considering the radiation produced by the oscillating dipole associated with the classical electron orbit for the various possible atomic orientations, as in Figure 14.1b. This is little more than a mnemonic device, since the nature of the radiation must be associated with the motion of the charge *during the transition* that produces it. Thus, the polarization can be understood by evaluating the time-dependent expectation value of the dipole moment $\mathbf{P}(t)$ during the transition, when the states can be expressed as a mixture of initial and final states ψ_i and ψ_f. $\mathbf{P}(t)$ can be expressed in terms of the quantity $\mathbf{P}_{fi}$ of equation 19:

$$\mathbf{P}(t) \propto \mathrm{Re}[\mathbf{P}_{fi}\exp(i\omega t)] \qquad (\mathrm{C}\cdot\mathrm{m}) \tag{24}$$

where $\omega \equiv (E_f - E_i)/\hbar$ is the frequency of the photon emitted in the transition between the two states of energy E_i and E_f. For π transitions, $\mathbf{P}$ oscillates along the z direction, so that radiation polarized along $\hat{z}$ would be viewed transversely, but no radiation is emitted longitudinally, along the dipole axis. For σ transitions, $\mathbf{P}$ rotates in the xy plane, producing circularly polarized light viewed along $\hat{z}$; when viewed transversely $\mathbf{P}$ appears as a linear dipole, perpendicular to $\hat{z}$, and so produces an ϵ perpendicular to the magnetic field.

EXERCISE 9

Verify that the transverse intensities for the Hg 4046-Å line that are indicated in Figure 14.8b correctly correspond to the transitions of Figure 14.8a.

EXERCISE 10

Make a sketch, similar to Figure 14.8, for the Hg 5461-Å line, which is produced by a $^3S_1 \rightarrow {}^3P_2$ transition. Indicate, for the allowed transitions, the relative transverse intensities and the energy spacings relative to the zero-field energy.

EXPERIMENT

Listed in Table 14.1 are a few transitions suitable for Zeeman effect observations in Hg, Na, and He—three atoms for which the above discussion of LS coupling is appropriate. For the Hg investigations, a sample enriched with ^{198}Hg (available commercially) should be used since its nucleus has zero spin; thus, the complications of the hyperfine structure associated with the magnetic moment of the nucleus are avoided.

The high resolution required for the observation of the Zeeman splittings may be obtained by use of either an Ebert-mount spectrograph or a Fabry–Perot interferometer; recording of the spectra may be done either photographically or with a phototube. The Ebert-mount spectrograph is described in Appendix D; below we briefly describe the operation of the Fabry–Perot device and an experiment that utilizes it with the phototube.

TABLE 14.1 SOME TRANSITIONS SUITABLE FOR ZEEMAN EFFECT OBSERVATIONS

Atom	Color	Wavelength (Å)	Initial Configuration		Final Configuration	
Hg	Violet	4046	6s 7s $^3S_1$		6s 6p $^3P_0$	
	Yellow	5791	6s 6d $^1D_2$		6s 6p $^1P_1$	
	Green	5461	6s 7s $^3S_1$		6s 6p $^3P_2$	
	Violet	4358	6s 7s $^3S_1$		6s 6p $^3P_1$	
	Yellow	5770	6s 6d $^3D_2$		6s 6p $^1P_1$	
Na	Yellow	5896	3p	$^2P_{1/2}$	3s	$^2S_{1/2}$
	Yellow	5890	3p	$^2P_{3/2}$	3s	$^2S_{1/2}$
He	Blue	4471	4d	$^3D_3$	2p	$^3P_2$
	Green	5016	3p	$^1P_1$	2s	$^1S_0$
	Yellow	5875	3d	$^3D_3$	2p	$^3P_2$

The Fabry–Perot Etalon

Consider a ray of monochromatic light, of vacuum wavelength λ_0, incident on an etalon consisting of a pair of parallel glass plates separated by a gap of width d filled with a medium with index of refraction n_f, as illustrated in Figure 14.10. The inner surfaces of the plates are partially mirrored so that, if the angle of incidence θ is small, the ray is multiply reflected. The optical path for each of the rays emerging at the right is different, so that beams will exhibit interference. The phase difference between adjacent emerging rays can be calculated by considering $\delta\ell$, the difference in path lengths for the rays emerging at B and D. Relative to the ray emerging at B, the ray reflected at B has an additional path length, measured from B to the wave front at D', of $\delta\ell = BC + CD' = 2d\cos\theta$. If we neglect the relatively small additional phase changes due to the metallic film, the difference in phase, $\delta\phi$, for these two rays is

$$\delta\phi = 2\pi\frac{\delta\ell}{\lambda} = 4\pi n_f\cos\theta\frac{d}{\lambda_0} \qquad \text{(rad)} \qquad (25)$$

where $\lambda = \lambda_0/n_f$ is the wavelength in the medium between the plates. The intensity I_t of the radiation transmitted through the Fabry–Perot will be a maximum if the beams interfere

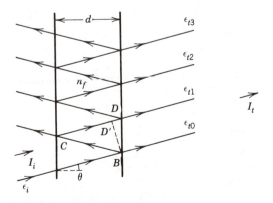

FIGURE 14.10 Multiple reflections of a beam between the plates of a Fabry–Perot etalon.

constructively, that is, if, for some integer m,

$$\delta\phi = 2\pi m \qquad \text{(rad)} \tag{26}$$

EXERCISE 11

Verify trigonometrically the relationship $\delta\ell = 2d\cos\theta$, which leads to the expression for $\delta\phi$ of equation 25.

The amplitude of the transmitted electric field ϵ_t^0 can be found in terms of the incident amplitude c_i^0 by superposing, with appropriate phase factors, the time-dependent field strengths of the transmitted beams. As in Figure 14.10, the time-varying strength of the electric field ϵ_{tk} of the kth transmitted beam can be written as $\epsilon_{tk}^0 \exp(i\omega t)$, where $k = 0$ corresponds to the first transmitted beam, which has not been reflected. The field transmission and reflection coefficients for the inner surfaces of the plates, t and r, can then be used to write $\epsilon_{tk}^0 = t^2 r^{2k} \epsilon_i^0$. The phase of the kth beam lags behind that of the $k = 0$ beam by an amount $k\,\delta\phi$. The field strength ϵ_t is then expressed as

$$\epsilon_t \simeq \sum_{k=0}^{\infty} \epsilon_{tk}^0 \exp[i(\omega t - k\,\delta\phi)]$$

$$= \left[\sum_{k=0}^{\infty} (t^2 r^{2k}) \exp(-ik\,\delta\phi) \right] \epsilon_i^0 \exp(i\omega t) \qquad \text{(N/C)} \tag{27}$$

where the number of reflections has been assumed so large that ϵ_t may be approximated by extending the series to an infinite number of terms. Evaluating this geometrical series yields

$$\epsilon_t = \left[\frac{T}{1 - R\exp(-i\,\delta\phi)} \right] \epsilon_i^0 \exp(i\omega t) \qquad \text{(N/C)} \tag{28}$$

where $T = t^2$ is the transmittance and $R = r^2$ the reflectance of the metal film. Since the intensity of the emerging beam is proportional to $\epsilon_t \epsilon_t^*$, equation 28 gives the ratio of transmitted and incident intensities as

$$\frac{I_t}{I_i} = \left(\frac{T}{1-R} \right)^2 \frac{1}{1 + F\sin^2(\delta\phi/2)} \tag{29}$$

where the abbreviation $F \equiv 4R/(1-R)^2$ and the identity $\cos\delta\phi = 1 - 2\sin^2(\delta\phi/2)$ have been used. This ratio is plotted in Figure 14.11a as a function of the phase difference $\delta\phi$ for

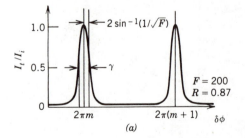

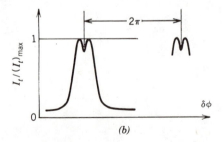

(a)   (b)

FIGURE 14.11 (a) The ratio of transmitted and incident intensities for a Fabry–Perot etalon with $R = .87$. (b) Intensity pattern from a Fabry–Perot for two barely resolved peaks.

a reflectivity of $R = 0.87$, assuming that the reflecting film absorbs no energy (i.e., that $T + R = 1$). Note, as mentioned above, that the peaks in the transmission occur for $\delta\phi = 2\pi m$.

Two ways in which the etalon can be used to measure wavelength are illustrated in Figure 14.12. If an extended source is configured with the etalon and two converging lenses, as in Figure 14.12a, rays enter the first plate of the Fabry–Perot with a range of incident angles θ. Each value of θ corresponds to a value of $\delta\phi$ given by equation 25 and to a radial position on a screen or photographic plate to which rays with this value of θ are focused; a series of bright rings are formed with radii corresponding to values of θ for which $\delta\phi = 2\pi m$. Measurements of these radii may then be used along with equations 25 and 26 to calculate the wavelength differences between the spectral components of the light from the source. As is discussed in reference 8, the wavelength separation for two adjacent fringes (corresponding to the same order m) with diameters D and D' is given by

$$\Delta\lambda = \lambda(D^2 - D'^2)/8f^2 \qquad \text{(m)} \tag{30}$$

where f is the focal length of the lens between the screen and the etalon.

A second method for determining spectral splittings, discussed further below in connection with the Zeeman effect measurements, is called central spot scanning, illustrated by Figure 14.12b. In this technique, the index of refraction n_f or the etalon spacing d is varied, thus changing $\delta\phi$. Peaks in I_t are observed for the light allowed through the pinhole, for which $\theta = 0$. From the curve of I_t versus n_f (or d), one can deduce the separations in λ for the various components of the incident beam.

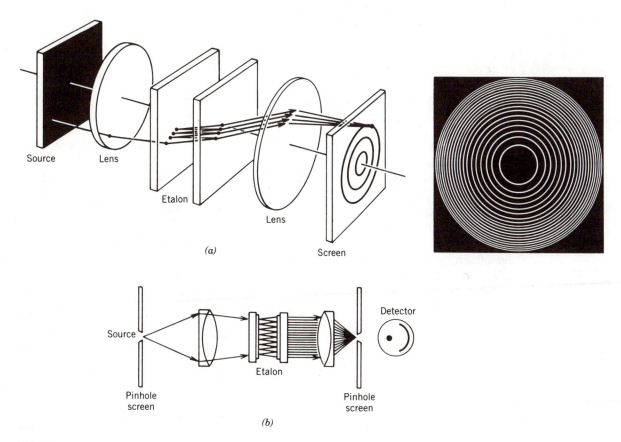

FIGURE 14.12 (a) Ring pattern produced by a Fabry–Perot with an extended source. (b) Central spot scanning with a Fabry–Perot.

Two parameters that determine the suitability of the Fabry–Perot for a particular set of measurements are the **free spectral range** $\Delta\lambda_{0f}$ and the **finesse** F_0. The free spectral range is that difference in wavelength between two spectral components of the source whose intensity maxima just overlap, that is, $\delta\phi(\lambda_0) - \delta\phi(\lambda_0 + \Delta\lambda_{0f}) = 2\pi$. Using equation 25 for $\delta\phi$ gives $\Delta\lambda_{0f}$ for $\theta = 0$:

$$\Delta\lambda_{0f} \simeq \frac{\lambda_0^2}{2n_f d} \quad (m) \tag{31}$$

EXERCISE 12

Verify that this expression for $\Delta\lambda_{0f}$ is nearly exact for visible wavelengths and an etalon having $d = 0.500$ cm and $n_f = 1$. What is $\Delta\lambda_{0f}$ in this case if $\lambda_0 \simeq 4046$ Å?

The finesse F_0 is the ratio of the separation, in $\delta\phi$, of two adjacent maxima in Figure 14.11a to the full width at half-maximum (FWHM) of the peaks, labeled γ. F_0 can be related to the reflectivity of the etalon surface by noting that equation 29 implies that, for large values of the parameter F (defined below, equation 29), I_t falls to one half its maximum value when $\delta\phi$ deviates by $2/F^{1/2}$ radians from $2\pi m$, its value at the mth peak. The full width γ is just double this amount, so $\gamma = 4/F^{1/2}$. Since two adjacent peaks are separated by a phase difference of 2π, $F_0 \equiv 2\pi/\gamma = \pi F^{1/2}/2$. Substituting the definition of F equation (29) gives F_0 in terms of R:

$$F_0 = \frac{\pi\sqrt{R}}{1 - R} \tag{32}$$

EXERCISE 13

Verify that I_t falls to half of its maximum value when $\delta\phi$ deviates from a peak value by $2/F^{1/2}$ radians. What approximation must be made to obtain this result?

We will say that two wavelengths λ_1 and $\lambda_2 \equiv \lambda_1 + \Delta\lambda_0$ are resolvable with the etalon if the phase differences $\delta\phi_1$ and $\delta\phi_2$ to which they correspond (for fixed values of n_f, θ, and d near a transmission peak) different by at least the FWHM, γ, of the I_t/I_i versus $\delta\phi$ curve: $\delta\phi_1 - \delta\phi_2 \gtrsim 2\pi/F_0$ for resolution of the two wavelength components. Figure 14.11b illustrates the intensity pattern due to two closely spaced wavelength components whose peaks are just barely resolved. To express the above resolution criterion directly in terms of wavelength separation, an approximation for $\delta\phi_1 - \delta\phi_2$ in terms of $\Delta\lambda_0$ may be obtained by differentiating equation 25 with respect to λ_0, giving us $\delta\phi_1 - \delta\phi_2 \simeq \delta\phi(\lambda_0)(\Delta\lambda_0/\lambda_0)$, where $\lambda_0 \simeq \lambda_1 \simeq \lambda_2$. Since $\delta\phi(\lambda_0) \simeq 2\pi m$ near a peak, this means that two wavelengths may be resolved if $\Delta\lambda_0 \gtrsim \lambda_0/mF_0$. The ratio of λ_0 to the minimum resolvable wavelength difference, $(\Delta\lambda_0)_{min}$, is the **chromatic resolution** R_0 and is given by

$$R_0 \equiv \frac{\lambda_0}{(\Delta\lambda_0)_{min}} = mF_0 = \frac{\pi m\sqrt{R}}{1 - R} \tag{33}$$

EXERCISE 14

For the situation described in Exercise 12, estimate the chromatic resolution of an etalon with surfaces that are 87 percent reflecting. Is this resolution sufficient for the study of the Zeeman pattern of, for example, the 4046-Å line of Hg using the magnetic fields available to you in your laboratory? What other factors will determine your ability to resolve the individual components of the Zeeman pattern?

Zeeman Effect Measurements

Figure 14.13a is a block diagram of the apparatus that can be used for the transverse observation of the Zeeman effect through the use of the central spot scanning technique described above. Light from the source, located between the magnet pole pieces, passes through converging lens 1 (and, optionally, through a polaroid sheet) so that it is incident on the Fabry–Perot etalon as a parallel (linearly polarized) beam at normal incidence. The etalon is located in a vacuum chamber, which can be evacuated and refilled with an adjustable leak (e.g., with a needle valve control); the index of refraction n_f can thus be varied by varying the pressure in the chamber. Lens 2 focuses light exiting the etalon onto the slit of a constant-deviation prism spectrometer (discussed in reference 6, Section 5.5); the prism can be rotated so that light from the spectral line under investigation can be viewed at the eyepiece. The data are taken by replacing the eyepiece with a pinhole so that

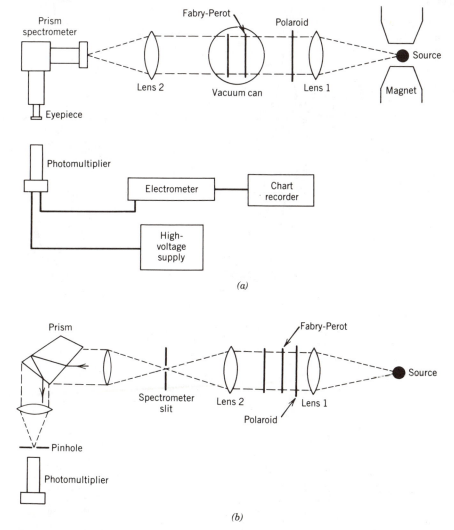

FIGURE 14.13 (a) Apparatus for the transverse observation of the Zeeman effect.
(b) Optical arrangement for the observation of the Zeeman effect.

light from the center of the Fabry–Perot pattern enters the photomultiplier tube. The photocurrent is detected with an electrometer, the output of which is connected to a chart recorder. Some further details of the optical system for this mode of operation are included in Figure 14.13b.

With the Fabry–Perot "crossed" with the prism spectrometer in the fashion described above, scanning of the Zeeman pattern is accomplished by slow readmission of gas into the evacuated vacuum chamber, which results in an n_f that varies in an approximately linear way with time. If the output of the phototube is then recorded as a function of time, the result will be a scaled version of the intensity versus n_f curve. For each wavelength component, the phase difference $\delta\phi$ is, according to equation 25, proportional to n_f, so that for a single wavelength (e.g., for a spectral line observed with $B = 0$), a portion of the I_t versus n_f curve derived from a pressure scan would appear as in Figure 14.14a; for two closely spaced wavelengths, the scan would produce a double-peak pattern, shown in Figure 14.14b for the case of unequal intensities. The splitting $\lambda_2 - \lambda_1$ can be estimated in terms of n_{f1}, n_{f2}, the free spectral range $\Delta\lambda_{0f}$, and Δn_{f0}, the change in n_f between successive orders:

$$\frac{\lambda_2 - \lambda_1}{\Delta\lambda_{0f}} \simeq \frac{n_{f2} - n_{f1}}{\Delta n_{f0}} \tag{34}$$

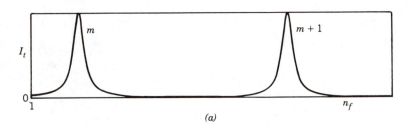

(a)

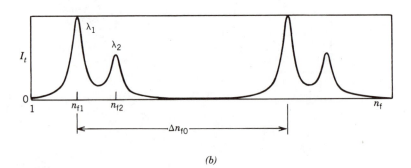

(b)

FIGURE 14.14 (a) I_t versus n_f curve derived from a pressure scan.
(b) Appearance of the I_t versus n_f curve in the case of two closely spaced peaks.

EXERCISE 15

For the wavelength and etalon in Exercise 12, estimate the number of intensity peaks that would be seen in a scan from zero pressure up to atmosphere, if $n_f \simeq 1.000277$ at 1 atm pressure.

With the Fabry–Perot and its vacuum can out of the optical path, collimate the light from the source with lens 1 so that it does not spread along the path. Use lens 2 to focus the light on the spectrometer slit. With the eyepiece in place, adjust the spectrometer and lens 2 to obtain the brighest spectral line images. Place the etalon, with its vacuum can, in the optical path and adjust it so that the ring pattern is centered in the eyepiece; the light reflected from the etalon should be directed back toward the source. Adjust the pressure for a peak in the intensity of the light entering the spectrometer and replace the eyepiece with the pinhole and phototube assembly. Maximum signal can be obtained by adjustments of the focus and the prism orientation.

The strength of the magnetic field should be measured for each set of observations. To eliminate error due to hysteresis in the magnet iron, the field measurements should be taken after the scan and before changing the magnet current.

For the spectral line(s) selected, predict the Zeeman patterns and produce pressure scans at a few different values of B. The central component may be filtered out with a polarizer to facilitate measurements of the splittings. As is suggested above, these splittings may be estimated as fractions of the free spectral range.

EXERCISE 16

The superposition of intensities from two closely spaced peaks will cause the measured separation in wavelength to be different from its actual value. Will your measurements then yield overestimates or underestimates of the splittings? How would you attempt to correct for this error? Note in this connection that the best data are obtained for splittings that amount to about one half of the free spectral range.

EXERCISE 17

Investigate the linearity of the splittings as functions of the field and derive, for each observed pattern, a value of μ_0 for comparison with its accepted value. Repeat enough of the measurements to derive an estimate of the associated uncertainties. Do your observations support the theory of LS coupling? If you can estimate your relative intensities, are they what you expect?

Summary

In atoms of intermediate atomic number for which the structure of the optical electrons is adequately described by the Russell–Saunders coupling scheme, the states (wave functions) for these electrons may be described by the quantum numbers s', ℓ', j', and $m_{j'}$. The energy of these states may be expressed as $E = E'(n, s', \ell', j') + E_Z$, where E_Z is the Zeeman energy, given by

$$E_Z = \mu_0 B g_{j'} m_{j'} \qquad (\text{J}) \tag{17'}$$

where the Landé g factor is

$$g_{j'} = 1 + \frac{j'(j'+1) + s'(s'+1) - \ell'(\ell'+1)}{2j'(j'+1)} \tag{18'}$$

The energy levels E' are illustrated schematically for a few multiplets of Hg in Figure 14.7, and Figure 14.8 illustrates the effect of a magnetic field on these levels, that is, the Zeeman effect. For spectral lines resulting form the transitions between pairs of states in different multiplets, calculations of the dipole matrix elements leads to selection rules, intensities, and polarizations for the radiation emitted in various directions.

The Zeeman effect may be investigated quantitatively by measuring the splitting, produced by an externally applied magnetic field, of the spectral lines corresponding to transitions between states of a multiplet. The high resolution required can be obtained by use of either an Ebert-mount spectrograph or by a Fabry–Perot interferometer. The central spot scanning technique for the Fabry–Perot etalon, illustrated in Figure 14.13, produces intensity peaks, as in Figure 14.14, from which measurements of the Bohr magneton may be derived.

COMPUTER-ASSISTED EXPERIMENTATION (OPTIONAL)

Prerequisite

Experiment 6, Introduction to Computer-Assisted Experimentation.

Introduction

The relatively simple method of acquiring and analyzing the Zeeman effect data suggested above depends for its success on the linearity of the relationships (a) between n_f (the index of refraction of the air between the reflecting surfaces of the interferometer) and p (the pressure), and (b) between p and the time elapsed since beginning the readmission of air to the can, which is measured on the time axis of the chart recorder. The linearity between pressure and time may be improved by using a needle valve to impose a very low flow rate for the air; this also minimizes effects due to local temperature changes and turbulence during the pressure scan. The magnitude of the problem posed by the nonlinearity of n_f versus p can be assessed by examining the quantity $\partial n_f/\partial p$, which is given by reference 7 for a constant temperature of $T = 15\,°C$ as

$$\left(\frac{\partial n_f}{\partial p}\right)_{T=15\,°C} = (n_f^0 - 1)(1.3149 \times 10^{-3} + 1.626 \times 10^{-9}p) \tag{35}$$

where p is in torr and n_f^0 is the index for air at $p = 760$ torr and $T = 15\,°C$. The effect of the nonlinearity given by this relationship depends on the range of pressures used in the scan, but is small for most cases of practical interest.

These nonlinearities can be effectively compensated for by a computer interface that utilizes software to linearize these data. Additionally, the computer affords a convenient method for averaging data (from the same scan or from different scans) in the event that low signal levels make this appropriate.

EXERCISE 18

Using equation 35, derive an expression for the index of refraction as a function of p at $15\,°C$. For the pressure range required to scan one free spectral range at optical wavelengths with your etalon, estimate the percentage by which the corresponding actual change in n_f deviates from the change calculated under the assumption of complete linearity.

Experiment

The configuration we suggest for computer-assisted data acquisition from central spot scans is shown in Figure 14.15. The output of the pressure transducer is an analog voltage that is linearly related to the pressure within the vacuum can; this signal is input, via a conditioning circuit (if required), to one of two input channels of the ADC that are to be polled for data by the software. The output of the electrometer is fed to a second channel via another conditioning circuit.

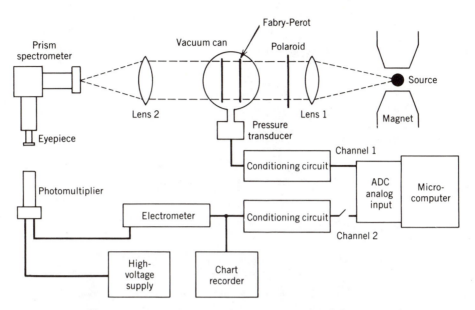

FIGURE 14.15 Configuration for computer-assisted data acquisition.

If the output of the pressure transducer matches the ADC input range, then no conditioning circuit is required. If conditioning is required, a modification of the circuit shown in the upper half of Figure 6.5 (consisting of op-amps 1 and 2) can be used to provide the appropriate gain and offset. The transducer output ground can be directly connected to pin 3 of op-amp 1 and resistors R_3 and R_4 can be eliminated. R_2 and V_{off} should be adjusted so that the full range of the ADC input is utilized for maximum precision.

The conditioning circuit for the electrometer output may be of the same design as that for the transducer, but since the input signal range during the scan will be determined by the light intensity and the electrometer scale setting, it is a good idea to adjust the conditioning gain and offset by setting up and observing some scans on the chart recorder before switching the output of the conditioning circuit into channel 2 of the ADC.

Measurements and Software   Both pressure and intensity data may be taken "on the fly" by polling channels 1 and 2 of the ADC input periodically as air is slowly readmitted to the vacuum can containing the Fabry–Perot. Write a program to take the data and to produce values of n_f and intensity at intervals during a scan; produce a graph of these results. You may wish to develop averaging schemes and/or attempt to process this data to produce values of the Zeeman splittings directly.

Condensed Matter Physics

INTRODUCTION TO MAGNETIC RESONANCE

Historical Note

The 1952 Nobel prize in Physics was awarded jointly to
  Felix Bloch, the United States, and Edward Mills Purcell, the United States
  For their development of new methods for nuclear magnetic precision measurements (nuclear magnetic resonance) and discoveries in connection therewith.

KEY CONCEPTS

Angular momentum	Magnetic dipole selection rules
Magnetic moment	Magnetic susceptibility
g factor	Spin–spin relaxation time T_2
Larmor frequency	Spin–lattice relaxation time T_1
Maxwell–Boltzmann statistics	Dc line shape
Fermi's golden rule No. 2	

REFERENCES

1. R. Norberg, Resource Letter NMR-EPR-1 on Nuclear Magnetic Resonance and Electron Paramagnetic Resonance, *Am. J. Phys.* **33** (Feb. 1963). This is a list of articles that were originally published in various journals. Each article is specified as elementary, intermediate, or advanced.

2. G. Pake, Fundamentals of Nuclear Magnetic Resonance Absorption I, *Am. J. Phys.* **18**, 438 (1950). An excellent paper. Essentially every topic presented in the introduction of this experiment is covered in Pake's article.

3. E. Andrew, *Nuclear Magnetic Resonance*, Cambridge University Press, New York, 1955. General discussion of theory, experimental methods, and applications of NMR. See especially Chapters 1, 2, and 3.

4. A. Abragam, *Principles of Nuclear Magnetism*, Oxford University Press, New York, 1961. Detailed discussion of theory, experimental methods, and applications of NMR.

5. R. Eisberg and R. Resnick, *Quantum Physics of Atoms, Molecules, Solids, Nuclei, and Particles*, Wiley, New York, 1985. Transition probability is discussed in Appendix K. The Boltzmann distribution as an approximation to quantum distributions is discussed on pages 391–392.

6. K. Symon, *Mechanics*, 3d ed., Addison-Wesley, Reading, MA, 1971. Rotating coordinate systems are treated.

INTRODUCTION

This is a prerequisite for Experiment 15, Nuclear Magnetic Resonance (NMR), and Experiment 16, Electron Spin Resonance (ESR).

The interactions among the nucleons of a nucleus give rise to a ground state and a constellation of excited states. Under the conditions of an NMR experiment, the nucleons occupy the ground state and, hence, we ignore the excited states. With the nucleons in the ground state, the nucleus has some total angular momentum **J** and a corresponding magnetic moment **μ**. Therefore, we may treat each such nucleus as *a particle having angular momentum* **J**.

We may make a similar statement about the atomic electrons of an atom. There exist a ground state and a constellation of excited states, and we assume that the excited states are not occupied. In the ground state the atomic electrons have total angular momentum **J** and a magnetic moment **μ**. Hence, we may treat each such atom as *a particle having angular momentum* **J**.

The introduction includes (1) a classical description of a magnetic moment **μ** in a uniform magnetic field **B**$_0$ and a time-varying magnetic field **B**$_1(t)$ which varies in the radio-frequency range for NMR and in the microwave range for ESR, (2) a quantum description of the same system, and (3) a discussion of the magnetic susceptibility of the sample.

Classical Description

For a distributed mass, the total angular momentum **J** is obtained by integration

$$\mathbf{J} = \int (\mathbf{r} \times \mathbf{v})\rho_m(\mathbf{r})\, dV \qquad (\text{kg} \cdot \text{m}^2/\text{s}) \tag{1}$$

where ρ_m is the mass density and **r** is the radius vector with respect to the center of the distributed mass. The units for **J** are typically expressed as joules · second. Similarly, the magnetic moment **μ** involving distributed charge is

$$\boldsymbol{\mu} = \frac{1}{2}\int (\mathbf{r} \times \mathbf{v})\rho_e(\mathbf{r})\, dV \qquad (\text{C} \cdot \text{m}^2/\text{s}) \tag{2}$$

where ρ_e is the charge density. If the ratio of charge to mass $\rho_e\, dV/\rho_m\, dV$ is constant and equal to Q/M, where Q and M are the total charge and mass, then $\rho_e\, dV = (Q/M)\rho_m\, dV$ and equation 2 becomes

$$\boldsymbol{\mu} = \frac{Q}{2M}\int (\mathbf{r} \times \mathbf{v})\rho_m(\mathbf{r})\, dV = \frac{Q}{2M}\mathbf{J} \qquad (\text{inadequate}) \tag{3}$$

Equation 3 is not in agreement with experimental results. It is necessary to replace Q with the charge q multiplied by a pure number, not too far from unity, called the g factor. Equation 3 then becomes

$$\boldsymbol{\mu} = \frac{gq}{2M}\mathbf{J} \qquad (\text{C} \cdot \text{m}^2/\text{s}) \tag{4}$$

For the electron system of an atom, $q = -e$, M is the mass of an electron, and g depends on the quantum state of the electrons. For a single electron having spin angular momentum only, g is 2.0023. For a given nucleus, g depends on the quantum state of the nucleus, $q = +e$, and M is the mass of the proton. Some g factors and nuclear spins are listed in Table M.1 for nuclei in the ground state.

TABLE M.1 g FACTORS AND SPINS FOR NUCLEI IN THE GROUND STATE

Nucleus	g	Spin
^1_0n	-3.826	$\frac{1}{2}$
^1_1H	$+5.586$	$\frac{1}{2}$
^2_1H	$+0.8576$	1
^3_1H	$+5.9594$	$\frac{1}{2}$
^3_2He	$+4.26$	$\frac{1}{2}$
^6_3Li	$+0.8223$	1
^7_3Li	$+2.1505$	$\frac{3}{2}$
$^{19}_9F$	$+5.2592$	$\frac{1}{2}$

We first consider the motion of a magnetic moment $\mathbf{\mu}$ in the presence of a static field $\mathbf{B}_0$. $\mathbf{B}_0$ will produce a torque on $\mathbf{\mu}$ of amount $\mathbf{\mu} \times \mathbf{B}_0$. The equation of motion of $\mathbf{\mu}$ is found by equating the torque with the rate of change of the angular momentum $\mathbf{J}$

$$\frac{d\mathbf{J}}{dt} = \mathbf{\mu} \times \mathbf{B}_0 \qquad (\mathrm{J}) \tag{5}$$

where typical units for $\mathbf{\mu}$ were used, namely, joules per tesla. Using equation 4 to eliminate $\mathbf{J}$, we find that

$$\frac{d\mathbf{\mu}}{dt} = \frac{gq}{2M}(\mathbf{\mu} \times \mathbf{B}_0) \qquad (\mathrm{J/T \cdot s}) \tag{6}$$

This equation says that at any instant the rate of change of $\mathbf{\mu}$ is perpendicular to both $\mathbf{\mu}$ and $\mathbf{B}_0$. The relation of $\mathbf{\mu}$ to $\mathbf{B}_0$ is shown in Figure M.1. Consider the tail of $\mathbf{\mu}$ as fixed; the tip of the vector moves in a circular path centered on $\mathbf{B}_0$ and the angle θ between $\mathbf{\mu}$ and $\mathbf{B}_0$ does not change. The precessional frequency of $\mathbf{\mu}$ about $\mathbf{B}_0$, obtained from solving equation 6, is called the Larmor frequency and is given by

$$\omega_0 = -\frac{gq}{2M}\mathbf{B}_0 \equiv -\gamma\mathbf{B}_0 \qquad (\mathrm{rad/s}) \tag{7}$$

where the defined γ is called the gyromagnetic ratio.

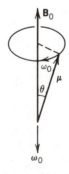

FIGURE M.1 The magnetic moment $\mathbf{\mu}$ precesses about the static field $\mathbf{B}_0$ with angular frequency $\mathbf{\omega}_0$.

EXERCISE 1

For a proton in a 0.5-T field, show that the precessional frequency v_0 is 21.3 MHz (radio frequency). For a free electron spin in a 0.5-T field, show that the precessional frequency v_0 is 14.0 GHz (microwave frequency).

We may use the torque $\boldsymbol{\mu} \times \mathbf{B}_0$ to calculate the work done in rotating the magnetic moment from $\theta = 0$ to some angle $\theta > 0$:

$$U = \int_0^\theta \mu B_0 \sin \theta \, d\theta = -\mu B_0 \cos \theta + \text{constant}$$

$$= -\boldsymbol{\mu} \cdot \mathbf{B}_0 + \text{constant} \quad \text{(J)} \tag{8}$$

Increasing θ increases the energy U; hence, increasing θ corresponds to an absorption of energy.

Suppose now that a circularly polarized magnetic field $\mathbf{B}_1(t)$ having angular frequency $\boldsymbol{\omega}$ and amplitude B_1 is applied at right angles to $\mathbf{B}_0$. If the rotation of the circularly polarized field is as shown in Figure M.2, then in terms of x and y components it may be written

$$\mathbf{B}_1(t) = B_1 \cos \omega t \, \hat{x} - B_1 \sin \omega t \, \hat{y} \quad \text{(T)} \tag{9}$$

The motion of $\boldsymbol{\mu}$, when both $\mathbf{B}_0$ and $\mathbf{B}_1(t)$ are present, is treated most simply by use of a rotating coordinate system. The rotating coordinate system of most use for this system is one in which $\mathbf{B}_1(t)$ is constant, that is, one rotating at $\boldsymbol{\omega}$. From classical mechanics

$$\frac{d\boldsymbol{\mu}}{dt} = \frac{d'\boldsymbol{\mu}}{dt} + \boldsymbol{\omega} \times \boldsymbol{\mu} \quad \text{(J/T} \cdot \text{s)} \tag{10}$$

where d/dt and d'/dt signifies differentiation with respect to the laboratory frame and the rotating frame, respectively. The equation of motion for $\boldsymbol{\mu}$ in the lab frame is

$$\frac{d\boldsymbol{\mu}}{dt} = \gamma(\boldsymbol{\mu} \times \mathbf{B}) \quad \text{(J/T} \cdot \text{s)} \tag{11}$$

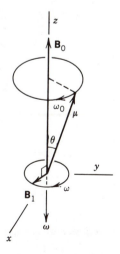

FIGURE M.2 Diagram of $\boldsymbol{\mu}$ precessing about $\mathbf{B}_0$ in the presence of a time-varying field of amplitude B_1 and frequency $\boldsymbol{\omega}$.

where $\mathbf{B} = \mathbf{B}_1(t) + \mathbf{B}_0$. Substituting equation 11 into equation 10 and solving for $d'\boldsymbol{\mu}/dt$ gives

$$\frac{d'\boldsymbol{\mu}}{dt} = \boldsymbol{\mu} \times (\gamma\mathbf{B} + \boldsymbol{\omega}) = \boldsymbol{\mu} \times \gamma \left[B_1 \cos \omega t \, \hat{x} - B_1 \sin \omega t \, \hat{y} + \left(B_0 - \frac{1}{\gamma}\omega \right)\hat{z} \right] \qquad (\text{J/T} \cdot \text{s}) \quad (12)$$

Equation 12 is the equation of motion for $\boldsymbol{\mu}$ as viewed from the frame rotating at $\boldsymbol{\omega}$. We define $\mathbf{B}_e = \mathbf{B}_0 + \boldsymbol{\omega}/\gamma$, the effective magnetic field in the frame rotating at $\boldsymbol{\omega}$. $\mathbf{B}_e$ is shown in the rotating frame in Figure M.3, where it is assumed that $\mathbf{B}_1$ is parallel to the x' axis in the rotating frame.

Magnetic resonance occurs when ω, the angular frequency of the time varying field, equals ω_0, the Larmor frequency:

$$\omega = \omega_0 = \gamma B_0 \qquad (\text{rad/s}) \tag{13}$$

Hence, at magnetic resonance the z component of $\mathbf{B}_e$ vanishes (see equation 12). In the rotating frame, the motion of $\boldsymbol{\mu}$ at magnetic resonance is precession about $\mathbf{B}_e$, which is assumed to lie along the x' axis. In the lab frame, the motion of $\boldsymbol{\mu}$ is precession about the x' axis, which rotates with angular frequency ω_0 about the z axis, see Figure M.4. Note that in this classical description of the motion of $\boldsymbol{\mu}$ at magnetic resonance, the angle θ continuously changes; that is, the magnetic moment is continuously exchanging energy with the time varying field.

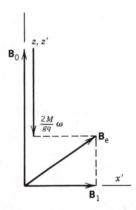

FIGURE M.3 Diagram of the effective field $\mathbf{B}_e$ in the rotating frame.

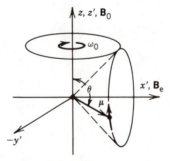

FIGURE M.4 In the rotating frame the tip of the vector $\boldsymbol{\mu}$ precesses about $\mathbf{B}_e$. The complete motion, in the laboratory frame, is then obtained by rotating the x' axis about $\mathbf{B}_0$ at the angular frequency ω_0.

Quantum Description

Each particle has total angular momentum $\mathbf{J}$, whose magnitude is $\hbar[j(j+1)]^{1/2}$. The interaction $-\boldsymbol{\mu} \cdot \mathbf{B}_0$ is a small perturbation on the ground state and the Hamiltonian H for this interaction is

$$H = -\boldsymbol{\mu} \cdot \mathbf{B}_0 = -\gamma \mathbf{J} \cdot \mathbf{B}_0 = -\gamma B_0 J_z \quad \text{(J)} \tag{14}$$

where the direction of $\mathbf{B}_0$ is taken as the z direction. The eigenfunctions of a particle having angular momentum $\mathbf{J}$ are, in Dirac notation, $|jm_j\rangle$, where $m_j = -j, -j+1, \ldots, j$. J^2 and J_z operating on these eigenfunctions yield

$$J^2|jm_j\rangle = \hbar^2 j(j+1)|jm_j\rangle \quad ((\mathbf{J} \cdot \mathbf{s})^2/(\text{steradians})^{1/2})$$

$$J_z|jm_j\rangle = \hbar m_j|jm_j\rangle \quad ((\mathbf{J} \cdot \mathbf{s}/(\text{steradians})^{1/2}) \tag{15}$$

These simplified eigenfunctions are useful to describe the ground state in the presence of $\mathbf{B}_0$; however, it should be pointed out that the real eigenfunctions are complicated functions. Such simplified eigenfunctions are discussed more fully in Experiment 16, Electron Spin Resonance. The time-independent Schrödinger equation for the ground state is

$$H|jm_j\rangle = E(m_j)|jm_j\rangle \quad (\text{J}/(\text{steradians})^{1/2}) \tag{16}$$

EXERCISE 2

Assuming $j = 2$ (arbitrarily), solve equation 16 for the five eigenvalues $E(m_j)$.

In the presence of $\mathbf{B}_0$, the ground state splits into $2j + 1$ levels. This is shown in Figure M.5 for $j = 2$ and positive charge q.

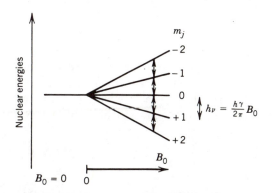

FIGURE M.5 For $B_0 = 0$, the level having $j = 2$, $m_j = 2$, 1, 0, -1, -2 is a single level. For $B_0 \neq 0$, it splits into five levels. At some value of B_0 the separation of neighboring levels equals the photon energy $h\nu$, that is, resonance occurs.

EXERCISE 3

For a proton in a 0.5-T field, what is the frequency of radiation that causes a transition between the two levels? For a proton $j = \frac{1}{2}$ and $g = +5.586$ (see Table M.1). Answer this question for a free electron spin ($g = 2.0023$). Compare your answers with those from Exercise 1.

The Hamiltonian given by equation 14 assumes independent, noninteracting particles, which implies a spatial separation and spatial distinguishability of particles. For such a system in thermal equilibrium with its surroundings, the probability that the energy level $E(m_j)$ is occupied is given by Maxwell–Boltzmann statistics:

$$P[E(m_j)] \propto \exp\left[-\frac{E(m_j)}{k_B T}\right] \tag{17}$$

If $N(m_j)$ and $N(m_j')$ are the number of particles per unit volume having energies $E(m_j)$ and $E(m_j')$ and N is the total number of particles per unit volume, then in equilibrium the ratio $N(m_j)/N(m_j')$ is given by

$$\frac{N(m_j)}{N(m_j')} = \frac{NP[E(m_j)]}{NP[E(m_j')]} = \exp\left[-\frac{E(m_j) - E(m_j')}{k_B T}\right] \tag{18}$$

where equation 17 was used.

EXERCISE 4

For protons at 300 K in a 0.5-T field, show that $N(+\frac{1}{2})/N(-\frac{1}{2})$ equals $1 + 3.4 \times 10^{-6}$. (We shall see that this small, but finite, excess of population in the lower energy state causes a net absorption of energy from the radio-frequency field in NMR.) For free electron spins in a 0.5-T field at 300 K, show that $N(-\frac{1}{2})/N(+\frac{1}{2})$ equals 1.002.

NMR (ESR) involves applying a time varying radiowave (microwave) magnetic field to the sample, which induces transitions between energy levels such as those shown in Figure M.5. For reasons that will become obvious, this magnetic field must be perpendicular to the static field $B_0\hat{z}$. Let this field be along the x axis and be given by

$$\mathbf{B}_1(t) = B_1 \cos \omega t \, \hat{x} \qquad \text{(T)} \tag{19}$$

where B_1 is the amplitude (small compared with B_0) and ω is the angular frequency. It is worthwhile to relate this linear polarized field to the previously mentioned circularly polarized field. The field given by equation 19 is equivalent to two circularly polarized fields rotating in opposite directions and of amplitude $B_1/2$:

$$B_1 \cos \omega t \, \hat{x} = \frac{B_1}{2}(\cos \omega t \, \hat{x} + \sin \omega t \, \hat{y}) + \frac{B_1}{2}[\cos(-\omega t) \, \hat{x} + \sin(-\omega t) \, \hat{y}] \qquad \text{(T)} \tag{20}$$

The component rotating in the same direction as the precessing magnetic moment will be in resonance when $\omega = \omega_0$, see Figure M.5. The other rotating component is completely out of phase with the precessing magnetic moment and has no effect on it.

The Hamiltonian or interaction between the magnetic moment $\boldsymbol{\mu}$ and $\mathbf{B}_1$ is given by

$$H_1 = -\boldsymbol{\mu} \cdot \mathbf{B}_1 = -\gamma \mathbf{J} \cdot \mathbf{B}_1 = -\gamma B_1 J_x \cos \omega t \qquad \text{(J)} \tag{21}$$

The probability per unit time that a transition occurs from an initial state m_j to a final state m_j' is given from Fermi's golden rule No. 2 as

$$W(m_j \to m_j') = \frac{2\pi}{\hbar^2} |\langle jm_j'|H_1|jm_j\rangle|^2 f(v)$$

$$= \frac{2\pi}{\hbar^2} |\gamma B_1 \cos \omega t|^2 |\langle jm_j'|J_x|jm_j\rangle|^2 f(v) \qquad (1/s) \qquad (22)$$

where $f(v)$ is a line-shape function (units of seconds), peaked at or near $v_0 = \gamma B_0/2\pi$. The origin and detailed nature of $f(v)$ are not important for the present discussion. To evaluate the matrix elements in equation 22 it is convenient to introduce raising and lowering operators, J_+ and J_-, where

$$J_+ = J_x + jJ_y \quad \text{and} \quad J_- = J_x - jJ_y \qquad (\text{J} \cdot \text{s}) \qquad (23)$$

and $j^2 = -1$. The operation of $J_\pm$ on the eigenfunction $|jm_j\rangle$ yields

$$J_\pm|jm_j\rangle = \hbar\sqrt{j(j+1) - m_j(m_j \pm 1)}|jm_j \pm 1\rangle \qquad (\text{J} \cdot \text{s/(steradian)}^{1/2}) \qquad (24)$$

where the upper sign is for J_+ and the lower sign is for J_-. Solving (23) for J_x in terms of J_+ and J_-, and then writing equation 22 in terms of J_+ and J_- we have

$$W(m_j \to m_j') = \frac{2\pi}{\hbar^2} |\gamma B_1 \cos \omega t|^2 |\langle jm_j'| \frac{J_+ + J_-}{2} |jm_j\rangle|^2 f(v) \qquad (1/s) \qquad (25)$$

EXERCISE 5

The selection rule for magnetic dipole transitions is $\Delta m_j = \pm 1$. Use equations 24 and 25 to show that the only transitions that will occur are those indicated by vertical lines in Figure M.5, where hv is the energy of a photon. Hence, verify that the selection rule is satisfied.

Note in Figure M.5 that for a field $B_1(t)$ of fixed frequency v the four transitions occur at the same value of B_0. The fixed frequency v is determined by the oscillator (NMR) or klystron (ESR).

EXERCISE 6

Use equation 25 to show that the transition probabilities per second, $W(m_j \to m_j - 1)$ and $W(m_j - 1 \to m_j)$, are equal.

Consider two energy levels, $E(m_j)$ and $E(m_j - 1)$, where the energy separation is hv, the energy of a photon. Let $N(m_j)$ and $N(m_j - 1)$ be the number of particles per unit volume of sample that occupy these levels. Then, for positive charges, the power radiated per unit volume of sample, due to stimulated emission, is $N(m_j - 1)hv W(m_j - 1 \to m_j)$, and the power absorbed, due to stimulated absorption, is $N(m_j)hv W(m_j \to m_j - 1)$. Using the result from Exercise 6, the instantaneous net power absorbed per unit volume of sample for transitions between these two levels is

$$P(t) = [N(m_j) - N(m_j - 1)]hv W(m_j \to m_j - 1) \qquad (\text{J/m}^3\text{s}) \qquad (26)$$

We do not include spontaneous emission in equation 26 because the transition probability for spontaneous emission is proportional to the cube of the frequency; therefore, it is small for both radio and microwave frequencies. (See equation 11, Introduction to Laser Physics.)

EXERCISE 7

Use equation 18 to eliminate $N(m_j - 1)$ in terms of $N(m_j)$, and then expand the exponential (the Boltzmann factor), keeping first-order terms only, to obtain

$$P(t) = N(m_j) \frac{\gamma \hbar}{k_B T} B_0 h\nu W(m_j \to m_j - 1) \qquad \text{(J/m}^3\text{s)} \tag{27}$$

EXERCISE 8

Show that the time-averaged power absorbed $\langle P \rangle$ by the sample is given by

$$<P> = N(m_j) \frac{\gamma \hbar}{k_B T} B_0 h\nu \frac{2\pi}{\hbar^2} (\gamma B_1)^2 \frac{\hbar^2}{8} [j(j+1) - m_j(m_j - 1)] f(\nu) \qquad \text{(J/m}^3\text{s)} \tag{28}$$

To obtain this result evaluate the matrix elements in the transition probability $W(m_j \to m_j - 1)$ and then calculate the time average of $P(t)$ over one period. Note that there are three independent variables in equation 28: T, B_0, and B_1. The frequency is not independent of B_0, since at magnetic resonance it is related to B_0, as shown in Figure M.5.

Magnetic Susceptibility

(One of the fundamental principles of magnetostatics is that, when a magnetic field **H** (called the magnetic intensity) is applied to matter, the matter becomes magnetized. The resulting magnetization of the matter **M** is proportional to **H**, and the defining equation for the dimensionless magnetic susceptibility χ is

$$\mathbf{M} = \chi \mathbf{H} \qquad \text{(A/m)} \tag{29}$$

The magnetic field **B** (called the magnetic flux density) is related to χ and **H** by

$$\mathbf{B} = \mu_0 (1 + \chi) \mathbf{H} \qquad \text{(T)} \tag{30}$$

where μ_0 is the vacuum permeability (μ_0 has no connection to the magnetic dipole moment **μ**). If the matter is subjected to both a static field $\mathbf{B}_0$ and a time-varying field $\mathbf{B}_1(t)$, then the magnetization given by (29) is described by both a static and a frequency dependent susceptibility, χ_0 and $\chi(\omega)$.

The magnetization **M** that results from the application of the magnetic field to the matter is due to the alignment of the magnetic moments of atomic electrons or, in the case of nuclear magnetism, the alignment of the nuclear magnetic moments. **M** is defined as the magnetic moment per unit volume of matter:

$$\mathbf{M} = \frac{1}{V} \sum_i \boldsymbol{\mu}_i \qquad \text{(A/m)} \tag{31}$$

where the sum is over the magnetic moments in the volume V.

We have seen that an assembly of nuclear (or electronic) magnets in a static magnetic field absorb power from a suitably applied radio-frequency (or microwave frequency) field.

We describe this power absorption in terms of the frequency-dependent susceptibility $\chi(\omega)$ of the assembly of magnets. We write this susceptibility as a complex quantity, having a real part $\chi'(\omega)$ and an imaginary part $\chi''(\omega)$:

$$\chi(\omega) = \chi'(\omega) - j\chi''(\omega) \qquad \text{(dimensionless)} \qquad (32)$$

As we will see below, it is the imaginary part $\chi''(\omega)$ that is responsible for power absorption.

To understand the absorption of power in terms of $\chi(\omega)$, we examine the impedance of the LC circuit shown in Figure M.6a for NMR spectroscopy and the impedance of the circuit shown in Figure M.6b for ESR spectroscopy. The NMR sample is inserted into the coil of the inductor shown in Figure M.6a. The radio frequency (rf) current I flowing through the coil provides the rf field $\mathbf{B}_1(t)$, which is perpendicular to the field $\mathbf{B}_0$ of the electromagnet. An ESR sample is placed inside the transmission microwave cavity shown in Figure 2.4, Experiment 2. The magnetic field strength lines $\mathbf{H}$ are shown in Figure 2.5, Experiment 2. The circuit shown in Figure M.6b is the equivalent circuit of a transmission cavity, where the coupling holes are represented by transformers.

For both circuits, when the field B_0 is off magnetic resonance, that is, $(h\gamma/2\pi)B_0 \neq h\nu$, the inductance L_0 is that of the coil only. When B_0 is at the magnetic resonance value, the inductance L is that of the coil plus the sample,

$$L = L_0(1 + \eta\chi) \qquad \text{(H)} \qquad (33)$$

where η is the filling factor or the fraction of the volume (of the coil or cavity) filled by the sample $(0 < \eta < 1)$.

The RLC circuit that is equivalent to an ESR transmission cavity is analyzed first. If we ignore the coupling transformers, then the impedance is that of a series RLC circuit with impedance

$$Z = R + j\left(\omega L - \frac{1}{\omega C}\right) \qquad (\Omega) \qquad (34)$$

Defining $\omega_r = 1/\sqrt{L_0 C}$, which is the resonance frequency when B_0 is off magnetic resonance, and substituting equation 33 into 34 yields

$$Z = R + \frac{jL_0(\omega^2 - \omega_r^2)}{\omega} + jL_0\omega\eta\chi \qquad (\Omega) \qquad (35)$$

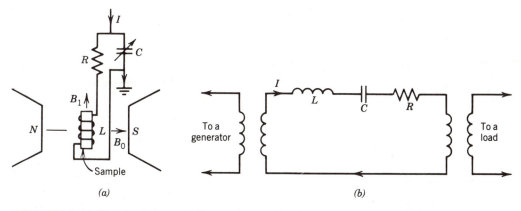

FIGURE M.6 (a) RLC circuit in an NMR experiment. R is the wiring resistance. (b) RLC circuit equivalent to an ESR transmission cavity.

Substituting equation 32 into 35 gives

$$Z = R + \eta\omega L_0 \chi'' + j\left[\eta\omega L_0 \chi' + \frac{L_0}{\omega}(\omega^2 - \omega_r^2)\right]$$

$$= R\left\{1 + \eta\frac{\omega L_0}{R}\chi'' + j\left[\eta\frac{\omega L_0}{R}\chi' + \frac{\omega L_0}{R}\left(\frac{\omega^2 - \omega_r^2}{\omega^2}\right)\right]\right\}$$

$$\equiv R\left\{1 + \eta Q\chi'' + j\left[\eta Q\chi' + Q\left(\frac{\omega^2 - \omega_r^2}{\omega^2}\right)\right]\right\} \quad (\Omega) \tag{36}$$

The defined quantity Q is called the Q of the circuit (for quality). A high-Q circuit has low resistance, low fractional energy loss per cycle, and a sharp resonance. The resistive term, represented by the real part of Z, determines the energy dissipated; hence, the energy dissipated by the sample is determined by χ''.

We now obtain the power dissipated by the ESR sample in terms of χ''. Let the current I be given by $I_0 \cos \omega t$; then the instantaneous power dissipated per unit volume of sample is

$$P(t) = \frac{\eta R Q \chi''}{V}(I_0 \cos \omega t)^2 \quad (\text{W/m}^3) \tag{37}$$

where V is the volume of the sample. Equation 37 gives the power absorbed in terms of the macroscopic quantity χ'', whereas equation 27 gives the power absorbed in terms of the microscopic quantum states of the nucleus or atom.

Note in equation 36 that the effect of χ' is to shift the circuit resonance from ω_r due to its contribution to the effective inductance of the coil.

The analysis to determine the impedance of the circuit in Figure M.6a is algebraically clumsy. For this reason we analyze the equivalent circuit shown in Figure M.7. The circuit in Figure M.7 is equivalent to that in Figure M.6a provided that $Q = \omega L_0/R \gg 1$, which is usually the case. The fictitious resistance R' is related to R by $R' = Q^2 R$ and, hence, $Q = \omega L_0/R = R'/\omega L_0$. Since R', L, and C are in parallel, the reciprocal of the impedance of the equivalent circuit is given by

$$\frac{1}{Z} = \frac{1}{R'} - \frac{\omega C}{j} + \frac{1}{j\omega L} \quad (\Omega^{-1}) \tag{38}$$

Substituting equation 33 into equation 38 and using $\omega = \omega_r = 1/\sqrt{L_0 C}$ (the oscillator angular frequency ω will equal ω_r when B_0 is off magnetic resonance and they are

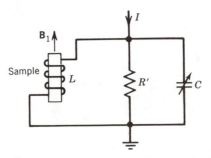

FIGURE M.7 Equivalent circuit to that in Figure M.6a provided $Q = \omega L_0/R \gg 1$.

approximately equal at magnetic resonance) and $Q = R'/\omega L_0$, we have

$$\frac{1}{Z} = \frac{1}{R'}\left[1 - \frac{R'}{j\omega L_0} + \frac{R'}{j\omega L_0(1 + \eta\chi)}\right] = \frac{1}{R'}\left[1 - \frac{Q}{j} + \frac{Q}{j(1 + \eta\chi)}\right] \quad (\Omega^{-1}) \qquad (39)$$

The magnitude of the susceptibility is small, that is, $|\chi| \ll 1$. Doing a binomial expansion of the term $(1 + \eta\chi)^{-1}$, and keeping linear terms in χ yields

$$\frac{1}{Z} = \frac{1}{R'}(1 + jQ\eta\chi) \quad (\Omega^{-1}) \qquad (40)$$

The impedance Z is given by

$$Z = R'(1 + jQ\eta\chi)^{-1} = R'(1 - jQ\eta\chi) \quad (\Omega) \qquad (41)$$

where again the binomial expansion was used and the very small quadratic and higher powers of χ were ignored. Substituting equation 32 into equation 41 yields

$$Z = R'(1 - Q\eta\chi'' - jQ\eta\chi') \quad (\Omega) \qquad (42)$$

It is the real part of Z that represents resistance and, hence, determines the dissipation of rf energy; therefore, the energy dissipated by the sample is determined by χ''.

To obtain the power dissipated by the NMR sample in terms of χ'', we simply replace R in equation 37 with $-R'$.

Both χ' and χ'' are frequency dependent, and they both depend on the spin–lattice and spin–spin relaxation times, T_1 and T_2, respectively, which we now discuss. Your answer for Exercise 3 shows that a system of protons or atomic electrons in thermal equilibrium with its surroundings has a slight excess of population in the lower energy state. If the time-varying field is applied at the resonance frequency with a large amplitude B_1, that is, $B_1^2 \simeq 1/\gamma^2 T_1 T_2$ (this condition on B_1 is discussed later), then the levels would become equally populated and net absorption of power would cease. The system is then said to be *saturated*.

A saturated system is in dynamic equilibrium in that the energy absorbed from the time-varying field is balanced by energy loss to vibrations of neighboring atoms, ions, or molecules. If the field is turned off, then the saturated system returns to equilibrium with its surroundings, where the relative population of levels is given by equation 18. The time required for the system to go from saturation to equilibrium with its neighbors is called the spin–lattice relaxation time T_1. The vibrating neighbors produce a time-varying electric field at the site of a particle. This electric field may produce transitions between energy levels of a particle. $1/T_1$ is defined to be the probability per second of such transitions occurring; hence, spin–lattice relaxation gives rise to lifetime broadening of the energy levels according to the uncertainty principle. For nuclei T_1 typically ranges from 10^{-4} to 10^4 s, depending on the sample. T_1 is temperature dependent for both nuclei and atomic electrons. A large value for T_1 means the transition probability is small; thus, the spin–lattice interaction (interaction of the spin with its neighbors) is weak.

There will be a small spread in frequencies for which absorption can occur. This spread is caused by (a) inhomogeneities in the field produced by the electromagnet, and (b) the magnetic field produced by the neighbors of each particle. This local field will differ from particle to particle depending on the relative orientation of the neighboring particles. We will estimate how the local field produced by a neighboring particle, particle 1, at the site of another particle, particle 2, affects the resonant absorption of energy by particle 2. The field produced by particle 1 at the site of particle 2 is a dipolar field given by

$$\mathbf{B}_d(\mathbf{r}) = \frac{\mu_0}{4\pi r^3}\left[\frac{3(\boldsymbol{\mu}_1 \cdot \mathbf{r})\mathbf{r}}{r^2} - \boldsymbol{\mu}_1\right] \quad (T) \qquad (43)$$

where μ_0 is the vacuum permeability, $\boldsymbol{\mu}_1$ is the magnetic moment of particle 1, and $\mathbf{r}$ is the radius vector from particle 1 to particle 2. This dipolar field is of the order of

$$|\mathbf{B}_{\mathrm{d}}(r)| \simeq \frac{\mu_0}{4\pi} \frac{|\boldsymbol{\mu}_1|}{r^3} \qquad (\mathrm{T}) \qquad (44)$$

The interaction of $\boldsymbol{\mu}_2$ with this field is the usual expression

$$H_{\mathrm{d}} = -\boldsymbol{\mu}_2 \cdot \mathbf{B}_{\mathrm{d}} \simeq -\frac{\mu_0}{4\pi} \frac{|\boldsymbol{\mu}_1||\boldsymbol{\mu}_2|}{r^3} \qquad (\mathrm{J}) \qquad (45)$$

where $\boldsymbol{\mu}_1$ and $\boldsymbol{\mu}_2$ are given by equation 4. This interaction gives rise to a spread in resonant frequencies over a range

$$\Delta v = \frac{1}{h} H_{\mathrm{d}} \simeq \frac{\mu_0}{4\pi h} \frac{|\boldsymbol{\mu}_1||\boldsymbol{\mu}_2|}{r^3} \qquad (\mathrm{Hz}) \qquad (46)$$

Classically, if two particles have precession frequencies differing by Δv and are initially in phase, they will be out of phase in a time $T_2 \simeq 1/\Delta v$. T_2 is called the spin–spin relaxation time and it is a measure of the microscopic field inhomogeneity.

EXERCISE 9

Use equation 46 to estimate Δv and, hence, T_2 for two protons separated by 2 Å, where you may assume $|\mathbf{J}_1| \simeq |\mathbf{J}_2| \simeq \hbar$. Also do the calculation for two free spin electrons.

In 1946 Felix Bloch calculated χ' and χ''. The results of his calculations are

$$\chi' = \frac{1}{2} \chi_0 \omega_0 T_2 \left[\frac{(\omega_0 - \omega)T_2}{1 + (\omega_0 - \omega)^2 T_2^2 + \gamma^2 B_1^2 T_1 T_2} \right] \qquad (\text{dimensionless}) \qquad (47)$$

$$\chi'' = \frac{1}{2} \chi_0 \omega_0 T_2 \left[\frac{1}{1 + (\omega_0 - \omega)^2 T_2^2 + \gamma^2 B_1^2 T_1 T_2} \right] \qquad (\text{dimensionless}) \qquad (48)$$

where χ_0 is the static magnetic susceptibility of the system, ω_0 is the resonant frequency of the system, γ is the gyromagnetic ratio, and B_1 is the amplitude of the time-varying field. If B_1 is small, then saturation of the sample will not occur. This condition to avoid saturation is usually stated as

$$\gamma^2 B_1^2 T_1 T_2 \ll 1 \qquad (\text{dimensionless}) \qquad (49)$$

Then χ' and χ'' become

$$\chi' \simeq \frac{1}{2} \chi_0 \omega_0 T_2 \left[\frac{(\omega_0 - \omega)T_2}{1 + (\omega_0 - \omega)^2 T_2^2} \right] \qquad (50)$$

$$\chi'' = \frac{1}{2} \chi_0 \omega_0 T_2 \left[\frac{1}{1 + (\omega_0 - \omega)^2 T_2^2} \right] \qquad (51)$$

These susceptibilities are plotted in Figure M.8 as a function of the dimensionless product $(\omega_0 - \omega)T_2$. The curve for χ'' shows the resonant character of the absorption, since the time-averaged power absorbed is, from equation 37, given by

$$\langle P \rangle = \frac{\eta R Q I_0^2}{2V} \chi'' \qquad (\mathrm{J/m^3 s}) \qquad (52)$$

Also, the half-width at half-maximum power absorbed occurs when

$$|\omega_0 - \omega| T_2 = 1 \qquad (53)$$

and thus $1/T_2$ is the half-width expressed as an angular frequency.

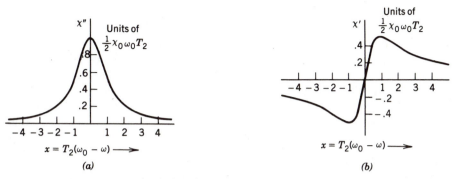

FIGURE M.8 Imaginary and real parts of the magnetic susceptibility $\chi(\omega)$ are plotted in (a) and (b), respectively.

Summary

Note the similarities between NMR and ESR. In both cases a sample containing particles having a magnetic moment $\boldsymbol{\mu}$ is placed in a uniform magnetic field B_0, and a time-varying field $B_1 \cos 2\pi vt$ is applied at right angles to B_0. At magnetic resonance, that is, when $hv = (h\gamma/2\pi)B_0$, the sample absorbs power from the time-varying field. The time-averaged power absorbed is given by both equations 28 and 52. Since $\langle P \rangle \propto \chi''$, the absorption curve (called the **dc line shape**) is proportional to χ'' (see Figure M.8). The width of the absorption curve is determined by T_1 and T_2.

---

15. NUCLEAR MAGNETIC RESONANCE: RADIO-FREQUENCY SPECTROSCOPY

APPARATUS

NMR spectrometer, see Figure 15.1. (Lock-in amplifier is optional.)

Oscilloscope

Gaussmeter

Equipment to measure radio frequency: grid-dip meter, frequency counter, or an oscilloscope with a sweep rate of 0.1 μs/cm or less

Samples: distilled water, Kel F vacuum grease, Teflon, copper sulfate or ferric nitrate (for dissolving in distilled water)

OBJECTIVES

To observe the NMR signal of several samples.

To observe the line width of the NMR signal of hydrogen nuclei in distilled water doped with paramagnetic ions.

To observe "wiggles" and measure the amplitude of these decaying oscillations to determine T_2.

To introduce you to NMR as a technique to probe the quantum states of the nucleus and to gain information about the atomic environment surrounding the nucleus.

KEY CONCEPTS

Radio-frequency (rf) oscillator	Modulated line shape
rf detection	Line width
rt signal	T_1, T_2
Audio-frequency (af) signal	Wiggles
Magnetic field modulation	Modulation spectroscopy
	Derivative of the dc line shape

REFERENCES

1. F. N. H. Robinson, J. Phys. E. **20**, 502 (1987). Dr. Robinson, Clarendon Laboratory, Oxford University, has designed several novel NMR spectrometers, one of which is described in this paper. He also designed and constructed the spectrometer shown in Figure 15.2, which is the spectrometer recommended for this experiment.

2. Appendix A, Modulation Spectroscopy: The Lock-in Amplifier. In this experiment you are asked to record the derivative of the NMR spectral lines. The technique for doing this is discussed in this appendix.

3. See the references listed in Introduction to Magnetic Resonance.

4. The *1989* ARRL *Handbook for the Radio Amateur*, published by the American Radio Relay League, Newington, CT.

INTRODUCTION

Read the Introduction to Magnetic Resonance, which precedes this experiment.

NMR Spectrometer

A block diagram of the NMR spectrometer is shown in Figure 15.1. Figure 15.2 is a photograph of the rf head and the probe. The aluminum tubing (0.750-in. o.d., 0.035-in. wall) has been removed to better show the coil (inductor) and the Teflon ™ coil form, which is also the sample holder. Figure 15.3 is the circuit diagram of the rf head. The circuit is discussed first.

In constructing the circuit use a printed circuit board and 1 percent metal-film resistors. Also use the smallest available resistors ($\frac{1}{8}$ W preferably, but $\frac{1}{4}$ W will do) and capacitors. Monolithic ceramic capacitors are ideal except where electrolytics are indicated. Keep all leads very short and think especially carefully where return currents are flowing in the ground leads. The inductance of a straight nonmagnetic wire with a length much greater than its diameter is approximately given by (ref. 4)

$$L = 0.0002b \left[\left(\ln \frac{2b}{a} \right) - 0.75 \right] \quad (\mu H) \tag{1}$$

where L is the inductance in microhenries, b is the length of the wire in millimeters, a is the wire diameter in millimeters, and ln is the natural logarithm.

EXERCISE 1

Calculate the inductive reactance of 150 mm of No. 24 A.W.G. (American wire gauge) wire for a 25-MHz signal frequency. Note that when laying circuits for use at radio-frequency every lead must be considered as an inductor.

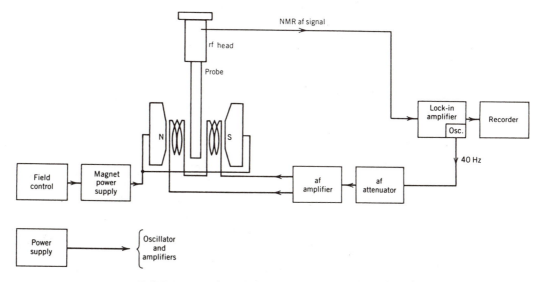

FIGURE 15.1 Block diagram of an NMR spectrometer.

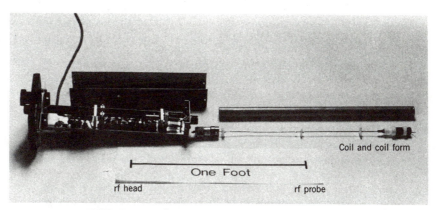

FIGURE 15.2 Photograph of the rf head and probe.

T_1 and T_2 are matched J309 FETs (field-effect transistors) that are configured to form an rf oscillator. The rf signal is fed back to the LC resonance circuit via C_1, a 10-pF capacitor, which sustains the rf oscillations. C_2, a 0-35-pF trimmer capacitor, permits some variation of the rf level. Increasing C_2 decreases the level by reducing the feedback. C_0, a 100-pF maximum variable capacitor, allows the resonance frequency to be tuned from about 15 to 30 MHz. Part of the rf signal output of the oscillator is coupled to T_3 (2N 3904 transistor) via two 1-nF capacitors. This rf signal is available for monitoring, with a frequency meter or oscilloscope, at the BNC connector. T_1 detects (half-wave rectifies) the rf signal. (In addition to the detected rf signal, on the "drain" of T_1 a low-level rf signal is also present. This low-level rf signal is available at the collector of T_3.) The 10-kΩ resistor and two 1-nF capacitors at the base pin of T_4 (2N 2484 transistor) filter the rf signal, but pass the af signal resulting from the low frequency (40 Hz) magnetic field modulation. (T_1 and the RC filter form the rf detector and filter.) T_5 (2N 3906) together with T_4 gives an af gain of 100, and the amplified signal comes off of the collector via the 1-kΩ resistor and 0.1-μF capacitor. T_6 (2N 3904) filters the applied +24 to +30-V dc.

The NMR af (40 Hz) signal shown in Figure 15.1 is the input to the lock-in amplifier, and the output of the lock-in amplifier will be the derivative of the absorption curve. (See Appendix A, Modulation Spectroscopy: The Lock-in Amplifier.) The 40-Hz signal drives

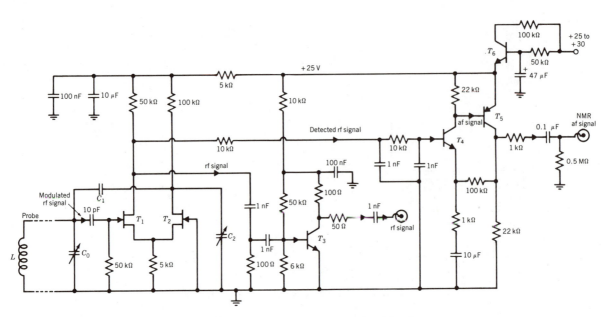

FIGURE 15.3 Circuit diagram of the rf head.

the modulation coils mounted on the poles of the electromagnet, producing a modulation magnetic field $B_m(t)$, which is alternately parallel and antiparallel to B_0, where

$$B_m(t) = B_m \cos 2\pi 40t \qquad (T) \qquad\qquad (2)$$

When one observes the NMR signal on the oscilloscope, it is convenient to modulate the magnetic field with a 60-Hz signal. A circuit for 60-Hz modulation is shown in Figure 15.4a. The amplitude of the modulated field B_m is adjustable by changing the variac setting. It is desirable to drive the horizontal oscilloscope sweep with a 60-Hz signal that has an adjustable phase. A circuit for such a sweep drive is shown in Figure 15.4b.

Remark: If you do not plan to observe the NMR signal by using a lock-in amplifier, then the magnetic field modulation may be done with 60-Hz, that is, the 40-Hz circuit is not needed.

EXERCISE 2

What is the theoretical dependence of B_m on the output voltage of the variac? To answer this question first calculate B_m as a function of the modulation coil current, and then relate the coil current to the primary voltage of the filament transformer. You will need the specifications for your modulation coils: number turns, coil radius, and coil separation.

The rf probe is considered next. The wire that connects the rf head to the coil is No. 16-gauge A.W.G. wire. The wire and aluminum tube form a rigid, air-filled coaxial transmission line. Three Teflon spacers keep the wire centered in the tubing, and two spots of solder on each side of a spacer hold it in place. A Teflon spacer is shown in Figure 15.5a. The Teflon sample holder and the inductor are shown in Figure 15.5b. Note that one end of the sample holder has a conducting pin mounted in it and the No. 16-gauge wire solders to the pin.

The inductor is 12 turns of closely spaced enameled No. 20 A.W.G. wire. The inductor is 11 mm in length, and the wire turns touch each other. The end of the sample holder is closed with a small rubber stopper. (For a liquid sample place a small diameter wire along side the stopper as you insert the stopper. The wire allows the compressed air to escape.

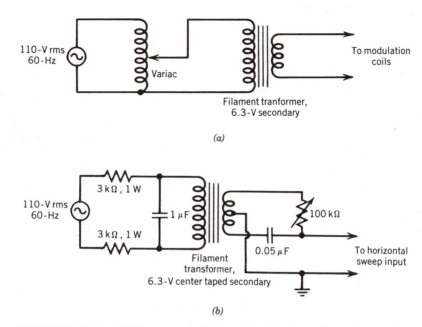

FIGURE 15.4 (a) A 60-Hz magnetic field modulation circuit. (b) A 60-Hz circuit to drive the horizontal sweep of the oscilloscope.

Remove the wire after the stopper is set.) Mount a ground lug on the inside of the tubing for grounding the inductor.

EXERCISE 3

Calculate the inductance L_0 of the coil and, hence, show that L_0 equals 1.6×10^{-6} H.

Figure 15.5c shows machined brass for mounting the ultra high frequency (UHF) connector and the aluminum tubing. The connector solders to the brass, and the tubing is attached with three 2-56 screws.

EXPERIMENT

We first discuss two methods of observing the resonance signal.

1. *Modulated Line Shape.* This method does not use the lock-in amplifier. Connect the signal output of the rf head to the vertical input of the oscillosope. Set B_0 at the resonance value $2\pi v_0/\gamma$, and adjust the amplitude B_m of the 60-Hz magnetic field modulation such that B_m is about 10 times ΔB, the line width. The 60-Hz modulation causes the sample to pass in and out of resonance 120 times per second. Each time the field passes through the resonance value the sample absorbs rf power. Figure 15.6a shows the amplitude-modulated rf signal on the "gate" of T_1. Figure 15.6b shows the filtered signal on the base of T_4, and Figure 15.6c is the modulated line shape output of T_5; however, the amplifications of T_4 and T_5 are not shown. The field $B_m(t)$ is zero at the center of each line shape in Figure 15.6c; one line shape corresponds to $B_m(t)$ sweeping through resonance with a positive slope, whereas the other line·shape corresponds to one half of a period later when $B_m(t)$ has a negative slope.

For clarity, decaying oscillations, called *wiggles*, were intentionally omitted from Figure 15.6c. The wiggles are the oscillations that appear on the side of the resonance curve, *following* the passage through resonance. These decaying oscillations are discussed in references 3 and 4, Introduction to Magnetic Resonance. Analysis of the oscillations shows

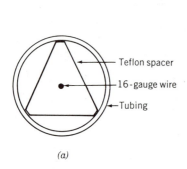

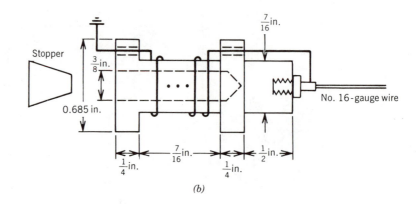

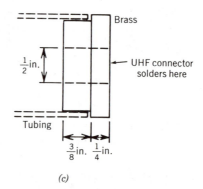

FIGURE 15.5 (a) Teflon spacer. (b) Teflon coil form and sample holder. (c) Brass for mounting the UHF connector.

that the time dependence is given by

$$e^{-t/T_2} \cos\left(\frac{\gamma}{2} t^2 \frac{dB_m}{dt}\right) \tag{3}$$

Suppose the left line shape in Figure 15.6c corresponds to $B_m(t)$ traversing the resonance condition with a negative slope, and the right line shape corresponds to the positive slope of $B_m(t)$. Then the wiggles that follow the passage through resonance are as shown in Figure 15.6d.

What appears as two separate resonances in Figure 15.6 is the result of sweeping $B_m(t)$ through resonance with a positive slope and then a negative slope, which are separated in time by 1/120 s. A sketch of $\langle P \rangle$ versus the total magnetic field $B_0 + B_m(t)$ would show the two curves superimposed with a common origin.

2. *Derivative of the Dc Line Shape.* This method uses the apparatus as configured in Figure 15.1 to record the first derivative of the dc line shape. The technique for doing this is discussed in detail in Appendix A.

Each experiment consists of the following:

1. Placing the sample in a uniform field B_0 and a modulation field $B_m(t)$, which is parallel to B_0.
2. Applying a rf field $B_1 \cos 2\pi v_0 t$ perpendicular to B_0.
3. Detecting the power loss by the rf field as the magnetic field $B_0 + B_m$ is brought to the resonance value $(2\pi/\gamma)v_0$.
4. Finally, measuring B_0, B_m, and v_0 as accurately as possible.

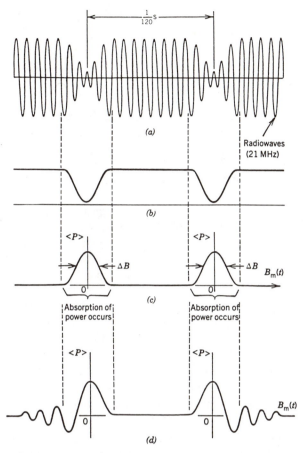

FIGURE 15.6 Refer to Figure 15.3. (a) Modulated signal on the gate of T_1. (b) Detected and filtered (af) signal on the base of T_4. (c) Modulated line shape output of T_5. (d) Wiggles follow the passage through resonance.

Three methods to measure the frequency ν_0 are

1. Use a "grid-dip" meter.
2. Use an oscilloscope with a sweep rate of 0.1 μs/cm or less.
3. Use a frequency counter.

The third method will yield the greatest precision.

A gaussmeter can be used to measure the magnetic field B_0. The gaussmeter should be placed as near the sample position as possible. Also, the poles faces of the electromagnet should be as close together as possible. This has two advantages: The field will be more uniform over the sample volume, and the magnet will regulate to higher values of the field.

In Figure 15.1, B_m is controlled by the af attenuator. A possible af attenuator is a 10-turn 100-kΩ potentiometer, as shown in Figure 15.7. The amplitude B_m may then be calibrated as a function of the potentiometer setting.

EXERCISE 4

If a calibration curve of B_m versus variac setting does not exist, then construct a coil with approximately the following characteristics: 100 turns of No. 28 wire and an area of

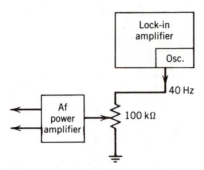

FIGURE 15.7 Audio-frequency attenuator or amplitude controller.

3.14 cm^2. Place the coil at the sample position and use an oscilloscope to measure the induced voltage as a function of the variac setting. Then calculate B_m as a function of variac setting. If you plan to use a lock-in amplifier, carry out a similar calibration for the 40-Hz circuit.

Some possible samples are the following:

1. Protons of hydrogen in distilled water. At room temperature $T_1 = 2.3$ s; thus, there is little lifetime broadening of the energy levels and the line width. A low-intensity rf field is necessary to avoid saturation. This signal will not be easy to observe, but you should be able to observe it by using a fairly large value of magnetic field modulation.

2. Distilled water doped with copper sulfate or ferric nitrate. Doping reduces T_1 from 2.3 s to 10^{-3}–10^{-4} s, depending on the concentration of paramagnetic ions. In this instance, T_1 contributes significantly to the line width and, hence, the line width will be greater than that of protons in distilled water.

3. Fluorine (^{19}F nucleus) in Teflon. The Teflon coil form may be used as a sample; however, the line shape will be wide and, therefore, weak. A stronger and narrower line shape can be obtained with Kel F vacuum grease.

Two methods of observing the resonance have been described. It is suggested that you use both methods to observe the resonance of protons in a doped sample of distilled water. This is also a good sample to use first because the greater line width (as compared with that of distilled water) makes the resonance easier to find. Using method 1 do the following.

1. Observe the resonance on the oscilloscope with $B_m \simeq 10\Delta B$.
2. Knowing ν_0 and B_0, calculate the nuclear g value and compare it with the accepted value.
3. The time dependence of the wiggles is given in expression 3. Measure the amplitude of the wiggles as a function of time, then plot your data and determine T_2.
4. Observe the resonance signal as you vary the amplitude of the modulation field B_m.
5. Prepare four samples with distilled water, each having a different concentration of copper sulfate or ferric nitrate. With B_m set as in Step 1, observe the line width of each sample. As was previously pointed out, the presence of paramagnetic ions in distilled water reduces the relaxation time T_1 of the protons of hydrogen. T_1 due to paramagnetic ions is given by (see reference 3, Introduction to Magnetic Resonance)

$$\left(\frac{1}{T_1}\right)_{\text{ions}} = \frac{9\pi^2 \gamma^4 \hbar^2 \eta N}{10 k_B T} \qquad (1/S) \tag{9}$$

where γ is the gyromagnetic ratio, η is the viscosity of the liquid, N is the number of paramagnetic ions per cubic centimeter, k_B is the Boltzmann constant, and T is the absolute temperature.

Remark: Unless the field of the electromagnet is very uniform, the line width observed may be due to the variation in the magnetic field over the sample rather than to the intrinsic width due to the sample. For such a nonuniform magnetic field it would be difficult to observe the dependence of the line width on the concentration of paramagnetic ions.

EXERCISE 5

Explain how decreasing T_1 causes the line width ΔB to increase. Do you qualitatively observe the expected change in line width as N changes for your four samples?

16. ELECTRON SPIN RESONANCE: MICROWAVE SPECTROSCOPY

APPARATUS

A block diagram of the apparatus is shown in Figure 16.9.

Microwave spectrometer (see Figure 2.15, Experiment 2; the slotted section of waveguide is not needed)

Electromagnet

Modulation coils

Oscilloscope

XY recorder

DPPH(α'-diphenyl-β-picryl hydrazyl) powder

Potassium chrome alum crystal

Gaussmeter

Lock-in amplifier (optional)

OBJECTIVES

To observe the ESR absorption lines by three methods.

To observe an absorption line as a function of microwave power.

To measure the absorption linewidth and to calculate the spin–spin relaxation time T_2.

To determine the electronic g value of the nearly free electrons of DPPH.

To determine the electronic g value and the axial crystalline field parameter D for the potassium chrome alum crystal.

To introduce you to ESR as a technique to probe quantum states of an atom and to gain information about the atomic environment surrounding the atom.

To increase your familiarity with the applications of quantum mechanics.

KEY CONCEPTS

Hamiltonian of a multielectron atom	Electron spin resonance (ESR)
LS coupling	Electronic g value
Term states	Axial crystalline field parameter D
Crystalline electric field	Linewidth
Spin Hamiltonian	Spin–spin relaxation time T_2

REFERENCES

1. Experiment 2, Waveguide. The microwave spectrometer assembled and studied in Experiment 2 is used in this experiment; therefore, Experiment 2 is a prerequisite for this experiment.

2. B. Bleaney and K. Stevens, Paramagnetic resonance, *Rep. Prog. Phys.* **16**, 108–159 (1953). Paramagnetic resonance and ESR are synonymous. This paper includes a discussion of the Hamiltonian of a multielectron atom in a crystalline electric field. Some experimental results are given for ESR studies of Cr^{3+} ion.

3. K. Bowers and J. Owen, Paramagnetic resonance, II, *Rep. Prog. Phys.* **18**, 305 (1955). The spin Hamiltonian and magnetic resonance for Cr^{3+} ion are discussed.

4. C. Whitmer, R. Weidner, J. Hsiang, and P. Weiss, *Phys. Rev.* **74**, 1478 (1948). This paper reports on electron spin resonance absorption in potassium chrome alum, which is a crystal you are asked to investigate in this experiment.

5. R. Eisberg and R. Resnick, *Quantum Physics of Atoms, Molecules, Solids, Nuclei, and Particles*, 2d ed., Wiley, New York, 1985. *LS* coupling is discussed on pages 356–361, spin–orbit interaction is discussed on pages 278–281 and in Appendix O, and the angular dependence of electron probability density is presented on pages 249–252.

6. A. Abragam and B. Bleaney, *Electron Paramagnetic Resonance of Transition Metal Ions*, Clarendon, Atlanta, 1970. A comprehensive text. Energy levels of Cr^{3+} are discussed on many pages.

7. A. Holden and P. Singer, *Crystals and Crystal Growing*, Doubleday, New York, 1960. Instructions are given for growing alum crystals.

8. Appendix A, Modulation Spectroscopy: The Lock-in Amplifier. An optional part of this experiment is recording the derivative of the spectral lines. The technique for doing this is described in Appendix A.

Also see the references listed under Introduction to Magnetic Resonance, which precedes Experiment 15.

INTRODUCTION

The Introduction to Magnetic Resonance, which precedes Experiment 15, is a corequisite for this experiment. This experiment also uses the microwave spectrometer assembled in Experiment 2; therefore, Experiment 2 is a prerequisite.

In this introduction the ESR spectra arising from Cr^{3+} ions in a potassium chrome alum crystal is discussed. The topics are (1) the Hamiltonian for the trivalent chromium ion, (2) *LS* coupling and the resulting term states, (3) spin–orbit interaction, (4) crystalline electric field, (5) spin Hamiltonian, (6) microwave spectrometer, and (7) growing potassium chrome alum crystals.

Hamiltonian of the Cr^{3+} Ion

In the ground state the electron configuration of Cr^{3+} is $1s^2, 2s^2\, 2p^6, 3s^2\, 3p^6\, 3d^3$. The general notation is $n\ell^q$, where $q =$ the number of electrons; n is the principal quantum number, where $n = 1, 2, 3, 4, \ldots$; and ℓ is the orbital angular momentum quantum number, and for a given value of n, $\ell = 0(s), 1(p), 2(d), 3(f), 4(g), \ldots, n-1$, where the spectroscopic notation for each value of ℓ is enclosed in parentheses. The magnitude L of the orbital angular momentum of an electron is given by

$$L = \hbar\sqrt{\ell(\ell + 1)} \quad \text{(Js)} \tag{1}$$

Also, m_ℓ is the azimuthal quantum number, and for a given value of ℓ,

$m_\ell = -\ell, -\ell + 1, \ldots, \ell - 1, \ell$. The z component L_z of the orbital angular momentum is quantized:

$$L_z = \hbar m_\ell \qquad \text{(Js)} \tag{2}$$

m_s is the spin quantum number, and for a single electron $m_s = +\frac{1}{2}$ or $-\frac{1}{2}$. The z component S_z of the spin angular momentum is quantized:

$$S_z = \hbar m_s \qquad \text{(Js)} \tag{3}$$

Also, the magnitude S of the spin angular momentum of a single electron is given by

$$S = \hbar\sqrt{s(s + 1)} \qquad \text{(Js)} \tag{4}$$

where $s = \frac{1}{2}$.

The four quantum numbers n, ℓ, m_ℓ, m_s specify the quantum state or energy of noninteracting electrons and the Pauli exclusion principle states that no two electrons may have identical quantum numbers.

For a given value of n, electrons with higher ℓ have a higher orbital angular momentum (a more positive kinetic energy) and on the average are farther from the nucleus, which means they are shielded from the nucleus by the other electrons and have a less negative Coulomb potential energy (a more positive total energy). Thus, for a given value of n, energy states with higher ℓ have a higher (more positive) energy. This is shown in Figure 16.1, where the electron configuration of Cr^{3+} in the ground state is indicated. The interactions that produce the set of energy levels in the figure, electron kinetic energy K plus electron-shielded nucleus potential energy V, are indicated at the bottom of the figure.

We want to determine the possible energy levels of the $3d^3$ electrons in some detail. We will not solve the Schrödinger equation for the eigenfunctions and eigenvalues; instead, we discuss the eigenvalues of each interaction qualitatively. We first make the useful approximation that the nucleus and the electrons in closed shells (argon core) form a spherically symmetric potential with charge $Z_{eff}e$, where Z_{eff} is an effective atomic number. The nonrelativistic Hamiltonian for an infinitely massive nucleus of the $3d^3$ electrons in an external static field $\mathbf{B}_0 = B_0\hat{z}$ and a crystalline electric field is given by

$$H = \sum_{i=1}^{3} \left(\underbrace{-\frac{\hbar^2}{2m}\nabla_i^2}_{K} \underbrace{- k\frac{Z_{eff}e^2}{r_i}}_{V} + \underbrace{\frac{1}{2}\sum_{\substack{j=1 \\ j \neq i}}^{3} k\frac{e^2}{|\mathbf{r}_i - \mathbf{r}_j|}}_{H_{e\text{-}e}} \underbrace{- \sum_{j=1}^{6} \frac{eP_j}{|\mathbf{r}_i - \mathbf{R}_j|^2}\cos\theta_{ij}}_{V_c} + \underbrace{\zeta(r_i)\mathbf{L}_i \cdot \mathbf{S}_i}_{H_{so}} \underbrace{- \boldsymbol{\mu}_i \cdot \mathbf{B}_0}_{H_Z} \right)$$

$$\tag{5}$$

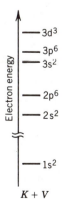

FIGURE 16.1 Electron configuration of Cr^{3+} in the ground state.

where the interactions are listed in order of decreasing strength, $k \equiv 1/(4\pi\varepsilon_0)$, and

K is the kinetic energy of the $3d^3$ electrons;

V is the Coulomb interaction of the $3d^3$ electrons with the core;

$H_{\text{e-e}}$ is the Coulomb interaction among the $3d^3$ electrons, where $\mathbf{r}_i$ is the position of the ith electron with respect to the nucleus;

V_c is the interaction of the $3d^3$ electrons with the electric dipole moments of the six nearest neighboring water molecules having dipole moment $\mathbf{P}_j$ and position $\mathbf{R}_j$ with respect to the Cr^{3+} nucleus, where θ_{ij} is the angle between $\mathbf{P}_j$ and a line from $\mathbf{P}_j$ to the electron located at $\mathbf{r}_i$ (see Figure 16.2);

H_{so} is the spin–orbit interaction;

H_Z is the Zeeman interaction, where $\boldsymbol{\mu}_i$, the total magnetic moment of the ith electron, is the sum of the orbital $\boldsymbol{\mu}_L$ and spin $\boldsymbol{\mu}_S$ magnetic moments, which may be written in terms of $\mathbf{L}_i$ and $\mathbf{S}_i$:

$$\boldsymbol{\mu}_i = \boldsymbol{\mu}_L + \boldsymbol{\mu}_S = -\frac{e}{2m}\mathbf{L}_i - \frac{ge}{2m}\mathbf{S}_i \tag{6}$$

where g is the electron g factor, 2.0023 for a free spin.

The abundance of stable Cr nuclei having a nuclear angular momentum $\mathbf{I}$ that is zero is 90.45 percent; therefore, the hyperfine and nuclear Zeeman interactions are not included in equation 5.

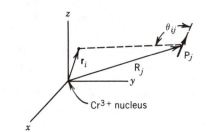

FIGURE 16.2 The ith 3d electron of Cr^{3+} and the jth water molecule, with electric dipole moment $\mathbf{P}_j$, of the chrome alum crystal are shown relative to the Cr^{3+} nucleus.

LS Coupling and Term States

In equation 5, note that the first two interactions are both sums of single particle interactions; therefore, for these interactions the $3d^3$ electrons move independently of each other. Since their motion is independent of each other, the spatial quantum states that these electrons occupy are labeled with spatial quantum numbers: n_i, ℓ_i, m_{ℓ_i}, where $i = 1, 2, 3$. The electron–electron interaction is an interaction between pairs of electrons and therefore the energy of each electron depends on the relative position of the other two electrons. This interaction and the requirement that electrons satisfy the Pauli exclusion principle give rise to energies of the $3d^3$ electrons that depend on the total orbital angular momentum $\mathbf{L}$ and the total spin angular momentum $\mathbf{S}$, where

$$\mathbf{L} = \mathbf{L}_1 + \mathbf{L}_2 + \mathbf{L}_3 \quad \text{and} \quad \mathbf{S} = \mathbf{S}_1 + \mathbf{S}_2 + \mathbf{S}_3 \tag{7}$$

The magnitudes L and S are given by

$$L = \hbar\sqrt{\ell(\ell + 1)} \quad \text{and} \quad S = \hbar\sqrt{s(s + 1)} \quad \text{(Js)} \tag{8}$$

where the possible quantum numbers ℓ and s are determined by ℓ_i and s_i, $i = 1, 2, 3$ (in this case $\ell_1 = \ell_2 = \ell_3 = 2$ and $s_1 = s_2 = s_3 = \frac{1}{2}$). This coupling to form $\mathbf{L}$ and $\mathbf{S}$ is called LS or Russell–Saunders coupling.

EXERCISE 1

Carry out LS coupling for the $3d^3$ electrons. Show that the possible values of s are $\frac{1}{2}(2)$, $\frac{3}{2}(1)$ and that the possible values of ℓ are $0(1)$, $1(3)$, $2(5)$, $3(4)$, $4(3)$, $5(2)$, $6(1)$, where the number in parentheses is the number of times the value occurs; for example, $\ell = 2$ occurs five times.

Obtaining the possible values of ℓ and s in Exercise 1 is primarily an exercise in vector addition and does not involve physical reasoning. We now require the values of ℓ and s to be physically correct; that is, we require that pairs of values of ℓ and s satisfy the Pauli exclusion principle. The Pauli exclusion principle can be stated in two equivalent ways: (1) In a multielectron atom there can never be more than one electron in the same quantum state, and (2) a multielectron atom must be described by an antisymmetric eigenfunction, where the eigenfunction is a product of a space function and a spin function. The theoretical details are beyond what we want to do in this experiment; however, the pairs of values of ℓ and s that satisfy the Pauli exclusion principle are given in Table 16.1. Spectroscopic notation is used for ℓ values in the table; that is, $\ell = 0(S)$, $1(P)$, $2(D)$, $3(F)$, $4(G)$, $5(H)$, $6(I)$, ..., and the *term* designation is $^{2s+1}\ell$. Often texts will use L and S to denote the total orbital and spin quantum numbers and the term designation becomes ^{2S+1}L. A term or term state is a group of states specified by ℓ and s (or L and S).

Hund's rule for determining the ground term of a configuration is the following: The ground term of a configuration is the term with the highest allowed value of the total spin and the highest value of the total orbital angular momentum consistent with the first requirement. Hence, for the $3d^3$ configuration the ground term will be the ^4F term.

In Figure 16.3 each interaction in equation 5, except for H_Z, and its effect on the energy levels of the $3d^3$ electrons is shown. The interaction labeled H_0 is defined to be $K + V$, the first two interactions in equation 5. The interaction $H_{e\text{-}e}$ separates the term energy levels designated by $^{2s+1}\ell$ as shown in Figure 16.3b. As specified by Hund's rule, the ground term is the ^4F term. The units used in the figure, cm^{-1} or inverse centimeters, are the units of wave number or reciprocal wavelength, which are standard in spectroscopy. Wave number is

TABLE 16.1 TERM STATES FOR THE $3d^3$
ELECTRONS OF Cr^{3+}

s	ℓ	$^{2s+1}\ell$
$\frac{1}{2}$	$1(P)$	^2P
$\frac{3}{2}$	$1(P)$	^4P
$\frac{1}{2}$	$2(D)$	^2D
$\frac{1}{2}$	$3(F)$	^2F
$\frac{3}{2}$	$3(F)$	^4F
$\frac{1}{2}$	$4(G)$	^2G
$\frac{1}{2}$	$5(H)$	^2H

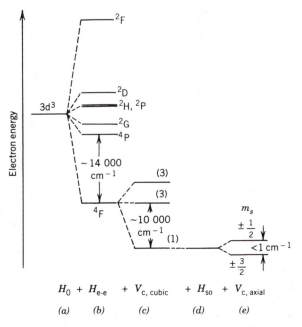

FIGURE 16.3 Each interaction in equation 5, with the exception of H_z, and its effect on the energy levels of the 3d³ electrons is shown.

defined as 1/wavelength or energy/hc, where h and c are Planck's constant and the velocity of light.

EXERCISE 2

If an electron, initially in the ⁴F level, absorbs a photon and makes a transition to the ⁴P level, then what is the frequency of the photon? Would the study of such transitions be infrared, optical, or UV spectroscopy?

Spin–Orbit Interaction

It is perhaps worthwhile to point out the physical basis of the $\mathbf{L} \cdot \mathbf{S}$, spin–orbit interaction. In the frame of reference of the nucleus there exists an electrostatic field $\mathbf{E}$ arising from the shielded nucleus, and a 3d electron moves with some velocity $\mathbf{v}$ through this field. In the frame of reference of the electron spin, the shielded nucleus orbits the electron. The electron spin then sees an effective magnetic field $\mathbf{B}' = -(\mathbf{v}/c^2) \times \mathbf{E}$ and has the ordinary magnetic interaction $-\boldsymbol{\mu}_S \cdot \mathbf{B}'$, modified by the Thomas factor, $\frac{1}{2}$ (see reference 5). For a spherically symmetric potential $\phi(r) = kZ_{\text{eff}}e/r$, then $\mathbf{E} = -\nabla\phi(r) = -(\partial\phi/\partial r)(\mathbf{r}/r)$, and $\mathbf{B}'$ is readily shown to be proportional to $\mathbf{r} \times m\mathbf{v} = \mathbf{L}$, giving finally the interaction $\zeta(r)\mathbf{L} \cdot \mathbf{S}$, where $\zeta(r)$ is defined to be the r dependence and constant that multiply $\mathbf{L} \cdot \mathbf{S}$.

EXERCISE 3

Carry out the steps to show that the ordinary magnetic interaction, modified by the Thomas factor of $\frac{1}{2}$, $-\boldsymbol{\mu}_S \cdot \mathbf{B}'/2$ gives the spin–orbit interaction. Estimate the magnitude of the spin–orbit interaction by assuming: $Z_{\text{eff}} = 3$, $r = 3^2 a_0$ ($a_0 = $ 1st Bohr orbit $=$ 0.529×10^{-10} m), $\mathbf{S} \cdot \mathbf{L} = \hbar^2$. Express your answer in joules and in electron-volts.

Crystalline Electric Field

The potassium chrome alum crystal contains both trivalent aluminum Al^{3+} and chromium Cr^{3+}. Each positively charged trivalent ion is at the center of a nearly regular octahedron of water molecules. (A regular octahedron has eight equilateral triangular surfaces.) This is shown in Figure 16.4, where the trivalent ion is at the origin O and each dot represents a water molecule. The six neighboring water molecules are located at $(\pm p, 0, 0)$, $(0, \pm q, 0)$, and $(0, 0, \pm r)$, where p, q, and r are approximately 2 Å. The negative end of the electric dipole moment of each water molecule points inward toward the positively charged trivalent ion.

The octhedral complex is not regular, rather it is stretched or compressed along a body diagonal of the complex. This distortion along an axis is called an *axial* distortion, and the axis is labeled as the z axis. (The z axis is no longer arbitrary.) The unit cell of an alum crystal contains four nonequivalent octahedral complexes. The complexes are nonequivalent in that the axial distortions are along different body diagonals of each complex.

To determine the effect of the crystal on the trivalent ion we use the following model. We assume the six water molecules surrounding the trivalent ion give rise to an electrostatic potential energy V_c. The potential energy V_c is assumed to arise from the electronic and nuclear charges of the six molecules. To calculate this potential energy we use the electric dipole approximation, where each water molecule is replaced by an electric dipole moment. Note that this model assumes that the $3d^3$ electrons of Cr^{3+} are localized on the ion itself, and that the atomic electrons of each water molecule are localized on the molecule. That is, the wave functions of these atomic electrons do not overlap; hence, the model assumes there is no covalent bonding.

The electric field produced by the water molecules at the trivalent ion site is called the crystalline electric field. Since this field originates from the water molecules making up the crystal, it must have the same point symmetry as the ion site. Of course, V_c must have the same point symmetry, and this point symmetry is used to determine V_c. The crystalline electric field produced by the six water molecules at the site of the Cr^{3+} ion can be expressed as a superposition of fields, a dominant one possessing cubic symmetry and a small component possessing axial (trigonal) symmetry. The interaction of the d electrons with these fields is written as

$$V_c = V_{c,\text{cubic}} + V_{c,\text{axial}} \tag{9}$$

where $V_{c,\text{cubic}}$ is the interaction of the 3d electrons with the field having cubic symmetry, and so on.

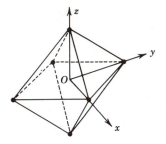

FIGURE 16.4 A positively charged trivalent ion, located at the origin O, is surrounded by six water molecules, represented by dots.

To understand the effect of the crystalline electric field on the 3d orbitals we simplify; that is, we consider an ion at the center of the octahedral complex that has a single 3d electron with $\ell = 2$, $s = \frac{1}{2}$, and we ignore the spin–orbit interaction. The ground term of an ion with a single 3d electron is ^2D, and each of the five angular momentum states corresponding to the five values of $m_\ell = -2, -1, 0, 1, 2$ have the same energy. The angular part of the orbitals are usually written in terms of spherical harmonics $Y_\ell^{m_\ell}(\theta, \phi)$, or in Dirac notation as $|\ell, m_\ell\rangle$. We express the five degenerate orbitals of the ^2D term state as functions of (θ, ϕ) and (x, y, z):

$$|2, 2\rangle = \frac{1}{4}\sqrt{15/2\pi}\, e^{j2\phi} \sin^2\theta = \frac{1}{4}\sqrt{15/2\pi}\, \frac{(x + jy)^2}{r^2}$$

$$= \frac{1}{4}\sqrt{15/2\pi}\,[(x^2 - y^2) + j2xy]\frac{1}{r^2}$$

$$|2, 1\rangle = -\frac{1}{2}\sqrt{15/2\pi}\, e^{j\phi} \sin\theta \cos\theta = -\frac{1}{2}\sqrt{15/2\pi}\, \frac{z(x + jy)}{r^2}$$

$$= -\frac{1}{2}\sqrt{15/2\pi}\,(zx + jzy)\frac{1}{r^2}$$

$$|2, 0\rangle = \frac{1}{2}\sqrt{5/4\pi}\,(3\cos^2\theta - 1) = \frac{1}{4}\sqrt{5/\pi}\, \frac{3z^2 - r^2}{r^2} \qquad (1/\sqrt{\text{steradians}}) \qquad (10)$$

$$|2, -1\rangle = \frac{1}{2}\sqrt{15/2\pi}\, e^{-j\phi} \sin\theta \cos\theta = \frac{1}{2}\sqrt{15/2\pi}\, \frac{z(x - jy)}{r^2}$$

$$= \frac{1}{2}\sqrt{15/2\pi}\,(zx - jyz)\frac{1}{r^2}$$

$$|2, -2\rangle = \frac{1}{4}\sqrt{15/2\pi}\, e^{-j2\phi} \sin^2\theta = \frac{1}{4}\sqrt{15/2\pi}\, \frac{(x - jy)^2}{r^2}$$

$$= \frac{1}{4}\sqrt{15/2\pi}\,[(x^2 - y^2) - j2xy]\frac{1}{r^2}$$

The angular probability density $Y_\ell^{m_\ell} \cdot Y_\ell^{m_\ell *}$ (units of probability per steradian) is sketched in Figure 16.5 for $\ell = 2$. The $\pm$ sign in $|2, \pm m_\ell\rangle$ implies opposite rotations about the z axis. Equations 10 show the ϕ dependence of $Y_\ell^{m_\ell}$ is $e^{jm_\ell \phi}$; therefore, the angular probability density will not be a function of ϕ, and we define $Y_\ell^{m_\ell}(\theta, \phi) \cdot Y_\ell^{m_\ell *}(\theta, \phi) \equiv \Theta_\ell^{m_\ell}(\theta) \cdot \Theta_\ell^{m_\ell *}(\theta)$. Each sketch in Figure 16.5 is valid for all angles ϕ, and the three-dimensional probability density is obtained by rotating about the z axis. The angular probability density is a plot of $\Theta_\ell^{m_\ell}(\theta) \cdot \Theta_\ell^{m_\ell *}(\theta)$ in the radial direction. This is shown for one value of θ in the sketch of $||2, \pm 2\rangle|^2$.

The spherical harmonics are appropriate with the spherical symmetry of a free atom or

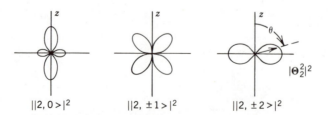

$$||2, 0\rangle|^2 \qquad\qquad ||2, \pm 1\rangle|^2 \qquad\qquad ||2, \pm 2\rangle|^2$$

FIGURE 16.5 Angular probability densities corresponding to the five 3d orbitals, where each orbital is a single spherical harmonic.

ion; however, in crystals it is more convenient to replace the spherical harmonics with the following linear combinations:

$$|2, xy\rangle = -\frac{j}{\sqrt{2}}|2, 2\rangle + \frac{j}{\sqrt{2}}|2, -2\rangle = \frac{1}{2}\sqrt{15/\pi}\,\frac{xy}{r^2}$$

$$|2, yz\rangle = \frac{j}{\sqrt{2}}|2, 1\rangle + \frac{j}{\sqrt{2}}|2, -1\rangle = \frac{1}{2}\sqrt{15/\pi}\,\frac{yz}{r^2}$$

$$|2, zx\rangle = -\frac{1}{\sqrt{2}}|2, 1\rangle + \frac{1}{\sqrt{2}}|2, -1\rangle = \frac{1}{2}\sqrt{15/\pi}\,\frac{zx}{r^2} \quad (1/\sqrt{\text{steradians}}) \qquad (11)$$

$$|2, x^2 - y^2\rangle = \frac{1}{\sqrt{2}}|2, 2\rangle + \frac{1}{\sqrt{2}}|2, -2\rangle = \frac{1}{4}\sqrt{15/\pi}\,\frac{x^2 - y^2}{r^2}$$

$$|2, 3z^2 - r^2\rangle = |2, 0\rangle = \frac{1}{4}\sqrt{5/\pi}\,\frac{3z^2 - r^2}{r^2}$$

Probability densities are sketched in Figure 16.6. The dots represent the water molecules of the octahedral complex.

What happens to the ^2D ground term in the presence of a cubic crystalline field? The six water molecules, each located equidistant from the central ion shown in Figure 16.4, give rise to a cubic field. The d electron in the orbital $|2, x^2 - y^2\rangle$ or $|2, 3z^2 - r^2\rangle$ has a higher probability of being near the negative charge of the water dipole moment, and, hence, has a more positive electrostatic potential energy than in the orbitals $|2, xy\rangle$, $|2, yz\rangle$, or $|2, zx\rangle$. Thus, the cubic field splits the fivefold orbital degenerate ^2D ground term into a higher energy twofold level and a lower energy threefold level. This is shown in Figure 16.7b, where the orbital degeneracy is given in parentheses.

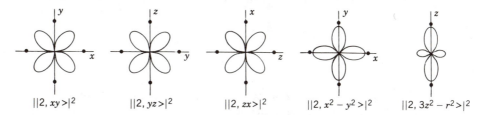

$||2, xy\rangle|^2 \qquad ||2, yz\rangle|^2 \qquad ||2, zx\rangle|^2 \qquad ||2, x^2 - y^2\rangle|^2 \qquad ||2, 3z^2 - r^2\rangle|^2$

FIGURE 16.6 Angular probability densities corresponding to 3d orbitals that are linear combinations of spherical harmonics. Each dot represents a water molecule of the octahedral complex.

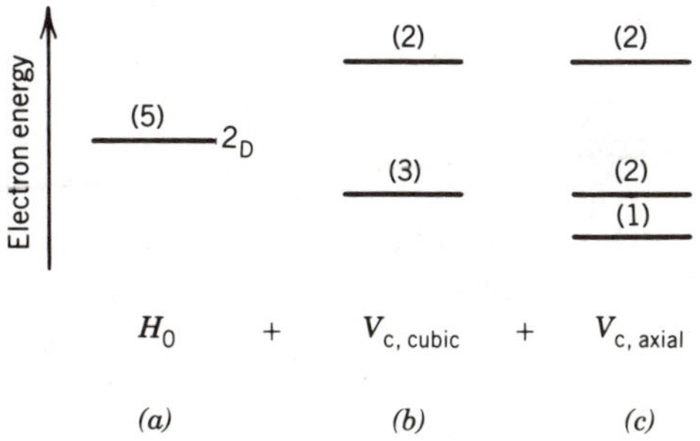

FIGURE 16.7 The splitting of the ground term of a 3d^1 electron by a cubic and then an axial crystalline field is shown. The degeneracy of each level is in parentheses. The spin−orbit interaction has been ignored, and the splittings are not to scale.

If the octahedral complex is distorted slightly from cubic symmetry, additional splittings may occur. Suppose an axial distortion causes the water molecules on the x axis to be slightly farther from the central ion than the other four molecules. In this case the electron in the $|2, zx\rangle$ orbital will experience the least dipole electric field, and, hence, have a less positive electrostatic potential energy, and, thus, a lower energy. This is shown in Figure 16.7c. *Summary*: A cubic crystalline field plus an axial field split the fivefold orbital degenerate ^{2}D term into a ground orbital singlet, the $|2, zx\rangle$ orbital, and two excited orbital levels, each twofold degenerate.

The interaction of the $3d^3$ electrons with the crystalline electric is more complex than that of a single 3d electron, and, although the basic mechanism is the same, we only summarize the results for this case.

The ^{4}F ground term of Cr^{3+} is sevenfold degenerate in orbital angular momentum since $m_\ell = -3, -2, -1, 0, 1, 2, 3$. The interaction $V_{c,cubic}$ splits the ^{4}F level into an orbital singlet and two orbital triplets, where the singlet energy level has the lowest energy. This is shown in Figure 16.3c, where the orbital degeneracy of each level is given in parentheses. In general the singlet level and the two triplet levels are all linear combinations of the spherical harmonics, $Y_\ell^{m_\ell}$, where $\ell = 3$. We will be interested in only the ground singlet since at the temperature of this experiment (room temperature) the probability of either triplet state being occupied is small.

EXERCISE 4

The Hamiltonian for a single Cr^{3+} ion, equation 5, assumes independent noninteracting Cr^{3+} ions, which implies a spatial separation and spatial distinguishability of ions. This system of ions is in thermal equilibrium with the alum crystal lattice of temperature T, and the probability that the energy level E_a is occupied is given by Maxwell–Boltzmann statistics:

$$P(E_a) \propto \exp\left(-\frac{E_a}{k_B T}\right)$$

What is the ratio of the probabilities that the first triplet orbital level is occupied to that of the ground singlet level at 300 K?

The spin–orbit interaction is smaller than $V_{c,cubic}$ and larger than $V_{c,axial}$; hence, we consider it next. What effect does the spin–orbit interaction have on the ground singlet level? Well, for a term state, the spin–orbit interaction may be written

$$H_{so} = \lambda \mathbf{L} \cdot \mathbf{S} \tag{12}$$

where, for a given term state, λ is a constant and the magnitudes L and S are, for the ^{4}F term state, $L = \hbar[3(3 + 1)]^{1/2}$ and $S = \hbar[\frac{3}{2}(\frac{3}{2} + 1)]^{1/2}$. A theorem by Van Vleck states that the expectation value of an operator, linear in orbital angular momentum, vanishes for an orbital singlet state. Hence, H_{so} does not affect the ground state, as shown in Figure 16.3d.

Historical Remark

The 1977 Nobel prize in Physics was divided equally between
    Philip W. Anderson, the United States; Nevill F. Mott, Great Britain, and John H. Van Vleck, the United States

For their fundamental theoretical investigations of the electronic structure of magnetic and disordered systems.

Note that these recipients of the Nobel prize were not cited for a single outstanding achievement. Usually the recipients of the prize are cited for a single outstanding piece of work.

The ground state shown in Figure 16.3d has $s = \frac{3}{2}$ and $m_s = \frac{3}{2}, -\frac{1}{2}, \frac{1}{2}, \frac{3}{2}$; hence, it is fourfold degenerate in spin. The axial field splits the ground state into two spin doublets as shown in Figure 16.3e.

Spin Hamiltonian

The splitting due to the axial field can be described by the operator H_c:

$$H_c = -D\left(\frac{1}{\hbar^2} S_z^2 - \frac{5}{4}\right) \tag{13}$$

where D, called the axial crystalline field parameter, depends on the strength of the axial field and has units of energy. The eigenvalues of H_c, E_c, are given by

$$E_c(m_s) = -D(m_s^2 - \tfrac{5}{4}) \qquad \text{(J)} \tag{14}$$

The application of an applied magnetic field will lead to the Zeeman splitting of the two doublets. Assuming the magnetic field is along the axial field or z direction, the Zeeman interaction is given by

$$H_Z = \frac{e}{2m} g_\parallel B_0 S_z \tag{15}$$

where $g_\parallel$, called *g parallel*, is the spectroscopic g factor when the static field $\mathbf{B}_0$ is parallel to the axial field. (Compare equation 15 with equation 9 in the Introduction to Magnetic Resonance.) The spectroscopic g factor differs from the free spin value (2.0023) in that the crystalline field modifies the g value through the indirect effect of the spin–orbit interaction, changing the magnitude of the g factor and converting it to a tensor quantity with different values depending on whether B_0 is parallel or perpendicular to the axial field.

For the purpose of this experiment each Cr^{3+} ion is *a spin-$\frac{3}{2}$ particle* in an axial electric field and a static magnetic field. The Hamiltonian for such a system is called the spin Hamiltonian H_s and is the sum of equations 13 and 15:

$$H_s = \frac{e}{2m} g_\parallel B_0 S_z - D\left(\frac{1}{\hbar^2} S_z^2 - \frac{5}{4}\right) \tag{16}$$

The eigenfunctions of a spin-$\frac{3}{2}$ particle are, in Dirac notation, $|s, m_s\rangle$, where $s = \frac{3}{2}$ and $m_s = -\frac{3}{2}, -\frac{1}{2}, \frac{1}{2}, \frac{3}{2}$. The spin Hamiltonian and these simplified eigenfunctions are useful tools to describe the low-lying energy states; however, it should be recognized that the real eigenfunctions for these states are complicated mixtures of the spin and orbital functions of $3d^3$ electrons.

The time-independent Schrödinger equation for these low-lying states is

$$H_s |s, m_s\rangle = E(m_s)|s, m_s\rangle \tag{17}$$

EXERCISE 4

Show that the general eigenvalue $E(m_s)$ is given by

$$E(m_s) = \frac{e\hbar}{2m} g_\parallel B_0 m_s - D\left(m_s^2 - \frac{5}{4}\right)$$

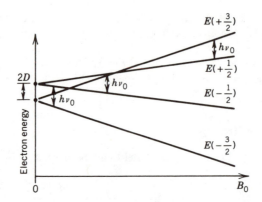

FIGURE 16.8 Splitting of the two spin doublets as a function of the magnetic field B_0. Magnetic dipole transitions ($\Delta m_s = \pm 1$) occur when the photon energy $h\nu_0$ equals the energy level separation.

Show that the spacing between the two spin doublets in zero magnetic field is $2D$, as shown in Figure 16.8. Also convince yourself that the general behavior of the eigenvalues with increasing magnetic field strength is as indicated in Figure 16.8.

EXERCISE 6

For the chrome alum crystal you are asked to study, reference 4 reports $g_{\parallel} = 1.98$ and $2D = 0.091$ cm^{-1} (their δ is the same as $2D$). Knowing these values and the resonance frequency ν_0 for your cavity, use conservation of energy, $h\nu_0 = E(m_s) - E(m_s - 1)$, to calculate the field B_0 in teslas for each of the three absorption lines. (In Table 1, reference 4, δ is mistakenly reported as 0.91 and it should be 0.091.)

The absorption of microwave power by the sample was discussed in the Introduction to Magnetic Resonance and will not be repeated here. We do point out that the interaction of the magnetic moment of Cr^{3+} with the microwave field $\mathbf{B}_1(t)$ is

$$H_1 = -\boldsymbol{\mu}_s \cdot \mathbf{B}_1 = \frac{e}{2m} g_{\perp} \mathbf{S} \cdot \mathbf{B}_1 = \frac{e}{2m} g_{\perp} S_x B_1 \cos \omega t \qquad (18)$$

where $g_{\perp}$, called *g perpendicular*, is the spectroscopic *g* factor when the magnetic field (here $\mathbf{B}_1$) is perpendicular to the axial field. (Compare equation 18 with equation 21, Introduction to Magnetic Resonance.)

The interaction H_1 will produce transitions between the energy levels shown in Figure 16.8. The calculation of transition probability, power absorption, and so on, is done in the Introduction to Magnetic Resonance. Transitions between energy levels produce magnetic dipole radiation, and the selection rule is $\Delta m_s = \pm 1$. See your answer for Exercise 5, Introduction to Magnetic Resonance.

Microwave Spectrometer

The microwave or ESR spectrometer is shown in Figure 16.9. The 40-Hz reference signal from the lock-in amplifier drives the modulation coils via the audio-frequency (af) attenuator and amplifier. (If you are not using a lock-in amplifier, a 40-Hz sine wave oscillator may be used to drive the coils via the attenuator and amplifier.) The field produced by the coils

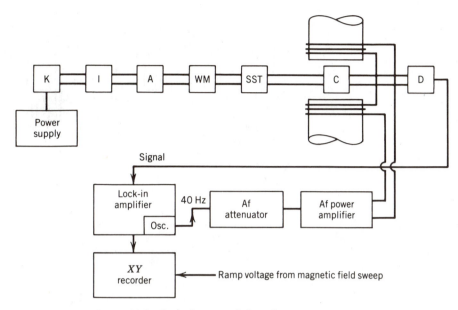

FIGURE 16.9 Block diagram of the microwave spectrometer.

is alternately parallel and antiparallel to B_0 (field of the electromagnet); hence, it *modulates* B_0. This modulation field is given by

$$B_m(t) = B_m \cos 2\pi 40t \qquad (T) \tag{19}$$

Three methods of observing the magnetic resonance signal will be discussed. Methods 1 and 2 do not use a lock-in amplifier, and in these two cases it is necessary to have an op-amp with a gain of about 25 following the diode D in Figure 16.9. (Operational amplifiers, op-amps, are discussed in reference 5, Experiment 6.) The three methods are as follows:

1. *Oscilloscope Observation of the Magnetic Resonance Signal.* Set B_0 and the frequency v at magnetic resonance values, adjust B_m until it is about five times ΔB (the linewidth), and connect the diode to the oscilloscope via the op-amp shown in Figure 16.10. (In place of the lock-in amplifier, drive the modulation coils with a 40-Hz sine wave oscillator.) The 40-Hz modulation causes the sample to pass in and out of resonance 80 times per second. Each time the sample passes through resonance it absorbs power from the microwave field. Figure 16.11*a* shows the microwave signal incident on the diode and Figure 16.11*b*

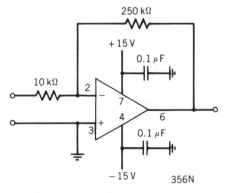

FIGURE 16.10 Operational amplifier having a gain of 25.

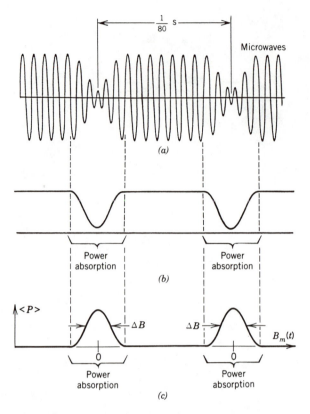

FIGURE 16.11 (a) With the modulation field amplitude B_m set to five times the linewidth ΔB, then resonance is swept through every 1/80 of a second, absorbing microwave power. (b) Input signal to the oscilloscope, after being rectified by the diode. (c) Inverted absorption signal.

shows the signal input to the oscilloscope. Figure 16.11c shows the inverted absorption signal, where the average power absorbed $\langle P \rangle$ by the sample is sketched versus $B_m(t)$. What appears as two separate resonances in 16.11c is the result of sweeping $B_m(t)$ through resonance with a positive slope and then with a negative slope, which are separated in time by 1/80 s. The two signals wll be superimposed on the oscilloscope. A good method to display the signal on the oscilloscope is to also apply the 40-Hz modulation signal to the horizontal input of the scope.

2. *Recording the Dc Line Shape on the XY Recorder.* Disconnect the lock-in amplifier from the spectrometer; hence, the magnetic field will not be modulated. Connect the Y axis of an XY recorder to the diode via the op-amp in Figure 16.10. Drive the X axis with the ramp voltage of the electromagnet power supply. The magnetic field sweep is a linear function of the ramp voltage, and the voltage should be calibrated in field units. (If such a voltage is not available from your power supply, an Xt recorder may be used with the X input connected to the diode. It will be necessary to calibrate the time axis in field units.) With the klystron set at the resonance frequency of the cavity, v_0, sweep the field through magnetic resonance, recording the signal on the recorder. Figure 16.12a shows the microwave signal incident on the diode D as the magnetic field is swept through resonance. The diode rectifies the microwave voltage and the dc voltage input to the recorder is shown in Figure 16.12b. The inverted dc voltage is shown in Figure 16.12c, where the average power absorbed $\langle P \rangle$ by the sample is sketched versus B_0.

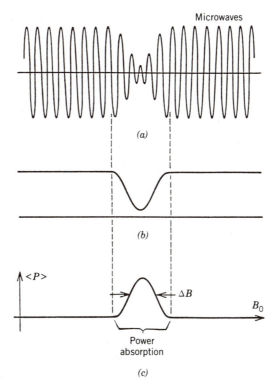

FIGURE 16.12 (a) Microwave signal incident on the diode. (b) Signal after diode rectification. (c) Average power absorbed by the sample as a function of B_0.

3. *Recording the Derivative of the Dc Line Shape.* An optional method is to connect the system as shown in Figure 16.9, and record the derivative of the dc line shape. The lock-in amplifier and the technique for doing this are discussed in Appendix A.

Growing Potassium Chrome Alum Crystals

The chemicals required to grow the crystals are potassium aluminum sulfate dodecahydrate $(KAl(SO_4)_2 \cdot 12H_2O)$ or potassium aluminum alum, potassium chromium sulfate dodecahydrate $(KCr(SO_4)_2 \cdot 12H_2O)$ or potassium chrome alum, and distilled water. The resulting crystal is potassium alumimum–chromium alum. We are interested in the chromium ions rather than the aluminum; thus, we call such crystals potassium chrome alums or just chrome alums.

If the concentration of Cr^{3+} ions is too small, then the magnetic resonance signal will be weak. (Equation 28, Introduction to Magnetic Resonance, shows that the time average power absorbed by the sample is proportional to the number of particles (here Cr^{3+} ions) per unit volume $N(m_j)$ (here m_s) in the initial or lower state.) On the other hand, a high concentration of Cr^{3+} ions gives rise to strong magnetic dipole–magnetic dipole interactions between neighboring Cr^{3+} ions, producing broadened spectral lines.

To grow chrome alum crystals, place five parts by weight of potassium aluminum alum to one part potassium chrome alum in a 25-mL beaker. (Reference 4 reports that a quantitative analysis indicates this ratio of weights yields crystals having a 8.5 to 1 ratio of aluminum to chromium ions.) Add enough distilled water to dissolve the two alums. Heating with a Bunsen burner will accelerate dissolution. The resulting crystals will be

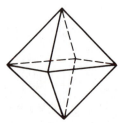

FIGURE 16.13 Chrome alum crystals are octahedrons.

octahedrons, as shown in Figure 16.13. Grow the crystals to a thickness of about 1 cm between opposite triangular faces, corresponding to a mass of about 2 g.

Growing the crystals may take a few weeks; therefore, they should be grown well ahead of the time that they will be needed. Reference 6 is useful in crystal preparation.

EXPERIMENT

The experiment consists of placing a sample containing the paramagnetic ions to be studied in a position inside the cavity where the microwave magnetic field $B_1 \cos \omega_0 t$ is strong. Place the cavity in a uniform field B_0, detect the power lost by the microwave field as B_0 is brought to the resonance value, and then measure B_0 and v_0 as accurately as possible.

v_0 may be measured with the wavemeter WM and B_0 may be measured with a gaussmeter. The gaussmeter should be placed as near the sample as possible. Also the pole faces of the electromagnet should be placed as close together as possible. This has two advantages. One is that the field should be more uniform over the sample, and the other is that the magnet will regulate to higher values of the field.

It is useful to calibrate and control the modulation field amplitude B_m. In Figure 16.9, B_m is controlled by the af attenuator. A possible af attenuator is a 10-turn 100-kΩ potentiometer as shown in Figure 16.14. The amplitude B_m may then be calibrated as a function of the potentiometer setting.

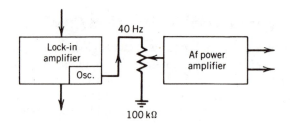

FIGURE 16.14 A possible af attenuator is a 10-turn, 100-kΩ potentiometer.

EXERCISE 7

If a calibration curve, for example, B_m versus potentiometer setting, does not exist for your modulation coils, then construct a *search* coil with a small diameter (about 1 cm) and having many turns of wire, and measure the induced voltage as a function of the potentiometer setting. Knowing the characteristics of your search coil, calculate B_m for each setting.

DPPH

The sample α'-diphenyl-β-picryl hydrazyl (DPPH) is good to use first in order to become familiar with ESR. DPPH has almost free spin electrons, that is, $g = 2$ and $s = \frac{1}{2}$, and it is a strong absorber of microwaves, giving a strong resonance absorption signal. The Hamiltonian for DPPH is given by equation 15.

DPPH is a powder and it must be placed in an evacuated glass tube. About 10 mg of DPPH is adequate.

A suggested procedure is the following:

1. Place the DPPH sample inside the cavity where B_1 is strong. The glass tube holding the sample can be held in position with vacuum grease.
2. Position the sample in the center of the pole faces with $B_0 = 0$.
3. Connect the diode D of the spectrometer to the vertical input of an oscilloscope, and tune the klystron to the resonant frequency of the cavity.
4. Turn on the electromagnet and observe the magnetic resonance signal by each of the three methods described previously.

Following equation 28 in the Introduction to Magnetic Resonance, it was pointed out that there are three independent variables in the equation for the time-averaged power absorbed $\langle P \rangle$: T, B_0, and B_1. For a given sample, B_0 is fixed by the microwave frequency and we are working at room temperature; hence, the only variable is B_1.

Method 1 of observing the signal allows you to see the entire absorption curve on the scope screen. The linewidth of the DPPH absorption curve is about 2 gauss (G); hence, set $B_m \simeq 10$ G. Once you have obtained the resonance signal on the scope, then observe the signal as B_1 is varied from small to large values.

EXERCISE 8

Do you find qualitative agreement between equation 28, Introduction to Magnetic Resonance, and your observation of signal amplitude as a function of B_1? Try to observe *saturation* of the sample. Does equation 28 predict saturation? See the discussion of saturation following equation 48, Introduction to Magnetic Resonance.

Using methods 2 and/or 3, measure the linewidth ΔB and then calculate the spin–spin relaxation time T_2. Use your magnetic resonance data to calculate the g value, and compare it to the free electron-spin value 2.0023.

Remark: To record a symmetrical line shape such as that shown in Figure 16.12c, the klystron must be stabilized to the cavity frequency. Since we are not doing this, you will not observe such a symmetrical line shape; hence, you will have to estimate where to measure ΔB.

Cr³⁺ in Chrome Alum Crystal

An important question is how do we orient the crystal relative to the field B_0? Well, if B_0 is parallel to the axial crystalline field then the spin Hamiltonian is given by equation 16, and the time-independent Schrödinger equation, equation 17, is readily solved for the eigenvalues $E(m_s)$. For other orientations of B_0, the spin Hamiltonian, and, hence, the eigenvalues, are more complex.

It was previously indicated that the unit cell of the alum crystal contains four nonequivalent octahedral complexes; therefore, at most, B_0 may be parallel to any one of the four axial fields of the unit cell. We choose to orient the crystal relative to B_0 such that one axial field is parallel to B_0 and all of the other axial fields make the same angle with respect to B_0. The orientation is shown in Figure 16.15, where a triangular surface of the crystal is flat against the narrow wall of the cavity. (In terms of Miller indices, B_0 is perpendicular to a (111) face of the crystal. Miller indices are discussed in Experiment 10, Diffraction of X Rays and Microwaves by Periodic Structures: Bragg Spectroscopy.)

The position of the crystal shown in Figure 16.15 assumes the cavity mode is a TE_{102}, with the center of the crystal one-quarter wavelength from a cavity iris. Vacuum grease may be used to hold the crystal in place.

In the orientation shown in Figure 16.15, B_0 is parallel to one of the axial fields and makes an angle of 70°33′ with the other three. You obtained the eigenvalues for the parallel case in Exercise 4. The eigenvalues for the angle of 70°33′ cannot be solved for exactly (depending on the strength of B_0, one must use second-order perturbation theory or do a numerical computation); hence, we will not be concerned with calculating these eigenvalues and the field values of their corresponding absorption lines.

There will be a total of six absorption lines, three corresponding to the parallel case and three for the 70°33′ case. However, the transition from $m_s = -\frac{1}{2}$ to $m_s = +\frac{1}{2}$ occurs at the same value of B_0 for the two cases; hence, a maximum of five distinct absorption lines are observed. The spectrum reported in reference 4, for your crystal orientation, is shown in Figure 16.16 at a resonance frequency of 9375 MHz. The solid curve is for a crystal grown from potassium chrome alum; that is, all of the trivalent ions are Cr^{3+}. The dashed curve is for a crystal with the same concentration as yours. Note the broadening of the solid curve relative to the dashed curve, caused by the magnetic dipole–magnetic dipole interactions among neighboring Cr^{3+} ions. Also their experimental apparatus included stabilizing the klystron to the cavity; hence, your spectrum will be somewhat different from theirs.

For the alum crystal follow the four procedural steps suggested for DPPH. The magnetic resonance signal of the Cr^{3+} ions will not be as strong as the DPPH; the amplitude will be less with broader linewidths, $\Delta B \simeq 150$ G. You may not be able to observe the resonance signals by method 1, but try it!

In doing Exercise 5 you calculated B_0 for the three absorption lines in the parallel case. Use these values to identify these absorption lines for the spectrum recorded by methods 2 and/or 3. Knowing B_0 for each of the three spectral lines in the parallel case and v_0, use

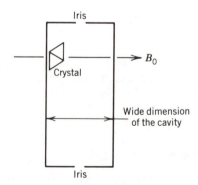

FIGURE 16.15 Orientation of the chrome alum crystal relative to the direction of B_0. B_0 is perpendicular to the (111) face of the crystal.

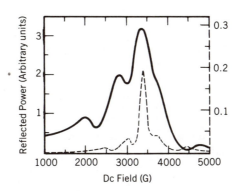

FIGURE 16.16 Spectrum reported in reference 4 for B_0 perpendicular to the (111) face of the crystal. The scales for the solid line curve (left) and dashed line curve (right) refer to the same input power.

conservation of energy (see Exercise 5) to calculate D and $g_\parallel$. Compare your values with those given in Exercise 5. Also measure the linewidth of the most distinct of the three lines and calculate T_2.

17. ELECTRICAL CONDUCTIVITY AND THE HALL EFFECT

Historical Note

The 1985 Nobel prize in Physics was awarded to
  Klaus von Klitzing, Germany
  For his discovery of the quantized Hall resistance.

APPARATUS [Optional Equipment in Brackets]

Cryostat

[Oven]

Laboratory electromagnet with power supply and gaussmeter

Rectangular semiconducting crystal (e.g., a low-impurity piece of germanium or other semiconductor with dimensions of about $1 \times 2 \times 10$ mm and $\rho > 0.5$ Ω-cm at 300 K)

Electrometer or potentiometer for measuring Hall voltages

Ammeter

Voltmeter

6-V battery

Assorted resistors

Copper–constantan thermocouple with potentiometer

OBJECTIVES

To measure the Hall voltage and conductivity for a semiconductor as functions of temperature and to interpret these data qualitatively.

To derive the sign of the extrinsic charge carriers, the Hall mobility, and the energy gap from the data.

To observe and understand qualitatively the phenomenon of magnetoresistance.

To understand the basic principles underlying the Hall effect and its relationship to charge transport mechanisms in semiconductors.

KEY CONCEPTS

Hall voltage   Hall coefficient

Hall resistance   Free electron model

Conductivity   Mobility

Fermi energy   Energy gap

Valence/conduction band   Holes

Effective mass   Intrinsic/extrinsic carriers

n-type/p-type semiconductor   Magnetoresistance

Thermomagnetic/thermoelectric
   effects

REFERENCES

1. C. Kittel, *Introduction to Solid State Physics*, 6th ed., Wiley, New York, 1985. Chapter 6 presents a simple model for electrical conduction in a free electron Fermi gas and discusses the Hall effect for the case of a single type of charge carrier. Chapter 8 contains a discussion of charge transfer in semiconductors and contains data on energy gaps, donor and acceptor ionization energies, and mobilities for some semiconductors.

2. C. Kittel and H. Kroemer, *Thermal Physics*, 2d ed., Freeman, San Francisco, 1980. Chapter 13 is a thorough discussion of semiconductor statistics which includes calculations of carrier concentrations.

3. O. Lindberg, *Proc. IRE* **40**, 1414 (1952). A brief discussion of the Hall effect in semiconductors and the thermomagnetic effects that complicate measurements of Hall voltages.

4. F. E. Martin and P. H. Sidles, *Am. J. Phys.* **41**, 103 (1973). A description of a low-cost cryostat suitable for conductivity and Hall effect measurements over a wide range of temperatures.

5. R. C. Weast (ed.), *Handbook of Chemistry and Physics*, CRC Press, Boca Raton, FL, any edition. Thermocouple calibration tables.

6. P. H. Bligh, J. J. Johnson, and J. M. Ward, *Phys. Educ.* **20**, 246 (1985). Example of a computerized experimental setup for making Hall effect measurements that is suitable for use in the undergraduate laboratory.

7. E. H. Putley, *The Hall Effect and Semi-conductor Physics*, Dover, New York, 1968. Discusses theoretical and experimental aspects of charge transport processes in semiconductors. Includes band theory, Hall and magnetoresistance effects, and thermal effects.

8. B. I. Halperin, *Sci. Am.* **254**, 52 (1986). A descriptive account of the integral and fractional quantized Hall effects.

9. Experiment 6, Introduction to Computer-Assisted Experimentation. In this experiment the Hall sample is interfaced to the computer, and the computer is used to measure the Hall voltage as a function of both the magnetic field and the applied current.

INTRODUCTION

If a block of conducting material that carries a current I along the x direction is placed in a magnetic field $\mathbf{B} = B\hat{z}$, as in Figure 17.1, an electrostatic potential difference, referred to as the Hall voltage V_H, is produced between faces 3 and 4 along the y direction. The production of this voltage, first observed by E. H. Hall in 1879, is called the Hall effect and is due to the deflection along the y axis of the moving charge carriers in the material. The ratio V_H/I, called the Hall resistance, is observed to increase with increasing B.

The Hall effect involves several independent variables and so is the operating principle for a rather wide variety of practical instrumentation. For the experimental physicist, the Hall-effect gaussmeter, which provides convenient and accurate measurements of laboratory magnetic fields, is perhaps the most well known of these devices. Other applications include, but are by no means limited to, displacement indicators, tachometers, ammeters, modulators, D/A converters, multipliers, and spectrum analyzers.

There are two regions in which experimental studies of the Hall effect yield useful results. For low fields, defined by the condition $\omega_c \tau \ll 1$ ($\omega_c \equiv eB/m$ is the cyclotron frequency and τ is the mean time between collisions for the charge carriers), scattering processes are of importance in determining the motion of charge carriers. In this region, the behavior of the Hall voltage, taken together with electrical conductivity measurements, is a rich source of information regarding charge transport properties of metals and semiconductors. Concentrations, mobilities, and signs of charge carriers can be determined under the low-field conditions, which are appropriate to all of the measurements described in this experiment.

For the high-field region, defined by $\omega_c \tau \gg 1$, the motion of the charge carriers is determined by the externally applied fields and the behavior of the Hall voltage can yield information about the quantum-mechanical band structure of the material under investigation. If, additionally, the temperature is close enough to absolute zero, the effect of the magnetic field on the charge carrier states is manifested in the quantum behavior of the Hall resistance. In 1985, Klaus von Klitzing won the Nobel prize in physics for his observation of the integral quantized Hall effect, in which the Hall resistance for a two-dimensional system does not increase linearly with B as it does in the high-field region, but rather exhibits plateaus as B increases. Electrical resistance vanishes for values of B for which the Hall resistance exhibits a plateau; the reciprocals of the values of the Hall resistance (the Hall conductances) on these plateaus are given with great precision as integer multiples of e^2/h, thus defining a basic unit of conductance. Such measurements can establish precise

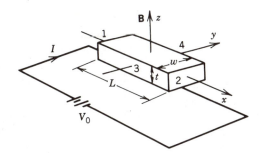

FIGURE 17.1 Geometry for measuring the Hall voltage for a conducting sample in a magnetic field.

standards of resistance, provide precise values for fundamental constants, and lead to more stringent tests of quantum electrodynamics.

Electrical Conductivity

For simple metals (e.g., the alkalis), many properties, including charge transport, are understood quite well in terms of the *free electron* model. In this picture, current is carried by electrons (charge $-e$ and number density n) that are accelerated by an applied electric field $\mathbf{E}$ as they undergo collisions with lattice imperfections, impurities, and phonons (thermal vibrations of the lattice). The current density $\mathbf{J} = ne\mathbf{v}$ can be calculated by noting that if the mean time between collisions is τ, then the effect of collisions can be approximated as a retarding force of magnitude mv/τ, where $\mathbf{v}$ is the average drift velocity acquired by the electrons between collisions. Newton's second law for these electrons reads, in this approximation

$$m\left(\frac{d}{dt} + \frac{1}{\tau}\right)\mathbf{v} = -e\mathbf{E} \qquad (\mathrm{N}) \tag{1}$$

In the steady state the retarding force just balances the force due to $\mathbf{E}$, so that the time derivative vanishes, yielding

$$\mathbf{v} = -\frac{e\tau}{m}\mathbf{E} \qquad (\mathrm{m/s}) \tag{2}$$

The electron density, n, is determined only by the density of atoms in the crystal and the number of valence electrons per atom. The current density produced by $\mathbf{E}$ is thus given by

$$\mathbf{J} = \frac{ne^2\tau}{m}\mathbf{E} \qquad (\mathrm{A/m^2}) \tag{3}$$

and the conductivity σ, defined by $\mathbf{J} = \sigma\mathbf{E}$, is just

$$\sigma = \frac{ne^2\tau}{m} \qquad (\Omega\cdot\mathrm{m})^{-1} \tag{4}$$

The result is also conveniently expressed in terms of the mobility μ of the charge carriers, defined by

$$\mu \equiv \frac{|\mathbf{v}|}{|\mathbf{E}|} \qquad (\mathrm{m^2(V\cdot s)^{-1}}) \tag{5}$$

In terms of μ, the conductivity for a metal is obtained from equations 2 and 3:

$$\sigma = ne\mu \qquad ((\Omega\cdot\mathrm{m})^{-1}) \tag{6}$$

In the quantum-mechanical description of a metal, the electrons are arranged in energy bands, as depicted in Figure 17.2a. Electrons occupy states within the allowed bands up to the Fermi energy $\mathscr{E}_F$. In a metal, $\mathscr{E}_F$ falls below the top of a band, so that an electric field can accelerate electrons by promoting them to unoccupied states of higher energy within this same band.

In a semiconducting material the situation is quite different. Examples of semiconductors are the crystalline forms of elements from group IV of the periodic table (e.g., Ge or Si) or the compounds of elements from groups III and V (e.g., GaAs or AlSb). In the pure forms

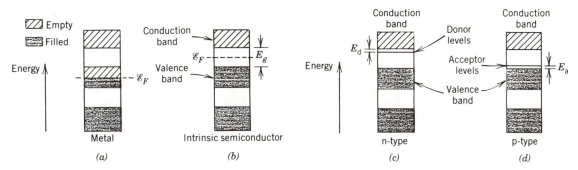

FIGURE 17.2 (*a*) Electronic energy bands for a metal. (*b*) Electronic energy bands for an intrinsic semiconductor. (*c*) Electronic energy bands for an n-type semiconductor. (*d*) Electronic energy bands for a p-type semiconductor.

of these materials at $T = 0$ K, there are no unoccupied states near the top of the electron energy distribution; that is, the **valence band** is filled, as indicated by Figure 17.2*b*. The unoccupied states of lowest energy are located in the **conduction band** at an energy E_g (the width of the *energy gap*) above the top of the valence band. Thus, an applied electric field cannot accelerate any of the electrons, and no current can flow. As T is raised above 0 K, some of the electrons near the top of the valence band are thermally excited into the conduction band, where they behave as free carriers of negative charge. The vacancies in the valence band left behind by these excited electrons are known as **holes** and behave as free carriers of positive charge. These thermally produced charge carriers are referred to as **intrinsic** carriers. The density of thermally excited electrons (n_e) and holes (n_h) will, along with their respective mobilities, determine the conductivity of the material. For a pure semiconductor, the intrinsic carrier densities are calculated from considerations of statistical mechanics (see reference 1 or 2):

$$n_e = n_h = 2\left(\frac{k_B T}{2\pi\hbar^2}\right)^{3/2}(m_e m_h)^{3/4}\, e^{-E_g/2k_B T} \qquad (\text{m}^{-3}) \qquad (7)$$

where k_B is Boltzmann's constant and m_e and m_h are the *density-of-states effective masses* of the electrons and holes in the crystal as determined from the band structure.

EXERCISE 1

Why are the hole and electron densities equal in a "pure" semiconductor?

Semiconductors are never pure. The presence of impurities, whether intended or not, markedly affects the carrier concentration and, hence, the conductivity of the material. Addition of group V atoms, which have an "extra" valence electron, creates an **n-type** semiconductor in which there are occupied donor levels located at an energy E_d below the conduction band, as illustrated by Figure 17.2*c*. Electrons occupying these levels are easily excited into the conduction band at temperatures much lower than that required to produce intrinsic carriers. On the other hand, a **p-type** semiconductor results from the addition of group III impurities. These atoms, which have one less valence electron than the atoms of the host crystal, create vacant acceptor levels located at an energy E_a above the valence band, as illustrated by Figure 17.2*d*. Electrons from the top of the valence band can readily be excited into these levels, leaving holes in the valence band to act as free charge carriers. Charge carriers created in this way are termed **extrinsic**, or impurity carriers. The calculation of the extrinsic carrier densities is somewhat involved, but if only one type of impurity is present, its associated carrier density n_{ex} can be expressed in closed form. If we neglect the concentration of intrinsic carriers, then

$$n_{ex} = \frac{n^*}{4}\left\{\sqrt{1 + \frac{8n_I}{n^*}} - 1\right\} \qquad (m^{-3}) \qquad (8)$$

In this expression the parameter n^* is defined by

$$n^* \equiv 2\left(\frac{m^* k_B T}{2\pi\hbar^2}\right)^{3/2} e^{-E_I/K_B T},$$

n_I is the impurity atom concentration, and we set $E_I = E_d$, $m^* = m_e$ for n-type materials, and $E_I = E_a$, $m^* = m_h$ for p-type materials. The energy E_I is called the **impurity ionization energy**.

EXERCISE 2

Show that in the low-temperature limit, $k_B T \ll E_I$, the concentration of extrinsic carriers has a temperature dependence given by

$$n_{ex} \propto T^{3/4} e^{-E_I/2k_B T} \qquad (m^{-3}) \qquad (8')$$

Note that this expression is good only in the temperature range indicated, and that as T is raised much beyond $E_I/2k_B$, the extrinsic carrier concentration saturates so that $n_{ex} \cong n_I$. For still higher temperatures, in the intrinsic region, intrinsic carrier production dominates charge transport processes in accordance with equation 7.

Carrier mobilities $\mu_e \equiv |v_e|/|E|$ and $\mu_h \equiv |v_h|/|E|$ for electrons and holes, respectively, are defined by analogy with the definition for metals stated in equation 5. These quantities are related, by comparison with equation 2, to the mean times between collisions for electrons and holes, τ_e and τ_h, by the relations

$$\mu_e = \frac{e\tau_e}{m_e} \qquad \mu_h = \frac{e\tau_h}{m_h} \qquad (m^2(V \cdot s)^{-1}) \qquad (9)$$

Mobilities are limited by scattering processes involving impurities and thermal phonons. In the temperature range where phonon scattering is of primary importance, carrier mobilities are expected to vary as $T^{-3/2}$. Hole mobilities are typically smaller than those for electrons because the distribution of the valence states allows a greater variety of scattering mechanisms.

In a semiconductor, both types of charge carriers contribute to the conduction of current. Following the same reasoning as for equations 1–6 above, we extend the expression for the conductivity to include the case in which both electrons and holes may be present:

$$\sigma = n_e e\mu_e + n_h e\mu_h \qquad ((\Omega \cdot m)^{-1}) \qquad (10)$$

Hall Effect

For a metal or for an n- or p-type semiconductor in the extrinsic temperature range, electrical conduction is due to a single carrier type with charge q and mobility μ. If a magnetic field **B** is applied along $\hat{z}$, as in Figure 17.1, then a carrier of charge q experiences a Lorentz force $\mathbf{F} = q(\mathbf{E} + \mathbf{v} \times \mathbf{B})$. If the definition of carrier mobility is expanded to include magnetic forces, then the drift velocity is given by

$$\mathbf{v} = \pm\mu(\mathbf{E} + \mathbf{v} \times \mathbf{B}) \qquad (m/s) \qquad (11)$$

where the plus sign is used for holes, and the minus sign is used for electrons. For this configuration of fields, the drift velocity components are related by

$$v_x = \pm\mu(E_x + v_y B_z) \qquad v_y = \pm\mu(E_y - v_x B_z) \qquad v_z = \pm\mu E_z \qquad \text{(m/s)} \qquad (12)$$

The components of the current density $\mathbf{J} = nq\mathbf{v}$ are thus

$$J_x = \pm\mu(nqE_x + J_y B_z) \qquad J_y = \pm\mu(nqE_y - J_x B_z) \qquad J_z = \pm\mu nqE_z \qquad \text{(A/m}^2) \quad (13)$$

If the current is being supplied by a battery attached to faces 1 and 2 of the sample, then in the steady state we must have $J_y = J_z = 0$. This implies that there is a component of $\mathbf{E}$ transverse to the current flow:

$$E_y = \frac{J_x B_z}{nq} \qquad \text{(V/m)} \qquad (14)$$

This is accounted for physically by the buildup of a static charge distribution on faces 3 and 4 which produces an electric field along the y axis that compensates for the force of $\mathbf{B}$ on the carriers. The **Hall coefficient** R_H is defined by

$$R_H = \frac{E_y}{J_x B_z} \qquad \text{(m}^3/\text{C)} \qquad (15)$$

We can restate this definition in terms of experimentally measured quantities by referring to Figure 17.1 and noting that the Hall voltage $V_H = V_3 - V_4$ is expressed in terms of E_y as $V_H = wE_y$, and that $J_x = I/wt$. Thus, the experimental Hall coefficient is expressed in terms of measurable quantities as

$$R_H = \frac{V_H t}{I B_z} \qquad \text{(m}^3/\text{C)} \qquad (16)$$

By comparing equations 14 and 15, it can be seen that the theoretical Hall coefficient for the case of a single carrier type is expressed in terms of the carrier density as

$$R_H = \frac{1}{nq} \qquad \text{(m}^3/\text{C)} \qquad (17)$$

Note that the carrier density in this case can be calculated directly from a measurement of R_H, and that the sign of R_H can be used to determine the sign of the charge carriers. If both R_H and σ are measured, then the **Hall mobility** μ_H is determined by

$$\mu_H = |R_H|\sigma \qquad \text{(m}^2(\text{V} \cdot \text{s})^{-1}) \qquad (18)$$

EXERCISE 3

Using Figure 17.1, along with the definition of equation 15 and the right-hand rule for the Lorenz force, verify that R_H is positive for extrinsic p-type semiconductors.

In semiconductors where both electrons and holes contribute to conduction, the expression for R_H involves the densities and mobilities of both carriers. The calculation of R_H for this case is a straightforward extension of the one above for a single species of carrier. The current density is the sum of the electron and hole current densities, and thus has a y component that is

$$J_y = n_h e v_{hy} - n_e e v_{ey} \quad \text{(A/m}^2) \tag{19}$$

where v_{hy} and v_{ey} are y components of the hole and electron velocities, respectively. Substituting v_y given in equation 12 for each carrier into this expression for J_y yields

$$J_y = (n_h \mu_h + n_e \mu_e)eE_y + (n_e \mu_e^2 - n_h \mu_h^2)eB_z E_x \quad \text{(A/m}^2) \tag{20}$$

Here we assume that we are working in the low-field region, as discussed in the introduction, so that we can neglect terms containing B^2 in comparison with those containing B. As for the single carrier calculation, we demand $J_y - 0$ in the steady state. Writing $J_x = \sigma E_x$, with σ given approximately by equation 10, gives us for R_H

$$R_H \equiv \frac{E_y}{J_x B_z} = \frac{n_h \mu_h^2 - n_e \mu_e^2}{e(n_e \mu_e + n_h \mu_h)^2} \quad \text{(m}^3/\text{C)} \tag{21}$$

EXERCISE 4

In light of the observation that hole mobilities are typically much smaller than electron mobilities, what is the expected sign of the Hall coefficient for an intrinsic semiconductor? What can you conclude about the impurities in a semiconductor for which R_H changes sign as the temperature is raised?

Note that for a material with a single type of charge carrier with a well-defined mobility, the steady-state condition $J_y = 0$ is equivalent to requiring that the y component of the Lorentz force on the carriers be zero, so that the carrier trajectories are straight lines along the x axis. In this instance it is expected that the sample in Figure 17.1 would have a measured resistance between faces 1 and 2 that is not affected by the applied magnetic field B. If, however, the carriers have a distribution of mobilities (e.g., in the two-carrier case), then $J_y = 0$ is not equivalent to zero Lorentz force for all carriers, and some will develop drift velocities with nonzero y components. This will result in curved carrier trajectories and an accompanying increase in the measured resistance of the sample. The resistance measured in the presence of B is termed the **magnetoresistance**, and is important for the determination of electronic structures. This same mechanism explains why R_H has a slight dependence on B.

EXPERIMENT

Cryogenics

Exploration of the temperature behavior of the Hall coefficient and conductivity of a sample requires a means of measuring and slowly varying its temperature as voltage and current measurements are made. Aside from the use of commercially available cryostats, there are several convenient methods for accomplishing this which vary in cost, sophistication, and range of temperatures available. Some that are suitable for this experiment are suggested below.

1. A mixture of dry ice and alcohol used as a cryogenic fluid allows exploration of temperatures from about 190 K to around room temperature. The alcohol should be added to a Dewar containing the dry ice and, after the mixture has reached a fairly

quiescent state, the alcohol should be poured off (with a minimum of dry ice) into a second Dewar or beaker in which thermal contact is made with the sample. Measurements should be taken as the temperature rises. When the rate of temperature increase becomes inconveniently slow, higher temperature data may be taken with the use of an oven.

2. The apparatus shown schematically in Figure 17.3a allows measurements down to about 80 K by utilizing a thermally conducting rod (e.g., of copper) as a "cold finger," which thermally links the sample and a liquid nitrogen (LN$_2$) bath. The overall efficiency of the link may be optimized by replacing a portion of the rod with material of lower thermal conductivity. Operation of the heating coil when liquid nitrogen is in the Dewar allows variation of the sample temperature. As in method 1, measurements can be taken "on the fly" as the temperature rises rather than awaiting stabilization at each point.

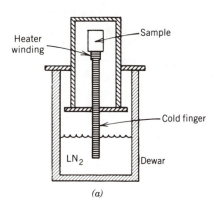

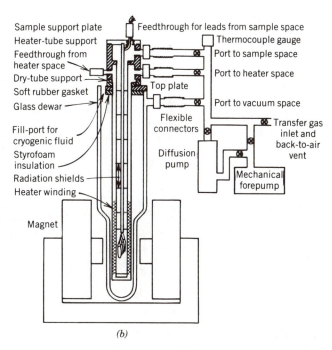

FIGURE 17.3 (a) Cold finger cryostat. (b) Low-cost cryostat described by Martin and Sidles, reference 4.

3. The construction of the cryostat shown in Figure 17.3*b* is discussed in detail by Martin and Sidles (reference 4) as a low-cost option for the instructional laboratory. The apparatus shown is suitable for operation down to liquid nitrogen temperature, but is readily modified for operation with liquid helium.

In all of the above systems, a copper–constantan thermocouple in thermal contact with the sample may be used for temperature measurements. One of the junctions should be maintained in an ice-water reference bath at 0 °C so that the calibration tables (reference 5) may be used. If desired, a linear correction may be made to these tables based on the thermocouple emf measured when the junction is dipped in a cryogenic liquid of known temperature (e.g., liquid netrogen at 77 K). For the temperature range available in method 1, a low-temperature toluol thermometer may be used instead of a thermocouple.

Electrical Connections

Electrical leads may be soldered to the sample, but care must be taken to wet the sample with the solder and to avoid rectifying contacts. A convenient and effective alternative to soldering is the use of the sample holder described in reference 4 and shown in Figure 17.4, where part a of the figure is a front view of the holder and part b is a cross-sectional side view. Contacts for measuring Hall potentials are made with segments of 20-mil tungsten wire, the ends of which are sharpened by anodic oxidation in a potassium hydroxide solution with a graphite cathode. Faces of the electrodes making contact with the ends of the sample are coated with indium solder; making the end faces equipotentials in this way will not significantly "short out" the Hall voltage measured across the middle as long as

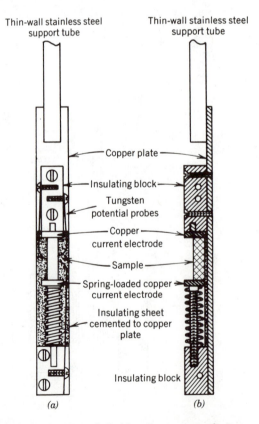

FIGURE 17.4 (*a*) Sample holder (front view). (*b*) Sample holder (cross-sectional side view).

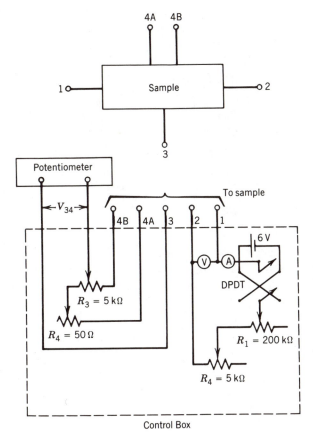

FIGURE 17.5 Circuit for conductivity and Hall voltage
measurements.

$L/w > 4$ for the sample. Spring loading of one current electrode eliminates difficulties caused
by differential thermal expansion and contraction. The base of the holder is a thin copper
plate to which the thermocouple may be attached and which is in thermal contact with but
electrically insulated from the sample and electrode supports. The surface of the sample
should be cleaned of grease and oxides before insertion into the holder; after insertion an
ohmmeter may be used to verify that nonrectifying contacts have been made.

The circuit for the conductivity and Hall voltage measurements is shown in Figure 17.5.
Connections from the control box to the sample may be made conveniently with a Jones
strip. Resistances R_1 and R_2 control the current through the sample. If a single probe were
used on each side of the sample to measure V_H, there would be an "IR drop" contribution
to the measured Hall voltage if the probes were not precisely located along a line
perpendicular to the direction of current flow. To eliminate this effect, two probes, 4A and
4B, are used, and adjustment of resistances R_3 and R_4 is equivalent to sliding a single probe
between these two contact points. Thus, these resistances should be adjusted at the
beginning of the measurements with the sample current flowing and with $B = 0$ until
$V_{34} = 0$.

Because of several thermomagnetic effects, the measured voltage V_{34} may not be the
"true" Hall voltage V_H. Reference 3 gives a concise description of the Nernst effect, the
Righi–Leduc effect, and the Ettingshausen effect, along with suggestions for reducing their
respective contributions to measured voltages. The Ettingshausen effect cannot be separated
from the Hall effect in the present experiment, but it will be small in samples with good bulk
thermal conduction. The Nernst and Righi–Leduc effects, along with any residual IR drop,

can be compensated for by averaging V_{34} over both directions of current and field. Thus, the Hall voltage corresponding to a current I and field B should be taken as an average of four separate measurements:

$$V_H = \tfrac{1}{4}[V_{34}(I, B) - V_{34}(I, -B) - V_{34}(-I, B) + V_{34}(-I, -B)] \quad (V) \qquad (22)$$

Thermoelectric effects can also affect V_{12}, the voltage measured between points 1 and 2 on the sample, if there are temperature gradients. Thus, the conductivity should be calculated using an average voltage V_0, given by

$$V_0 = \tfrac{1}{2}[V_{12}(I) - V_{12}(-I)] \quad (V) \qquad (23)$$

Measurements

Measure the dimensions of the sample. With the sample removed from the pole pieces of the magnet, zero the IR drop as discussed above. Place the sample between the pole pieces and, with a current of a few milliamperes flowing in the sample, rotate it until V_H is maximized; this procedure aligns the apparatus for future measurements. Note how the directions of **B** and I in the sample correlate with the polarity of the meter readings for the magnet and sample current.

With the sample at room temperature, select a convenient sample current and obtain a few values of V_H as a function of B at constant I. Do the same for V_H at constant B as a function of I.

EXERCISE 5

Do plots of V_H against B and V_H against I verify the linear relationships implied by equation 16?

Bring the sample to the lowest temperature in the range to be investigated and set the sample current at a convenient value, to be maintained throughout the experiment. Take measurements every few degrees as the temperature rises, reading the thermocouple emf both before and after the voltage measurements at each temperature. For each temperature do the following:

(a) With the magnetic field at a constant magnitude of about 1 kG, record V_{12} and V_{34} for current flow in both directions.
(b) Record V_{12} with $B = 0$, for both current directions.
(c) With the magnetic field reversed but equal in magnitude to that in (a), record V_{34} for both current directions.

From measurements of V_{12}, V_{34}, and I, determine the Hall coefficient and electrical conductivity for each temperature.

EXERCISE 6

Plot both R_H and σ against T and explain the behavior of these curves. Can you identify an extrinsic temperature region? An intrinsic temperature region?

EXERCISE 7

Plot $\ln(R_H T^{3/2})$ against $1/T$. From the linear portion of this curve in the intrinsic region, determine E_g, the energy gap in the semiconductor. Assume for this calculation that $n_e = n_h$ in this temperature range. What assumption(s) do you need to make about the carrier mobilities? How does your value for E_g compare with that given by reference 1? Bear in mind that E_g is a function of temperature.

EXERCISE 8

Estimate the impurity ionization energy E_I from the behavior of the low-temperature portion of the curve plotted in response to Exercise 7, if possible. What is the sign of the impurity charge carrier?

EXERCISE 9

Determine the energy gap and, if possible, the impurity ionization energy from the conductivity data. What assumptions are you making? How do these values compare with those obtained from the Hall coefficient data?

EXERCISE 10

Calculate the Hall mobility μ_H given by equation 18 for each temperature. From a plot of $\ln \mu_H$ versus $\ln T$, can you identify a temperature range in which a simple power law fits the data? If such a range exists, does the exponent in the power law suggest lattice scattering (discussed following equation 9) as the primary limiting factor for the mobility? It should be noted in this connection that the Hall mobility, defined experimentally by equation 18, is in general not equal to the mobility derived from conductivity measurements alone. Why might this be so?

EXERCISE 11

Calculate, for each temperature, the ratio $(R_m - R)/R$, where R_m is the resistance measured in the presence of a magnetic field and R is the resistance in zero magnetic field. Can you explain qualitatively the behavior of this ratio as the temperature is increased?

Nuclear Physics

18. INTERACTIONS OF GAMMA RAYS WITH MATTER

APPARATUS

Scintillation counter: scintillator, photomultiplier tube (PMT), multichannel analyzer (MCA) (see Figure B.1, Appendix B)

1-μCi radioactive sources: $^{137}_{55}$Cs, $^{60}_{27}$Co, and $^{22}_{11}$Na.

OBJECTIVES

Use the scintillation counter to record the complete spectra of two or more γ-ray sources.

To identify the Compton edge, the total energy peak, and other peaks that occur in the spectra of γ rays.

To become familiar with the scintillation counter.

To understand three interactions of γ rays with matter: the photoelectric effect, Compton scattering, and pair production.

KEY CONCEPTS

Photoelectric effect   Compton wavelength
Compton scattering   Compton edge
Pair production   Total energy peak
Cross section   Scintillation counter
Linear attenuation coefficients

REFERENCES

1. R. Eisberg and R. Resnick, *Quantum Physics of Atoms, Molecules, Solids, Nuclei, and Particles*, 2d ed., Wiley, New York, 1985. The photoelectric effect, Compton scattering, and pair production and pair annihilation are treated in Chapter 2. Selection rules for γ decay are discussed on pages 578–584.

2. R. Evans, *The Atomic Nucleus*, of 14th printing (1972), Krieger, Melbourne, FL, 1982. Compton scattering is presented in Chapter 23. Photoelectric effect, pair production, and pair annihilation are presented in Chapter 24. The attenuation and absorption of electromagnetic radiation are discussed in Chapter 25.

3. C. Lederer and V. Shirley, *Table of Isotopes*, 7th ed., Wiley, New York, 1978. The decay of radionuclides is given in detail.

Also see the references listed under Appendix B, Sorting and Counting Particles: The Scintillation Counter. Appendix B is a corequisite for this experiment.

INTRODUCTION

Gamma rays are the very high-frequency electromagnetic radiations accompanying nuclear transitions. The decay scheme of three radionuclides, $^{137}_{55}Cs$, $^{60}_{27}Co$, and $^{22}_{11}Na$, are shown in Figure 18.1. Spin and parity for each state are given in parentheses. The energy range of γ rays is typically from 0.01 to 10 MeV. The interaction of γ rays with matter is dependent on the γ-ray energy $h\nu$. In the energy range given above, there are just three major interactions: (1) Compton scattering, (2) the photoelectric effect, and (3) pair production. In (2) and (3) the γ ray is completely absorbed, whereas in (1) it is scattered.

For each interaction, the probability of a γ ray interacting with matter is specified by a quantity called the **cross section**. To define this concept consider a parallel beam of γ rays passing through the slab of matter shown in Figure 18.2. I_0 γ rays are incident on the slab and I primary or unmodified γ rays pass through. The probability that a γ ray will be, for example, absorbed due to the photoelectric interaction in passing an atom of the slab is expressed in terms of the photoelectric cross section σ_{PE}. The number N_{PE} of γ rays absorbed due to the photoelectric interaction is proportional to I_0 and N_a, the number of

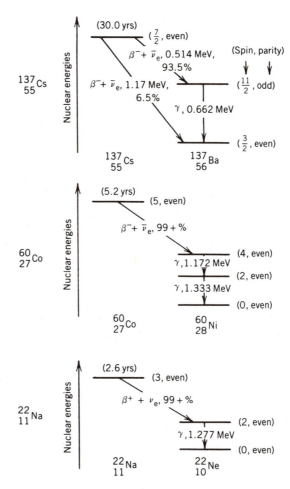

FIGURE 18.1 Decay scheme of three radionuclides.

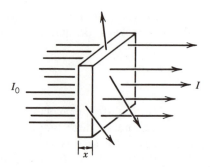

FIGURE 18.2 Interactions reduce the number of γ rays
that emerge from the slab of thickness x.

atoms per unit area:

$$N_{PE} \propto I_0 N_a \tag{1}$$

Writing expression (1) as an equality, we obtain the defining equation for that cross section:

$$N_{PE} \equiv \sigma_{PE} I_0 N_a \tag{2}$$

where σ_{PE} must have units of cm^2/atom.

The defining equations for the Compton scattering and pair production cross sections, σ_C and σ_{PP}, are analogous to (2):

$$N_C \equiv \sigma_C I_0 N_e \tag{3}$$

where N_C is the number of primary γ rays that are scattered because of Compton scattering, N_e is the number of electrons per unit area, and σ_C has units of cm^2/electron. Also

$$N_{PP} \equiv \sigma_{PP} I_0 N_n \tag{4}$$

where N_{PP} is the number of absorbed γ rays due to pair production, N_n is the number of nuclei per unit area, and σ_{PP} has units of cm^2/nucleus.

Compton scattering, the photoelectric effect, and pair production involve a γ-ray interacting with an electron, an atom, and a nucleus, respectively; hence, the cross sections are an area per unit particle of interaction. A cross section is a measure of the probability that a certain type of particle (electron, atom, or nucleus) will cause a γ ray of a particular energy to undergo a particular interaction. A large cross section implies a high probability that a certain interaction will occur.

Linear attenuation coefficients for each interaction are defined in terms of their cross sections:

$$\mu_{PE} \equiv \sigma_{PE} \rho_a \quad (1/\text{cm}) \tag{5}$$

$$\mu_C \equiv \sigma_C Z \rho_a \quad (1/\text{cm}) \tag{6}$$

$$\mu_{PP} \equiv \sigma_{PP} \rho_a \quad (1/\text{cm}) \tag{7}$$

where ρ_a is the number of atoms/cm^3 and Z is the number of electrons/atom. The three linear attenuation coefficients depend on the γ-ray energy $h\nu$ and the atomic number Z of the absorber. The relative importance of μ_{PE}, μ_C, and μ_{PP} is shown graphically in Figure 18.3. The lines show the values of Z and $h\nu$ for which two neighboring interactions are just

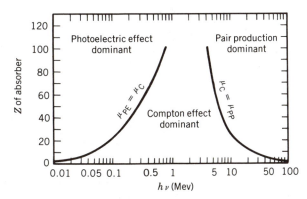

FIGURE 18.3 The γ-ray–absorber interaction depends
on the atomic number Z of the absorber
and the γ-ray energy $h\nu$.

equal; for example, for $Z = 13$ (aluminum) the energy ranges in which each effect dominates
are approximately

Photoelectric effect	$h\nu < 0.05$ MeV
Compton scattering	0.05 MeV $< h\nu < 10$ MeV
Pair production	$h\nu > 10$ MeV

The total linear attenuation coefficient μ is given by the sum

$$\mu = \mu_{\text{PE}} + \mu_{\text{C}} + \mu_{\text{PP}} \qquad (1/\text{cm}) \tag{8}$$

It is usually shown in lower-division texts that the intensity of a beam of photons, specified
by the number I of primary or unmodified photons, after passing a slab of thickness x is
given by

$$I(x) = I_0\, e^{-\mu x} \qquad \text{(number of unmodified photons)} \tag{9}$$

The intensity is often written in terms of the total mass attenuation coefficient, which is the
total linear attenuation coefficient divided by the density ρ (g/cm³) of the slab:

$$I(x) = I_0\, e^{-(\mu/\rho)\rho x} \qquad \text{(number of unmodified photons)} \tag{10}$$

where μ/ρ has units of cm²/g. Mass attenuation coefficients for sodium iodide versus γ-ray
energy are shown in Figure 18.4. It is pointed out in Appendix B that a sodium iodide
crystal is often used as a scintillator for γ-ray detection.

We now examine the three interactions individually. Our approach will be to apply the
conservation of energy and linear momentum before and after the interaction, and to bypass
the complicated physics of the interaction, because it obscures our purpose here.

Compton Scattering

We consider the electron struck by the γ ray to be initially at rest. The general case can be
obtained from this special case by a Lorentz transformation. The electron is also considered
to be free. This approximation is reasonable provided that the γ-ray energy $h\nu$ is much
greater than the binding energy B of the electron. If B is about equal to $h\nu$, then the

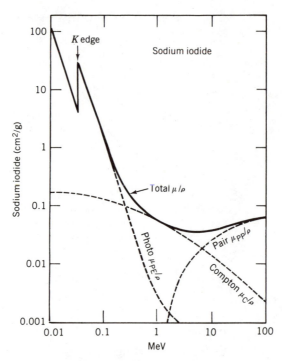

FIGURE 18.4 Mass attenuation coefficients for photons in sodium iodide as a function of photon energy.

photoelectric cross section greatly exceeds the Compton scattering cross section so that Compton scattering becomes of minor importance.

Compton scattering is shown in Figure 18.5, where v' is the frequency of the scattered γ, and $\mathbf{P}$ and K are the momentum and kinetic energy of the electron. Conservation of momentum yields two equations:

$$\frac{hv}{c} = \frac{hv'}{c} \cos \theta + P \cos \phi \qquad (\text{kg} \cdot \text{m/s}) \qquad (11)$$

$$0 = \frac{hv'}{c} \sin \theta - P \sin \phi \qquad (\text{kg} \cdot \text{m/s}) \qquad (12)$$

Conservation of energy for an elastic collision gives

$$hv = hv' + K \qquad (\text{J}) \qquad (13)$$

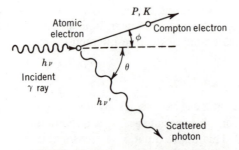

FIGURE 18.5 Compton scattering.

K and P are related by the relativistic relationship

$$E = K + m_0 c^2 = \sqrt{c^2 P^2 + m_0^2 c^4} \qquad \text{(J)} \tag{14}$$

where m_0 is the rest mass of the electron.

A number of useful relationships follow from algebraic combinations of the three conservation equations. These include the following, in which $\alpha \equiv h\nu/m_0 c^2$.

1. The Compton wavelength shift:

$$\lambda' - \lambda = \frac{c}{\nu'} - \frac{c}{\nu} = \frac{h}{m_0 c}(1 - \cos \theta) \qquad (\text{cm}^{-1}) \tag{15}$$

where $h/m_0 c = 2.426 \times 10^{-10}$ cm is called the Compton wavelength. It is the wavelength of a γ ray whose energy is $m_0 c^2$ or 0.511 MeV. Note that $0 \leq \theta \leqq \pi$ implies $0 \leq \lambda' - \lambda \leqq 2h/m_0 c$.

2. The energy of the scattered γ ray:

$$h\nu' = \frac{m_0 c^2}{1 - \cos \theta + 1/\alpha} \qquad (\text{MeV}) \tag{16}$$

Note for $\alpha \gg 1$ and $\theta = 180°$ the energy of the backscattered γ ray is

$$h\nu' = \frac{m_0 c^2}{2} = 0.256 \text{ MeV} \tag{17}$$

while for $\theta = 90°$,

$$h\nu' = m_0 c^2 = 0.511 \text{ MeV} \tag{18}$$

3. The kinetic energy of the struck electron:

$$K = h\nu - h\nu' \qquad (\text{MeV}) \tag{19}$$

$$K = h\nu \frac{2\alpha \cos^2 \phi}{(1 + \alpha)^2 - \alpha^2 \cos^2 \phi} \qquad (\text{MeV}) \tag{20}$$

$$K = h\nu \frac{\alpha(1 - \cos \theta)}{1 + \alpha(1 - \cos \theta)} \qquad (\text{MeV}) \tag{21}$$

The maximum energy transferred $K_{\max}$ occurs when $\theta = 180°$:

$$K_{\max} = \frac{h\nu}{1 + \dfrac{1}{2\alpha}} < h\nu \qquad (\text{MeV}) \tag{22}$$

This maximum energy is known as the **Compton edge**.

EXERCISE 1

Derive equations 15, 16, and 21.

Note that the Compton shift in wavelength is independent of $h\nu$.

The Compton shift in energy or the energy of the electron is strongly dependent on hv; hence, low-energy γ rays are scattered with only a moderate energy change, but high-energy γ rays suffer a large change in energy.

EXERCISE 2

At $\theta = 90°$, if $hv = 10$ keV, then show $hv' = 9.8$ keV (a 2 percent change); but if $hv = 10$ MeV, then $hv' = 0.49$ MeV (a 20-fold change).

In Figure 18.6, the Compton cross section per unit energy interval versus Compton electron energy K is shown for three values of α, or γ-ray energy. Since the probability of the interaction occurring is proportional to the cross section, the most probable electron energy for each value of α corresponds to the Compton edge.

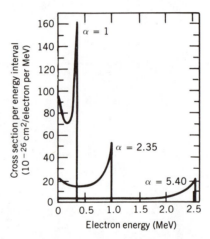

FIGURE 18.6 Energy distribution of Compton electrons produced by three primary photons.

EXERCISE 3

For each value of α given in Figure 18.6, use equation 22 to calculate the energy corresponding to the Compton edge. Compare your values with Figure 18.6.

Photoelectric Effect

In this case the γ ray is totally absorbed. An incident photon cannot be totally absorbed by a free electron because momentum is not conserved. If the electron is bound, then momentum is conserved by the recoil of the entire atom. Since the entire atom participates, the photoelectric effect may be visualized as an interaction of the γ ray with the entire electron cloud in which the entire γ-ray energy is absorbed and an electron is ejected. The process is shown in Figure 18.7 and conservation of momentum and energy gives

$$\frac{hv}{c} = P \cos \theta + P_a \cos \phi \qquad \text{(kg} \cdot \text{m/s)} \qquad (23)$$

$$0 = P \sin \theta - P_a \sin \phi \qquad \text{(kg} \cdot \text{m/s)} \qquad (24)$$

$$hv = K + K_a + B \qquad \text{(J)} \qquad (25)$$

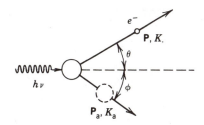

FIGURE 18.7 Photoelectric effect.

where B is the binding energy of the electron (the energy required to remove the electron from the atom), $\mathbf{P}$ and K are the electron's momentum and kinetic energy, and $\mathbf{P}_a$ and K_a are the atom's momentum and kinetic energy.

When we ignore the small kinetic energy of the atom, then from equation 25 the kinetic energy of the ejected electron is given by

$$K = h\nu - B \quad \text{(J)} \tag{26}$$

Since the atom is left with a vacancy in an atomic shell, it will be immediately filled by an outer electron and an x ray will be emitted.

Pair Production

If $h\nu \geqq 1.02$ MeV, then pair production becomes an important interaction of γ rays with matter. In this interaction, the γ ray is completely absorbed and a positron–electron pair are created whose total energy is just equal to $h\nu$. Thus,

$$h\nu = (K_- + m_0 c^2) + (K_+ + m_0 c^2) \quad \text{(J)} \tag{27}$$

where K_- and K_+ are the kinetic energies of the electron and positron, respectively, and $m_0 c^2 = 0.511$ MeV is the electronic rest-mass energy. Pair production occurs only in the field of charged particles, mainly in the field of the nucleus. The presence of the particle is necessary to conserve momentum. Pair production is shown in Figure 18.8.

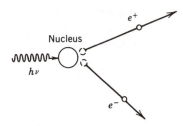

FIGURE 18.8 Pair production.

Scintillation Counter

In this experiment and in Experiment 19, Compton Scattering, the scintillation counter is used to detect γ rays and measure their energy. Read Appendix B, Counting and Sorting Particles: The Scintillation Counter.

EXPERIMENT

Become familiar with the equipment (read the manufacturer's description of operating voltages, etc.) before starting the experiment.

Start with a monoenergetic source of γ rays, such as $^{137}_{55}$Cs, connect the output of the linear amplifier to the oscilloscope, and observe the pulses. Vary the gain of the amplifier until saturation occurs, then set the gain so that the output pulses from the amplifier are not saturated.

For each source available observe the pulses on the oscilloscope to determine a gain setting such that saturation does not occur for any source. It may be necessary to reduce the high voltage of the PMT and the gain of the amplifier to avoid saturation. Reducing the high voltage will reduce the amplitude of the output pulses from the PMT. Check the manufacturer's minimum high-voltage setting and do not go below it. Leave the gain and high voltage at a level that does not create saturation.

Connect the amplifier to the MCA. Observe the complete spectra for each source. Using the known decay energies (see Figure 18.1 and/or reference 3), plot energy versus channel number for several total energy peaks to obtain an energy calibration curve for the amplifier and MCA.

EXERCISE 4

Determine the equation for your energy calibration curve. What is the error in both the slope and vertical intercept?

EXERCISE 5

For each source, calculate the expected channel number for the Compton edge, the single escape peak, and the other expected peaks that you identified in Exercise 4 of Appendix B.

EXERCISE 6

Explain the origin of all peaks that you find in each spectra. Sketches similar to those in Figures B.7 and B.8 of Appendix B are suggested. The spectra of ^{22}Na will include a large peak that is not related to the 1.277-MeV γ ray shown in Figure 18.1c. Carefully determine the position of this peak and explain its origin. Note that a positron is emitted by ^{22}Na.

EXERCISE 7

For the large peaks, what is the pulse height resolution of the scintillation counter?

19. COMPTON SCATTERING: GAMMA-RAY SPECTROSCOPY

Historical Note

The 1927 Nobel prize in Physics was divided equally between
  Arthur Holly Compton, the United States
  For his discovery of the effect (Compton scattering) named after him

and Charles Thomson Wilson, Great Britain
  For his method of making the paths of electrically charged particles visible by condensation of vapour.

APPARATUS

Scintillation counter: scintillator, photomultiplier tube (PMT), multichannel analyzer (MCA) (see Figure B.1, Appendix B)

1-mCi ^{137}Cs source

Scattering apparatus (see Figure 19.1)

Calibration sources, 1-μCi each of ^{137}Cs, ^{22}Na, ^{133}Ba, and ^{203}Hg

Apparatus to record the output of the multichannel analyzer: oscilloscope and camera or *XY* recorder or computer-driven data acquisition system.

OBJECTIVES

To obtain an energy calibration curve for the scintillation counter using several low-intensity γ-ray sources of known decay energies.

To use the calibrated scintillation counter to investigate Compton scattering and to compare experimental results with theory.

To understand Compton scattering.

To become familiar with the scintillation counter.

KEY CONCEPTS

Compton scattering   Background counts
Pair annihilation   Compton wavelength shift
Total energy peak

REFERENCES

1. A. Compton, *Am. J. Phys.* **29**, 817 (1961). Compton reviews the experimental evidence and the theoretical considerations that led to the discovery and interpretation of x rays acting as particles.

2. A. Bartlett, *Am. J. Phys.* **32**, 120 (1964). This paper is a historical review of the experiments that were later explained by Compton's discovery of the Compton effect.

3. A. Burns and R. Singhal, *Am. J. Phys.* **46**, 646 (1978). A Compton scattering expeirment is described and the experimental results are presented in this paper.

Also see the references listed under Experiment 18, Interaction of Gamma Rays with Matter, and Appendix B, Sorting and Counting Particles: The Scintillation Counter.

INTRODUCTION

The Introduction of Experiment 18, Interaction of Gamma Rays with Matter, where Compton scattering is discussed, is a prerequisite for this experiment.

Scintillation Counter

Appendix B, Sorting and Counting Particles: The Scintillation Counter, is a corequisite for this experiment.

Radiation Safety

Radionuclide samples that are used in the laboratory with "minimal" safety standards are one-microcurie (1-μCi) sources. One millicurie (1 mCi) is 1000 times this value; therefore, such a source must be handled in a safe manner. Mounting the 1-mCi source in a $4 \times 4 \times 4$-in. lead shield meets all radiation safety requirements. The lead shield is shown in Figure 19.1.

The following procedures are recommended in handling the 1-mCi source:

1. Mount the source in the lead shield such that it is securely held in place. A method for doing this is discussed below.
2. Measure the radiation on all faces of the shield using a calibrated survey meter. The results of such measurements are reported later.

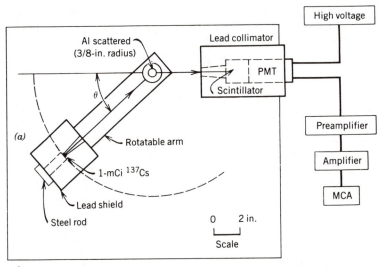

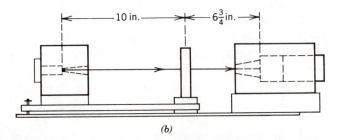

FIGURE 19.1 (*a*) Top view of the apparatus. (*b*) Side view of the apparatus.

3. Place the apparatus shown in Figure 19.1 in a corner of the laboratory with the opening in the lead shield directed primarily toward the corner of concrete walls.

4. Keep a lead brick at the station and place it against the opening in the shield before reaching in to adjust the scintillator or other equipment. Place the brick over the opening when the equipment is not in use.

5. For long-term storage place the lead shield in an appropriate lead vault.

We now discuss a method of mounting the source in the lead shield and measurement of the emerging radiation. A $\frac{3}{8}$-in. hole was drilled from face to face and passing through the center of the shield. A stainless steel sleeve, shown in Figure 19.2, fits inside the hole. The 1-mCi source mounts inside the sleeve and it is held in place by a stainless steel bolt that threads into the sleeve. A conical opening for radiation to escape is approximated by drilling holes of increasing diameter ranging from 0.3 to 0.6 cm.

A calibrated survey meter was placed against the five surfaces of the shield and against the head of the bolt that holds the source in place. The measured values are given in Table 19.1, where the *front* surface has the opening for radiation to escape, the *back* surface is the surface with the protruding bolt head, and the other four surfaces are the four *side* surfaces.

The units in Table 19.1, milliroentgen per hour (mR/h), are units of radiation exposure. Exposure indicates the production of ions in a material by radiation, and it is defined as the amount of ionization produced in a unit mass of dry air at standard temperature and pressure. The roentgen is the conventional unit for exposure, where

$$\text{radiation exposure unit: } 1 \text{ roentgen} = 1 \text{ R} = 2.58 \times 10^{-4} \text{ coulomb per kilogram}$$

Thus, 1 R of radiation produces 2.58×10^{-4} C of positive ions in a kilogram of air at standard temperature and pressure, and an equal charge of negative ions.

Radiation safety standards are expressed in units of *roentgen equivalent mammal* per year (rem/yr). We now relate roentgens to rems via the *rad* unit and the *RBE* or *QF*, discussed below.

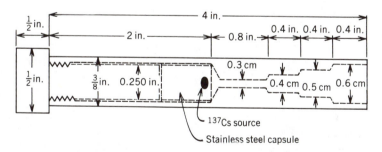

FIGURE 19.2 ^{137}Cs source-holder sleeve.

TABLE 19.1 MEASURED RADIATION EXPOSURE

Surface	Exposure (mR/h)
Front	6
Back	0.7
Four sides	0.5

The absorbed dose is the radiation energy absorbed per kilogram of absorbing material. It is measured in rads, where

absorbed dose unit: 1 rad = 0.01 joule per kilogram

In animal tissue it takes about 30 eV or 4.8×10^{-18} J to produce an ion pair, and assuming the magnitude of the charge of each ion is 1.60×10^{-19} C, then for animal tissue

$$1\,\text{R} \times \frac{4.8 \times 10^{-18}\,\text{J}}{1.6 \times 10^{-19}\,\text{C}} = 2.58 \times 10^{-4}\,\frac{\text{C}}{\text{kg}} \times 30\,\frac{\text{J}}{\text{C}} = 0.008\,\frac{\text{J}}{\text{kg}} \simeq 1\,\text{rad}$$

Hence, a 1-R exposure to x rays or γ rays produces an animal tissue absorbed dose of approximately 1 rad.

The effects of radiation on biological systems depend on the type of radiation and its energy. The *relative biological effectiveness* (**RBE**) or *quality factor* (**QF**) of a particular radiation is defined by comparing its effects to those of a standard kind of radiation, which is usually taken to be 200-keV x rays. The RBE or QF is the ratio of the dose in rads of a particular kind of radiation to a 1-rad dose of 200-keV x rays, where the particular radiation produces the same biological effect as the x rays. Note that RBE or QF is dimensionless.

$$\text{RBE or QF:} \quad \frac{\text{Number of rads of a particular kind of radiation}}{1\,\text{rad of 200-keV x rays}}$$

In animal tissue the RBE is about 1.0 for x rays, γ rays, and β rays, and it ranges from about 2 to 20 for protons, neutrons, and α particles.

The rem is defined as

$$\text{rem} \equiv \text{dose in rads} \times \text{RBE}$$

For animal tissue a 1-rad dose of γ rays is equivalent to an exposure of 1 R, and the RBE is about 1 for γ rays in animal tissue; therefore, the dose in rems and the exposure in roentgens are equivalent. Thus, the measurements in Table 19.1 may be expressed in mR/h or mrem/h.

Radiation standards adopted by the United States Government are the following:

1. For workers employed around nuclear facilities: 5 rem/yr, which would be 2.5 mrem/h continuously while at work for those on a 40-h week. For comparison, 300 to 600 rem of acute whole-body radiation is fatal to humans.
2. For the general population living near a nuclear facility: 0.5 rem/yr.
3. For worldwide population: 5 rem total up to age 30 (0.17 rem/yr), in addition to natural background radiation, which is about the same intensity. Primary concern is for genetic damage. It is estimated, rather uncertainly, that 0.17 rem/yr may produce 5000 extra deaths and 5000 birth defects in the United States per year.

Assuming that a student spent two laboratory periods on this experiment at 3 hours per lab period, then the exposure from the back surface of the lead shield would be 4.2 mrem. This is low-level exposure; however, the following is recommended: NEVER POSITION YOURSELF UNNECESSARILY CLOSE TO THE LEAD SHIELD.

We now show that the theoretical number of γ rays striking the aluminum scatterer per second is 2.7×10^4. The total number of emitted γ rays per second, n, by a radionuclide is given by

$$n = K \cdot S \cdot 3.666 \times 10^{10} \qquad (\gamma \text{ rays/s}) \tag{1}$$

where K is the average number of emitted γ rays per decay ($K = 1$ for ^{137}Cs) and S is the source activity in curies ($S = 10^{-3}$ Ci), where 1 Ci $= 3.666 \times 10^{10}$ distintegrations/s. Approximating the sample as a point source and the channel in the lead block as a cone, the γ-ray flux at a distance of 25 cm (source–target distance, see Figure 19.1a) is spread over a circular area of radius R:

$$\frac{0.6 \text{ cm}}{5 \text{ cm}} = \frac{R}{25 \text{ cm}} \quad \text{or} \quad R = 3 \text{ cm} \tag{2}$$

where the first equality follows from similar triangles. The aluminum rod or target has a vertical intersection of the γ-ray beam of 3 cm and a horizontal intersection of its width, 1.9 cm. The number of γ rays per second striking the target n_1 is given by

$$n_1 = n \frac{\Omega}{4\pi} = n \frac{(1.9 \times 3)/25^2}{4\pi} = 2.7 \times 10^4 \quad (\gamma \text{ rays/s}) \tag{3}$$

where Ω is the solid angle subtended by the area of the target that intersects the γ-ray beam.

The measured number of γ rays striking a target of area 1.9×3.0 cm located 25 cm from the source was 1.2×10^4 γ rays/s, which is of the predicted order of magnitude.

EXPERIMENT

Figure 19.1a shows a top view of the apparatus at an arbitrary scattering angle θ, and Figure 19.1b shows a side view corresponding to $\theta = 0°$. The scatterer or target is a $\frac{3}{4}$-in. aluminum rod mounted 10 in. from the Cs source. The lead collimator, shown in Figures 19.1 and 19.3, is tapered to better define the scattering geometry. Also the collimator in Figure 19.3 is for a 2-in.-diameter scintillator. Lead bricks may be necessary for additional shielding of the scintillator and PMT, especially for small scattering angles.

Become familiar with the equipment before starting the experiment; for example, read the manufacturer's operating instructions for both the PMT and the MCA.

We first use several low-intensity sources of known decay energies to obtain an energy calibration curve for the linear amplifier and MCA.

Remark: Many MCAs will generate the energy calibration curve for you. Check the specs of your MCA on this point.

For these sources shielding is not necessary. Starting with the 1-μCi ^{137}Cs source, connect the output of the linear amplifier to the oscilloscope and observe the pulses. The arrangement of the apparatus is shown in Figure B.1, Appendix B. Set the gain of the amplifier so that the output pulses are not saturated. It may be necessary to reduce the high voltage of

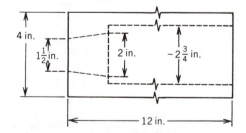

FIGURE 19.3 Lead collimator to shield a 2-in.-diameter scintillator and the photomultiplier tube.

the PMT and the gain of the amplifier to avoid saturation. Reducing the high voltage will reduce the amplitude of the output pulses from the PMT. Disconnect the oscilloscope and connect the amplifier to the MCA. Record the complete spectrum of the ^{137}Cs source in the MCA with the amplifier gain adjusted such that the total energy peak, which occurs at 0.662 MeV, is approximately located in the channel number given by $0.9 \times$ maximum channel, for example, channel number 475 for a 528-channel analyzer. (See Figure 18.1, Experiment 18, for the decay scheme of ^{137}Cs.) Leave this gain setting fixed for the remainder of the experiment.

Replace the ^{137}Cs source with the ^{22}Na source and record its spectrum. The total energy peak of ^{22}Na occurs at 1.277 MeV (see Figure 18.1, Experiment 18, for the decay scheme of ^{22}Na), which, for your gain setting, exceeds the maximum channel number. ^{22}Na decays by β^+ emission, and pair annihilation occurs in the source when the β^+ combines with an atomic electron, producing two 0.511-MeV γ rays. Hence, a peak will occur in the spectrum of ^{22}Na corresponding to a 0.511-MeV γ ray.

EXERCISE 1

Since pair annihilation produces two 0.511-MeV γ rays, why do we not expect a peak corresponding to an energy of 1.022 MeV?

The other two suggested calibration sources are ^{133}Ba and ^{203}Hg, which emit γ rays of energy 0.356 and 0.279 MeV, respectively; see Figure 19.4. Record the spectrum of each source. Using the data for the four sources, plot known peak energy versus channel number.

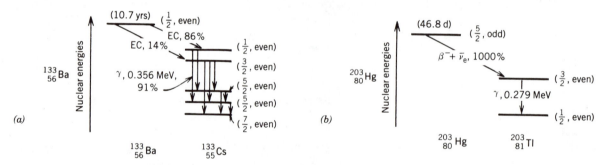

FIGURE 19.4 (a) Decay scheme of $^{133}_{56}$Ba. (b) Decay scheme of $^{203}_{80}$Hg.

EXERCISE 2

Using the method of least squares, what is the equation of your calibration curve?

Figure 19.5 is a photograph of the apparatus, where the spectrum shown on the MCA face was recorded over a 10-min time interval at a scattering angle of 50°. The cursor (vertical line) marks the channel number of the total energy peak.

Assemble the apparatus shown in Figure 19.1. Adjust the rotatable arm so $\theta = 45°$, and, with the target in place, record the spectrum from the 1-mCi cesium source for about 15 min. Then block the opening to the source with a lead brick, set the MCA on the subtract option, and accumulate background counts for the same length of time. The resulting spectrum is due to Compton-scattered photons at $\theta = 45°$. Record the spectrum in the MCA by connecting the output of the MCA to an available device. Use your energy calibration curve to determine the energy of the scattered γ rays in the total energy peak.

FIGURE 19.5 Photograph of the apparatus, including a typical spectrum.

Increase θ by 15° and repeat the measurement. The largest possible angle, ∼120°, is determined by the geometry of the apparatus.

Record the data at 0, 15, and 30°. For these angles additional shielding is likely to be required while recording the Compton-scattered γ rays to eliminate the direct observation of the spectrum.

From your data you have the energy of the total energy peak at each angle θ. Knowing this energy, calculate the wavelength λ' of the scattered γ ray and the Compton wavelength shift $\lambda' - \lambda$, where λ is the wavelength corresponding to the energy of the unscattered γ rays: 0.662 MeV. The theoretical Compton wavelength shift is given by equation 15, Experiment 18.

Plot $\lambda' - \lambda$ versus some function of θ such that a straight line is expected. Determine the equation of the line by the method of least squares.

EXERCISE 3

What value do you obtain for the Compton wavelength? Compare your value with the accepted value. See the line following equation 15, Experiment 18.

20. IONIZATION OF GASES BY ALPHA PARTICLES

APPARATUS

Brass pieces and Teflon for ionization chamber (detailed in Figure 20.8*b*)

$\frac{1}{4}$-mil mylar sheet

Electrodeposited ^{241}Am α source

Collimator (e.g., as shown in Figure 20.7)

Compressed argon supply

90-V battery with potentiometer

Voltmeter (0–100 V)
Electrometer
Pulse amplifier (e.g., as shown in Figure 20.9)
Scaler

OBJECTIVES

To construct and test an integral/differential dc ionization chamber.

To measure the range and stopping power of α particles in air and in argon and to compare these results, where possible, with theory.

To understand the energy loss mechanism for charged particles passing through matter.

To understand how the stopping power and the range for a heavy charged particle depend on the particle energy and on material parameters.

KEY CONCEPTS

Stopping power   Range
Extrapolated range   Straggling
Straggling parameter   Integral ionization chamber
Differential ionization chamber   Specific ionization
Bragg curve   Single-particle ionization curve
Saturation curve   Residual energy

REFERENCES

1. E. Fermi, *Nuclear Physics*, University of Chicago Press, Chicago, 1950. Chapter II contains an elementary discussion of the theory of energy loss by charged particles in matter.

2. E. Bleuler and G. J. Goldsmith, *Experimental Nucleonics*, Rinehart, New York, 1958. Contains a concise discussion of the range, Bragg curve and straggling for α particles.

3. R. D. Evans, *The Atomic Nucleus*, McGraw-Hill, New York, 1955. Chapter 22 discusses the passage of heavy charged particles through matter.

4. E. Segre (Ed.), *Experimental Nuclear Physics*, Vol. 1, Wiley, New York, 1960. Part I contains a discussion of the physics of particle detectors, the ionization chamber in particular. Part II discusses energy loss by heavy particles in matter.

5. W. J. Price, *Nuclear Radiation Detection*, McGraw-Hill, New York, 1964. Contains a discussion of scaling laws and empirical expressions for the range.

6. R. G. Marclay, *Am. J. Phys.* **29**, 845 (1961). Describes the construction of an ionization chamber suitable for the undergraduate laboratory, although much more sophisticated than the one described in this experiment.

INTRODUCTION

A heavy charged particle passing through a material loses energy primarily by means of the Coulomb interaction between the moving charge and the electrons of the material. The energy deposited per unit length by a projectile of energy E passing through a material, $-(dE/dx)$, is called the **stopping power** and depends on the particle energy and on the density and excitation energies of the electrons in the atoms of the material. An estimate of

dE/dx can be obtained from classical arguments like those given by Fermi (reference 1), which are outlined briefly here.

We calculate the energy deposited per unit length by a particle with charge $+ze$ (depicted in Figure 20.1a) as it passes through a medium containing electrons of mass m assumed to be initially at rest and which are assumed not to move appreciably during their interaction with the particle. The impulse **J** acquired, during the passage, by a single electron situated a distance b from the particle trajectory can be calculated by writing

$$\mathbf{J} = \int_{-\infty}^{+\infty} \mathbf{F}\, dt \qquad (\text{kg} \cdot \text{m/s}) \qquad (1)$$

where **F** is the electrostatic force exerted by the particle on the electron. If the electric field at the location of the electron due to the particle is $\boldsymbol{\epsilon}$, then $\mathbf{F} = -e\boldsymbol{\epsilon}$. The time differential dt is related to the increment dx along the particle path: $dt = dx/v$, where v is the instantaneous speed of the particle as it passes the electron. The expression for the impulse delivered to a single electron is thus

$$\mathbf{J} = -\frac{e}{v} \int_{-\infty}^{+\infty} \boldsymbol{\epsilon}(x)\, dx \qquad (\text{kg} \cdot \text{m/s}) \qquad (2)$$

if we take v as approximately constant during the nonnegligible portion of the interaction. Note that only $J_\perp$, the component of **J** perpendicular to the particle trajectory, is nonzero: The electron receives no net impulse along x during the passage of the particle. The expression for $J_\perp$,

$$J_\perp = -\frac{e}{v} \int_{-\infty}^{+\infty} \epsilon_\perp(x)\, dx \qquad (\text{kg} \cdot \text{m/s}) \qquad (3)$$

can be evaluated easily by noting that while $\epsilon_\perp(x)$ in the integral refers to the perpendicular or y component of the electric field at the location (assumed fixed) of the electron due to the particle when it is instantaneously at x, the desired result can also be obtained by considering the electric field at the changing location of the electron as viewed in the rest frame of the particle. Then Gauss's law states

$$\int_S \epsilon_\perp\, dA = \frac{ze}{\epsilon_0} \qquad (\text{N} \cdot \text{m}^2/\text{C}) \qquad (4)$$

where now the integral is over the surface of the very (infinitely) long cylinder of radius b, pictured in Figure 20.1b, that encloses the charge $+ze$, located on the symmetry axis. Using

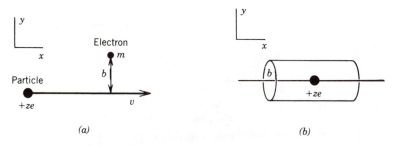

FIGURE 20.1 (a) Particle with charge $+ze$ passing an electron with mass m. (b) Gaussian surface for the calculation of $J_\perp$.

$dA = (2\pi b)\, dx$, the impulse delivered to the electron is evaluated as

$$J_\perp = -\frac{ze^2}{2\pi b v \epsilon_0} \qquad (\text{kg}\cdot\text{m/s}) \tag{5}$$

since the integral in equation 4 is just $2\pi b$ times the integral in equation 3. Because $J_\perp$ is just the momentum acquired by the electron, the energy acquired is approximately $E_1 = J_\perp^2/2m$, or, upon substitution of the expression in equation 5,

$$E_1 = \frac{z^2 e^4}{8\pi^2 b^2 v^2 \epsilon_0^2 m} \qquad (\text{J}) \tag{6}$$

Using this expression for the energy given to a single electron by the passing particle, we can express the change in the particle projectile energy, dE, due to the energy it loses to all of the electrons located within a cylindrical shell of thickness db and in a slab of thickness dx, as

$$dE = -\left(\frac{z^2 e^4}{8\pi^2 b^2 v^2 \epsilon_0^2 m}\right) N(2\pi b\, db)\, dx \qquad (\text{J}) \tag{7}$$

where N is the number density of electrons in the material. (Note that a minus sign has been inserted because the energy *gained* by the electrons equals the energy *lost* by the projectile particle.) The expression for dE/dx is now obtained by integrating over b, the distance from the particle trajectory:

$$\frac{dE}{dx} = -\frac{Nz^2 e^4}{4\pi m v^2 \epsilon_0^2} \int_{b_{\min}}^{b_{\max}} \frac{db}{b} \qquad (\text{J/m}) \tag{8}$$

Note that we cannot integrate over all values of b, since the integral would then diverge; integration of equation 8 should be performed only over a range of distances $b_{\min} < b < b_{\max}$ where the classical expression for dE in equation 7 is approximately valid. Below we summarize Fermi's arguments for obtaining estimates for $b_{\min}$ and $b_{\max}$.

The upper limit $b_{\max}$ can be estimated by noting that electrons with orbital motions characterized by frequencies $f > \tau^{-1}$, where τ is the duration of the field pulse at the location of an electron during passage of the particle, will not be able to absorb energy from it. Classically, these motions correspond to orbits in which the electrons reverse direction too rapidly to gain any net energy from the field. The time τ is approximately b/v, but is further shortened by the factor $(1 - \beta^2)^{1/2}$ (where $\beta \equiv v/c$) because of the relativistic contraction of the electric field in the direction of the motion. We include in the range of integration, therefore, only those values of b for which $f < (v/b)(1 - \beta^2)^{-1/2}$, so that for $b_{\max}$ we have

$$b_{\max} \simeq \frac{v}{f}(1 - \beta^2)^{-1/2} \qquad (\text{m}) \tag{9}$$

For $b_{\min}$, the relevant concern is that the Coulomb field of the particle be uniform over a region that has dimensions on the order of $\lambda/2\pi$, where λ is the de Broglie wavelength of the electron as seen from the particle rest frame. Thus, we consider only distances $b \gtrsim \lambda/2\pi$ to be within the range of validity. For particle speeds much greater than the speeds associated with the electron orbitals, the observed wavelength is $\lambda = h/p = (h/mv)(1 - \beta^2)^{1/2}$, where the relativistic expression for the momentum has been used. Thus, we have

$$b_{\min} \simeq \frac{\hbar\sqrt{1 - \beta^2}}{mv} \qquad (\text{m}) \tag{10}$$

Performing the definite integral in equation 8 with the limits of equations 9 and 10 yields an approximation for the energy lost per unit length in a material by a fast, heavy charged particle:

$$\frac{dE}{dx} = -\frac{Nz^2 e^4}{4\pi m v^2 \epsilon_0^2} \ln \frac{mv^2}{\hbar f (1 - \beta^2)} \quad \text{(J/m)} \qquad (11)$$

A better approximation for dE/dx is given by Bethe as

$$\frac{dE}{dx} = -\frac{Nz^2 e^4}{4\pi m v^2 \epsilon_0^2} \left\{ \ln\left[\frac{2v^2 m}{I(1 - \beta^2)} \right] - \beta^2 \right\} \quad \text{(J/m)} \qquad (12)$$

where I is an empirically determined average of the electronic excitation potentials.

The measured stopping power for protons and alpha particles in air, given as the *negative* x derivative of the *particle* energy, is plotted versus particle energy in Figure 20.2. For energies $E_\alpha \gtrsim 5$ MeV this curve is in agreement, at least qualitatively, with equation 12. The $1/v^2$ dependence is evident for the midrange of particle energies; for higher energies (Figure 20.2b), the effects of relativistic corrections (i.e., of the terms containing β) can be seen as a slight rise in stopping power as v approaches c.

For energies below a few MeV, however, the behavior of the stopping power deviates substantially from the prediction of the above semiclassical calculation, which assumed particle velocities much greater than the orbital speeds of all Z of the electrons in the atoms of the material. For particle velocities less than the velocities of the K-shell electrons, an

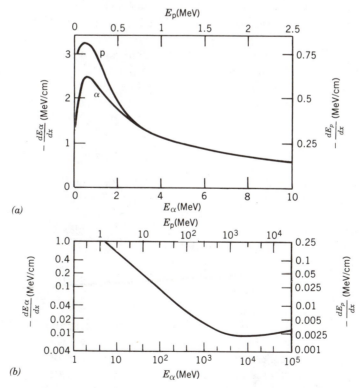

FIGURE 20.2 (a) Stopping power versus energy for protons and α particles in air (low energies). (b) Stopping power versus energy for protons and α particles in air (higher energies).

energy-dependent correction term is typically subtracted from the stopping power given above to account for the nonparticipation of these electrons in the energy absorption process. The corrected stopping power can be written, according to reference 3, as

$$\frac{dE}{dx} = -\frac{Nz^2e^4}{4\pi mv^2\epsilon_0^2}\left\{\ln\left[\frac{2v^2m}{I(1-\beta^2)}\right] - \beta^2 - \frac{C(v)}{Z}\right\} \qquad \text{(J/m)} \qquad (13)$$

where $C(v)$, a dimensionless number of order unity, depends on both the particle velocity and the nature of the binding of the innermost electrons to the atoms of the material. For even smaller particle velocities, the particle may capture one or more electrons, thus reducing its effective charge below $+ze$ and, hence, reducing the stopping power even further.

While the semiclassical calculation outlined above provides only a rough prediction of the stopping power for a given particle/material combination, the result of equation 13 allows comparisons of stopping powers for different combinations. For example, if the stopping power for α particles in air is measured for a particular energy, then the stopping power for α particles in another material may be calculated approximately from its dependence on the electron number density N for the material.

EXERCISE 1

Calculate the approximate thickness, in centimeters, of a sheet of aluminum in which a 6-MeV α particle would lose the same amount of energy as it would in 1 cm of air at standard temperature and pressure. Assume the effective average number of electrons per atom in the air molecules, Z_{air}, is 7.22, $I_{air} \simeq 80.5$ eV, and $I_{Al} \simeq 164$ eV. What approximations do you need to make? Express your answer also in units of mg/cm^2 of Al. Compare your calculated answer to the measured result of 1.51 mg/cm^2.

The distance traveled by a particle before all of its initial energy E is transferred to the material is referred to as the **range** R of the particle. If dE/dx is known as a function of E, then the distance dx corresponding to an energy dE delivered to the material is just $dx = dE/(-dE/dx)$, and the relationship between the range and the stopping power is

$$R(E) = \int_0^E \frac{dE'}{-(dE'/dx)} \qquad \text{(m)} \qquad (14)$$

If dE/dx is known analytically (e.g., from equation 13) or is measured, then the range can be computed by performing the above integral directly. The energy dependence of R for a given type of particle in a given type of material may often be represented as a simple function of E over a range of energies. For example, the range of α particles in air, in the energy interval from 4 to 11 MeV, is approximated (see reference 5) to within 1 percent by

$$R(E) = 5 \times 10^{-5}E^{5/2} + 2.85 \times 10^{-3}E^{3/2} \qquad \text{(m)} \qquad (15)$$

where E is the α energy in MeV.

Because of statistical fluctuations in the interactions between particles and material electrons, α particles with the same initial energy will have slightly different ranges, a phenomenon called **straggling**. For a large number of incident particles, the ranges are normally distributed about a **mean range** R_0, as indicated by Figure 20.3a. The probability distribution for the ranges is given, as in reference 2, by

$$p(R) = \frac{\alpha}{\pi^{1/2}}\exp[-\alpha^2(R - R_0)^2] \qquad \text{(m}^{-1}) \qquad (16)$$

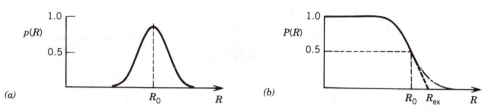

FIGURE 20.3 (a) Distribution of ranges for charged particles. (b) Probability of a particle having a range larger than R.

where α is called the **straggling parameter**. Since $p(R)\,dR$ is the probability that the range will be within an interval dR about R, the probability $P(R)$ that a particle will have a range larger than R can be found by integrating the above distribution:

$$P(R) = \int_{R}^{\infty} p(R')\,dR' \qquad (17)$$

This is plotted as a function of R in Figure 20.3b. As expected, $P(R)$ dips to $\frac{1}{2}$ at $R = R_0$. This curve, unlike the one for $p(R)$, is easy to obtain experimentally by noting the number of particles reaching a detector as the distance from a reference point is increased. In addition to R_0, the **extrapolated range** R_{ex} is often measured from the $P(R)$ curve by finding the intersection with the abscissa of a line tangent to $P(R)$ at $R = R_0$, as is illustrated in Figure 20.3b. The quantity $R_{ex} - R_0$ is directly related to the straggling parameter:

$$R_{ex} - R_0 = \frac{P(R_0)}{-dP/dR} = \frac{\sqrt{\pi}}{2\alpha} \qquad \text{(m)} \qquad (18)$$

where the derivative dP/dR is evaluated at $R = R_0$. For α particles in air, for example, this amounts to about 1 percent of R_0.

EXERCISE 2

Verify the relationship between R_{ex}, R_0, and α given above by equation 18.

If the particle source has nonnegligible thickness, particles emitted from different locations within the source will have different ranges; the $P(R)$ curve in Figure 20.3b will thus be modified accordingly. The true mean range R_0 for particles emitted from such a source is no longer the same as $R_{1/2}$, which is the measured half-intensity distance. R_0 may still be calculated from a measurement of R_{ex} and $R_{1/2}$; the technique is described concisely in reference 2, page 282.

EXERCISE 3

Indicate qualitatively, by means of a rough sketch, how Figure 20.3b would be modified for a source of nonnegligible thickness.

Measurements of the range and stopping power for charged particles emitted from radioactive sources may be accomplished either with a surface barrier detector in conjunction with a pulse analysis system or with an ionization chamber and an electrometer. In

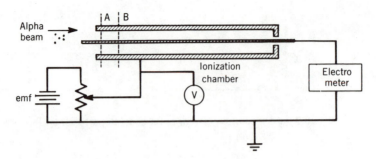

FIGURE 20.4 Configuration for ionization chamber measurements.

what follows we describe measurements on α particles taken with simply constructed integral and differential ionization chambers configured as shown schematically in Figure 20.4. In this arrangement, α particles enter the region enclosed by the conducting cylinder. Some of the energy lost to gas molecules in this region goes to the production of ion pairs, most of which are collected by the cylinder and the coaxial wire as a result of the strong electric field produced by the application of the emf between these two conductors. The current produced by the collected ion pairs is measured by the electrometer. If the length of the region within the chamber from which ions are collected exceeds the range of the entering α particles, then the device is configured as an **integral** chamber and can be used to measure the total energy of the entering beam. If ions are collected only from a small segment of the chamber (e.g., the region between the dashed lines A and B in the figure), then the device is a **differential** chamber and can be used to measure the stopping power $-(dE/dx)$ for the α particles as a function of distance from the source.

   The number of ion pairs produced per unit path length due to the passage of a single particle is called the **specific ionization**, shown as the solid curve in Figure 20.5 for the case of Po α particles in air. The *average* specific ionization, $B(x)$, for a collection of many particles with the same initial energy, as will be measured in the present experiment, is called the **Bragg curve** and includes the effects of straggling. Where the Bragg curve differs from the single-particle ionization curve in Figure 20.5, it is represented by a dashed curve. The Bragg curve can be obtained directly from the experimental setup in Figure 20.4 by reading the electrometer current from the differential chamber segment AB.

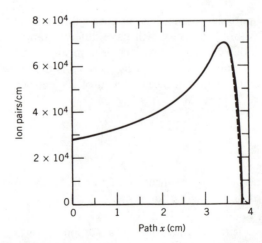

FIGURE 20.5 Specific ionization (—) and Bragg curve
(– – –) for Po α particles in air.

EXERCISE 4

Use equation 15 to estimate the range of the Po α particles that produced the data of Figure 20.5.

The experimentally obtained Bragg curve can be simply related in an approximate way to the stopping power $-dE/dx$. The energy E delivered to the stopping material may be distributed to its atoms in various ways: (1) It may go directly to the production of an ion pair; (2) it may eject fast electrons which themselves produce additional ion pairs; (3) it may excite an atom without ionizing it. Although the detailed process by which the charged particle transfers its energy may be quite complicated, the remarkable fact is that the average energy, w, expended in the production of a single ion pair varies only by about a factor of two for all common gases and for all heavy incident particles over a large energy range. Of particular relevance for this experiment is that w for α particles depends only on the gas used as the stopping material and is virtually independent of the α-particle energy E. Thus, the stopping power may be obtained from the Bragg curve $B(x)$ by writing

$$\frac{dE}{dx} = -wB(x) \qquad (\text{J/m}) \tag{19}$$

Values for w found in the literature vary slightly and have been shown to depend somewhat on the purity of the gas. Measurements made for α particles indicate that for air and Ar, values of w are 35.5 and 26.4 eV/ion pair, respectively.

EXERCISE 5

Explain qualitatively the behavior of the specific ionization curve of Figure 20.5 for regions on both sides of the peak, using the above discussion of the dependence of dE/dx on E. Account for the difference between the Bragg and single-particle curves.

EXERCISE 6

For many common gases, w is substantially greater than the ionization potential. Explain why this should be so.

EXPERIMENT

The Source

The source recommended for this set of measurements consists of electrodeposited ^{241}Am, which has a half-life of 458 years with respect to alpha decay to ^{237}Np. The decay scheme described by Figure 20.6a indicates that the majority of α particles are emitted with two closely spaced energies: 5.476 MeV (84%) and 5.433 MeV (14%). These two energy peaks can be resolved by use of a surface barrier detector. Figure 20.6b shows these two peaks in an Am α-particle energy spectrum measured with such a device.

The source emits α particles in all directions. Thus, if the experimental configuration in Figure 20.4 is to be used, it is clearly advantageous to collimate the beam of α particles emitted from the source so that the trajectories of the particles as they enter are approximately parallel. If the beam were left uncollimated, not only would the number of α particles entering the chamber be a complicated geometrical function of source–chamber distance,

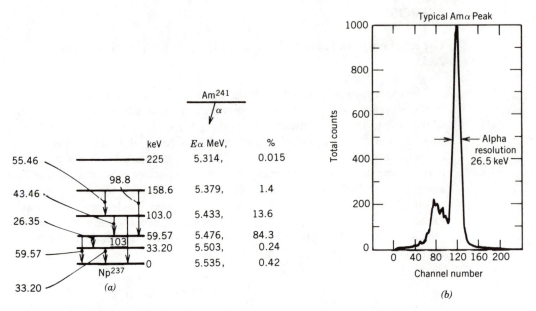

FIGURE 20.6 (a) Decay scheme for ²⁴¹Am. (b) Am α-particle energy spectrum.

but many of these α particles would impact the electrodes, reducing the extent of ionization, thus making energy measurements very difficult. The required collimation can be achieved by the simple device pictured in Figure 20.7, which is mounted so that it can be moved independently of the ionization chamber. The source material is mounted on a movable piston in an evacuated chamber sealed on one end with a $\frac{1}{4}$-mil mylar window. The angular divergence of the α beam is then determined by the effective area of the source, the area of the mylar window, and the distance between them. Observe that although the collimation produced by this device greatly simplifies the interpretation of the data, the intensity of the beam is thereby greatly reduced.

Note: Never touch the source material directly!

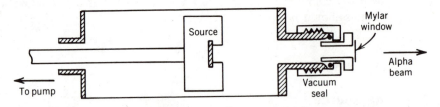

FIGURE 20.7 Collimator assembly.

EXERCISE 7

Roughly estimate, for your collimator arrangement, the distance between the source and the window required to insure that α particles entering the chamber do not collide with the cylindrical electrode. (Use equation 15 to estimate the range.) If you have done Experiment 21, Rutherford Scattering, can you make an educated guess as to whether the elastic scattering of the α particles by the nuclei of the gas molecules will substantially affect the collimation of the beam after it enters the chamber?

Ionization Chamber

The assembled ionization chamber is shown with the collimator in Figure 20.8a; the numbered parts are detailed in Figure 20.8b. The insulator mount (3) and gas inlet (2) may be soldered to the main tube (1), but the differential collar (6) should slip on and off of it. The inner electrode is a No. 18 wire which is held by the feed-through (4) and insulated from the main tube, which serves as the outer electrode, by a Teflon gasket (5). The beam enters the chamber through a screen mesh that covers the opening to the collar; the mesh reduces the fringing field outside the chamber, thus defining the collection volume for the ions. Leave a gap of about 2 mm between the end of the wire and the screen mesh. The

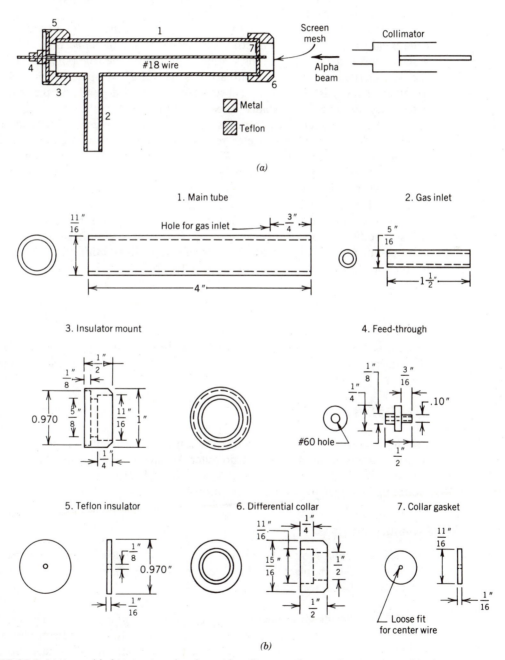

FIGURE 20.8 (a) Assembled ionization chamber with collimator. (b) Parts for ionization chamber.

Teflon collar gasket (7) at the front stops the α particles and thus separates the differential collar region from the rest of the chamber. With the gasket in place, only the collar region collects ions and the device functions as a differential ionization chamber; with the gasket removed, all of the ions along the α trajectory within the chamber are collected, and the device functions as an integral ionization chamber.

Construct the circuit shown in Figure 20.4, in which the chamber is represented schematically in cross section. For the chamber voltage supply, use the smallest potentiometer consistent with low battery drain. To avoid ground loops, use leads as short as possible and connect all grounded leads to a common ground at the electrometer. A shielded cable should be used to connect the center electrode to the electrometer, which measures the ionization current. Good, solid connections are essential here; poor connections can result in large fluctuations in the current readings.

The bias voltage applied by the battery between the inner electrode (wire) and the cylindrical shell provides the electric field necessary for the collection of ions. This voltage is read approximately as V on the voltmeter; to obtain the true potential difference between the electrodes one needs to take into account the voltage drop across the electrometer input. Whether or not this represents a substantial correction depends on the scale setting, which determines the effective resistance between the input terminals. This resistance will be quite high, and can be determined from inspection of the electrometer instruction manual.

The reading of the electrometer represents the approximate average of the ionization current i over a time $\tau \simeq RC$, where R and C are, respectively, the resistance and capacitance of the electrometer input. This average current is related to the rate of ion-pair production: $i = naq$, where n is the average number of α particles arriving each second at the chamber entrance, a is the number of ions produced by each particle, and q is the charge per ion. Even if the circuit is configured in such a way as to maximize the signal-to-noise ratio, there will still be fluctuations in i because of the statistical fluctuations in n. If the variations in n are assumed to be well described by the Poisson distribution, then the magnitude of the fluctuations in i should be given approximately by

$$\frac{\Delta i}{i} \simeq (n\tau)^{-1/2} \tag{20}$$

Thus, it is possible to reduce statistical fluctuations either by increasing n (e.g., by making observations on a stronger source) or by increasing τ by changing scales on the electrometer.

EXERCISE 8

Use the discussion of the Poisson distribution in Experiment 6, Introduction to Computer-Assisted Experimentation (under Computer-Assisted Counting and Data Analysis) to verify the above estimate for $\Delta i/i$. (*Hint*: Assume the electrometer produces its value for i by "counting" charges for a time τ and then dividing the result by τ.)

Measurements

All measurements should be made with the collimator chamber evacuated, with the differential collar in place on the front end of the chamber, and with the axes of the collimator and chamber aligned.

1. For both the integral and differential chamber configurations, measure the ionization current i as a function of the potential difference between the electrodes of the chamber while it is filled with air. Be sure to take measurements at voltages in the saturation region, that is, at voltages high enough so that further increases in voltage produce no

further increases in ionization current. Plot this saturation curve of ionization current against interelectrode voltage. Using this curve, determine a convenient voltage to be used for all subsequent measurements made with air that guarantees that all ions produced in the chamber will be collected.

EXERCISE 9

Explain the behavior of your saturation curve in both the low- and high-voltage regions.

2. Measure i versus x, the distance from the source, for both integral and differential chambers. Plot the data from both curves. Check to see that your integral and differential curves are consistent with each other.

3. Correct the data obtained above from the differential chamber for the presence of the mylar collimator window, since it absorbs some of the α-particle energy. This can be done by placing an *additional* $\frac{1}{4}$-mil piece of mylar over the screened entrance to the chamber (held in place by a rubber band, e.g.) and remeasuring the i versus x curve., which should be a shifted version of the original curve. Use these data to predict an i versus x curve for α particles in air without *any* mylar present.

EXERCISE 10

What assumption must you make about the ratio of the stopping powers of mylar and air to produce the corrected i versus x curve?

4. From the corrected i versus x curve for the differential chamber, produce the dE/dx versus x curve for the α particles in air. The vertical scale can be determined by noting that the integral of this curve over the range must equal the energy with which a single α particle is emitted from the source. Estimate the range of the α particles.

5. From the dE/dx versus x curve, produce a dE/dx versus E_{res} curve, where E_{res} is the residual α-particle energy. E_{res} can be obtained from the data by writing

$$E_{res} = \int_x^R -\frac{dE}{dx'}\,dx' \qquad \text{(J)} \tag{21}$$

How does this curve compare with the ones discussed above?

6. The air in the chamber may be replaced with flowing argon by means of the gas inlet and a piece of mylar (perforated if necessary) over the front collar. Repeat enough of these measurements with argon to produce a dE/dx versus E_{res} curve for this gas. Can you explain the comparison with the results for air?

7. Pulses of ionization due to individual α particles may be observed and counted by using the chamber (or surface barrier detector, if preferred) in conjunction with a pulse amplifier, scaler, and oscilloscope. The circuit in Figure 20.9, for example, produces voltage pulses suitable for counting from the pulses of ionization current input from the chamber. Using the pulse counting equipment, produce a number versus distance curve, as discussed following equation 17. Use this curve to produce estimates of the extrapolated range, the mean range, and the straggling parameter.

8. Make an estimate, from the scaler output and the collimation geometry, of the source strength. Compare it with the manufacturer's specification, if available. (You may need to take into account the age of the source.) Make an estimate of the α-particle energy by observing the pulse heights on the scope and estimating the gain and input capacitance of your amplifier.

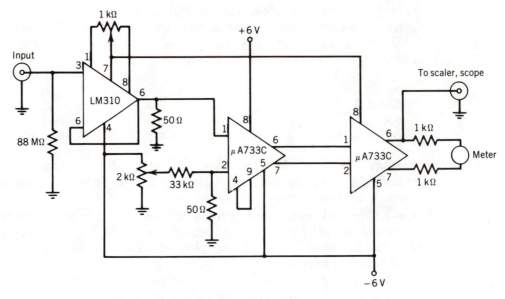

FIGURE 20.9 Pulse amplifier circuit.

21. RUTHERFORD SCATTERING

APPARATUS [Optional Equipment in Brackets]

Mechanical vacuum pump

Vacuum hose

[2 Vacuum valves]

Binocular microscope (10–50×) or slide projector

[Thermocouple vacuum gauge]

Aquarium heater

Thermometer

Sodium hydroxide (2.5N solution)

EN-20 Rutherford scattering apparatus (available from Daedalon Corporation) or a similar device composed of

  Vacuum chamber

  Alpha-particle source (10 μCi ^{241}Am recommended)

  Collimating device (e.g., two plates with holes at centers)

  Thin gold foil

  Kodak LR-115 Type 2 cellulose nitrate film (available from Kodak–Pathe in France, or from Daedalon Corp.)

  Counting mask (see Figure 21.8)

OBJECTIVES

To observe the distribution of α-particle scattering angles and compare it with the predictions of the Rutherford model over a range of small angles.

To understand the concepts of differential and total cross section and their relationship to the results of scattering experiments.

To understand how the differential cross section for α-particle scattering is derived from the Rutherford model of the atom.

KEY CONCEPTS

Elastic scattering

Differential cross section

Total cross section

Solid angle

REFERENCES

1. R. Eisberg and R. Resnick, *Quantum Physics of Atoms, Molecules, Solids, Nuclei, and Particles*, 2d ed., Wiley, New York, 1985. Sections 4-1 and 4-2 contain a discussion of the Thomson and Rutherford models of the atom. The differential cross section for Rutherford scattering is derived using the details of the α-particle trajectory developed "from scratch" in Appendix E.

2. P. A. Tipler, *Modern Physics*, Worth, New York, 1978. At the end of Chapter 4 is a simple derivation, similar to the one outlined in this experiment, of the relationship between the deflection angle and the impact parameter for the α particle which does not utilize the details of the trajectory.

3. L. Basano and A. Bianchi, *Am. J. Phys.* **48**, 400 (1980). An elegant derivation of the Rutherford scattering differential cross section using only the conservation of the Runge–Lenz vector.

4. H. A. Enge, *Introduction to Nuclear Physics*, Addison-Wesley, Reading, MA, 1966. Section 3-2 is a brief but general discussion of scattering cross sections.

5. J. Walker, *Sci. Am.* **254**(2): 114 (1986). An interesting narrative account of a Rutherford scattering experiment performed on homemade apparatus similar to that recommended for the present experiment. Also contains an idea for a computer interface to record the data.

INTRODUCTION

The elastic scattering of α particles (^{4}He nuclei) by heavy atoms, referred to as Rutherford scattering, is at the heart of one of the most dramatic illustrations of the interplay between theory and experiment that is the essence of the scientific method. Rutherford and his colleagues, around 1911, studied the distribution of angular deflections of α particles scattered from the atoms of a thin gold foil and concluded that, contrary to the Thomson model of atomic structure, the positive charge of an atom is concentrated in a nucleus at its center. This conclusion was based primarily on the quite unexpected occurrence of very large angle scatterings of the α particles, a result that Rutherford himself described by saying that it was "as if you fired a 15-inch naval shell at a piece of tissue paper and the shell came right back and hit you." The resulting picture of an extremely dense, positively charged nucleus at the center of the atom remains today at the heart of our understanding of atomic structure.

In what follows we describe the classical theory of α-particle scattering and a relatively simple technique for obtaining and analyzing the scattering pattern for small angles.

Scattering Cross Section

The scattering cross section for a process involving the collision of an incident beam of particles and a target is an experimentally measurable quantity that can be compared with theoretical predictions concerning the interaction between particle and target. The idea can be illustrated for the macroscopic system depicted in Figure 21.1a, in which an incident beam of intensity I particles per second · area is projected at a hard sphere of cross-sectional area σ. The total cross section of the sphere can be measured by noting the fraction, f, of the incident particles that are scattered by impact with the target. If the total area of the beam is A, then one expects a fraction $f = \sigma/A$ of the beam to be scattered, so that the **total cross section** of the target can be defined operationally as $\sigma \equiv fA$. If N particles per second are observed as scattered from the target, then we can express σ as

$$\sigma \equiv fA = \frac{N}{IA} A = \frac{N}{I} \quad (\text{m}^2) \tag{1}$$

We naturally expect the cross section of this sphere, or of any macroscopic object, calculated from scattering observations using equation 1, to correspond to its size, that is, the geometrical area it presents to the incident beam; but the "size" of atomic and subatomic targets is not so easily defined. Figure 21.1b, for example, depicts an incident positively charged particle scattered "at a distance" from a stationary target nucleus by its Coulomb field, which extends its effective size for scattering far beyond the "boundary" of the nucleus. For scattering on this scale, however, the scattering cross section can still be defined and measured operationally using equation 1. It should be noted that, for example, every particle will be scattered by the bare nucleus through some nonzero angle θ; for a given particle energy, the larger the impact parameter b (defined in Figure 21.1b), the smaller is θ. For this case, the total cross section for scattering at all angles will be infinite; the quantity of practical importance is therefore the total cross section for scattering at angles greater than some angle θ_0.

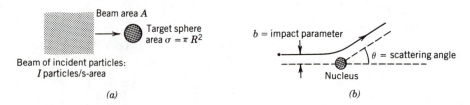

FIGURE 21.1 (a) A beam of particles incident on a target sphere. (b) Positively charged particle scattered from a stationary positive nucleus.

EXERCISE 1

Verify that the total cross section for a target, $\sigma \equiv N/I$, has dimensions of area.

The detailed distribution of scattering angles can give, as in the present experiment, details concerning the interaction between the target and the incident particles that the measurement of the total cross section alone could not supply. If, as in the case of Rutherford scattering, the interaction potential of the target is spherically symmetric, the scattering angle for particles of a given type and energy depends only on the impact parameter b, which measures the distance by which a particle "misses" a head-on collision with the target. As is indicated in Figure 21.2a, particles incident with impact parameters between b and $b + db$ will have deflection angles between θ and $\theta + d\theta$. The relationship

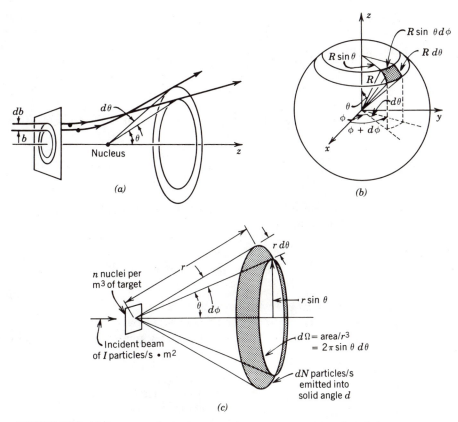

FIGURE 21.2 (a) Correspondence between the impact parameter b and the scattering angle θ. (b) Area on a sphere subtended by $d\theta$ and $d\phi$. (c) Azimuthally symmetric solid angle corresponding to $d\theta$.

between b and θ is derived from the known (or postulated) target–particle interaction law. The cross-sectional target area enclosed by the annulus of radius b and of width db is just $d\sigma = 2\pi b \, db$; particles with impact parameters in this range db will have deflections in a range $d\theta$ given by $d\theta = (d\theta/db) \, db$. If $b(\theta)$ is known, we can associate the differential area $d\sigma$ with the angular range $d\theta$ by writing

$$d\sigma = 2\pi b \left| \frac{db}{d\theta} \right| d\theta \qquad (\text{m}^2) \qquad (2)$$

where the absolute value has been used to ensure a positive differential area. Substituting in the $b(\theta)$ appropriate to a given target and particle then gives $d\sigma$ in terms of θ and $d\theta$. It is often customary and convenient to express $d\sigma$ in terms of the differential solid angle $d\Omega$ instead of the polar angle $d\theta$. The solid angle, expressed in units of steradians (sr), corresponding to a polar angular interval $d\theta$ and an azimuthal angular interval $d\phi$, is just the corresponding area on a unit sphere centered at the origin, that is, the scattering center. As illustrated in Figure 21.2b, we can write the area on the sphere of radius R subtended by $d\theta$ and $d\phi$ at the origin as $dA = (R \, d\theta)(R \sin \theta \, d\phi) = R^2 \sin \theta \, d\theta \, d\phi$. Setting $R = 1$ for the unit sphere then yields

$$d\Omega = \sin \theta \, d\theta \, d\phi \qquad (\text{sr}) \qquad (3)$$

as the general expression for $d\Omega$. For the symmetric scattering problem, as shown in Figure

21.2*a* and *c*, the annular area $d\sigma$ given by equation 2 corresponds to all values of ϕ from 0 to 2π. Thus, the solid angle that corresponds to $d\theta$ (the shaded portion in Figure 21.2*c*) is found by integrating $d\Omega$ in equation 3 over ϕ to obtain

$$d\Omega = 2\pi \sin \theta \, d\theta \qquad (\text{sr}) \tag{4}$$

Using equation 4 to express $d\sigma$, given by equation 2, in terms of $d\Omega$ gives

$$d\sigma = \frac{b}{\sin \theta} \left| \frac{db}{d\theta} \right| d\Omega \qquad (\text{m}^2) \tag{5}$$

Dividing both sides of the above expression by $d\Omega$ yields the **differential cross section** of a single scattering center as

$$\frac{d\sigma}{d\Omega} = \frac{b}{\sin \theta} \left| \frac{db}{d\theta} \right| \qquad (\text{m}^2/\text{sr}) \tag{6}$$

The behavior of $d\sigma/d\Omega$ with angle and incident particle energy reflects the nature of the interaction between the target and the incident particle.

The total cross section for scattering angles greater than θ_0 is readily expressed in terms of $d\sigma/d\Omega$:

$$\sigma = \int_{\theta_0}^{\pi} \left(\frac{d\sigma}{d\Omega} \right) d\Omega \qquad (\text{m}^2) \tag{7}$$

In the present scattering experiment, the measured quantity is the rate of scattering, $N(\theta)$, per unit polar angle θ. This can be expressed simply in terms of $d\sigma/d\Omega$ for a single target atom for the situation depicted in Figure 21.3, in which a thin foil (thickness t) composed of atoms (number density n) with *nonoverlapping* target areas, scatters a particle beam of intensity I. The area of the foil is assumed to be larger than or equal to the total cross-sectional area A of the incident beam, as is the case in our experiment. The effective area for scattering into a differential solid angle $d\Omega$ is $(d\sigma/d\Omega) \, d\Omega \times$ (number of scatterers in the beam) $= nAt(d\sigma/d\Omega) \, d\Omega$. The number of scatterings per second into $d\Omega$ is then just $InAt(d\sigma/d\Omega) \, d\Omega$. Finally, using equation 4 for $d\Omega$ gives $N(\theta) \, d\theta$, the rate of scattering into $d\theta$, as

$$N(\theta) \, d\theta = 2\pi InAt \frac{d\sigma}{d\Omega} \sin \theta \, d\theta \qquad (\text{s}^{-1}) \tag{8}$$

Analogously, the total rate of scattering, N, for angles greater than θ_0 for this collection of

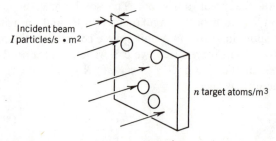

Incident beam
I particles/s $\cdot$ m^2

n target atoms/m^3

FIGURE 21.3 Particle beam incident on a thin foil.

scatterers is obtained by integration of $N(\theta)$:

$$N = \sigma I n A t \qquad (\text{s}^{-1}) \qquad (9)$$

where σ, as in equation 7, is the total cross section of a single scatterer for $\theta > \theta_0$.

Rutherford Scattering

The result of equation 6 for the differential cross section applies quite generally to any spherically symmetric single scatterer. To apply this discussion to the specific case of Rutherford scattering of α particles by atomic targets, we need an explicit expression for $b(\theta)$; that is, we need to know the relationship between the impact parameter of a particle in the incident beam and its angle of scattering. To facilitate the calculation, we assume the following:

1. Scattering by the atomic electrons can be neglected because of their small mass.
2. The target nuclei remain stationary during impact, since their mass is assumed much greater than that of the α particles.
3. The velocities of the α particles are such that the scattering is adequately described by Newtonian mechanics.
4. The energies of the α particles are not high enough to permit them to penetrate the nucleus, so that they interact with the nuclei predominantly via the Coulomb repulsion force.

As is illustrated by Figure 21.4a, the incident α particle has a charge $+ze$ and an initial momentum of magnitude $P_i = mv_0$ as it starts out far away from the nucleus of charge $+Ze$; for the momentum P_f after the scattering we must have $P_f = P_i$, since energy conservation applied to this elastic collision guarantees that the initial and final speeds are the same. Because the force is central and hence exerts no torque on the α particle, the angular momentum ℓ, which initially has a magnitude $\ell_i = mv_0 b$ with respect to the origin at the nucleus, remains constant during the interaction. We can calculte the scattering angle θ by using these conservation laws and Newton's second law without knowing the details of the α particle's hyperbolic trajectory.

The change $\Delta \mathbf{P}$ in the α-particle momentum during the scattering can be calculated by writing $\Delta \mathbf{P} = \int_{-\infty}^{+\infty} \mathbf{F} \, dt$, where the Coulomb force of repulsion is $\mathbf{F} = (kzZe^2/r^2)\hat{r}$, $\hat{r}$ being the radial unit vector. The final momentum $\mathbf{P}_f$ is then given by $\mathbf{P}_f = \mathbf{P}_i + \Delta \mathbf{P}$, as in Figure 21.4$b$. We use the expression for $\mathbf{F}$ in the above expression for $\Delta \mathbf{P}$ to write

$$\Delta \mathbf{P} = kzZe^2 \int_{-\infty}^{+\infty} \frac{\hat{r}}{r^2} \, dt \qquad (\text{kg} \cdot \text{m/s}) \qquad (10)$$

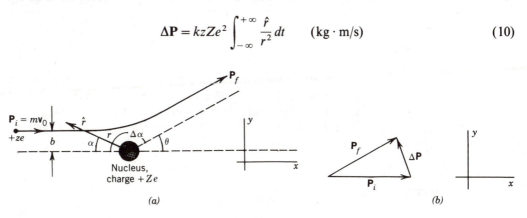

FIGURE 21.4 (a) Trajectory of an α particle scattered by a nucleus. (b) The change in the α-particle momentum during a scattering.

where we will use r and α to represent the polar coordinates of the α particle. The integration in equation 10 can be done by noting that the differential dt can be written in terms of $d\alpha$: $dt = (dt/d\alpha)\, d\alpha$. After this substitution is made, the integrand becomes $(\hat{r}/r^2)(dt/d\alpha)\, d\alpha = \hat{r}[r^2(d\alpha/dt)]^{-1}\, d\alpha = m\hat{r}[mr^2(d\alpha/dt)]^{-1}\, d\alpha$. The quantity in brackets may be recognized as the magnitude of the angular momentum ℓ, which is constant and thus may be replaced by its initial value $mv_0 b$, yielding for the change in momentum

$$\Delta\mathbf{P} = \frac{kzZe^2}{v_0 b}\int_0^{\Delta\alpha}\hat{r}\,d\alpha \qquad (\text{kg}\cdot\text{m/s}) \tag{11}$$

where $\Delta\alpha$ in the upper limit is the supplement of the desired scattering angle θ: $\Delta\alpha = \pi - \theta$. Inserting $\hat{r} = -\hat{x}\cos\alpha + \hat{y}\sin\alpha$ and integrating with respect to α, we have the final momentum in terms of b and θ:

$$\begin{aligned}\mathbf{P}_f &= \mathbf{P}_i + \Delta\mathbf{P}\\[4pt] &= mv_0\hat{x} + \frac{kzZe^2}{v_0 b}[(1+\cos\theta)\hat{y} - \sin\theta\hat{x}] \qquad (\text{kg}\cdot\text{m/s}) \end{aligned} \tag{12}$$

Finally, since the initial and final α-particle momenta have the same magnitude, we can set $P_f = mv_0$ and solve for b in terms of θ. The result is

$$b = \frac{kzZe^2}{mv_0^2}\cot\frac{\theta}{2} \qquad (\text{m}) \tag{13}$$

The differential cross section for Rutherford scattering is then obtained by substituting this result for $b(\theta)$ into equation 6, giving

$$\frac{d\sigma}{d\Omega} = \left(\frac{kzZe^2}{2mv_0^2}\right)^2\sin^{-4}\frac{\theta}{2} \qquad (\text{m}^2/\text{sr}) \tag{14}$$

It is interesting to note that even though this formula was derived by using purely classical considerations, it is valid quantum mechanically as well for nonrelativistic particles.

EXERCISE 2

Obtain the above results for b and $d\sigma/d\Omega$ from the discussion preceding equation 13.

EXERCISE 3

Justify assumption 4 above for the case of α particles emitted from ^{241}Am ($E_\alpha = 5.5$ MeV) and scattered off Au atoms ($Z = 79$). Use conservation of energy to calculate the distance of closest approach to the center of the nucleus, and use the approximate empirical expression for the nuclear radius R in terms of the atomic mass number A: $R \simeq (1.4 \times 10^{-15})A^{1/3}$ (m).

EXERCISE 4

Verify that the total Rutherford scattering cross section for *all* angles is infinite. Interpret this result in the context of a practical measurement.

EXERCISE 5

Use equations 8 and 14 to show that the dependence of the scattering angle distribution, $N(\theta)$, on θ and on the kinetic energy (K) of the α particles can be written as

$$N(\theta) \propto \frac{1}{K^2} \left[\frac{\sin \theta}{\sin^4(\theta/2)} \right]$$

EXPERIMENT

We describe below the use of a device, developed by C. W. Leming and D. H. Garrison and pictured schematically in Figure 21.5a, to study the Rutherford scattering of α particles from the atoms of a thin metallic foil. The angular dependence of the scattered intensity over angles ranging from about 3 to 13° can be studied and compared to the expected result calculated above using the Rutherford model of atomic structure.

The apparatus consists of a vacuum chamber with an α source attached to the center of the left wall. Two disks, each with a hole of diameter d at the center, are mounted coaxially a distance L apart; these serve to collimate the unscattered beam, that is, to reduce its angular divergence before scattering. A thin metal foil (gold, for example) attached to the right-hand collimator, scatters the α beam. When the chamber is evacuated sufficiently, the scattered particles impact the film mounted on the end plate at the right of the chamber.

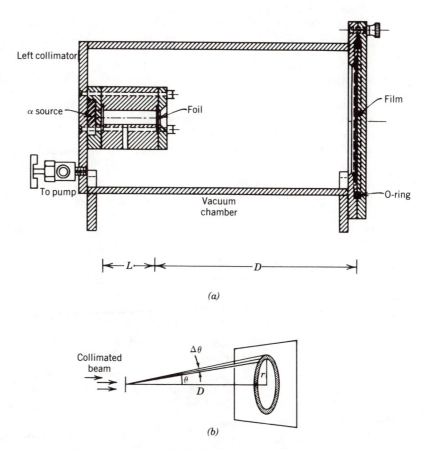

FIGURE 21.5 (a) Daedalon Corporation's Rutherford scattering apparatus. (Courtesy of Daedalon Corp.) (b) Experimental scattering geometry.

This film consists of a thin sensitive layer of cellulose nitrate on a polyester backing; etching the film with NaOH leaves clear spots at the points where α particles hit the film. These spots can be counted to determine the distribution of scattering angles. The experimental geometry is indicated in Figure 21.5b.

EXERCISE 6

From the values of D and L (given or measured) for your apparatus, determine the approximate angular divergence of the unscattered beam. What limits does this divergence place on your measurements?

EXERCISE 7

If you have read Experiment 20, Ionization of Gases by Alpha Particles, estimate, for your α source and for the source-to-film distance in your apparatus, the maximum air pressure inside the chamber that permits α particles to impact the film.

EXERCISE 8

If the thickness of your foil is known or can be measured, estimate the average number of nuclei per cm^2 in the target. From this number and from the expression for $b(\theta)$ in equation 13, estimate the range of angles for which the effects of multiple scattering (overlapping targets) would be important.

Procedure

Attach the mechanical vacuum pump to the vacuum chamber. A vacuum valve or pinch clamp installed between the pump and chamber will allow the pump to be isolated before being turned off so that oil is not sucked into the chamber. A bleeder valve may also be installed so that the pressure in the chamber can be changed gradually; even though a pump-out channel for the region between the two collimators may reduce the probability of sudden pressure differentials across the foil, abrupt changes in chamber pressure may rupture it. The use of hose clamps and Teflon tape, where appropriate, is also recommended.

The starting and ending times for the exposure may be recorded with a scribe near the edge of the film on the sensitive (concave) surface. Mount the film so that it is held symmetrically between the chamber vacuum flange and the end plate with the sensitive surface facing the α source. Check that the chamber is vacuum tight and then evacuate it to start the exposure; the pressure should be maintained at 15 Pa (0.1 Torr) or below during the exposure. The optimal exposure time, which will depend on the initial activity of the source, its half-life, and its age, can be determined by trial and error. The exposure is terminated by the gradual readmission of air into the chamber.

The film is processed by etching away, with a 2.5N NaOH solution, the portions of the emulsion that have been damaged by the α particles. One arrangement for doing this is shown in Figure 21.6. The solution in the canister is maintained at constant temperature, for example, by a water bath heated by an aquarium heater. After etching, films should be rinsed for at least 30 min in water.

The size of the etched spots is controlled by the temperature of the bath and the development time: 24 h at 40 °C has been reported as one satisfactory time/temperature combination. Other combinations are possible, but because development time is a very

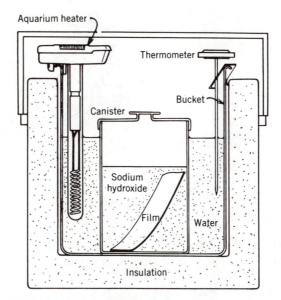

FIGURE 21.6 Film processing apparatus. *Source*: From "The Amateur Scientist," by Jearl Walker, copyright © 1986, Scientific American, Inc. All rights reserved.

sensitive function of the bath temperature, very frequent monitoring of the process is recommended. Charles Leming and Shane Thompson have studied the dependence of the development time on temperature. The results of their work, which may be used as a guide in selecting an appropriate time and temperature, are represented by the plot in Figure 21.7. Fine tuning of these variables may be expedited by dividing a single exposed film into sectors, then processing each one individually.

Analysis

To compare the distribution of scattering angles with theory, it is necessary to measure $N(\theta)$ as defined preceding equation 8. Referring to the geometry in Figure 21.5*b*, this means, for

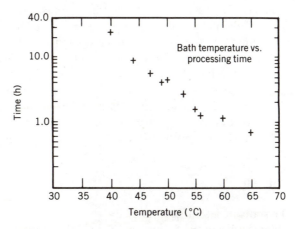

FIGURE 21.7 Film development time versus temperature. (Courtesy of C. Leming and S. Thompson.)

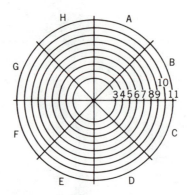

FIGURE 21.8 Counting mask.

each value of θ, counting the number of spots that fall within an annulus of average radius $r = D \tan \theta$; each annulus should correspond to the same fixed value of $\Delta\theta$. A convenient way to do this is to examine the etched film under a binocular microscope with illumination from beneath the microscope stage. A transparent mask, like the one pictured in Figure 21.8, may be placed over the film to facilitate the angle measurement. Note that it is not necessary to count spots over an entire annulus for each θ to obtain data proportional to $N(\theta)$; a fixed portion (sector) of each annulus will do.

It should be noted that the number of spots in the various sectors of a given annulus will be statistically distributed. It is therefore appropriate to associate an uncertainty with the count for each sector, which is just the standard deviation of this distribution. If the counts are distributed according to the Poisson distribution (discussed in Experiment 6, Introduction to Computer-Assisted Experimentation), then the percentage uncertainty in the number of spots increases as this number decreases.

EXERCISE 9

Strictly speaking, should the annuli of the mask in Figure 21.8 all have the same area if they are to correspond to the same angular interval $\Delta\theta$? Explain.

As with any particle-counting experiment, background counts are a potentially important source of systematic error. It is possible to determine the rate at which spots attributable to background accumulate on the film. If time permits, expose a sheet of film in the apparatus with a piece of aluminum covering the hole in the second collimator, cutting off the α beam. Make an estimate of the number of background counts per unit area, and correct your tabulated $N(\theta)$ accordingly. Note that since the background count should have a uniform area density on the film, the correction to $N(\theta)$ must be calculated for each annulus.

Using your corrected $N(\theta)$, make a plot of log N against θ (with error bars), for angles ranging from 3 to 13°. Using the same set of axes, plot $\log[\sin^{-4}(\theta/2)]$. To compare these two it will probably be necessary to shift one of these curves along the y axis by some constant to account for normalization factors. Does your $N(\theta)$ behave as expected?

EXERCISE 10

Count the spots in several sectors of the same annulus and calculate the standard deviation in this count. Utilizing the discussion of the Poisson distribution in Experiment 6, determine if your result is consistent with the assumption that the number of counts is Poisson distributed.

EXERCISE 11

What additional piece(s) of information would you need to estimate the thickness of your scattering foil from your data? If this information is available or can be easily obtained, make this estimate and, if possible, compare it with the thickness obtained by another means.

Physical Optics

22. THE FARADAY EFFECT

APPARATUS [Optional Equipment in Brackets]

Monochromatic collimated light source (laser or white light with monochromator)
Laboratory electromagnet (bored axially with at least 0.5-T capability)
Gaussmeter
Photodiode or phototube detector
Light and heavy flint glass (prisms and rectangular blocks)
Two polaroid filters (one of which can be rotated)
Prism spectrometer
Mercury vapor lamp
[Oscilloscope, motor]

OBJECTIVES

To observe and measure the rotation of the plane of polarization of light propagating in a medium in the presence of a magnetic field.

To measure the dispersion $dn/d\lambda$ of the medium and relate it to the observed rotations.

To observe qualitatively the rotation of the plane of polarization by a substance with natural optical activity

To use the relation between the measured dispersion of the medium and the observed Faraday rotation to determine a value for e/m, the charge-to-mass ratio of the electron.

To understand the physical basis of dispersion in materials and how it is related to the observed rotations.

KEY CONCEPTS

Polarization (linear and circular)   Optical activity
Levo- and dextrorotatory   Verdet constant
Circular birefringence   Faraday rotation
Susceptibility   Dielectric constant
(Complex) index of refraction   Dispersion
Absorption   Minimum deviation

REFERENCES

1. F. Jenkins and H. White, *Fundamentals of Optics*, 3d ed., McGraw-Hill, New York, 1957. The Faraday effect is discussed qualitatively on page 596. The Verdet constants for some common substances are given.

2. M. Klein, *Optics*, Wiley, New York, 1970. Chapter 10 discusses linearly and circularly polarized light.

3. J. Marion, *Classical Electromagnetic Radiation*, Academic, New York, 1965. A classical description of dispersion in solids, liquids, gases, and plasmas is contained in Chapter 9. The Zeeman effect is also treated classically here.

4. R. Eisberg and R. Resnick, *Quantum Physics of Atoms, Molecules, Solids, Nuclei, and Particles*, 2d ed., Wiley, New York, 1985. Chapter 10 contains a quantum-mechanical description of the Zeeman effect.

5. C. Kittel, *Introduction to Solid State Physics*, 6th ed., Wiley, New York, 1985. Contains a discussion of dielectric properties of solids in Chapter 13.

6. A. Portis, *Electromagnetic Fields: Sources and Media*, Wiley, New York, 1978. Chapter 12 contains discussions of optical activity, dispersion, circular polarization and Faraday rotation in plasmas.

INTRODUCTION

An optically active medium is one that will rotate the plane of polarization of linearly polarized light. Although optical activity may occur naturally as a result of the intrinsic helical character of the molecules composing the medium, Michael Faraday in 1845 observed that this property could be induced in glass by application of an external magnetic field (**B**) that is parallel to the direction of propagation of the wave. This is known as the **Faraday effect** and can be observed for almost any transparent medium. It produces a rotation of the plane of polarization by an angle ϕ given by

$$\phi = VBL \quad \text{(rad)} \tag{1}$$

where L is the length of the path traversed in the medium and V is known as the *Verdet constant* for the material. The Faraday effect has practical application wherever one requires a "one-way street" (irreversible path) for electromagnetic radiation, as in the microwave isolator used in Experiment 2.

Below we describe the physical basis of the rotation given by equation 1 and determine V in terms of the optical properties of the medium.

Polarization of Light; Circular Birefringence

Light is said to be *linearly polarized* if its associated electric field $\mathbf{E}(\mathbf{r}, t)$ and the propagation vector **k** are always in the same plane, called the plane of polarization. In Figure 22.1a, the

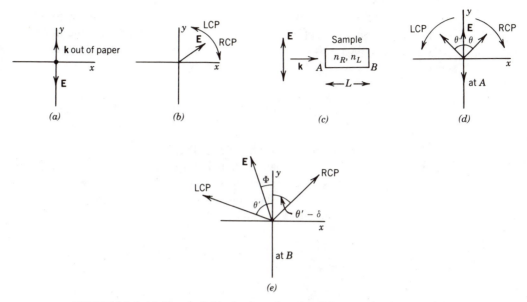

FIGURE 22.1 (a) Electric field of a linearly polarized wave. (b) Electric field of a circularly polarized wave. (c) Linearly polarized wave illuminating a sample of circularly birefringent material. (d) Linearly polarized wave as a linear combination of LCP and RCP waves. (e) Rotated electric field as the resultant of two circularly polarized waves out of phase.

wave propagates out of the paper along $\hat{z}$ and is polarized in the yz plane. An expression for such a wave would be

$$\mathbf{E}(z, t) = E_0 \sin(\omega t - kz)\hat{y} \qquad \text{(V/m)} \tag{2}$$

in which $k \equiv 2\pi/\lambda$ is the wave number and where we have arbitrarily chosen the y axis as the direction for $\mathbf{E}$. If the plane containing $\mathbf{E}$ and $\mathbf{k}$ rotates while $|\mathbf{E}|$ remains constant, as in Figure 22.1b, then the tip of the vector $\mathbf{E}$ traces out a circle, and the wave is said to be circularly polarized:

$$\mathbf{E}(z, t) = E_0' \sin(\omega t - kz)\hat{x} + E_0' \sin\left(\omega t - kz \pm \frac{\pi}{2}\right)\hat{y} \qquad \text{(V/m)} \tag{3}$$

The x and y components of $\mathbf{E}$ are thus $\pi/2$ radians out of phase. Using the plus sign in the phase of the y component describes a *right* circularly polarized (RCP) wave in which $\mathbf{E}$, as observed head-on at some fixed point, rotates clockwise with time at angular frequency ω. The minus sign corresponds to a *left* circularly polarized (LCP) wave.

We can understand how the plane of polarization of the wave in equation 2 can be rotated as it passes through a sample of length L as in Figure 22.1c if the sample is circularly birefringent, that is, if it exhibits slightly different indices of refraction n_R and n_L for RCP and LCP waves, respectively. We consider the linearly polarized wave of amplitude E_0 to be a linear superposition of a LCP wave and a RCP wave, each of amplitude E_0', as shown in Figure 22.1d just after the crest of the wave has entered the sample at $z = 0$ (point A). Since the indices of refraction are different for the LCP and RCP components of the wave, so are the speeds, wavelengths, and wave numbers k_L and k_R inside the sample. For the LCP wave, the wave number k_L will be given by $k_L = 2\pi/\lambda_L = n_L k$, where k is the wave number of both components outside the sample; similarly for the RCP wave, $k_R = n_R k$. At point B, where the waves exit the sample, $\mathbf{E}(t)$ for each circular component is given by equation 3, using the appropriate sign and the corresponding value of k. If we assume that $n_R > n_L$, then the phase of the RCP component at $z = L$ will lag that of the LCP component by δ, where

$$\delta = (n_R k)L - (n_L k)L = \frac{\omega L}{c}(n_R - n_L) \qquad \text{(rad)} \qquad (4)$$

As is indicated by Figure 22.1e, this means that the RCP component is an angle δ radians closer to its LCP partner than it would have been in the absence of the sample. The resultant $\mathbf{E}$, obtained as the instantaneous vector sum of the two components, is thus no longer oscillating along the y axis, but has been rotated by an angle $\phi = \delta/2$, so that the Faraday rotation is given by equation 4 as

$$\phi = \frac{\omega L}{2c}(n_R - n_L) \qquad \text{(rad)} \qquad (5)$$

EXERCISE 1

Write the linearly polarized wave of equation 2 explicitly in terms of LCP and RCP waves, each of which has a form like that of equation 3, to show that the above-mentioned decomposition is valid. How would your answer change if the original wave had been polarized with $\mathbf{E}$ along $\hat{x}$?

For materials that exhibit optical activity in the absence of a magnetic field, the difference between n_R and n_L arises because of the molecular symmetries involved, leading to the rotation given by equation 5. Figure 22.2 schematically depicts left-handed (levorotatory) and right-handed (dextrorotatory) stereoisomers (e.g., of glucose), which respond differently to LCP and RCP waves because of their lack of reflection symmetry. Although it may seem that the rotatory effect of these molecules on an incident light wave would be "averaged away" as all possible orientations of these molecules are considered, you should convince yourself that the "handedness" is characteristic of the molecular species itself, regardless of its orientation, and thus defines a unique direction of rotation for $\mathbf{E}$ with respect to the direction of propagation of the wave.

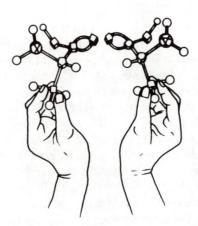

FIGURE 22.2 Left- and right-handed stereoisomers.

Frequency Dependence of the Index

To understand how the presence of **B** creates a situation in which $n_R \neq n_L$ for almost any transparent material, we first examine the frequency dependence of $n(\omega)$ in the absence of the field. We take a simple classical model in which the electrons of the medium are bound harmonically to noninteracting molecules. The polarization response of an individual molecule to an electric field $\mathbf{E} = \mathbf{E}_0 \, e^{-j\omega t}$ is then calculated by solving the classical equation of motion of the electron:

$$\ddot{\mathbf{r}} + 2b\dot{\mathbf{r}} + \frac{\kappa}{m}\, \mathbf{r} = \frac{e}{m}\, \mathbf{E} \qquad (\text{m/s}^2) \qquad (6)$$

where b and κ represent the *damping* and *spring* constants, respectively. The steady-state solution for $\mathbf{r}(t)$ is

$$\mathbf{r}(t) = \frac{(e/m)\mathbf{E}}{(\omega_0^2 - \omega^2) - 2jb\omega} \qquad (\text{m}) \qquad (7)$$

where $\omega_0 = \sqrt{\kappa/m}$ is the natural frequency of the system in radians per second. Assuming there are N molecules per unit volume with the same resonant frequency ω_0, the polarization (i.e., the induced dipole moment per unit volume, denoted **P**) then follows as

$$\mathbf{P} = Ne\mathbf{r} = \frac{(Ne^2/m)\mathbf{E}}{(\omega_0^2 - \omega^2) - 2jb\omega} \qquad (\text{C} \cdot \text{m}^{-2}) \qquad (8a)$$

and the susceptibility χ_E is

$$\chi_E \equiv \frac{P}{\epsilon_0 E} = \frac{Ne^2/m\epsilon_0}{(\omega_0^2 - \omega^2) - 2jb\omega} \qquad (8b)$$

where ϵ_0 is the permittivity of free space. The relative dielectric constant $\epsilon_r \equiv \epsilon/\epsilon_0$ now follows from

$$\epsilon_r = 1 + \chi_E = 1 + \frac{Ne^2/m\epsilon_0}{(\omega_0^2 - \omega^2) - 2jb\omega} \qquad (9)$$

The index of refraction $n(\omega)$ for all polarizations in the absence of **B** is then given approximately, for ϵ close to ϵ_0, as

$$n \equiv \sqrt{\epsilon_r} \cong 1 + \frac{Ne^2/2m\epsilon_0}{(\omega_0^2 - \omega^2) - 2jb\omega} \qquad (10)$$

Note that $n(\omega)$ calculated on this model is complex; its real and imaginary parts are plotted in Figure 22.3. We can interpret these readily by writing **E** in the medium for the propagating wave of equation 2, using complex exponential form

$$\mathbf{E} = \mathbf{E}_0 \, e^{-j(\omega t - kz)} \qquad (\text{V/m}) \qquad (11)$$

which becomes, on making the substitution $k = n\omega/c$ from Maxwell's equations,

$$\mathbf{E} = \mathbf{E}_0 \exp\left\{-j\left[\omega t - (\text{Re } n)\frac{\omega}{c}z\right]\right\} \exp\left[-\frac{\omega}{c}(\text{Im } n)z\right] \qquad (\text{V/m}) \qquad (12)$$

This represents a wave traveling along $+z$ with speed $c/\text{Re } n$ and an attenuation per unit

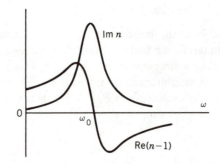

FIGURE 22.3 Real and imaginary parts of the index of refraction plotted against angular frequency.

length of sample that is proportional to Im n. The behavior of Re n (written simply as n in Snell's law) is thus referred to as the **dispersion** of the material, because of its relation to the separation of wavelengths (colors) with prisms; the plot of Im n versus ω describes the **absorption**, since it governs the attenuation of the wave. Note from the appearance of these curves in Figure 22.3 that the position of the absorption peak at ω_0 corresponds to the point where Re$(n-1)$ crosses the axis, and that a shift in the absorption peak will produce a corresponding shift in Re$[n(\omega)]$.

EXERCISE 2

Verify that Figure 22.3 represents Re$(n-1)$ and Im n by calculating these quantities from the approximate form given in equation 10 above. Note that the computation leading to this equation was simplified by expanding $n(\omega)$ to first order. Although this is a poor approximation for materials in which n is much different than 1, the argument serves, nevertheless, to illustrate the basic physics involved.

To understand and calculate the Faraday rotation in terms of the dispersion, we note that the application of a **B** field parallel to the direction of propagation of the wave creates, because of the Zeeman effect (Experiment 14), two different absorption curves for the left and right circularly polarized components of the plane wave, as is shown in Figure 22.4. This corresponds directly to the Zeeman splitting of excited electronic states having different values of L_z, the component of the orbital angular momentum in the direction of **B**. To calculate the separation, $\Delta\omega$, of the two absorption peaks (plotted as Im n), we note that, since the LCP and RCP photons carry an angular momentum of $+\hbar$ and $-\hbar$, respectively, along the direction of propagation, absorption of such a photon by the electrons of the material must involve, by conservation of angular momentum, a corresponding change in the angular momentum of the electronic system. Changing the angular momentum of an electron by an amount ΔL_z changes its energy in the presence of **B** by an amount

$$\Delta\mathscr{E} = \Delta(-\boldsymbol{\mu} \cdot \mathbf{B}) = \left[\frac{e}{2m}(\Delta L_z)\right]B \qquad \text{(J)} \qquad (13)$$

where we have assumed that the spin vector of the electron does not change during the absorption by using $(-e/m)\Delta L_z$ for the change in the component of the magnetic moment μ along the z axis. The difference in the peak energies of the LCP and RCP photons causing

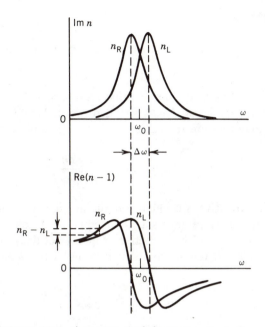

FIGURE 22.4 Absorption and dispersion curves for
LCP and RCP components of the plane
wave.

the transitions depicted schematically in Figure 22.5 is then given by

$$\Delta\mathscr{E} = \hbar\Delta\omega = \frac{e}{2m}(2\hbar)B \qquad (\text{J}) \qquad (14)$$

so that the frequency separation of the two absorption peaks is just

$$\Delta\omega = \frac{eB}{m} \qquad (\text{rad/s}) \qquad (15)$$

According to our simple model, which should be a good description near frequencies
such as ω_0 where the optical behavior is dominated by a resonant absorption, the dispersion
peaks [plotted as $\text{Re}(n-1)$] in Figure 22.4 should also be separated in frequency by the
amount given by equation 15.

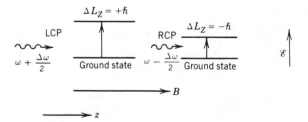

FIGURE 22.5 Schematic depiction of the transitions
caused by LCP and RCP photons.

EXERCISE 3

Calculate $\Delta\omega$ for an applied B of 0.5 T and show that it is small compared with ω for a visible light wave.

Finally, to calculate the Faraday rotation we refer to equation 5 and approximate $n_R - n_L \equiv \Delta n \cong (dn/d\omega)\,\Delta\omega$, where here we understand n to be the real part of the index. The Faraday rotation ϕ is then given from equations 5 and 15 as

$$\phi = \frac{\omega e}{2mc}\left(\frac{dn}{d\omega}\right)BL \quad \text{(rad)} \tag{16a}$$

where the derivative is evaluated at the angular frequency of the radiation. Note that this result is the same as that stated in equation 1, but that the Verdet constant has been expressed in terms of the refractive properties of the medium. It is sometimes convenient to express this result for the rotation in terms of the wavelength λ:

$$\phi = -\frac{e}{2mc}\lambda\frac{dn}{d\lambda}BL \quad \text{(rad)} \tag{16b}$$

EXERCISE 4

If a mirror were placed at point B in Figure 22.1c, what would be the polarization of the reflected wave at A in terms of ϕ given in equation 5? Does your answer depend on whether the rotation is due to the Faraday effect or to the natural optical activity of the material?

Equation 16b for the Faraday rotation (derived by H. Becquerel) was shown here to result from some rough semiclassical arguments that ignore the detailed nature of the material involved. The quantum-mechanical treatment of the rotation for a particular material is quite complicated, but could be used, in conjunction with experimental observations of the rotation, to study its electronic structure. The discussion presented here is intended to explain the physical basis for the rotation and to predict its dependence on magnetic field and wavelength.

EXPERIMENT

The apparatus for the investigation of the Faraday effect is shown schematically in Figure 22.6. If the source is not a laser, it should be collimated and the light should pass through some type of monochromator. Polaroid 1 determines the initial plane of polarization of the light passing through the sample; polaroid 2 analyzes the polarization of the light that

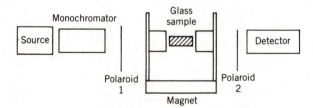

FIGURE 22.6 Apparatus for the investigation of the Faraday effect.

emerges and so must be able to rotate. These sheets should be placed as close as possible to the magnet itself to minimize interference from stray light.

Calibration of the Magnet

It will be necessary to use the gaussmeter to produce a magnet current (I) versus field (B) calibration plot. Note that hysteresis in the magnet iron may introduce substantial uncertainty in the field produced by a particular value of I. This effect can be minimized by always starting with the maximum allowable current (corresponding to saturation of the pole pieces) and working *downward*, both for calibration and for subsequent rotation measurements. This is good practice for magnetism experiments in general!

With the sample removed from between the pole pieces, measure B against I for a fixed position z_0 along the optical path in the sample region, starting with the maximum allowable magnet current and working downwards. Plot the B versus I curve for this position z_0. Because of the inhomogeneity of the field in the sample region, you will need to adjust this curve so that it represents $\bar{B}$ versus I, where $\bar{B}$ is the magnetic field averaged over the optical path through the sample. To do this, select a magnet current I_0 in the linear region of the B versus I curve. For this current, measure the value of B at z_0 again [call it $B_0(z_0)$]. For the same current measure B_0 against z along the optical path in the sample region and compute the adjustment factor f:

$$f = \frac{\bar{B}_0}{B_0(z_0)} = \left[\frac{1}{B_0(z_0)}\right] \frac{1}{L} \int_0^L B_0(z)\,dz \qquad (17)$$

using the rectangular rule to estimate the integral. Now calculate and plot a $\bar{B}$ versus I curve by multiplying each value of B in your B versus I curve by the factor f calculated above.

EXERCISE 5

By considering the infinitesimal Faraday rotation produced by a slab of sample with thickness dL, show that $\bar{B}$ should replace B in equations 16a,b for ϕ, which was derived assuming a uniform field.

Measurement of the Rotation

With light of fixed wavelength shining through the sample and with $\bar{B} = 0$, adjust polaroid 2 so that the transmission of light is at a maximum (or minimum). This may be done by eye (if the source is not a laser) or with the detector. Note the angular reading. For several values of $\bar{B}$ (again, starting at the high end) record the angular position of polaroid 2 for which the intensity is again a maximum (or minimum). For each field, the rotation is the difference between this reading and the zero field reading.

An alternative way to measure the rotation is to use a motor to rotate polaroid 2 while feeding the output of the detector into the input of an oscilloscope. A trigger signal may be derived from the rotating rim of the polaroid, so that the rotation can be measured by observing the phase relationship between the start of the sweep and the peak of the sinusoidal detector signal. Also, if your scope has them, the B intensified by A, time-delay multiplier and A delayed by B features will enhance this technique.

For each sample, produce a plot of rotation versus $\bar{B}$ for this wavelength, putting one or two error bars on the graph to indicate the uncertainties in your measurements.

EXERCISE 6

Is the rotation linear in $\bar{B}$? What is your value of the Verdet constant for your samples and how do they compare with values in the literature? See reference 1 for some typical values.

For one value of $\bar{B}$, measure the rotation against λ if more than one wavelength is available to you. Plot ϕ/λ versus λ.

EXERCISE 7

Assuming a single resonance is responsible for the dispersive properties of the materials, can you deduce, from the data described above, whether this resonance is above or below the range of visible wavelengths?

Measurement of Dispersion and Absorption

To further compare your results with theory, you need to measure $dn/d\lambda$ and the absorption versus wavelength. To measure the dispersion (i.e., Re n), use the prism spectrometer setup shown in Figure 22.7a, in which the prism is made of the same material as the sample for which the rotations were measured. Adjust the telescope and collimator as described in Experiment 4. For each visible line in the mercury spectrum, rotate the prism until the deviation D, defined in the figure, is a minimum. The index of refraction for the wavelength corresponding to the line is then given in terms of this minimum deviation, D_{min}, and A, the vertex angle of the prism shown in Figure 22.7b, as

$$n = \frac{\sin[(D_{min} + A)/2]}{\sin(A/2)} \tag{18}$$

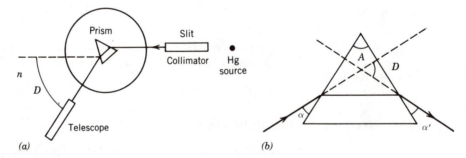

FIGURE 22.7 (a) Prism spectrometer setup for measuring the dispersion. (b) Geometry for the determination of the relationship between the index of refraction, the minimum deviation, and the vertex angle.

EXERCISE 8

Derive equation 18 using Snell's law. (*Hint*: Refer to Figure 22.7 and show that D takes on its minimum value when $\alpha = \alpha'$.)

Calculate n for each line in the visible Hg spectrum. By referring to the known wavelengths in Table 22.1, produce a plot of n versus λ. For wavelengths on the high side

TABLE 22.1 WAVELENGTHS OF THE MERCURY SPECTRUM

Color	Line (nm)
Violet (bright)	404.66
Violet (faint)	407.78
Blue (bright)	435.83
Green (moderate)	491.61
Green (faint)	—
Yel-gr (bright)	546.07
Yellow (bright)	576.96
Yellow (bright)	579.07
Red (faint)	—
Red (faint)	—
Red (bright)	629.13

Source: D. W. Preston, J. W. Kane, and M. M. Sternheim, *Experiments in Physics*, Wiley, New York, 1983. Copyright © 1983, John Wiley & Sons, Inc. Reprinted by permission.

of the absorption resonance, the dispersion can now be estimated by fitting your $n(\lambda)$ data to the Cauchy form for n in this region of the spectrum:

$$n = C + \frac{C'}{\lambda^2} \tag{19}$$

This expression is simply a low-frequency expansion of the real part of n given by equation 10.

EXERCISE 9

Use a least-squares fit to a straight line to determine C and C' for each material in the region where equation 19 is appropriate. Plot $dn/d\lambda$ as a function of λ (using C and C' wherever possible).

Remove polaroid 2 and measure the absorption versus λ for each material using a detector for which the output is linear in the input intensity. Measure the ratio $(I_{out} - I_{in})/I_{out}$, where I_{in} and I_{out} refer respectively to the light intensities detected with the sample in and out of the optical path.

EXERCISE 10

For each wavelength, a certain portion of the light incident on the sample will be reflected, causing the ratio measured above to be an overestimate of the fractional absorption. How would you correct for this, using your dispersion data? Produce a plot of absorption versus wavelength.

To compare your results with theory, do a least-squares fit to the linear portion of your rotation versus field (at constant wavelength) curves and derive a value of e/m for the electron using equation 16b.

EXERCISE 11

In view of your uncertainties, how does this value of e/m compare with the accepted value and with your value from Experiment 7 if you have done it.

EXERCISE 12

How does your absorption versus wavelength curve compare with your ϕ/λ versus λ data? What should be the relationship between these two curves, given our result of equation 16b and our simple model for the calculation of Re n and Im n?

Optically Active Materials

Make qualitative observations of the rotation of the plane of polarization by turpentine and sugar solutions both with and without a magnetic field.

EXERCISE 13

What effect does varying the concentration of the optically active molecules have on the rotation? Are these molecules dextrorotatory or levorotatory?

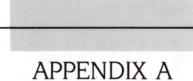

APPENDIX A

MODULATION SPECTROSCOPY: THE LOCK-IN AMPLIFIER

APPARATUS

Lock-in amplifier or a microcomputer and a lock-in amplifier card (the former is described here)
Circuit to generate square waves, see Figure A.9
Transmitter circuit, see Figure A.10a
Receiver circuit, see Figure A.10b.

OBJECTIVES

To test the operation of the lock-in amplifier using the circuit shown in Figure A.9.
To construct and use the circuits shown in Figure A.10 as an application of the lock-in amplifier.
To understand the operation of the lock-in amplifier.

KEY CONCEPTS

Signal-to-noise ratio   Phase difference
Reference signal   Phase detection
Signal modulation

REFERENCES

1. P. Temple, *Am. J. Phys.* **43**(9), 801 (1975). This paper is a good introduction to lock-in or phase-sensitive amplifiers. Several examples of the use of a lock-in amplifier are given. A circuit diagram for an inexpensive lock-in amplifier is included, along with construction and operating notes.

2. A. Martin and P. Quinn, *Am. J. Phys.* **52**(12), 1114 (1984). An experiment is described which uses a lock-in amplifier to record signals from a Franck–Hertz tube.

3. P. Horowitz and W. Hill, *The Art of Electronics*, Cambridge University Press, New York, 1980. The lock-in amplifier is discussed on pages 628–631.

4. M. Cardona, *Modulation Spectroscopy*, Academic, New York, 1969. This is a specialized advanced text pertaining to modulation spectroscopy of solids.

5. *Lock-in Applications Anthology*, edited by Douglas Malchow, EG&G Princeton Applied Research, Princeton, NJ, 1985. Twenty applications of the lock-in amplifier, with circuit diagrams, are discussed in some detail. The applications include solid state measurement, spectroscopy, electrochemistry, along with mechanical, instrumentation, and engineering applications.

INTRODUCTION

Modulation spectroscopy is an electronics technique for recording signals when the signal-to-noise ratio is small. This technique uses the lock-in or phase-sensitive amplifier.

Experimental investigations are usually carried out by recording a signal as a function of a single parameter while other parameters are held constant. Some examples of the signal and the single varied parameter are listed in Table A.1. For all cases, we denote the parameter by K and the signal by $e_s(K)$. An arbitrary curve of e_s versus K is shown in Figure A.1, where the parameter K is slowly swept over some range of interest determined by the experiment.

A signal that depends on a parameter can easily be made periodic by varying the parameter in a periodic manner. Let the periodic variation of the parameter K be $\Delta K \equiv k \cos \omega t$, where k is the amplitude (assumed small relative to K) and ω is the angular

TABLE A.1 EXAMPLES OF EXPERIMENTS, SIGNALS, AND VARIED PARAMETERS

Experiment	Signal	Parameter
Franck–Hertz	Transmission current	Accelerating voltage
NMR	Rf power absorbed	Dc magnetic field
ESR	Microwave power absorbed	Dc magnetic field
Photoelectric effect	Photocurrent	Voltage

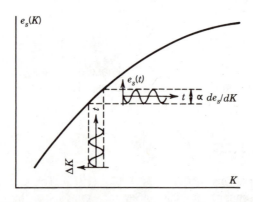

FIGURE A.1 Modulation of the e_s versus K curve with a small ac signal ΔK.

frequency. The modulation of the e_s versus K curve with the small ac component $k \cos \omega t$ is shown in Figure A.1, and the modulation permits measurement of de_s/dK and higher derivatives.

The signal depends on K and the modulation of K, that is, $e_s(K + \Delta K)$. Expanding e_s in a Taylor series about K gives

$$e_s(K + \Delta K) = e_s(K) + \frac{de_s}{dK} \Delta K + \frac{1}{2} \frac{d^2 e_s}{dK^2} (\Delta K)^2 + \cdots \tag{1}$$

where $\Delta K = k \cos \omega t$. Using the half-angle, trig identity, equation 1 becomes

$$e_s(K + \Delta K) = e_s(K) + \frac{de_s}{dK} k \cos \omega t + \frac{1}{2} \frac{d^2 e_s}{dK^2} k^2 \frac{1}{2} (1 + \cos 2\omega t) + \cdots \tag{2}$$

Hence, modulation produces a signal e_s that has a dc component, $e_s(K)$, and components that are periodic with angular frequencies $\omega, 2\omega, \cdots$

Figure 15.1, Experiment 15, Nuclear Magnetic Resonance, shows that the modulation of the dc magnetic field is accomplished with the 40 Hz shown driving the Helmholtz coils. (The modulation frequency is arbitrary.) A similar modulation process is shown in Figure 16.6, Experiment 16, Electron Spin Resonance. The modulation of the retarding voltage in the Franck–Hertz experiment and the modulation of the anode voltage in the photoelectric effect experiment are shown in Figure A.2a and b, respectively.

Note from equation 2 that the amplitude of the component of e_s having frequency ω is proportional to the first derivative of e_s. Hence, if we detect this frequency component of e_s, then the detected signal is proportional to de_s/dK. In a similar manner, if we detect the component of e_s having frequency 2ω, then the detected signal is proportional to $d^2 e_s/dK^2$.

Why bother to make a signal a periodic function for some parameter, and then to detect a frequency component of that signal? Well, in many experiments, the signal that we want to measure is small. In addition to this small signal, there will be present an additional experimental output, called noise, which tends to obscure the desired signal. By making the signal periodic, then the lock-in amplifier may be used and, as you will see, the lock-in amplifier reduces the experimental output identified as noise and it amplifies the selected frequency component of the signal.

A block diagram of the lock-in amplifier is shown in Figure A.3. For each arrow shown in Figure A.3, there is an adjustable knob on the front panel. The output from the experiment is the signal e_s given by equation 2 plus noise. The tuned amplifiers have a narrow bandwidth centered on a selectable frequency Ω. This frequency is selected by adjusting a knob on the lock-in amplifier; hence, a particular frequency component of e_s may be selected for amplification. Signals falling outside the passband of the amplifier are rejected. For example, if the modulation is 40 Hz and if the tuned amplifier is adjusted to 80 Hz, then the components of e_s having frequencies of 40 Hz, 120 Hz, ... are rejected. Also noise is rejected, except for the noise having frequencies in the narrow bandpass of the amplifier.

In Figure A.3, the oscillator that provides the modulation is shown as being external to the lock-in amplifier. In commercial units an oscillator is usually built in the lock-in amplifier and a switch on the front panel gives the experimentalist a choice of using the internal oscillator or connecting an external oscillator. In this appendix we show the oscillator as an external circuit. The oscillator voltage provides both the modulation of the parameter K and the reference input to the lock-in amplifier. The output of the harmonic generator is a superposition of sinusoidal waves of frequencies $\omega, 2\omega, 3\omega, 4\omega$, and so on; that is, the fundamental frequency and all of the harmonics are present. The tuned amplifier has a sinusoidal output at the selected frequency Ω, where Ω is the fundamental frequency ω or a harmonic $2\omega, 3\omega$, and so on. (Some older lock-in amplifiers have a square wave generator

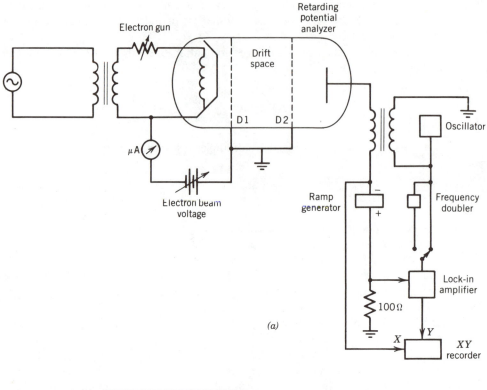

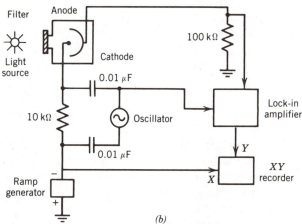

FIGURE A.2 Two circuits for carrying out modulation spectroscopy: (*a*) Modulation of the retarding voltage in the Franck–Hertz experiment. (*b*) Modulation of the anode voltage in the photoelectric effect experiment.

rather than a harmonic generator and the output of a square wave generator is a superposition of sinusoidal waves of frequencies ω, 3ω, 5ω, and so on, that is, the fundamental frequency and the odd harmonics.) The variable phase shifter allows us to shift the phase of the reference signal relative to the input signal e_s. The Schmitt trigger produces a square wave of frequency Ω.

We now examine the phase detector, which produces an output voltage proportional to the cosine of the phase difference between the input signal and the square wave reference signal. The phase detector circuit and a simple RC low-pass filter are shown in Figure A.4. The amplifier shown has a gain of -1; hence, the voltages on the input side of the switch,

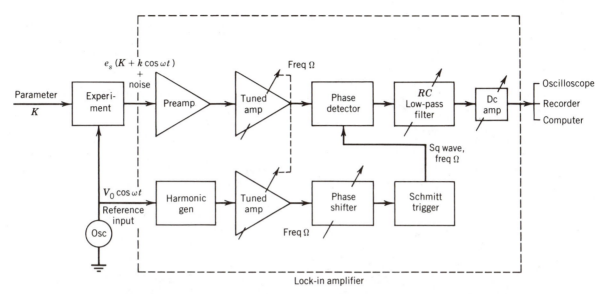

FIGURE A.3 Block diagram of the lock-in amplifier, reference oscillator, and experimental apparatus.

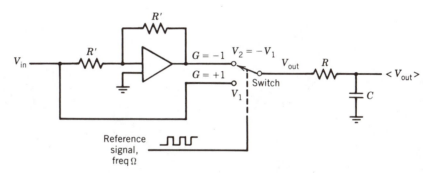

FIGURE A.4 Phase detector.

V_1 and V_2, are 180° out of phase. The switch is controlled by the reference voltage; that is, switching occurs each time the square wave makes a transition.

To analyze the phase detector, we assume the input signal is

$$V_{in} = V \cos(\Omega t + \phi) \quad (V) \tag{3}$$

where the amplitude V depends on the parameter K, ϕ is the phase difference between the input and reference signals, and Ω is the fundamental frequency ω or a harmonic frequency. We assume that the square wave output of the Schmitt trigger has transitions at the zeroes of $\cos \Omega t$, that is, at $t = \pi/2\Omega, 3\pi/2\Omega, 5\pi/2\Omega, \ldots$. Note in Figure A.4 that the voltage V_{out} will be V_1 or V_2, depending on the state of the switch, and that every half period the state of the switch changes.

Assuming the time constant τ of the low-pass filter is much longer than one period,

$$\tau = RC \gg T = \frac{2\pi}{\Omega} \quad (s) \tag{4}$$

then the filter averages the voltage V_{out}. The low-pass-filtered output is

$$\langle V_{out} \rangle = \frac{1}{T} \int_{\pi/2\Omega}^{3\pi/2\Omega} V_1 \, dt + \frac{1}{T} \int_{3\pi/2\Omega}^{5\pi/2\Omega} V_2 \, dt$$

$$= \frac{1}{T} \int_{\pi/2\Omega}^{3\pi/2\Omega} V \cos(\Omega t + \phi) \, dt + \frac{1}{T} \int_{3\pi/2\Omega}^{5\pi/2\Omega} -V \cos(\Omega t + \phi) \, dt \quad \text{(V)} \quad (5)$$

Evaluation of the integrals yields

$$\langle V_{out} \rangle = -\frac{2V}{\pi} \cos \phi \quad \text{(V)} \quad (6)$$

EXERCISE 1

Evaluate the integrals in equation 5 to obtain equation 6.

In Figure A.5, V_{ref}, V_1, V_2, V_{out}, and $\langle V_{out} \rangle$ are sketched for $\phi = 0°$. Similar graphs are sketched in Figure A.6 for $\phi = 90°$.

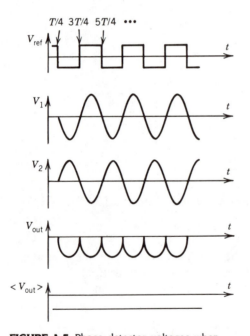

FIGURE A.5 Phase detector voltages when $\phi = 0°$.

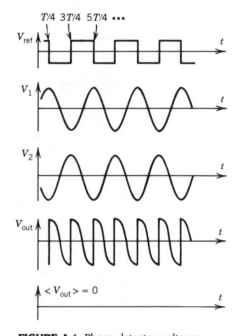

FIGURE A.6 Phase detector voltages when $\phi = 90°$.

EXERCISE 2

Sketch similar voltage versus time curves for $\phi = 45°$.

The signal $\langle V_{out} \rangle$ is maximized by adjusting the phase shifter in Figure A.3 until $\phi = 0$ or $180°$.

Note that the input signal to the phase detector V_{in} has a definite phase relation with the reference input V_{ref}; that is, the phase difference ϕ is fixed. We say the input signal is *phase locked* to the reference signal. In addition to the signal V_{in} at the input of the phase detector, there is also noise present that has the same frequency as V_{in}. (Recall that noise at other frequencies was rejected by the tuned amplifier.) The phase of the noise, relative to V_{ref}, is random. Since the noise has random phase, the noise output of the RC filter averages to zero; for example, if at an instant the phase difference between the noise and V_{ref} is 90°, then the noise output of the RC filter is zero as shown in Figure A.6.

Without modulation, the output of both the nuclear magnetic resonance and electron spin resonance spectrometers is the dc lineshape $e_s(B_0)$ versus magnetic field B_0, as shown in Figures 15.4d and 16.12c, Experiments 15 and 16, respectively. With modulation and the tuned amplifier frequency Ω set to ω, then V is proportional to de_s/dB_0. If the tuned amplifier is set to $\Omega = 2\omega$, then V is proportional to d^2e_s/dB_0^2. These three signals are shown in Figure A.7.

In Figure A.8a photocurrent I versus voltage across the phototube V_0, without modulation, is sketched for an ideal phototube, that is, a tube having zero reverse current. Here the voltage is labeled V_0 so as not to confuse it with the voltage V in equation 6. (Figure 9.6, Experiment 9, shows photocurrent versus voltage for a nonideal tube.) Figure A.8b and c show the output of the lock-in amplifier when tuned to frequencies ω and 2ω, respectively. The circuit for modulating the phototube is shown in Figure A.2b.

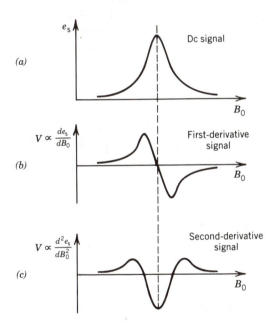

FIGURE A.7 Nuclear magnetic resonance or electron spin resonance signals. (a) Dc signal. (b) Lock-in amplifier output when $\Omega = \omega$. (c) Lock-in amplifier output when $\Omega = 2\omega$.

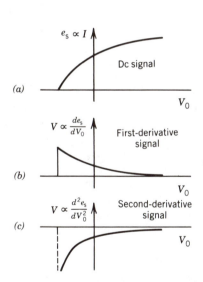

FIGURE A.8 Signals from a phototube. (a) Dc signal. (b) Lock-in amplifier output when $\Omega = \omega$. (c) Lock-in amplifier output when $\Omega = 2\omega$.

TEST OF THE LOCK-IN AMPLIFIER

The output of the lock-in amplifier is given by equation 6. To verify that the ouput is dependent on the phase difference ϕ, connect the circuit shown in Figure A.9. The oscillator, arbitrarily set to 40 Hz, drives the comparator and it provides the signal input to the lock-in amplifier. The output of the comparator is a square wave and, hence, it has

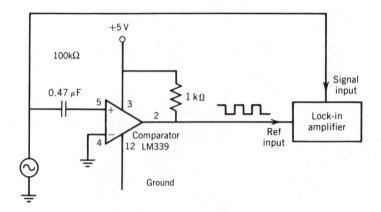

FIGURE A.9 Lock-in amplifier test circuit.

sinusoidal frequency components of 40 Hz, 120 Hz, 200 Hz, ... The output of the comparator connects to the reference input of the lock-in amplifier.

Usually lock-in amplifiers have BNC connectors for monitoring signals at various points in the circuit. Use an oscilloscope to monitor the signal at the output of the phase detector. This signal is labeled V_{out} in Figure A.4. The output of the lock-in amplifier, labeled $\langle V_{out} \rangle$ in Figure A.4, may be observed on the meter on the front panel of the amplifier.

EXERCISE 3

Set the frequency knob on the lock-in amplifier to 40 Hz. Observe V_{out} on the scope and $\langle V_{out} \rangle$ on the meter as functions of the phase angle. Compare your observations with V_{out} and $\langle V_{out} \rangle$ sketched in Figures A.5 and A.6.

Application of the Lock-in Amplifier

Connect the circuits shown in Figure A.10a and b. The LED shown in Figure A10a, the *transmitter* circuit, is of the kind used for panel indicators. The *receiver* circuit, Figure A.10b, should be located about a foot from the transmitter with the phototransistor looking in the direction of the LED. The output of the phototransistor is the input to the lock-in

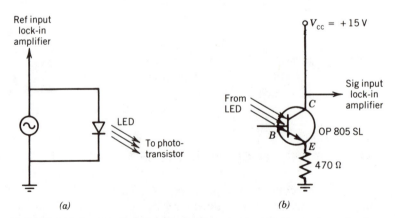

FIGURE A10 Lock-in amplifier application circuits. (a) Transmitter circuit.
(b) Receiver circuit.

amplifier. Also the oscillator in the transmitter circuit provides the reference input to the lock-in amplifier.

Disconnect the phototransistor output from the lock-in and connect it to the input of an oscilloscope. Set the oscillator to 500 Hz. With the room lights out, vary the LED–photo-transistor separation until the scope displays the signal from the phototransistor at the 500-Hz modulation frequency. There will also be noise mixed with the signal. Observe the signal on the scope while using your hand to block the light from the LED. Remove your hand and turn on the (fluorescent) room lights.

EXERCISE 4

With the room lights on, what is the frequency of the largest voltage observed on the scope? Is the 500-Hz signal observable?

Disconnect the output of the phototransistor from the scope and connect it to the signal input of the lock-in amplifier. Set the frequency of the lock-in to 500 Hz (this sets the tuned amplifiers to 500 Hz) and the time constant to a few seconds. Connect the output of the phase detector to the oscilloscope. Observe the output of the phase detector with the room lights on and then off.

EXERCISE 5

Do you observe any change in the signal at the output of the phase detector when the room lights are switched on or off?

With the room lights turned off, observe the output of the phase detector while using your hand to block the light from the LED.

EXERCISE 6

From your observations of the output of the phase detector can you conclude that the lock-in amplifier is working properly?

APPENDIX B

COUNTING AND SORTING PARTICLES: THE SCINTILLATION COUNTER

APPARATUS

None, since you are not asked to carry out measurements in this appendix.

OBJECTIVES

To understand the scintillation counter.

KEY CONCEPTS

Scintillator	Transit time
Photomultiplier tube (PMT)	Pulse height
Single-channel analyzer (SCA)	Total energy peak or photopeak
Multichannel analyzer (MCA)	Primary γ ray
Compton scattering	Secondary γ ray
Photoelectric effect	Full width at half-maximum (FWHM)
Pair production	Pulse height resolution
Quantum efficiency	Single escape peak
Compton edge	Double escape peak
Decay time	

REFERENCES

1. P. Ouseph, *Introduction to Nuclear Radiation Detectors*, Plenum, New York, 1975. Scintillators and the mechanism of scintillation are presented on pages 85–95. Photomultipliers are

discussed on pages 104–109. Efficiency of scintillation counters for γ-ray detection is given on pages 115–117. The interaction of γ rays with matter is discussed on pages 37–44.

2. R. Engstrom, RCA *Photomultiplier Handbook*, RCA, Lancaster, PA, 1980. The emphasis is primarily on the theory, design, and applications of photomultiplier tubes. Scintillators, scintillation counting, and the scintillation mechanism are treated on pages 69–74 and 92–97. The statistical theory of noise in photomultiplier tubes is treated in Appendix G, pages 160–174, and is summarized on pages 174–175.

3. J. Birks, *The Theory and Practice of Scintillation Counting*, Pergamon, Elmsford, NY, 1964. Unfortunately this book does not include a subject index; however, it does have a detailed table of contents. Gamma-ray interactions with matter are treated on pages 24–34. The scintillation process is presented in several chapters for a variety of scintillators. Photomultipliers are discussed on pages 113–147.

4. E. Bleuher and G. Goldsmith, *Experimental Nucleonics*, Holt, Rinehart, & Winston, Fort Worth, TX, 1960. Interaction of γ rays with matter is discussed on pages 175–178. Scintillation counters are discussed on pages 305–315.

SCINTILLATION COUNTER

In Experiments 18 and 19 a scintillation counter is used to study γ rays emitted by various radionuclides. A scintillation counter consists of five components:

1. A *scintillator*, for example, a NaI crystal doped with Tl, which converts the energy lost within it to many quanta of light (scintillations).
2. A *photomultiplier tube*, which converts the light quanta to photoelectrons, amplifies the number of photoelectrons, and converts the amplified pulse of charge to a voltage pulse.
3. A *preamplifier*, which amplifies the voltage pulse.
4. A *linear amplifier*, which further amplifies the voltage pulse.
5. *Circuitry* to measure the height of the voltage pulse.

A block diagram of a scintillation counter is shown in Figure B.1, where a γ ray is shown Compton scattering from an electron in the scintillator and the scattered γ ray is shown escaping from the scintillator. The components of the scintillation counter are discussed separately below.

SCINTILLATOR

A scintillator converts the energy lost within it into many photons of light. For a γ-ray incident on a scintillator the energy conversion is the following:

1. Some or all of the γ-ray energy is given to an electron in the scintillator by Compton scattering, photoelectric effect, or pair production.
2. Some fraction of the kinetic energy of the struck electron is transferred to activator atoms in the scintillator, raising their atomic electrons to excited states.
3. Some fraction of these excited atomic electrons decay to the ground state, emitting photons of light.

The typical quantum efficiency QE of scintillators is one photon of light produced per

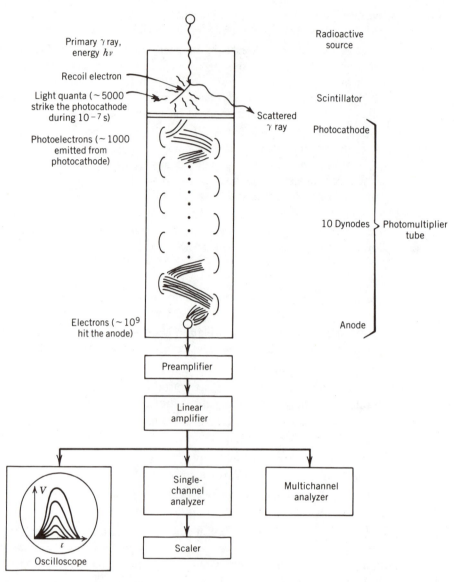

FIGURE B.1 Block diagram of a scintillation counter.

100 eV of energy deposited in the scintillator:

$$QE = \frac{1 \text{ photon}}{100 \text{ eV}} \qquad \text{(for scintillators)} \qquad (1)$$

The standard deviation of the number of photons is determined by the Poisson distribution and is the square root of the number of photons.

EXERCISE 1

If the energy absorbed by a scintillator corresponds to the Compton edge for a 0.662-MeV γ ray, then how many photons are produced? See equation 22, Experiment 18, where the energy corresponding to the Compton edge is discussed. What is the standard deviation of the number of photons?

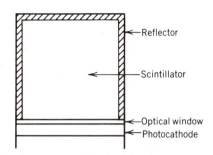

FIGURE B.2 Light created in the scintillator is ultimately reflected toward the optical window and, hence, the photocathode.

The **decay time**, the time interval between absorption of energy and emission of visible quanta, ranges from 10^{-9} to 10^{-6} s, depending on the scintillator material. The wavelength of maximum emission of the scintillator is typically 420 nm. Scintillator materials include NaI(Tl) crystal, anthracene, and certain plastics. A NaI(Tl) crystal is often used as the scintillator for γ-ray detection. The crystal is doped with Tl and light quanta are emitted by excited atomic electrons of Tl. Scintillators are essentially 100 percent transparent to the light quanta. The scintillator in a scintillation counter is partially enclosed with reflecting material and about 95 percent of the light quanta pass through an optical window and strike the photocathode of the photomultiplier, see Figure B.2.

PHOTOMULTIPLIER TUBE

The photocathode of a photomultiplier tube (PMT) is typically an alkali antimonide, having a spectral range of 300–660 nm. These materials are discussed in Experiment 9, Photoelectric Effect. The maximum quantum efficiency, the average number of electrons photoelectrically emitted from the photocathode per incident photon, is about 20 percent for alkali antimonides (see Figure 9.3, Experiment 9).

$$\text{QE}|_{\max} = \frac{1 \text{ photoelectron}}{5 \text{ photons}} \quad \text{(for photocathode)} \quad (2)$$

The standard deviation of the number of photoelectrons is the square root of the average number, and it is determined by Poisson statistics.

EXERCISE 2

Assuming that 95 percent of the photons calculated in Exercise 1 strike the photocathode, determine the number of photoelectrons emitted. What is the standard deviation of this number?

A PMT typically has 10 dynodes, a current gain of 10^6, and requires a cathode-to-anode voltage of about 1000 V. Secondary electron emission at each dynode gives a gain of about 4; hence, the overall gain is 4^{10}, which is approximately equal to 10^6. If all of the electrons are collected at each stage of the PMT, then 10^3 photoelectrons would create 10^9 electrons at the anode, as shown in Figure B.1. The collection efficiency at any stage is not 100 percent. The presence of a magnetic field, for example, reduces the efficiency. For this reason

PMTs are enclosed in a material having high permeability in order to shield against such fields.

The standard deviation of the number of electrons reaching the anode is not just the square root of the number. If we assume each dynode is identical and obeys Poisson statistics, then the standard deviation σ is given by

$$\sigma \simeq \sqrt{\frac{\eta n_p (\delta - 1)}{\delta}} \, m_{10} \tag{3}$$

where η is the photocathode quantum efficiency ($\simeq 0.20$), n_p is the number of photons produced in the scintillator, m_{10} is the gain of all 10 dynodes of the PMT ($\simeq 10^6$), and δ is the gain of a single dynode ($\simeq 4$).

EXERCISE 3

Using your value of n_p calculated in Exercise 1, what is the standard deviation of the number of electrons collected at the anode?

The dark or thermionic emission current from the photocathode is an additional source of PMT noise. (Thermionic emission is covered in Experiment 7, Electron Physics: Thermionic Emission and Charge-to-Mass Ratio.) Thermionic emission current in a PMT at 22 °C is typically 5 nA. Thermionic emission is temperature dependent and, hence, the resulting noise can be reduced by cooling the PMT.

The **transit time** of a PMT is defined to be the time interval between the arrival of a delta-function light pulse at the entrance window of the tube and the time at which the output pulse at the anode terminal reaches peak amplitude. The transit time is shown in Figure B.3. It is a function of anode-to-cathode voltage, and for an anode-to-cathode voltage of 1000 V the transit time ranges from 20 to 100 ns, depending on the PMT.

The number of electrons in the pulse of charge arriving at the anode of the PMT is proportional to the total energy deposited in the scintillator. This charge pulse is converted into a voltage pulse and its amplitude or **pulse height** is proportional to the total energy deposited in the scintillator.

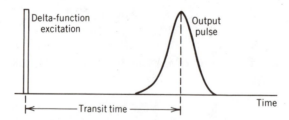

FIGURE B.3 Transit time of a PMT.

OSCILLOSCOPE

By connecting the output of the linear amplifier directly to the oscilloscope, pulses of various heights can be observed, as shown in Figure B.1. Each pulse is created by the interaction of a primary γ ray in the scintillator and the pulse height is proportional to the total energy deposited.

If the gain of the linear amplifier is too high, then saturation may occur, resulting in some output pulses being "clipped off." This is shown in Figure B.4. You should make sure the pulse heights you want to measure are not saturating the amplifier.

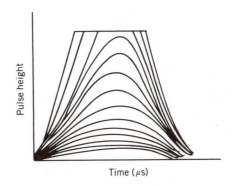

FIGURE B.4 Saturation of the amplifier causes pulse "clipping."

SINGLE-CHANNEL PULSE HEIGHT ANALYZER

The single-channel analyzer (SCA) can be used to count pulses, for a preselected time interval, that fall in a preselected range of heights. If a pulse whose voltage magnitude falls above a baseline voltage set by the baseline control of the SCA and in the range of "window" set by the channel width control of the SCA, then a standardized output pulse is emitted for each input pulse that falls in this range. The standardized output pulses can be counted with a scaler. Figure B.5 shows three pulses that are input to the SCA. Pulse 2 would be counted, but 1 and 3 would not be counted since their peaks are not inside the window.

A typical data-collection procedure using the scintillation counter shown in Figure B.1 with the output of the linear amplifier connected to the SCA would be to take 50 one-minute counts, where for each 1-min count the window voltage is the same but the baseline voltage is different. Then we plot the number of counts per minute per unit pulse height versus baseline voltage or pulse height, where the *unit pulse height* is determined by the window voltage setting.

Suppose we carried out such a set of measurements using a radioactive source that has monoenergetic γ rays of energy 1.5 MeV, say. What would a plot of counts per minute per unit pulse height versus pulse height look like? We first answer this question for a physically unrealistic case! Suppose that only Compton scattering occurs in the scintillator; that is, the photoelectric effect and pair production do not occur. In addition we assume that the scattered γ ray escapes from the scintillator and that it is not scattered by the laboratory

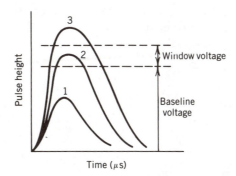

FIGURE B.5 Baseline voltage and window voltage of a SCA determines the pulses that are counted.

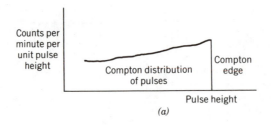

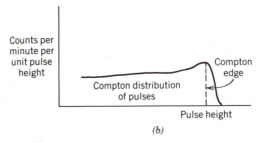

FIGURE B.6 Counts per minute per unit pulse height versus pulse height for a scintillation counter having (a) infinite resolution and (b) finite resolution.

walls, and so forth, back to the scintillator. In other words, the only energy deposited in the scintillator is that of the initially struck electron. It follows from Figure 18.6, Experiment 18, that our plot would be like that shown in Figure B.6a for a scintillation counter having infinite resolution and like that in Figure B.6b for a real counter with finite resolution.

It was previously pointed out that the delay time ranges from 10^{-9} to 10^{-6} s. Even for the fastest scintillators (10^{-9} s), a number of secondary events may occur that produce light quanta and contribute to the pulse height. For example, a secondary event could be the photoelectric absorption of a Compton-scattered γ ray. The **total energy peak**, also called the **photopeak**, is produced by γ rays for which all of their energy is absorbed in the scintillator by the primary event or by the primary event plus secondary events.

Figure B.7 shows events occurring in the scintillator that contribute to the pulse height distribution. The straight arrows shown in the scintillator, labeled with e$^+$ or e$^-$, represent the path of a positron or an electron, where the kinetic energy of the positron or electron is zero at the end of the arrow. A wiggly line leaving the scintillator represents a γ ray escaping from the scintillator. In parts (a), (b), and (d) only the primary event contributes to the pulse height distribution since the secondary γ rays in both (b) and (d) are shown escaping. In event (a) the following occurs: (1) the primary event is the photoelectric effect,

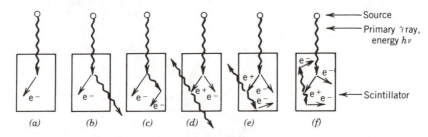

FIGURE B.7 Events occurring in the scintillator that contribute to the pulse height distribution.

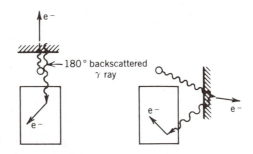

FIGURE B.8 Gamma rays scattered from surroundings contribute to the pulse height distribution.

(2) the energy absorbed is $h\nu$, the energy of the primary γ ray, (3) the pulse height is a maximum, producing a pulse in the photopeak. In event (e), (1) the primary event is pair production, followed by positron annihilation and the absorption of one annihilation γ ray, (2) the energy absorbed is $h\nu - 0.511$ MeV, (3) the pulse height is less than maximum and such pulses produce a peak in the Compton distribution of pulses.

Figure B.8 gives two contributions to the pulse height distribution by γ rays scattered from the surroundings into the scintillator.

For a monoenergetic source of γ rays, an ideal pulse height distribution is shown in Figure B.9a and a real pulse height distribution is shown in Figure B.9b. The abscissa is labeled in pulse height, which is the quantity measured and which is proportional to the energy deposited in the scintillator, shown on the upper scale. The total energy peak or photopeak occurs at energy $h\nu$, the Compton edge occurs at the maximum Compton–electron energy K_{max}, and event e in Figure B.7 produces the so-called single escape peak at energy $h\nu - m_0 c^2$. There will be other peaks in the spectrum.

The ideal pulse height spectrum, Figure B.9a does not take into account the previously discussed statistical fluctuations. These fluctuations broaden lines into peaks of finite width.

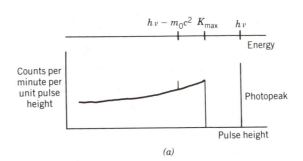

(a)

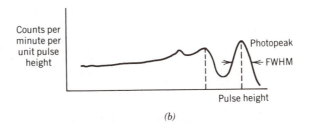

(b)

FIGURE B.9 Pulse height spectrum for (a) an ideal scintillation counter and (b) a real scintillation counter.

The full width at half-maximum (FWHM) of the total energy peak is shown in Figure B.9b. The **pulse height resolution** of the scintillation counter is defined to be the ratio of the FWHM (ΔE) to the energy E of the peak. For the total energy peak $E = h\nu$, the energy of the primary γ ray. For a 1-MeV γ ray the pulse height resolution is typically 10 percent.

EXERCISE 4

For each event given in Figures B.7 and B.8 specify (1) the absorption process, (2) the energy or the range of energy absorbed in the scintillator, and (3) the pulse height and where the pulse occurs in the pulse height distribution. Sketch a pulse height distribution curve for an ideal scintillator counter, similar to Figure B.9a, showing each expected peak. Label the "upper" abscissa scale in energy.

MULTICHANNEL PULSE HEIGHT ANALYZER

A multichannel analyzer (MCA) having N channels is equivalent to N single-channel analyzers that have a common input connection, identical window voltages, and successive baseline voltages spaced at exactly the window width. The channel number is proportional to the baseline voltage or the pulse height voltage, which is proportional to the energy deposited in the scintillator.

The MCA carries out pulse height analysis by converting the pulse height into a time interval, where the time interval is proportional to the pulse height, and the length of the time interval determines the channel number where the "event" is stored. The method of doing this is the following. An input pulse charges a capacitor to the peak voltage of the pulse, and the capacitor is discharged by a constant current supply. This is shown in Figure B.10a, where the voltage across the capacitor is plotted versus time and the input pulse is shown by the dashed curve. The time required for the capacitor voltage to reach zero is proportional to the pulse height, and the time interval is measured with a fast counter (a 200-MHz oscillator is typical). At the peak of the input pulse the oscillator is switched on and when the capacitor voltage reaches zero it is switched off. The oscillations are shown in Figure B.10b. The number of periods of the oscillator that occurs during the capacitor discharge time determines the channel number where the "event" is stored in memory. This has the disadvantage of giving the MCA dead time that depends on the height of the last pulse.

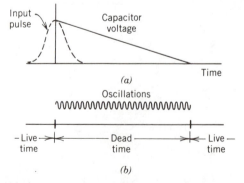

FIGURE B.10 The MCA sorts pulses by converting the input pulse height to a time interval, the time for a capacitor to be discharged by a constant current source.

When the measurements have been completed, the memory contains the number of counts sorted in each channel according to pulse height. To read out the information stored in the memory, the MCA puts out two voltages. One voltage is proportional to the channel number, and, hence, provides a horizontal energy reading, and the other voltage is proportional to the number of counts in the channel (or the log of this number as selected by a switch on the MCA). Thus, the output of a MCA, displayed on an oscilloscope or an *XY* recorder, would resemble Figure B.9*b*.

THE GENERAL PURPOSE
INTERFACE BUS (GPIB)

REFERENCES

1. S. Leibson, The Input/Output Primer. Part 3: The Parallel and HP-IB (IEEE-488) Interfaces, BYTE *Magazine* **7**(4), 186 (1982). Discusses the GPIB handshakes and bus configurations. Part 1 of the primer series contains a glossary of I/O terminology.

2. S. Leibson, *The Handbook of Microcomputer Interfacing*, TAB Books, Blue Ridge Summit, PA, 1983. Contains a discussion of serial and parallel interfacing. A brief discussion of the GPIB similar to the one cited in Reference 1 is found on pages 106–111.

3. D. J. Malcolme-Lawes, *Microcomputers and Laboratory Instrumentation*, 2d ed., Plenum, New York, 1988. Chapter 8 contains a discussion of the GPIB standard and its implementation on the IBM PC.

4. E. Fisher and C. W. Jensen, PET/CBM *and the* IEEE 488 Bus (GPIB), 2d ed., Osborne/McGraw-Hill, Berkeley, CA, 1982. Describes the implementation of the IEEE 488 specifically for applications involving PET/CBM machines.

5. *Tutorial Description of the Hewlett-Packard Interface Bus*, Hewlett-Packard, Sunnyvale, CA, revised Nov. 1987. Description of the technical fundamentals of the HP-IB. Available directly from HP.

Note: Copies of the IEEE 488 standard are available from IEEE.

In Experiment 6, Computer-Assisted Experimentation, three methods for interfacing a microcomputer to a laboratory experiment are discussed. The third method listed involves the use of the General Purpose Interface Bus (GPIB), which is discussed here. The GPIB, also known as the Hewlett-Packard Interface Bus (HP-IB), is a standard method by which various pieces of laboratory instrumentation can be conveniently connected to communicate with each other and with a *controller* (e.g., a microcomputer). The operation of the bus is specified by the IEEE 488 standard, the main features of which we briefly describe below. While this method of interfacing has the advantage of standardization, it can be less

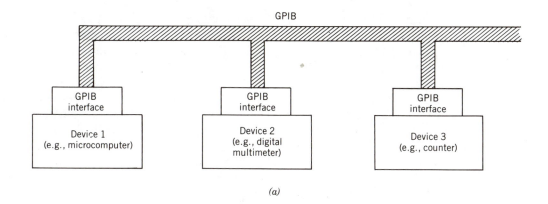

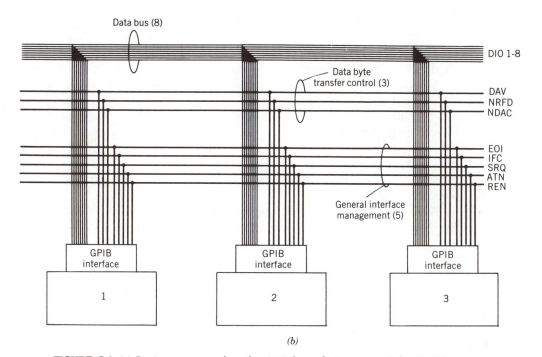

FIGURE C.1 (*a*) Devices connected to the GPIB bus. (*b*) Structure of the GPIB bus.

cost-effective than the various alternatives for the applications encountered in the upper-division physics laboratory.

As illustrated schematically in Figure C.1*a*, the GPIB allows bidirectional communication of data between devices connected to the bus. Among other things, the IEEE 488 standard specifies the following:

• Signals are carried along the bus in digital form, with TTL-compatible voltage levels and negative-true logic (lo = 1 or true, hi = 0 or false).

• The bus cable connecting any two devices may be as long as 4 m, while the total length of cabling in the system may be no more than 20 m. This limitation may be overcome by a bus extension technique involving, for example, telephone lines or serial optical fiber cables.

• The rate of data transfer may be as high as one megabyte per second.

• As many as 15 devices may be configured along the bus.

A device configured along the bus may be assigned the attribute of talker, listener, or controller. Devices configured as talkers can only place bytes on the bus (e.g., a simple ADC or counter), whereas listeners (e.g., a signal generator or printer) can only accept data bytes placed on the bus. The controller (e.g., a microcomputer) configures the bus, assigning attributes and controlling the data transactions. Devices may also possess some combinations of these three attributes: The controller can have all three, and "intelligent" devices (e.g., a programmable multimeter) can talk or listen, depending on which attributes have been made active by the controller for each device. While there can be only one active controller and one active talker at any time, more than one device can be an active listener. As a simple example, the Variable Dc Voltage exercise of Experiment 6 may be performed using a GPIB system with two devices: a microcomputer controller (listener) and a digital voltmeter or ADC configured as a talker.

STRUCTURE OF THE BUS

As indicated in Figure C.1*b*, the GPIB comprises a total of 16 lines divided into 3 subbusses: the data bus (8 lines), the data byte transfer control lines (3 lines), and the general interface management lines (5 lines). The ground wires associated with these signal lines are not shown.

The data bus (DIO 1–8) carries data (readings from instruments, e.g.) or control information (commands and addresses), depending on the state of the general interface management lines as determined by the controller. Information is transmitted along the data bus in 8-bit parallel, byte serial fashion, that is, 8 bits at a time, one byte after another. The transfer of each byte of data or control information is accompanied by a handshaking sequence, as described below.

The 3 lines in the data byte transfer control subbus perform the handshaking ritual and are assigned mnemonics appropriate to their functions:

DAV (data valid): Driven lo (true) by the active talker to indicate that data has been placed on the DIO lines.

NRFD (not ready for data): Driven hi (false) when all of the active listeners are ready to accept data.

NDAC (not data accepted): Driven hi (false) to indicate that all of the active listeners have accepted a data byte.

The 5 general interface management lines are used to send uniline messages:

ATN (attention): Driven low (asserted) by the controller to indicate to all devices on the bus that the bytes on the DIO lines represent commands or addresses rather than data.

SRQ (service request): May be asserted by any device requiring the attention of the controller. The identity of the device requesting service in this way must be determined by a polling process, since any device may, by itself, drive this line lo.

EOI (end or identify): May be asserted by an active talker to indicate that the current byte is the end of a data sequence. If asserted in conjunction with ATN by the controller, up to 8 devices may respond by sending 1 bit of information along each of the DIO lines in a *parallel poll* to identify a device requesting service.

IFC (interface clear): Asserted by the controller to force devices to leave the bus in some known, predetermined state.

REN (remote enable): Used by the controller to determine whether devices on the bus are controlled remotely by the GPIB (REN lo) or locally by the controls on the front panels (REN hi).

Note that not all of the above interface management lines are required on every system, but for systems with several devices, the ATN and the 3 transfer control lines constitute the minimum requirement for control of the system.

THE GPIB HANDSHAKE

The signaling protocol for transferring information between devices in a synchronized manner is referred to as a **handshake**. The GPIB handshake procedure, carried out by the 3 data byte transfer control lines for each byte conveyed by the bus, is designed so that the active listener with the slowest response time can participate in the exchange. This is accomplished by connecting the devices to the bus by means of an open collector circuit or a TRI-STATE® buffer. In the open collector scheme, each line is connected to the $+5$-V supply through a resistor and to the collector of an output transistor on each device. When a device turns its transistor on, current flows in the resistor, and the line voltage goes lo. Thus, any one device can, by itself, drive a line lo, but the line can be hi only by "unanimous consent" of all devices connected to it.

The relative order of events for one handshake cycle is shown in the timing diagram of Figure C.2, in which the voltage (V) is plotted against time (t) for each of the 3 byte transfer control lines. The cycle begins as the active listeners begin to indicate their individual readiness for data by turning off their output transistors for the NRFD line, as symbolized by the dashed lines; when *all* of the active listeners have so signaled their ready status, the NRFD line is driven hi (false). The active talker responds to NRFD = hi by placing the byte to be transmitted on the DIO lines and then pulling DAV lo, thus signaling active listeners that the data lines contain a valid byte. The listeners respond to DAV = lo by pulling NRFD back down and accepting the data. Each listener signals acceptance of the data by turning off its NDAC output transistor, as indicated again by the dashed lines. When all of the listeners have accepted the data, they release, by consensus, the NDAC line, making NDAC false. The talker acknowledges this transition by driving DAV back up; the listeners in turn pull NDAC low to start the next handshake cycle.

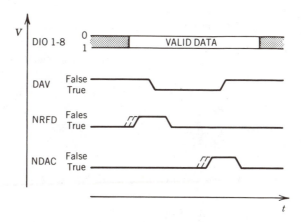

FIGURE C.2 Timing diagram for the GPIB handshake cycle.

COMMANDS AND ADDRESSES

As mentioned above, the DIO lines can carry data from instruments (talkers) as well as commands and addresses that configure the bus and control the devices. If the controller

TABLE C.1 COMMAND AND ADDRESS BYTES ON THE GPIB BUS

Type of Information	Bit Number							
	7	6	5	4	3	2	1	0
Bus command	X	0	0	C	C	C	C	C
Listen address	X	0	1	L	L	L	L	L
Talk address	X	1	0	T	T	T	T	T
Secondary address	X	1	1	S	S	S	S	S

brings ATN lo, all devices accept and interpret signals on the data bus as commands or addresses rather than as data. Table C.1 summarizes the meanings of the bytes carried by the data bus when ATN = lo. Note that bit 7 is represented by X, indicating that it is of no significance in this context, whereas bits 5 and 6, in combination, indicate whether bits 0–4 constitute a bus command (denoted by CCCCC in Table C.1), a listen address (LLLLL), a talk address (TTTTT), or a secondary address (SSSSS).

If bits 5 and 6 are both 0, the byte currently on the DIO lines is to be interpreted as a multiline bus command. Two kinds of bus commands are the *universal* commands, which cause every instrument so equipped to perform a specific interface function, and the *addressed* commands, which affect only those devices that have been previously addressed as a talker or a listener. Commands of both types are listed below, along with their three-letter mnemonics and decimal codes; these codes correspond to the 5-bit pattern (CCCCC) that sends the command.

The five universal commands are the following:

Device Clear (DCL, code 20): Devices that respond to this command go into some predetermined state (empty buffers, e.g.). The bit pattern for this command is CCCCC = 10100 (i.e., decimal code 20).

Local Lockout (LLO, code 17): This disables the "return-to-local" control on devices recognizing this command.

Serial Poll Enable (SPE, code 24): This establishes the serial poll mode for responding talkers, so that each device will, when addressed to talk, return a single 8-bit byte indicating its status.

Serial Poll Disable (SPD, code 25): This terminates the serial poll mode initiated by the SPE command described above.

Parallel Poll Unconfigure (PPU, code 21): All parallel poll devices are reset to a state in which they do not respond to a parallel poll.

The five addressed commands are the following:

Group Execute Trigger (GET, code 08): This causes the currently active listeners to take some predetermined action.

Selected Device Clear (SDC, code 04): This resets the devices currently addressed as listeners to a device-dependent state.

Go To Local (GTL, code 01): This causes currently addressed listeners to return to local control.

Parallel Poll Configure (PPC, code 05): This causes addressed listeners to be configured according to the Parallel Poll Enable command (a secondary command) that immediately follows PPC.

Take Control Talker (TCT, code 09): This causes the addressed talker to begin operating as the controller.

It should be noted that all GPIB-compatible devices do not support all of the standard commands.

If bits 5 and 6 are different, then the bit sequence LLLLL (or TTTTT) represents the address (in binary) of a device being configured to become an active listener (or talker). An address must be in the range 0–30 and correspond to the address code recognized by the device hardware. The address for a device may typically be set by the user by means of toggle switches on the device. Address 31 is used to deactivate talkers and listeners, so that the byte X0111111 (LLLLL = 11111 = 31) deactivates all active listeners (*unlisten*), while the byte X1011111 (*untalk*) deactivates all active talkers.

If bits 5 and 6 are both 1, then bits 0–4 contain a number in the range 0–31, which represents a secondary address, which is sometimes used to issue a command (change an instrument function, e.g.) to a device that has been previously addressed.

In laboratory applications, the controller is commonly a GPIB-equipped microcomputer, and many (but not all) of the functions supported by the GPIB interface in the acquisition of data and the control of instrumentation can be implemented using a high-level language, such as BASIC. In PET® BASIC, for example, commands such as OPEN, GET #, INPUT #, and PRINT # handle some of the possible transactions, so that knowledge of the bus conventions explained above is not always essential to the use of a GPIB system. For the implementation of interface functions that are not supported in BASIC on a particular microcomputer, it is usually possible to employ assembly language routines.

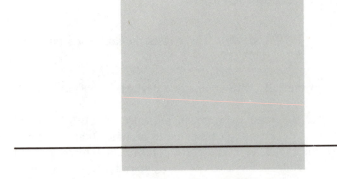

APPENDIX D

THE TWO-METER EBERT-MOUNT SPECTROGRAPH

APPARATUS

The apparatus required is listed in Experiment 12 or Experiment 13, depending on which experiment you are doing.

OBJECTIVES

To verify that the tilt adjustments of both the mirror and grating are adequate.

To obtain a sharp focused image of the slit on the film plate.

To become familiar with the Ebert-mount spectrograph and to understand two important characteristics of a spectrograph: dispersion and resolution.

KEY CONCEPTS

Resolution	Dispersion curve
Linear dispersion	Plate factor
Angular dispersion	Rayleigh's criterion

REFERENCES

1. C. Schreiber, E. Wong, and D. Johnston, *Am. J. Phys.* **39**, 1333 (1971). Three Ebert-mount spectrographs are described in this paper: a 1.5-m spectrograph, a 2-m spectrograph, and a 2.5-m spectrograph.
2. F. Jenkins and H. White, *Fundamentals of Optics*, 3d ed., McGraw-Hill, New York, 1957. Dispersion and resolution of a reflection grating are discussed in Chapter 17.

3. G. Harrison (ed.), M.I.T. *Wavelength Tables*, Wiley, New York, 1969. The wavelengths of the spectral lines of mercury can be found in this book.

4. A. Gatterer and J. Junkes, *Arc Spectrum of Iron*, Specola Vaticana, Citta del Vaticano (Vatican city), 1935. The arc spectrum of iron from 8388 to 2242 Å is reproduced on 21 photographic plates.

INTRODUCTION

Reference 1 describes the design, construction, and testing of three Ebert-mount spectrographs. The basic difference in the three instruments is their resolution and dispersion. These spectrographs are low-cost, high-resolution instruments that are designed for the undergraduate laboratory. The required optical components for the spectrographs are readily available and only modest shop facilities are needed to construct the various parts and cover box.

Machine shop drawings of the 2-m Ebert-mount spectrograph are available. Copies of the drawings may be obtained by contacting:

Daryl W. Preston
Physics Department
California State University   or
Hayward, CA 94542
415/881-3401

Eric Dietz
Physics Department
California State University
Chico, CA 95929

The optical path of the Ebert-mount spectrograph is shown in Figure D.1, where the focal length of the mirror f is equal to the distance s. All three of the spectrographs use mirrors with $f/8$ surfaces and plane reflection gratings that have 600 lines/mm and a blaze for 1 μ. Table D.1 is a partial summary of the three instruments. The length of the spectrum (measured on the photographic plate) and the resolving power are measured values reported in reference 1.

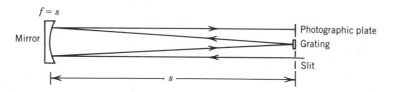

FIGURE D.1 Optical path of the Ebert-mount spectrograph.

TABLE D.1. SUMMARY OF THE 1.5-, 2-, AND 2.5-M, EBERT-MOUNT SPECTROGRAPHS

	Spectrograph		
Parameter	1.5 m	2 m	2.5 m
Mirror focal length (m)[a]	1.5	2	2.5
Mirror diameter (in.)	8	10	12.5
Grating dimension (mm)[b]	52 × 52	64 × 64	102 × 128
Length of spectrum (in.)	7	10	10
Resolving power	180 000	280 000	~450 000

[a]Mirrors are available from Edmund Scientific.
[b]Gratings are available from Milton Roy, Rochester, NY, (716) 248-4000.

The plane reflection grating is discussed in Experiment 4, where the equation for constructive interference is derived, namely,

$$d(\sin \theta_{in} - \sin \theta_{out}) = m\lambda \qquad \text{(m)} \tag{1}$$

where λ is the wavelength, m is the order number, d is the distance per reflecting surface of the grating, and θ_{in} and θ_{out} are the angles of the incident and diffracted rays, respectively. See equation 3 and Figure 4.5 of Experiment 4, and read the discussion of the reflection grating.

The principle of the Ebert-mount spectrograph is as follows. Light, consisting of multiple discrete wavelengths, is incident on the slit. This light is reflected by the mirror and incident on the grating at some angle θ_{in}, which is determined by the design of the spectrograph. The angle of the diffracted light, θ_{out}, depends on λ according to equation 1; hence, the different wavelengths are, in principle, separated. The separated wavelengths are incident on the photographic plate after being reflected by the mirror. Each spectral line on the plate is an image of the slit. Even if the slit is extremely narrow, its image is of appreciable width since it is a single-slit diffraction pattern. Two such images having wavelengths λ_1 and λ_2, and wavelength difference $d\lambda$, are shown in Figure D.2, where ℓ is the distance measured along the photographic plate and $d\ell$ is the separation of the intensity maxima.

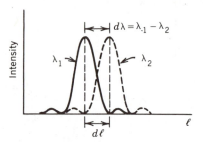

FIGURE D.2 Single-slit diffraction patterns that are barely resolved.

DISPERSION AND RESOLUTION

In this section some questions have two labels, Experiment 12 and Experiment 13. If you are doing Experiment 12 or Experiment 13, then answer that part of the question labeled Experiment 12 or Experiment 13, respectively.

Two important characteristics of a spectrograph are dispersion and resolution. Dispersion is discussed first.

The linear dispersion $d\ell/d\lambda$ gives the actual distance $d\ell$ of separation, measured on the photographic plate, in the spectrum of two close spectral lines differing in wavelength by $d\lambda$. For a given $d\lambda$, a large value of $d\ell$ implies high dispersion. Figure D.3a shows a spectrum taken with a spectrograph having higher dispersion than that in Figure D.3b, where $d\lambda = \lambda_1 - \lambda_2$, with λ_1 and λ_2 being the wavelengths at the center of each spectral line. We first discuss the "blazed" grating, and then determine the angular dispersion of a blazed grating, and then use the result to determine the linear dispersion of a grating used in the Ebert mount.

The blazed grating equation is the same as that for the plane grating, equation 1, where both θ_{in} and θ_{out} are positive when measured on opposite sides of the normal, and θ_{out} is negative when the two angles are measured on the same side of the normal. In Figure D.4a, which shows two spectral lines in first or higher order, θ_{in} is positive and θ_{out} is negative, and θ is the angular position of the grating. (It is not difficult to show that θ_{out} is negative for all orders higher than the zeroth for the two-meter Ebert-mount spectrograph.) For convenience of scale the spherical mirror is not shown in the figure, and the 4.56-inch light beam separation is that of the two-meter Ebert-mount spec-

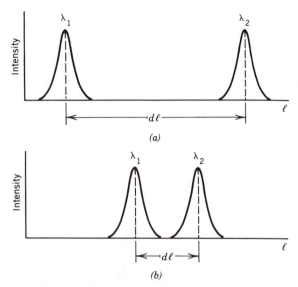

FIGURE D.3 The spectrum in (*a*) was recorded with a spectrograph having higher dispersion than that in (*b*).

trograph. (Figure D.1 shows zeroth order, where the normal to the grating is parallel to the mirror axis. Using s = 2 m and a beam separation of 4.56″, it follows that θ_{in} and θ_{out} are both 1.66°. As the grating is rotated the angle between the incident and diffracted rays will always be about 3.3°.)

Figure D.4b shows a magnified view of a "blazed" grating. The purpose of the blazed grating grooves is to diffract most of the incident light into a particular wavelength in each order, rather than to spread it out over many wavelengths in each order like a plane reflection grating. The wavelength that receives most of the light is determined by the shape of the grooves. Note: A plane reflection grating provides a spectrum of wavelengths (e.g., red through violet in the visible) in many orders, each weak in intensity; whereas, a blazed grating provides a bright spectrum peaked about one wavelength in each order.

A grating with a blaze for 1μ will produce a bright spectrum in first order centered at 1μ, or 10^4 Å, a bright spectrum in second order centered at 5000 Å, a bright spectrum in third order centered at 3333 Å, a bright spectrum in fourth order centered at 2500 Å, etc.

The angular dispersion of a grating $d\theta_{out}/d\lambda$ gives the angular separation $d\theta_{out}$ of two spectral lines differing in wavelength by $d\lambda$. Calculating the differential of both sides of equation 7 yields

$$-d \cos \theta_{out}\, d\theta_{out} = m\, d\lambda \qquad \text{(m)} \qquad\qquad (2)$$

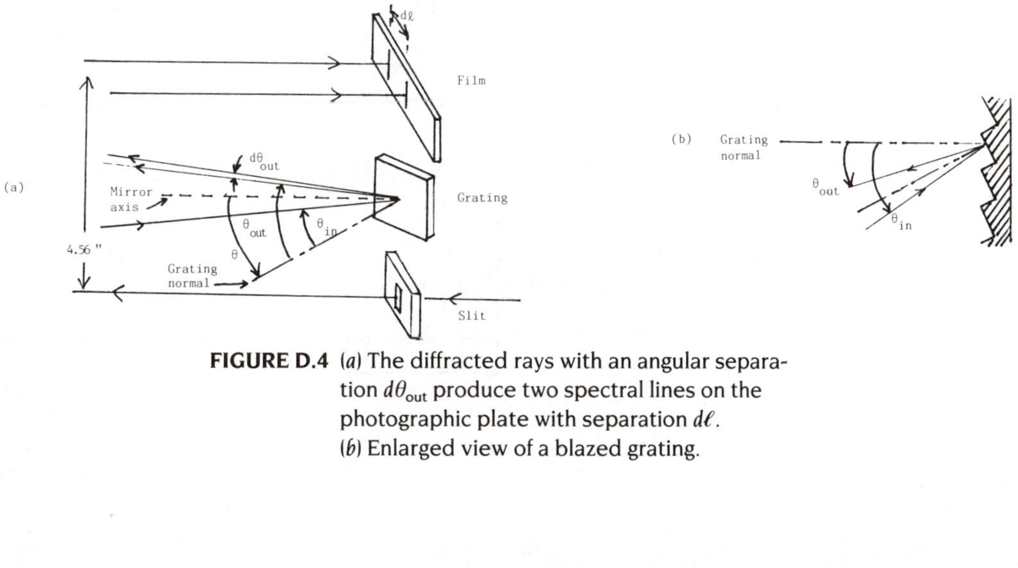

FIGURE D.4 (*a*) The diffracted rays with an angular separation $d\theta_{out}$ produce two spectral lines on the photographic plate with separation $d\ell$.
(*b*) Enlarged view of a blazed grating.

When we ignore the negative sign, the angular dispersion is

$$\frac{d\theta_{out}}{d\lambda} = \frac{m}{d \cos \theta_{out}} \qquad (rad/\text{Å}) \qquad (3)$$

where the typical units are radians per angstrom.

Figure D.4a shows two diffracted rays that differ in wavelength by $d\lambda$ with an angular separation $d\theta_{out}$. The separation of the two corresponding spectral lines on the photographic plate is $d\ell$. The separation $d\ell$ is related to the focal length of the mirror, s, by

$$d\ell \simeq s \, d\theta_{out} \qquad (mm) \qquad (4)$$

The equation for the linear dispersion is obtained by using equations 3 and 4 to eliminate $d\theta_{out}$. The result is

$$\frac{d\ell}{d\lambda} \simeq \frac{ms}{d \cos \theta_{out}} \, (mm/\text{Å}) \qquad (5)$$

where the typical units are millimeters per angstrom. For a particular order number m, large s, large θ_{out}, and/or small d imply high dispersion.

The reciprocal of the linear dispersion, $d\lambda/d\ell$, is often referred to in books and papers, and is called the *dispersion* or the *plate factor*.

$$\frac{d\lambda}{d\ell} \simeq \frac{d \cos \theta_{out}}{ms} \qquad (\text{Å}/mm) \qquad (6)$$

One of two methods is often used to determine unknown wavelengths of spectral lines recorded on the photographic plate. One method uses a light source of known wavelengths, for example, a mercury light source, to determine a *dispersion curve*, λ versus ℓ, for the photographic plate. Once the known wavelengths are recorded on film, then use the traveling microscope to measure ℓ, the position on the film, for each line. Plot λ versus ℓ and do a least-squares fit to determine the dispersion curve. (If you use this method, it is suggested you try both a linear fit, $\lambda = A\ell$, and a quadratic fit, $\lambda = C\ell + D\ell^2$, where the constants are determined from the fit. The quadratic curve will probably yield the better fit.)

The other method to determine unknown wavelengths is to record a spectrum of known wavelengths, for example, the spectrum of an iron arc, and the spectrum of unknown wavelengths on the same film with the unknown spectrum displaced relative to the known spectrum by means of a Hartman diaphram. (The Hartman diaphram will be discussed later.) The spectrum of the iron arc has a large number of closely spaced spectral lines; hence, there will be a spectral line of iron of known wavelength on each side of the spectral line of unknown wavelength and the unknown can be determined by interpolation. A circuit diagram for the iron arc is shown in Figure D.5. A convenient method of starting the arc is to spring load one iron electrode, and then with the power supply turned on, the spring-loaded electrode can be momentarily placed in contact with the other electrode, which will start the arc.

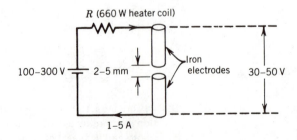

FIGURE D.5 Circuit for the iron arc.

CAUTION

THE IRON ELECTRODES SHOULD BE ENCLOSED IN A BOX WITH A SMALL APERTURE SO THAT THE RADIATION IS NOT OBSERVED DIRECTLY WITH THE EYE.

The iron arc method of determining the unknown wavelengths is more precise than the dispersion curve method.

EXERCISE 1

For each of the three spectrographs summarized in Table D.1 calculate the plate factor, $d\lambda/d\ell$, as a function of $\cos \theta_{out}$ in first, second, and third order. Express your answer in Å/mm. Assume the grating has 600 lines/mm.

EXERCISE 2

Experiment 12. The wavelength difference of the two lines of the fine structure of hydrogen is $d\lambda = 0.14$ Å. Calculate the separation $d\ell$ of the fine-structure lines of hydrogen for the two-meter spectrograph in first, second, and third order. Determine θ_{out} from equation 1, using $\lambda = 6563$ Å (H_α line), $\theta_{in} = |\theta_{out}| + 3.3°$, and $d = 1$ mm/600. Do not solve for θ_{out}, but rather use your calculator to find the value of θ_{out} that satisfies equation 1.

Experiment 13. The wavelength difference of two consecutive lines in the $C^3\Pi_u \rightarrow B^3\Pi_g$ transition of the N_2 molecule is approximately $d\lambda = 0.45$ Å. Calculate the separation $d\ell$ of two such lines for the two-meter spectrograph in first, second, and third order. Determine θ_{out} from equation 1, using $\lambda = 3226$ Å, $\theta_{in} = |\theta_{out}| + 3.3°$, and $d = 1$ mm/600. Do not solve for θ_{out}, but rather use your calculator to find the value of θ_{out} that satisfies equation 1.

EXERCISE 3

The maximum length of spectrum that can be recorded on the photographic plate of each spectrograph is summarized in Table D.1.

Experiment 12. Could all of the Balmer lines of hydrogen be recorded in third order within the length of spectrum of the two-meter spectrograph? To answer this question calculate θ_{out} for both the 6563 Å line and the 4105 Å line, and use the value of $\Delta\theta_{out}$ and $s = 2000$ mm to determine the $\Delta\ell$ required to record the spectrum.

Experiment 13. In reference 3, Experiment 13, photographic plate 3 includes the second positive spectrograph of N_2, which is the $C^3\Pi_u \rightarrow B^3\Pi_g$ transition, and it shows seven band origins between 2950 Å and 4340 Å. Could all seven of the band origins be recorded in third order within the length of spectrum of the two-meter spectrograph? To answer this question calculate θ_{out} for both 2950 Å and 4340 Å, and use $\Delta\theta_{out}$ and $s = 2000$ mm to determine the $\Delta\ell$ required to record the spectrum.

The resolution or resolving power R of an optical instrument is defined to be

$$R \equiv \frac{\lambda}{\Delta\lambda} \qquad \text{(for any optical instrument)} \qquad (7)$$

where $\Delta\lambda = \lambda_1 - \lambda_2$ is the wavelength difference of two lines that are barely resolved and $\lambda = (\lambda_1 + \lambda_2)/2$ is the average wavelength of the two lines. The value adopted for $\Delta\lambda$ is a matter of definition. The value of $\Delta\lambda$ that is generally used is based upon Rayleigh's criterion. Rayleigh's criterion, when applied to images from a slit, states that two lines are barely resolved when the central maximum of one line falls on the first diffraction minimum

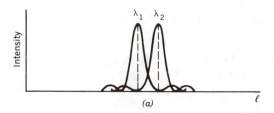

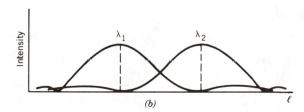

FIGURE D.6 The spectrum in (*a*) was recorded with a spectrograph having higher resolution than that in (*b*).

of the other. The two lines shown in Figure D.2 are barely resolved according to Rayleigh's criterion.

It should be pointed out that the resolution defined by equation 7 is the theoretical limit; that is, diffraction limits the resolution to the value determined by equation 7. Some mechanisms that may cause the resolution to be less than this value are (1) Doppler broadening in the light source, (2) collisional broadening in the light source, and (3) natural linewidth (uncertainty principle). Figure D.6*a* shows a spectrum recorded with a spectrograph having higher resolution than that in Figure D.6*b*. Figure D.7*a* and *b* show the same dispersion, but (*b*) indicates better resolution.

The resolution of a particular optical instrument is obtained by analysis. The result of the

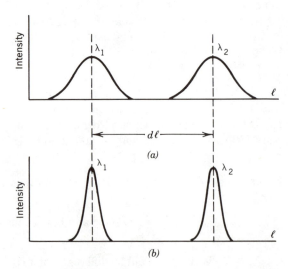

FIGURE D.7 The spectra in (*a*) and (*b*) show the same dispersion, but (*b*) shows better resolution.

analysis for a plane grating is (see reference 2)

$$R = mN \quad \text{(for a plane grating)} \tag{8}$$

where m is the order number and N is the total number of lines.

EXERCISE 4

Experiment 12. In Experiment 12 it is reported that the two spectral lines of the isotope structure of hydrogen have a wavelength difference of $\Delta\lambda = 1.79$ Å. How many ruled lines N on a grating would be required to resolve the isotope structure of hydrogen in first order? What value of N is required to resolve the fine structure of hydrogen in first order, where $\Delta\lambda = 0.14$ Å? In both cases assume the transitions are the H_α lines, where $\lambda = 6563$ Å.

Experiment 13. In answering Exercises 9 and 12, Experiment 13, you determined the frequency v and the wavelength difference $\Delta\lambda$ for $C^3\Pi_u \rightarrow B^3\Pi_g$ transitions of N_2. How many ruled lines N on a grating would be required to resolve such lines in first order?

EXERCISE 5

For each spectrograph in Table D.1 calculate the resolving power in first, second, and third order.

Experiment 12. What order will be required to resolve the fine structure of hydrogen?

Experiment 13. What order will be required to resolve the rotational structure of N_2? Assume the initial and final rotational quantum numbers are 4 and 3. (See your answer to Exercise 2, Experiment 13, where you calculated the frequency corresponding to this transition.)

SPECTROGRAPH ADJUSTMENTS

The optical path is shown in Figure D.8. The focal length of the 2-m spectrograph is actually 80 in. In Figure D.9 the photographic plate, slit, grating, and grating mount are shown. The calibrated disk shown in Figure D.9 is calibrated from $+90$ to $-90°$ and the angular position of the grating can be selected by rotating the disk. Figure D.10 shows the mount that supports the mirror, the mirror mask, and the incident and reflected rays. The light incident on the mirror from the slit reflects from the square region of the mirror that is not covered by the mask, and the light diffracted by the grating reflects from the trapezoid region that is not covered by the mask.

The adjustments of the spectrograph are made in the following order: (1) the tilt adjustment of the mirror, (2) the tilt adjustment of the grating, and (3) the focusing

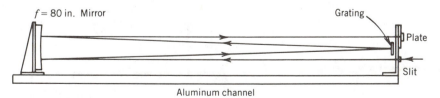

FIGURE D.8 Optical path for the 2-m, Ebert-mount spectrograph.

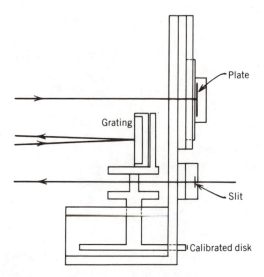

FIGURE D.9 Calibrated disk, slit, grating, and photographic plate.

adjustment of the mirror. The first two adjustments are, in principle, one time only adjustments that are made when the spectrograph is assembled.

We consider the tilt adjustments first. If we let the z axis be parallel to the light incident on the mirror from the slit, then by the tilt of the mirror or grating we mean rotations about the x and y axes as shown in Figure D.11 for the grating. Both the grating and the mirror have push and pull bolts that produce the rotations.

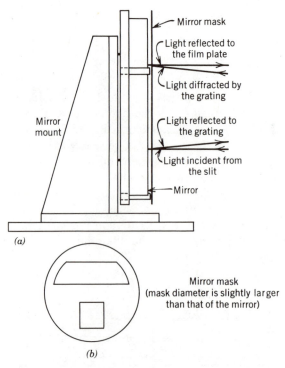

FIGURE D.10 (a) Side view of the mirror mount, mirror, and mirror mask. (b) Front view of
the mirror mask.

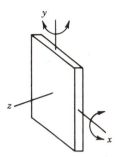

FIGURE D.11 Tilt adjust of the grating is carried out by a rotation about the *x* and/or *y* axes.

Before discussing the tilt adjustments we must make the following comments:

1. Setting the grating to 0° means the axes of the mirror and grating are then ideally parallel.
2. The Hartman diaphram is shown in Figure D.12. It is a thin metal plate with three holes that slides in front of the slit. If the center hole is in front of the slit, then the spectral lines will be in the center of the film. If the top or bottom hole is in front of the slit, then the spectral lines will fall on the upper or lower portion of the film, respectively.
3. The spectrograph has removable top covers at each end. Removing the front top cover exposes the grating and its mount, shown in Figure D.9, and removing the back top cover exposes the mirror and its mount, shown in Figure D.10.

The tilt adjustment of the mirror is made in the following way. Remove the front and back top covers. Rotate the grating to 0° and place the center hole of the Hartman diaphram in front of the slit; then adjust the push and pull bolts located on the mirror mount until the light reflected from the mirror falls on the center of the grating. Leave the bolts so adjusted.

With the grating set to 0° adjust the push and pull bolts located on the grating mount until the zeroth order spectrum of the mercury source is positioned on the center of the film plate. If this adjustment has been properly made, then rotation of the grating should not produce any vertical movement of the higher order spectra in the plane of the film plate. (It is pointed out in reference 1 that an aberration in this type of mounting is a rotation of the image of the slit as the plane of the grating is rotated away from the zero degree setting. A vertical image of the slit can be obtained by an appropriate rotation of the slit. See reference 1 on this point.)

We recommend that the above adjustments be checked by each student using the spectrograph to verify that the adjustments are correct.

The focusing adjustment produces a focused image of the slit on the film plate. This adjustment is first made approximately with the eye, and the final adjustment is made photographically. To focus the image of the slit with the eye, replace the film plate with the plate that has an attached eyepiece. Rotate the grating to 0°. Loosen the two bolts that fasten the mirror mount to the aluminum channel. This permits the mirror to be moved

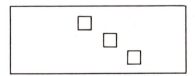

FIGURE D.12 Hartman diaphram.

toward or away from the film plate. Place an object, for example, a mark on paper, in the film plane and adjust the eyepiece until the object is in focus. Leave the eyepiece so set and remove the object. Using the mercury source rotate the grating until the two first-order yellow lines fall on the eyepiece. Slide the mirror toward or away from the eyepiece until focused images of the two yellow lines are observed.

The final focusing adjustment is made photographically. Replace the eyepiece with the film plate. Move the mirror 4 mm farther away from the film plate and then photograph the spectrum. (The actual time of the exposure depends on the source intensity, slit width (resolution desired), and the type of film. For both the CENCO hydrogen–deuterium discharge tube and the CENCO nitrogen spectrum tube, with a 20-μ slit and Panatomic-X 35-mm film, an exposure time of about 1 min is required. The mercury source will likely have a higher intensity than the above two tubes, and, hence, a shorter exposure time can be used. The exposure of the film is not a linear function of the exposure time. A large change in exposure time will produce a small effect on the exposed film. In obtaining the desired exposure of the film, it is suggested that the exposure time be changed by factors of ten; for example, go from a 1-min exposure to 0.1 or 10 min.) Move the mirror 2 mm toward the film plate and photograph the spectrum again. Continue to move the mirror toward the film plate in 2-mm increments, photographing the spectrum each time until the spectrum has been photographed about five times. Develop the film and use the data to determine the position of the mirror that gives the best image.

APPENDIX E

THE ELECTROMAGNETIC SPECTRUM: DETECTORS AND GENERATORS

Branch	Frequency (Hz)	Wavelength	Photon Energy in Typical Units	Detector	Generator
Low frequency	$<10^5$	>3 km	$<6.63 \times 10^{-29}$ J	Ammeter Voltmeter	Mechanical
Radiowaves	10^6-10^8	300–3 m	$6.63 \times 10^{-28}-6.63 \times 10^{-26}$ J	Antenna Diode	Tuned circuit Crystal
Microwaves	10^9-10^{11}	30 cm–3 mm	$6.63 \times 10^{-25}-6.63 \times 10^{-23}$ J $3.33 \times 10^{-2}-3.33$ cm^{-1a}	Antenna Crystal diode Bolometer	Klystron Magnetron Solid-state generator
Infrared	$10^{12}-3 \times 10^{14}$	300–1 μm	$4.14 \times 10^{-3}-1.24$ eV $3.33 \times 10^1-1 \times 10^4$ cm^{-1}	Bolometer Photodiode Phototransistor Photographic plate	Heat source Laser
Visible	$4 \times 10^{14}-7.5 \times 10^{14}$	700–400 nm	$1.66-3.11$ eV $1.43 \times 10^4-2.50 \times 10^4$ cm^{-1}	Photomultiplier tube Photodiode Phototransistor Photographic plate	Incandescent lamp Discharge tube Laser
Ultraviolet	$10^{15}-3 \times 10^{15}$	300–100 nm	$4.14-1.24 \times 10^1$ eV $3.33 \times 10^4-1.00 \times 10^5$ cm^{-1}	Same as visible	Same as visible
X rays	$10^{16}-10^{19}$	30–0.03 nm	$4.14 \times 10^{-1}-4.14 \times 10^1$ keV	Proportional counter Geiger–Mueller tube Scintillator Semiconductor detector	Heavy element bombardment K-shell transitions Accelerator
γ rays	$>10^{19}$	<0.3 pm	$>4.14 \times 10^{-2}$ MeV	Same as x rays	Nuclear energy level transitions Accelerator

[a] Units of inverse wavelength, cm^{-1}, are widely used in spectroscopy. The inverse of the wavelength is proportional to the energy.

Source: Charles P. Poole, Jr., Electron Spin Resonance: A Comprehensive Treatise on Experimental Techniques, Wiley, New York, 1983. Copyright © 1983, John Wiley & Sons, Inc. Reprinted by permission.

APPENDIX F

WRITING PROGRAMS

REFERENCES FOR THE IBM PC

BASIC

1. S. Gates with J. Becker, *Laboratory Automation Using the* IBM PC, Prentice-Hall, Englewood Cliffs, NJ, 1989.
2. L. Goldstein, IBM PC: *An Introduction to the Operating System, BASIC Programming, and Applications*, 3d ed., Brady, Englewood Cliffs, NJ, 1986.

Assembly language

3. J. Coffron, *Programming the 8086/8088*, Sybex, Alameda, CA, 1983.
4. P. Norton, *Programmer's Guide to the* IBM PC, Microsoft Press, Redmond, WA, 1985.

REFERENCES FOR THE APPLE II

BASIC

5. *Basic Programming Reference Manual*, Apple Computer, Cupertino, CA, 1978.
6. M. De Jong, *Apple II Applications*, Sams, Indianapolis, IN, 1983.

Assembly Language

7. C. Finley, Jr. and R. Myers, *Assembly Language for the Applesoft Programmer*, Addison-Wesley, Reading, MA, 1984.

INTRODUCTION

This appendix discusses writing programs in BASIC and in assembly language. It assumes you have prior experience in writing programs, and that your programming skills will be enhanced by studying the programs that accompany Experiment 6, Introduction to Computer-Assisted Experimentation. Program numbers referred to below, for example, program number 2, is the program number listed in the menu of the disks accompanying Experiment 6.

BASIC

The sample programs on the disks accompanying this book are written in interpreted BASIC (Applesoft, basica, gwbasic) because this language is the easiest for students and faculty to learn and to use, and with its easy access to assembly language it offers an excellent combination of simplicity, flexibility, and performance. BASIC is the closest we have to a universal language for computers, and experimenters would be well advised to familiarize themselves with it. The programs provided with Experiment 6 are illustrative and exemplary, and they are generously supplied with REM (remark) statements so that they may provide useful information for students who are writing programs.

Before starting to write your own programs, you should study program number 1, the A/D and D/A check. Operate it, introduce analog signals as input, and measure analog output signals. Then print out a listing. The program will be called MTN1 for the Mountain card, S-V1 for the Sunset-Vernier card, ML1 for the Multi-Lab card, or DACA1 for the IBM DACA card. (In BASIC, load the program and turn on the printer. Then for the PC type LLIST. For the Apple II, type PR # 1, LIST, and PR # 0.) This program includes a very complete set of REM statements to assist you in creating you own BASIC programs.

If you are using a Sunset-Vernier card, then you should also load and list program S-V0, which gives the peculiar initialization procedure in BASIC.

Such programs generally refer to offsets from a particular base address (BA). In the Apple II, each slot gets 16 numbered addresses, starting with 49280 for slot zero. Thus, slot 1 should start with 49296, so BA = 49296. (Your printer card should be in slot 1, and the interfacing card may be in any of slots 2 through 5.) On this disk, the HELLO program picks up the slot number and stores the corresponding slot offset in memory location 6.

In the IBM PC, for the ML-16 card, set the base address to hex 280 using the dip switches on the card, following the instruction manual for that card. For the DACA, set the base address to hex 2E2 similarly.

Using information gleaned from program 1, you should be able to write BASIC programs capable of accepting and sending out analog signals in times of a few tens of milliseconds. Thus, you will be able to interface your computer with most of the laboratory equipment you are likely to encounter at this level.

The IBM PC cards used here also contain counters and thus they can be utilized in BASIC programs for experiments like that of program 4, Nuclear Counting. For programming details, list program ML4 or DACA4 and study the REMs along with the manual for your card.

ASSEMBLY LANGUAGE

For various uses BASIC is simply not fast enough. A BASIC step typically takes 5 ms. A machine-language (or assembly-language) step typically takes only 5 μs, but it usually takes several of them to get as much done as one BASIC step. Consequently, one can usually increase speed by a factor of ten to a hundred by writing routines in assembly language. The strategy is to write the whole program in BASIC, and then substitute assembly language in subroutines where time is critical.

Assembly language is unjustly maligned as difficult; actually, it is as easy to learn as higher languages other than BASIC (such as Pascal and C), and its primary drawback is that it takes many steps to perform tasks that a higher language can do in one. We recommend purchase and study of books on assembly language.

The Apple II and the IBM PC use fundamentally different CPUs (central processing units): for the Apple, the 6502 and its descendants; for the IBM, the 8080 and its descendants. We will discuss their machine language characteristics and programs separately.

IBM

The system disks contain a program called Debug which is well suited to creation of simple routines. Such routines are normally interfaced to a BASIC program via an integer array. For example, program 5, Poisson Intervals, involves an assembly-language subroutine, which is listed below. The routine is encoded in pairs of hex bytes in a DATA statement starting at line 1160 of the BASIC program; you should examine it to see that the DATA statement does in fact correspond to the assembly-language program. The data are read into an integer array in lines 1200–1220. The initial element in the array is addressed with a CALL in line 1310 or 1320, and the array then constitutes the subroutine.

For the ML-16 card, the assembly-language routine is MLAL, listed in Table F.1. Although this card has good built-in counter–timers, BASIC is not fast enough to trigger them properly in the Poisson-Interval experiment. Counter (CTR) 1 is set in timer mode 0 (see REMs in the corresponding BASIC program); it receives timing information from the standard 14-MHz video signal, via CTR 2. This counter is set in flip-flop mode. To avoid missing part of an interval, the first loop, at byte 108, causes operation to pause until CTR 2 is off. The next loop waits until CTR 2 is on, that is, the next pulse from the detector. Then CTR 1 is loaded with hex FFFF and proceeds to count down. When CTR

TABLE F.1 MLAL: ASSEMBLY LANGUAGE ROUTINE, MULTI LAB CARD

0CD7:0100	BB8002	MOV	BX,0280	Base address
0CD7:0103	BA0300	MOV	DX,0003	Offset for CTR 2
0CD7:0106	01DA	ADD	DX,BX	
0CD7:0108	ED	IN	AX,DX	
0CD7:0109	A880	TEST	AL,80	CTR 2 off
0CD7:010B	74FB	JZ	0108	
0CD7:010D	ED	IN	AX,DX	
0CD7:010E	A880	TEST	AL,80	CTR 2 on
0CD7:0110	75FB	JNZ	010D	
0CD7:0112	BA0900	MOV	DX,0009	
0CD7:0115	01DA	ADD	DX,BX	
0CD7:0117	B0FF	MOV	AL,FF	
0CD7:0119	EE	OUT	DX,AL	Load CTR 1 with # FFFF
0CD7:011A	EE	OUT	DX,AL	
0CD7:011B	BA0300	MOV	DX,0003	
0CD7:011E	01DA	ADD	DX,BX	
0CD7:0120	ED	IN	AX,DX	
0CD7:0121	A880	TEST	AL,80	CTR 2 off
0CD7:0123	74FB	JZ	0120	
0CD7:0125	BA0B00	MOV	DX,000B	
0CD7:0128	01DA	ADD	DX,BX	
0CD7:012A	B040	MOV	AL,40	
0CD7:012C	EE	OUT	DX,AL	Latch CTR 1
0CD7:012D	CB	RETF		
0CD7:012E	90	NOP		
0CD7:012F	90	NOP		

TABLE F.2 DACAAL: ASSEMBLY LANGUAGE ROUTINE, IBM CARD

0CD7:0100	BBE202	MOV	BX,02E2	Base address
0CD7:0103	BA0020	MOV	DX,2000	Offset for CTR 2
0CD7:0106	01DA	ADD	DX,BX	
0CD7:0108	ED	IN	AX,DX	
0CD7:0109	A880	TEST	AL,80	CTR 2 off
0CD7:010B	74FB	JZ	0108	
0CD7:010D	ED	IN	AX,DX	
0CD7:010E	A880	TEST	AL,80	CTR 2 on
0CD7:0110	75FB	JNZ	010D	
0CD7:0112	BA0090	MOV	DX,9000	
0CD7:0115	01DA	ADD	DX,BX	
0CD7:0117	B0FF	MOV	AL,FF	
0CD7:0119	EE	OUT	DX,AL	Load CTR 1 with # FFFF
0CD7:011A	EE	OUT	DX,AL	
0CD7:011B	BA0020	MOV	DX,2000	
0CD7:011E	01DA	ADD	DX,BX	
0CD7:0120	ED	IN	AX,DX	
0CD7:0121	A880	TEST	AL,80	CTR 2 off
0CD7:0123	74FB	JZ	0120	
0CD7:0125	BA00B0	MOV	DX,B000	
0CD7:0128	01DA	ADD	DX,BX	
0CD7:012A	B040	MOV	AL,40	
0CD7:012C	EE	OUT	DX,AL	Latch CTR 1
0CD7:012D	CB	RETF		
0CD7:012E	90	NOP		
0CD7:012F	90	NOP		

2 turns back off, CTR 1 is latched (byte 12C of the listing). The elapsed time is then read in BASIC.

For the DACA card, the routine is DACAAL, listed in Table F.2. This program is similar to its ML-16 counterpart, except in base address and offsets. The card contains the same three counter–timers.

Apple II

You can enter the system monitor from BASIC with CALL-151. (To return to BASIC, ctrl-C.) You can then key in assembly-language routines directly in hexadecimal numbers, or via a mini-assembler. The routines we employ here begin at the usual hexadecimal 300, and they must end before 3D0 because of DOS requirements. To list, in the monitor type 300L. Since hex 300 is decimal 768, we invoke the routines from BASIC with CALL 768. How do you save your routine on disk? Suppose the routine you create is named MINE, for example, and is hex 37 bytes long. Then type BSAVE MINE, A$300, L$37. It will be saved on disk. To recall it, just type BLOAD MINE. (In a program, you have to put PRINT CHR$(13) CHR$(4) "BLOAD MINE".) To run it from BASIC, type CALL 768. You can stash a useful number like 17 in location 17 with POKE 7,17 in BASIC and then pick it up with assembly language with LDA 7. Similarly, if the assembly language places 17 (hex 11) in location 7 with STA 7, then that value can be picked up in BASIC with X=PEEK(7). (Safe locations: 6-9, EB-EF, FA-FF.)

For the Mountain card, the counting routine is called MCOUNT, listed in Table F.3. Since this card has no built-in counter, A/D channel 1 is used instead. It is read at byte 30C. If it is less than hex C0 ("off"), there is no count and the program branches directly to the

TABLE F.3 MCOUNT: ASSEMBLY LANGUAGE ROUTINE, MOUNTAIN CARD

0300-	A9	00		LDA	#$00	
0302-	85	EC		STA	$EC	Count, lo
0304-	85	ED		STA	$ED	Count, hi
0306-	85	EE		STA	$EE	Time, lo
0308-	85	EF		STA	$EF	Time, hi
030A-	A6	06		LDX	$06	Slot * 16
030C-	BD	81	C0	LDA	$C081,X	Read A/D channel 1
030F-	C9	C0		CMP	#$C0	
0311-	90	1B		BCC	$032E	Time while counter is off
0313-	AD	30	C0	LDA	$C030	Click
0316-	E6	EC		INC	$EC	Counter is on
0318-	D0	02		BNE	$031C	
031A-	E6	ED		INC]	$ED	
031C-	A9	03		LDA	#$03	
031E-	20	A8	FC	JSR	$FCA8	Wait
0321-	BD	81	C0	LDA	$C081,X	
0324-	20	3D	03	JSR	$033D	
0327-	C9	C0		CMP	#$C0	
0329-	B0	F6		BCS	$0321	Wait till counter is off
032B-	AD	30	C0	LDA	$C030	Click
032E-	EA			NOP		
032F-	20	3D	03	JSR	$033D	
0332-	20	3D	03	JSR	$033D	
0335-	E6	EE		INC	$EE	
0337-	D0	D3		BNE	$030C	
0339-	E6	EF		INC	$EF	
033B-	D0	CF		BNE	$030C	
033D-	60			RTS		

timer at 32E. The timing steps are such that the entire data-gathering process takes about 3 s.

If A/D channel 1 exceeds hex C0 ("on"), the counter is incremented and a click is produced. (A click requires that location C030 be addressed twice with an intervening delay.) To avoid counting one pulse twice, operation does not proceed until A/D channel 1 goes back below hex C0.

Counts are accumulated in the safe registers EC and ED, with each count in ED representing 256 counts in EC. Similarly, EE and EF are used for the time.

Also for the Mountain card, the timing routine is MTIME, listed in Table F.4. Since this card has no built-in timer, an assembly language loop is used instead. Data for a histogram are accumulated in an array at hex 9000; the array is first cleared with the routine at byte 300. Timing begins when A/D channel 1 goes below hex C0, and ends when it goes above. Timing steps are such that the resulting time is read directly in milliseconds.

For the Sunset-Vernier card, corresponding routines are STIME and SCOUNT, listed in Tables F.5 and F.6, respectively. These programs are similar to their Mountain card counterparts; but digital input User #8 is used in place of A/D channel 1.

TABLE F.4 MTIME: ASSEMBLY LANGUAGE ROUTINE, MOUNTAIN CARD

0300-	A2	7F		LDX	#$7F	
0302-	A9	00		LDA	#$00	
0304-	9D	00	90	STA	$9000,X	Clear array
0307-	CA			DEX		
0308-	10	FA		BPL	$0304	
030A-	60			RTS		
030B-	00			BRK		
030C-	A9	00		LDA	#$00	
030E-	85	EE		STA	$EE	Time, lo
0310-	85	EF		STA	$EF	Time hi
0312-	A4	06		LDY	$06	Slot * 16
0314-	B9	81	C0	LDA	$C081,Y	Read A/D channel 1
0317-	20	55	03	JSR	$0355	
031A-	C9	C0		CMP	#$C0	
031C-	B0	F6		BCS	$0314	Wait till counter is off
031E-	E6	EE		INC	$EE	
0320-	D0	02		BNE	$0324	
0322-	E6	EF		INC]	$EF	
0324-	EA			NOP		
0325-	20	55	03	JSR	$0355	
0328-	B9	81	C0	LDA	$C081,Y	
032B-	C9	C0		CMP	#$C0	
032D-	90	EF		BCC	$031E	Time until counter turns on
032F-	AD	30	C0	LDA	$C030	Click counter is on
0332-	A5	EF		LDA	$EF	
0334-	06	EE		ASL	$EE	
0336-	2A			ROL		
0337-	06	EE		ASL	$EE	
0339-	2A			ROL		
033A-	06	EE		ASL	$EE	
033C-	2A			ROL		
033D-	10	02		BPL	$0341	
033F-	A9	7F		LDA]	#$7F	
0341-	AA			TAX		X is time in milliseconds
0342-	FE	00	90	INC	$9000,X	
0345-	A9	03		LDA	#$03	
0347-	20	A8	FC	JSR	$FCA8	Wait
034A-	Ad	30	C0	LDA	$C030	Click
034D-	C6	EC		DEC	$EC	Count down
034F-	D0	BB		BNE	$030C	
0351-	C6	ED		DEC	$ED	
0353-	10	B7		BPL	$030C	
0355-	60			RTS		

TABLE F.5 STIME: ASSEMBLY LANGUAGE ROUTINE, SUNSET-VERNIER CARD

```
0300-   A2  7F          LDX    #$7F
0302-   A9  00          LDA    #$00
0304-   9D  00  90   ┌  STA    $9000,X      Clear array
0307-   CA           │  DEX
0308-   10  FA       └  BPL    $0304
030A-   60              RTS

030B-   00              BRK
030C-   A9  00       ┌  LDA    #$00
030E-   85  EE       │  STA    $EE          Time, lo
0310-   85  EF       │  STA    $EF          Time hi
0312-   A4  06       │  LDY    $06          Slot * 16
0314-   B9  86  C0   │┌ LDA    $C086,Y      Read User #8
0317-   30  FB       │└ BMI    $0314        Wait till counter is off
0319-   E6  EE       │┌ INC    $EE
031B-   D0  02       │  BNE    $031F
031D-   E6  EF       │  INC]   $EF
031F-   EA           │  NOP
0320-   EA           │  NOP
0321-   20  4F  03   │  JSR    $034F
0324-   B9  86  C0   │  LDA    $C086,Y      Time while counter is off
0327-   10  F0       │└ BPL    $0319
0329-   AD  30  C0   │  LDA    $C030        click; counter is on
032C-   A5  EF       │  LDA    $EF
032E-   06  EE       │  ASL    $EE
0330-   2A           │  ROL
0331-   06  EE       │  ASL    $EE
0333-   2A           │  ROL
0334-   06  EE       │  ASL    $EE
0336-   2A           │  ROL
0337-   10  02       │  BPL    $033B
0339-   A9  7F       │  LDA]   #$7F
033B-   AA           │  TAX                 X is time in milliseconds
033C-   FE  00  90   │  INC    $9000,X
033F-   A9  03       │  LDA    #$03
0341-   20  A8  FC   │  JSR    $FCA8        Wait
0344-   AD  30  C0   │  LDA    $C030        Click
0347-   C6  EC       │  DEC    $EC          Count down
0349-   D0  C1       │  BNE    $030C
034B-   C6  ED       │  DEC    $ED
034D-   10  BD       └  BPL    $030C
034F-   60              RTS
```

TABLE F.6 SCOUNT: ASSEMBLY LANGUAGE ROUTINE, SUNSET-VERNIER CARD

0300-	A9	00		LDA	#$00	
0302-	85	EC		STA	$EC	Count, lo
0304-	85	ED		STA	$ED	Count, hi
0306-	85	EE		STA	$EE	Time, lo
0308-	85	EF		STA	$EF	Time, hi
030A-	A6	06		LDX	$06	Slot * 16
030C-	BD	86	C0	LDA	$C086,X	Read User #8
030F-	10	16		BPL	$0327	Time while counter is off
0311-	AD	30	C0	LDA	$C030	Click
0314-	E6	EC		INC	$EC	Counter is on
0316-	D0	02		BNE	$031A	
0318-	E6	ED		INC]	$ED	
031A-	A9	03		LDA	#$03	
031C-	20	AB	FC	JSR	$FCA8	Wait
031F-	BD	86	C0	LDA	$C086,X	
0322-	30	FB		BMI	$031F	Wait till counter is off
0324-	AD	30	C0	[LDA	$C030	Click
0327-	EA			NOP		
0328-	EA			NOP		
0320-	20	37	03	JSR	$0337	
032C-	20	37	03	JSR	$0337	
032F-	E6	EE		INC	$EE	
0331-	D0	D9		BNE	$030C	
0333-	E6	EF		INC	$EF	
0335-	D0	D5		BNE	$030C	
0337-	60			RTS		

APPENDIX G

A SOFTWARE TUTORIAL FOR THE IBM PC

REFERENCES

See references 1 and 2 of Appendix F.

INTRODUCTION

This appendix is a tutorial, intended for students who have a little experience with microcomputers other than IBM PC type. Those students who have never operated any kind of computer may need substantial assistance over and above what is offered here.

The objective in this appendix is to learn several simple operations on the IBM PC or its clones. In particular, you will

1. Turn the machine on and prepare it for operation.
2. Learn some disk operations: format, copy, delete.
3. Read and write simple programs.
4. Observe how I/O ports are employed in the case of a particular interfacing (A/D and D/A) card.

BOOT

Place the DOS Master disk in drive a, and turn on the machine. Drive a is probably up or to the left. If the machine is already on, press ctrl-alt-del (all at once). Wait. If it asks for the date, just press the enter key marked ↵. Ultimately the system prompt comes on: A⟩ or such.

Place your disk in drive b. If it is not yet formatted, type format b:.

Press the enter key after each command. Capital versus lower case doesn't matter, but spaces often do.

To see what's on your disk, type dir b: for the directory for disk drive b. Then type dir.

413

BASIC

To get into BASIC, type basica for the IBM. For others, see particular instructions (for example, cd gwbasic followed by gwbasic). The prompt is ok. To return to the system, type system. Then go to BASIC again.

(To get rid of the bottom junk, type key off.)

Type print 5 + 7; the result will be 12. This is immediate execution of an instruction. Now type print "5 + 7"; what's inside of the quotes is just mindlessly reproduced as a string of characters.

Type new, to clear any programs that may be present. To clear the screen, press ctrl-Home both at once; or type the letters cls. Key in the following program. Note that a line that begins with a number becomes part of the resident program. This is deferred execution of instructions.

```
10   x = 10
20   y = 20
30   z = x + y
40   print z
```

Type list, and check the program. If there is an error, you can just retype the whole line, wherever your cursor may be; because of its line number, it will find its way into the program at the right place. To delete a line, just type the line number. Or you can set Num Lock and use the arrows to bring the cursor to the point of error, and then correct it using Ins, Del, etc. Continue until the list is correct. Then,

Type run; the result is 30.

Type new, and then list. The program is gone.

Now key in the program in a slightly different form:

```
5    x = 10:y = 20
10   z = x + y:REM add them
11   print z
```

Type List, and then Run. Note that the REM is an annotation and is not acted upon. Then type renum. List again; the line numbers are renumbered from ten, by tens.

(For a long list, to pause, press ctrl-numlock, of all things. Press anything else to continue.)

DISK

In BASIC: To save the above program, type save "b:sum". Then type: files "b:" to see that it's there on the disk.

Now, to remove the program from memory, type new, and then list. To get the program back from the disk, type load "b:sum". List it, and then run it.

To get rid of the program file on the disk, type kill "b:sum.bas". Then type files "b:" again to make sure it's gone.

Thus we have come full circle: we have created a program, run it, saved it, restored it, and deleted it.

In BASIC, put your disk in drive a, and type files. Observe that the b:'s above can be omitted when your disk is in drive a. Type system, and then type basica. This time it doesn't work, because basica is not a file on your disk. (If you boot from hard disk c:, then a:'s and b:'s may both be necessary.)

In the system: To copy a whole disk, put the DOS master in drive a and type diskcopy a: b:; then follow instructions. To copy the BASIC file "sum" from drive a to drive b, type copy a:sum.bas b:. To delete a file, type del sum.bas.

A file such as sum.com or sum.exe can be run by typing sum. A text file such as readme or read.me can be read by typing type readme, or type read.me; to pause, press ctrl-s. But all this doesn't work for your BASIC file sum.

To read a directory called DEMO, type cd demo. To get back, type cd\.

PRINTER

In BASIC, key in the above program, just to have something to work with. Now, to print out the program list, first be sure the printer is on, and then type llist, with two l's.

To print something like "Hello", type lprint "Hello".

To get the printer to print text on the printer while the program is running, press Ctrl-PrtSc (both together), before starting. Then type list. Type run. To stop printing everything, press Ctrl-PrtSc again.

To print a graphics screen, go back to the system and type graphics. Then type basica or whatever. After this, to print the graphics screen, press: shift-PrtSc.

To format your output, use print using "###.##";a. For example, key in the following and run it:

```
new
10   a = 3.14159
20   print a tab(10) using "##.##";a
```

SOUND

To beep, type beep.

To sound 500 Hz for 40/18.2 s, type sound 500,40. The lowest allowed frequency is 37 Hz.

To play notes, type play "cdef".

To click, type sound 37,1:sound 37,0. (The 0 argument turns off the sound immediately.)

USEFUL KEYS

Here are a few keys that make typing and editing programs easier. Try them.

```
f1    LIST
alt-1  LOCATE
alt-p  PRINT
alt-s  SCREEN
alt-w  WIDTH
home   cursor to top
end    cursor to end of line
esc    clear current line
```

INPUT AND LOOP

Key in the following program and run it. It takes the values of x and y that you input, adds them, and displays the result. Also, it loops: It goes on doing its thing until you press Esc, which is ASCII character Chr$(27).

```
new
10    Input "Input x,y:",x,y
20    Print x + y
30    If Input$(1)<>Chr$(27) then 10
```

Input$(1) waits for one keystroke. If it's not Esc, the program goes back to line 10. The symbol $ indicates that what is involved is a *string* of one or more characters, like letters, which do not necessarily have a particular quantitative or numerical significance in themselves. The variable x is a number, but x$ would be a string and x(17) is an array element (see below); they are all different variables.

Now change line 30 to Goto 10. Run the thing. There is no way out except to press ctrl-C or ctrl-break. You should always provide an escape for your users. Input a$ won't pick up the Esc key, but input$(1) will. So will inkey$, but it doesn't wait for you to press a key.

Inkey$ has three important uses:

1. To pick up a particular keystroke "on the fly". For example, the array program, below, continues to run unless the Esc key is pressed; but it does not stop and wait for a key to be pressed.

2. To prevent old keystrokes from getting picked up in the input of a program; see line 120 of the graphics program, below, which we will display here:

```
120 If inkey$ > ""then 120
```

where "" means the null string, that is, no key has been pressed. Each time inkey$ is invoked, it removes one character from the input buffer, until finally nothing is left, which is the null character.

3. To pick up extended or secondary ASCII codes; in particular, the arrow keys generate a string of two bytes of which the first is zero. To select the right-hand letter in the two-character string, use right$(a$,1). This is demonstrated in the A/D and D/A program in the I/O section, below.

GRAPHICS

In BASIC, type width 40. For graphics computers, the letters get twice as wide; for others, they are confined to the left half of the screen. Then type width 80.

To invoke graphics, type screen 1. The screen number is 0 for text, 1 for low-resolution graphics, and 2 for high-resolution graphics. In lo-res the screen is x, y = 0, 0 at upper left, and 199, 319 at lower right. In lo-res the text width is 40. To go back to text type screen 0. Then again type: screen 1.

(If you had wanted color, you could have typed screen 1,0. The second number is 0 for color and 1 for no color. To set colors, type color 1,1. The first number is background; 0 is black, 15 is white, and 1–14 are colors. The second number is foreground palette:

Palette 0		Palette 1	
0	background	0	background
1	green	1	cyan
2	red	2	magenta
3	brown	3	white.)

To plot a pixel at 200,100, type pset(200, 100). A tiny white point should duly appear at the middle right of the screen. (If there's no graphics, you get an error message at this point. In color, try pset(200,100),1 for cyan.) To erase it, type preset(200,100).

For a line, type line(20,50) − (80,100). For another, starting at the end of the first, type line-(100,80).

For a box, type line(50,100) − (90,160),,b. To fill the box, use bf instead of b. (The omitted variable is color.)

For a circle at 80,80, of radius 40 type circle(80,80),40. For aspect ratio .5, circle(80,80),40,,,,.5. (The omitted variables are color, and startangle and endangle for circular arc.)

To place the cursor at y,x type locate 20,5. For text the screen is 25 × 40. This program plots a sine wave. To see the effect of line 120, delete it by typing 120, and list to make sure it's gone. Then press a key while the sine is being plotted. Try it without STEP 4.

```
10    SCREEN 1
20    LOCATE 5,15:PRINT "y = sin(x)"
30    LOCATE 7,2:PRINT "1"
40    LOCATE 19,1:PRINT "−1"
50    LINE(16,100) − (319,100)
60    LINE(16,0) − (16,199)
70    PSET(16,100)
80    FOR X = 16 TO 319 STEP 4
90    Y = 50*SIN((X − 16)/20)
100   LINE − (X,100 − Y)
110   NEXT
120   IF INKEY$ > "" THEN 120
130   A$ = INPUT$(1)
140   SCREEN 0:WIDTH 80
```

MATH

Functions available: sqr(x), sin(x), cos(x), tan(x), atn(x), log(x), exp(x), int(x), fix(x) (which actually means truncate), abs(x), sgn(x).

For example, type print sqr(2).

Arithmetic commands: +, −, *, /. Exponentiation is also available; for example, type print 3^2.

Division can be confusing. Type print 8/2*2; the result is 8, because the second 2 is regarded as being in the numerator. Then type print 8/(2*2), which yields 2. Type print 8/2/2, which is also 2.

Scientific notation: 3.567E6.

RND produces a random number, 0 to 1. Type several times: print rnd.

If you don't seed it with different numbers, it always produces the same series of random numbers. To seed it, you could use a number, for example, randomize 129, at the beginning of the program. But instead, for a random seed, use val(right$(time$,2)). Time$ is the time function; to see it, type print time$.

Then type print right$(time$,2).

Since time is constantly changing, it will serve as a seed for random numbers. Right$ means string characters on the right-hand side of the time$ string; 2 are requested here. If we want simply to display them, we can print them as a string; however, when we want a numerical value, the val function evaluates them as numbers, as far as possible. For example, type

```
print  val("-3.1F7")
print  val("-3.1E7")
```

Note that E7 can be evaluated but F7 cannot.

ARRAYS

Type a(3) = 17. Then type print a(3). It's 17. Type print a(2); it's still zero, because nothing has been stored there. Arrays provide a means of organizing large numbers of data and variables. (The name could be b(3) or whatever.)

The following program handles the alphabet by a process that is instructive if not efficient. We begin by dimensioning a string array so that the computer will set aside space for it.

```
10    data a,b,c,d,e,f,g,h,i,j,k,l,m,n,o,p,q,r,s,t,u,v,w,x,y,z
20    clear
30    dim a$(26)                    Dimension the array.
40    restore
50    for i = 1 to 26
60    read a$(i)                    Put the alphabet in it.
70    next
80    for i = 1 to 26 step 2
90    print a$(i);                  Print every other letter.
100   next
110   print
120   if inkey$<>chr$(27) then 10   Loop.
```

Try the program. Then, to see what the commands do, try making some changes:

The semicolon in line 90 keeps the computer from doing a line-feed after every letter; try removing it, and run, and then replace it. Then try line 110.

Normally one would not need to use the clear and restore commands here, but we want to show what they do. Change the 10 in line 120 to 20; it still works. Change it to 30; now you commit the error of redimensioning an array without first clearing all variables.

Change the 10 (in line 120) to 50. You run out of data, because the program simply reads data in order as they appear in the program. But if you change it to 40, it's okay, because the restore command gets the program to start reading data from the beginning again.

Change the 10 to 80 and it will run faster.

(There are also multidimensional arrays a(x,y). For very large arrays of integers, memory space is saved by putting the data in integer arrays a%(x) if appropriate.)

I/O PORTS

The following sample programs are for use with an ML-16 general-purpose I/O (input/output) board. This board is particularly simple and flexible, and it is well suited to educational purposes. Even if you are using another board, the material presented here will assist you in developing an understanding of the methods and BASIC programs that are typically employed in this sort of interfacing.

The board has the following capabilities:

A/D (analog-to-digital conversion)
D/A (digital-to-analog conversion)
Counters (count-down, for internal or external sources)
DI and DO (digital input and output)

We will exercise all four of these options in the programs that follow, in the sections on A/D and D/A and Counter and Timer. Each step is explained briefly in the program. Some relevant information is presented below; for more, consult the ML-16 manual.

The base address (ba) is hexadecimal 280 (&H280), which is decimal 640. This is already set by switches on the card. Offsets are shown below. For example, for offset 3, the address is ba + 3 = 283 hex.

Offset	Read (inp)	Write (out)
0	A/D data	Start A/D conversion
1		(Reset interrupt)
2		A/D command byte
3	Digital input	Digital output
4		D/A data, ch.0
5		D/A data, ch.1
6 = 4		
7 = 5		
8	Counter 0	Load counter 0
9	Counter 1	Load counter 1
10	Counter 2	Load counter 2
11		Counter control byte

Information is written to the relevant address using the out command, which is analogous to the BASIC POKE. Information is read from the address using the inp command, which is analogous to the BASIC PEEK. Such numbers are limited to the range 0-255. (That represents one byte or character.)

Counters

Using internal jumpers, counter 0 has been connected in two ways:

1. The IBM PC video oscillator frequency of 14.32 MHz is divided by 16, and the resulting 0.895 MHz (or 1.12 μs) pulses go to the counter 0 clock.

2. The counter 0 output goes to the counter 1 clock; they are cascaded.

Counter 2 is not connected internally.

A/D and D/A

Key in this program:

```
10   ba = &H280
20   out ba + 2,&H30
```

Base address is hex 280.
Hex 30 is sent to address ba + 2, the A/D command byte. Hex 30 is binary 00110000. Displaying these bits vertically:
0   convert on "out" (to address ba + 0);
0   output format is offset binary;

	1 bipolar (positive and negative);
	1 high range (-5 to 5 V);
	0 ⎫
	0 ⎪
	0 ⎬ channel number 0 (of 16 possible).
	0 ⎪
	0 ⎭

```
30   out ba + 0,0
```
Start the A/D conversion, which takes 10 μs. It doesn't matter what you write to ba + 0; this step could be out ba,17 or out &H280,0.

```
40   locate 12,20:print "AD" inp(ba + 0)
```
Inp gets the result of the A/D conversion.

```
50   a$ = inkey$
```
Inkey$ can pick up the two-byte ASCII codes used for the arrows.

```
60   If right$(a$,1) = "K" and d > 0 then d = d − 1
```
Left arrow.

```
70   If right$(a$,1) = "M" and d < 255 then d = d + 1
```
Right arrow.
(The K and M must be capitalized.)

```
80   locate 15,20:print "DA" d
```
Print it.

```
90   out ba + 4,d
```
Send it to D/A channel 0.

```
100  If a$<>chr$(27) then 30
```
Loop.

List it, check it, then ctrl-Home and run. Touch a 1.5-V battery to connector pins 1 and 17 and watch AD. Attach a voltmeter to pins 19 and 21 and press the left and right arrows.

Counter and Timer

Key in the following program and run it. Clock 0 is already connected internally to the computer oscillator.

```
10   ba = &H280
```

```
20   out ba + 11,&H30
```
Offset 11 is the counter control byte. Hex 30 is binary 00110000. Displaying the bits vertically:
0
0   counter number 0 (of 3 counters);
1
1   read two bytes, in order low-high;
0 ⎫
0 ⎬ mode 0 (of 6): count;
0 ⎭
0   hexadecimal output.

```
30   out ba + 8,&Hff  ⎫
40   out ba + 8,&Hff  ⎭
```
Load counter 0 in order low-high, to a total of 65535.

```
50   for i = 1 to 3:next
```
Delay.

```
60   out ba + 11,0
```
Latch counter 0.

```
70   x1 = inp(ba + 8)  ⎫
80   xh = inp(ba + 8)  ⎭
```
Read counter 0.

```
90   print 256*xh + x1
```
Print it.

```
100  If input$(1)<>chr$(27) then 10
```
Loop.

At steps 30 and 40 the counter is loaded with hex ffff, which is decimal 65535. In hexadecimal numbers, A is decimal 10, b is 11, f is 15, and 10 is decimal 16. A *byte* is a number from 0 to 255, or hex 0 to ff. You can only output or input a byte at a time: first low byte, then high byte.

The counter counts down from 65535 during the delay in line 50. Each for-next loop takes around a millisecond, so line 50 introduces a delay of 3 ms. During that time the counter can count down by 3000, because the clock is about a MHz; this brings it to around 62000. Try various values instead of 3 in line 50. When you get up to 60 or 70, counter 0 will pass through zero and wrap-around to 65535.

Generally speaking, the counter is cycling all the time. Loading it resets it, and latching it saves a well-defined number to print out while the counter charges on.

Key in the following program. Connect the output of counter 1 to the digital input 7 (DI 7), which controls the msb (most significant bit) of the digital input byte at ba + 3.

10	ba = &H280	
20	out ba + 11,&H34	Counter 0, mode 2: divide by N.
30	out ba + 8,0 ⎤	
40	out ba + 8,5 ⎦	Load counter 0 in order low–high.
50	out ba + 11,&H70	Counter 1, mode 0: output starts low, goes high at the end.
60	out ba + 9,0 ⎤	
70	out ba + 9,8 ⎦	Load counter 1 in order low–high.
80	out ba + 11,&Hb8	Counter 2, mode 4: count. Mode 0 would also be okay here.
90	out ba + 10,&Hff ⎤	
100	out ba + 10,&Hff ⎦	Load counter 2 to 65535.
110	wait ba + 3,&H80	Hex 80 is 10000000; its msb is 1. Wait until the msb's match. At the end, the counter 1 output drives DI 7 high, viz. to 1.
120	out ba + 11,&H80	Latch counter 2.
130	xl = inp(ba + 10) ⎤	
140	xh = inp(ba + 10) ⎦	Read counter 2.
150	print 65535-256*xh-xl	Print the change.
160	sound 37,1:sound 37,0	Click.
170	If inkey$<>chr$(27) then 10	Loop.

Send 5-V pulses into the clk (clock) of counter 2. (Be careful with the input: five volts peak-to-peak; check it with a scope.) Run. The program will count pulses for about 3 s, display the count, and then repeat. Change the pulse frequency and observe the effect.

THE FUNDAMENTAL PHYSICAL CONSTANTS

The 1986 recommended values of the fundamental physical constants

Quantity	Symbol	Value	Units	Relative uncertainty ppm
Universal constants				
Speed of light in vacuum	c	299 792 458	m sec^{-1}	exact
Permeability of vacuum	μ_0	$4\pi \times 10^{-7}$	N A^{-2}	
		$= 12.566\,370\,614\ldots$	10^{-7} N A^{-2}	exact
Permittivity of vacuum	ϵ_0	$1/\mu_0 c^2$		
		$= 8.854\,187\,817\ldots$	10^{-12} F m^{-1}	exact
Newtonian constant of gravitation	G	6.672 59(85)	10^{-11} m^3 kg^{-1} sec^{-2}	128
Planck constant	h	6.626 075 5(40)	10^{-34} J sec	0.60
in electron volts, $h/|e|$		4.135 669 2(12)	10^{-15} eV sec	0.30
$h/2\pi$	$\hbar$	1.054 572 66(63)	10^{-34} J sec	0.60
in electron volts, $\hbar/|e|$		6.582 122 0(20)	10^{-16} eV sec	0.30
Planck mass, $(\hbar c/G)^{1/2}$	m_P	2.176 71(14)	10^{-8} kg	64
Planck length, $\hbar/m_P c = (\hbar G/c^3)^{1/2}$	l_P	1.616 05(10)	10^{-35} m	64
Planck time $l_P/c = (\hbar G/c^5)^{1/2}$	t_P	5.390 56(34)	10^{-44} sec	64
Electromagnetic constants				
Elementary charge	e	1.602 177 33(49)	10^{-19} C	0.30
	e/h	2.417 988 36(72)	10^{14} A J^{-1}	0.30
Magnetic flux quantum, $h/2e$	Φ_0	2.067 834 61(61)	10^{-15} Wb	0.30
Josephson frequency–voltage quotient	$2e/h$	4.835 976 7(14)	10^{14} Hz V^{-1}	0.30
Quantized Hall conductance	e^2/h	3.874 046 14(17)	10^{-5} S	0.045
Quantized Hall resistance, $h/e^2 = \mu_0 c/2\alpha$	R_H	25 812.805 6(12)	Ω	0.045
Bohr magneton, $e\hbar/2m_e$	μ_B	9.274 015 4(31)	10^{-24} J T^{-1}	0.34
in electron volts, $\mu_B/|e|$		5.788 382 63(52)	10^{-5} eV T^{-1}	0.089
in hertz, μ_B/h		1.399 624 18(42)	10^{10} Hz T^{-1}	0.30
in wavenumbers, μ_B/hc		46.686 437(14)	m^{-1} T^{-1}	0.30
in kelvins, μ_B/k		0.671 709 9(57)	K T^{-1}	8.5
Nuclear magneton, $e\hbar/2m_p$	μ_N	5.050 786 6(17)	10^{-27} J T^{-1}	0.34
in electron volts, $\mu_N/|e|$		3.152 451 66(28)	10^{-8} eV T^{-1}	0.089
in hertz, μ_N/h		7.622 591 4(23)	MHz T^{-1}	0.30
in wavenumbers, μ_N/hc		2.542 622 81(77)	10^{-2} m^{-1} T^{-1}	0.30
in kelvins, μ_N/k		3.658 246(31)	10^{-4} K T^{-1}	8.5
Atomic constants				
Fine-structure constant, $\mu_0 c e^2/2h$	α	7.297 353 08(33)	10^{-3}	0.045
inverse fine-structure constant	α^{-1}	137.035 989 5(61)		0.045
Rydberg constant, $m_e c \alpha^2/2h$	R_∞	10 973 731.534(13)	m^{-1}	0.0012
in hertz, $R_\infty c$		3.289 841 949 9(39)	10^{15} Hz	0.0012
in joules, $R_\infty hc$		2.179 874 1(13)	10^{-18} J	0.60
in eV, $R_\infty hc/|e|$		13.605 698 1(40)	eV	0.30

The 1986 recommended values of the fundamental physical constants

Quantity	Symbol	Value	Units	Relative uncertainty ppm
Bohr radius, $\alpha/4\pi R_\infty$	a_0	0.529 177 249(24)	10^{-10} m	0.045
Hartree energy, $e^2/4\pi\epsilon_0 a_0 = 2R_\infty hc$	E_h	4.359 748 2(26)	10^{-18} J	0.60
in eV, $E_h/\{e\}$		27.211 396 1(81)	eV	0.30
Quantum of circulation	$h/2m_e$	3.636 948 07(33)	10^{-4} m^2 sec^{-1}	0.089
	h/m_e	7.273 896 14(65)	10^{-4} m^2 sec^{-1}	0.089
Electron				
Mass	m_e	9.109 389 7(54)	10^{-31} kg	0.59
		5.485 799 03(13)	10^{-4} u	0.023
in electron volts, $m_e c^2/\{e\}$		0.510 999 06(15)	MeV	0.30
Electron–muon mass ratio	m_e/m_μ	4.836 332 18(71)	10^{-3}	0.15
Electron–proton mass ratio	m_e/m_p	5.446 170 13(11)	10^{-4}	0.020
Electron–deuteron mass ratio	m_e/m_d	2.724 437 07(6)	10^{-4}	0.020
Electron–α-particle mass ratio	m_e/m_α	1.370 933 54(3)	10^{-4}	0.021
Specific charge	$-e/m_e$	$-$1.758 819 62(53)	10^{11} C kg^{-1}	0.30
Molar mass	$M(e)$	5.485 799 03(13)	10^{-7} kg mol^{-1}	0.023
Compton wavelength, $h/m_e c$	λ_C	2.426 310 58(22)	10^{-12} m	0.089
$\lambda_C/2\pi = \alpha a_0 = \alpha^2/4\pi R_\infty$	λ_C	3.861 593 23(35)	10^{-13} m	0.089
Classical radius, $\alpha^2 a_0$	r_e	2.817 940 92(38)	10^{-15} m	0.13
Thomson cross section, $(8\pi/3)r_e^2$	σ_e	0.665 246 16(18)	10^{-28} m^2	0.27
Magnetic moment	μ_e	928.477 01(31)	10^{-26} J T^{-1}	0.34
in Bohr magnetons	μ_e/μ_B	1.001 159 652 193(10)		1×10^{-5}
in nuclear magnetons	μ_e/μ_N	1838.282 000(37)		0.020
Magnetic moment anomaly, $\mu_e/\mu_B - 1$	a_e	1.159 652 193(10)	10^{-3}	0.0086
g-factor, $2(1 + a_e)$	g_e	2.002 319 304 386(20)		1×10^{-5}
Electron–muon magnetic moment ratio	μ_e/μ_μ	206.766 967(30)		0.15
Electron–proton magnetic moment ratio	μ_e/μ_p	658.210 688 1(66)		0.010
Muon				
Mass	m_μ	1.883 532 7(11)	10^{-28} kg	0.61
		0.113 428 913(17)	u	0.15
in electron volts, $m_\mu c^2/\{e\}$		105.658 389(34)	MeV	0.32
Muon–electron mass ratio	m_μ/m_e	206.768 262(30)		0.15
Molar mass	$M(\mu)$	1.134 289 13(17)	10^{-4} kg mol^{-1}	0.15
Magnetic moment	μ_μ	4.490 451 4(15)	10^{-26} J T^{-1}	0.33
in Bohr magnetons	μ_μ/μ_B	4.841 970 97(71)	10^{-3}	0.15
in nuclear magnetons	μ_μ/μ_N	8.890 598 1(13)		0.15
Magnetic moment anomaly, $[\mu_\mu/(e\hbar/2m_\mu)] - 1$	a_μ	1.165 923 0(84)	10^{-3}	7.2
g-factor, $2(1 + a_\mu)$	g_μ	2.002 331 846(17)		0.0084
Muon–proton magnetic moment ratio	μ_μ/μ_p	3.183 345 47(47)		0.15
Proton				
Mass	m_p	1.672 623 1(10)	10^{-27} kg	0.59
		1.007 276 470(12)	u	0.012
in electron volts, $m_p c^2/\{e\}$		938.272 31(28)	MeV	0.30
Proton–electron mass ratio	m_p/m_e	1836.152 701(37)		0.020
Proton–muon mass ratio	m_p/m_μ	8.880 244 4(13)		0.15
Specific charge	e/m_p	9.578 830 9(29)	10^7 C kg^{-1}	0.30
Molar mass	$M(p)$	1.007 276 470(12)	10^{-3} kg mol^{-1}	0.012
Compton wavelength, $h/m_p c$	$\lambda_{C,p}$	1.321 410 02(12)	10^{-15} m	0.089
	$\lambda_{C,p}/2\pi$	2.103 089 37(19)	10^{-16} m	0.089
Magnetic moment	μ_p	1.410 607 61(47)	10^{-26} J T^{-1}	0.34
in Bohr magnetons	μ_p/μ_B	1.521 032 202(15)	10^{-3}	0.010
in nuclear magnetons	μ_p/μ_N	2.792 847 386(63)		0.023
Diamagnetic shielding correction for protons (H_2O, spherical sample, 25 °C), $1 - \mu_p'/\mu_p$	σ_{H_2O}	25.689(15)	10^{-6}	—
Shielded proton moment (H_2O, spherical sample, 25 °C)	μ_p'	1.410 571 38(47)	10^{-26} J T^{-1}	0.34
in Bohr magnetons	μ_p'/μ_B	1.520 993 129(17)	10^{-3}	0.011
in nuclear magnetons	μ_p'/μ_N	2.792 775 642(64)		0.023
Gyromagnetic ratio	γ_p	26 752.212 8(81)	10^4 sec^{-1} T^{-1}	0.30
	$\gamma_p/2\pi$	42.577 469(13)	MHz T^{-1}	0.30
uncorrected (H_2O, spherical sample, 25 °C)	γ_p'	26 751.525 5(81)	10^4 sec^{-1} T^{-1}	0.30
	$\gamma_p'/2\pi$	42.576 375(13)	MHz T^{-1}	0.30
Neutron				
Mass	m_n	1.674 928 6(10)	10^{-27} kg	0.59
		1.008 664 904(14)	u	0.014
in electron volts, $m_n c^2/\{e\}$		939.565 63(28)	MeV	0.30
Neutron–electron mass ratio	m_n/m_e	1838.683 662(40)		0.022
Neutron–proton mass ratio	m_n/m_p	1.001 378 404(9)		0.009

The 1986 recommended values of the fundamental physical constants

Quantity	Symbol	Value	Units	Relative uncertainty ppm
Molar mass	$M(n)$	1.008 664 904(14)	10^{-3} kg mol^{-1}	0.014
Compton wavelength, $h/m_n c$	$\lambda_{C,n}$	1.319 591 10(12)	10^{-15} m	0.089
$\lambda_{C,n}/2\pi$	$\lambda_{C,n}$	2.100 194 45(19)	10^{-16} m	0.089
Magnetic moment[a]	μ_n	0.966 237 07(40)	10^{-26} J T^{-1}	0.41
in Bohr magnetons	μ_n/μ_B	1.041 875 63(25)	10^{-3}	0.24
in nuclear magnetons	μ_n/μ_N	1.913 042 75(45)		0.24
Neutron–electron magnetic moment ratio	μ_n/μ_e	1.040 668 82(25)	10^{-3}	0.24
Neutron–proton magnetic moment ratio	μ_n/μ_p	0.684 979 34(16)		0.24
Deuteron				
Mass	m_d	3.343 586 0(20)	10^{-27} kg	0.59
		2.013 553 214(24)	u	0.012
in electron volts, $m_d c^2/\{e\}$		1875.613 39(57)	MeV	0.30
Deuteron–electron mass ratio	m_d/m_e	3670.483 014(75)		0.020
Deuteron–proton mass ratio	m_d/m_p	1.999 007 496(6)		0.003
Molar mass	$M(d)$	2.013 553 214(24)	10^{-3} kg mol^{-1}	0.012
Magnetic moment[a]	μ_d	0.433 073 75(15)	10^{-26} J T^{-1}	0.34
in Bohr magnetons	μ_d/μ_B	0.466 975 447 9(91)	10^{-3}	0.019
in nuclear magnetons	μ_d/μ_N	0.857 438 230(24)		0.028
Deuteron–electron magnetic moment ratio	μ_d/μ_e	0.466 434 546 0(91)	10^{-3}	0.019
Deuteron–proton magnetic moment ratio	μ_d/μ_p	0.307 012 203 5(51)		0.017
Physicochemical constants				
Avogadro constant	N_A, L	6.022 136 7(36)	10^{23} mol^{-1}	0.59
Atomic mass constant, $m(C^{12})/12$	m_u	1.660 540 2(10)	10^{-27} kg	0.59
in electron volts, $m_u c^2/\{e\}$		931.494 32(28)	MeV	0.30
Faraday constant	F	96 485.309(29)	C mol^{-1}	0.30
Molar Planck constant	$N_A h$	3.990 313 23(36)	10^{-10} J sec mol^{-1}	0.089
	$N_A hc$	0.119 626 58(11)	J m mol^{-1}	0.089
Molar gas constant	R	8.314 510(70)	J mol^{-1} K^{-1}	8.4
Boltzmann constant, R/N_A	k	1.380 658(12)	10^{-23} J K^{-1}	8.5
in electron volts, $k/\{e\}$		8.617 385(73)	10^{-5} eV K^{-1}	8.4
in hertz, k/h		2.083 674(18)	10^{10} Hz K^{-1}	8.4
in wavenumbers, k/hc		69.503 87(59)	m^{-1} K^{-1}	8.4
Molar volume (ideal gas), RT/p (at 273.15 K, 101 325 Pa)	V_m	22 414.10(19)	cm^3 mol^{-1}	.8.4
Loschmidt constant, N_A/V_m	n_0	2.686 763(23)	10^{25} m^{-3}	8.5
Stefan–Boltzmann constant, $(\pi^2/60)k^4/\hbar^3 c^2$	σ	5.670 51(19)	10^{-8} W m^{-2} K^{-4}	34
First radiation constant, $2\pi hc^2$	c_1	3.741 774 9(22)	10^{-16} W m^2	0.60
Second radiation constant, hc/k	c_2	0.014 387 69(12)	m K	8.4
Wien displacement law constant, $\lambda_{max} T = c_2/4.965\ 114\ 23\ldots$	b	2.897 756(24)	10^{-3} m K	8.4
Conversion factors and units				
Electron volt, $(e/C)J = \{e\}J$	eV	1.602 177 33(49)	10^{-19} J	0.30
Atomic mass unit (unified), $m_u = m(C^{12})/12$	u	1.660 540 2(10)	10^{-27} kg	0.59
Standard atmosphere	atm	101 325	Pa	exact
Standard acceleration of gravity	g_n	9.806 65	m sec^{-2}	exact
'As–maintained' electrical units				
BIPM-maintained ohm, $\Omega_{69\text{-}BI}$ as of 1 Jan 1985	Ω_{BI85}	$1 - 1.563(50)\times10^{-6}$	Ω	
		$= 0.999\ 998\ 437(50)$	Ω	0.050
Drift rate of $\Omega_{69\text{-}BI}$	$d\Omega_{69\text{-}BI}/dt$	$-0.0566(15)$	$\mu\Omega$/yr	—
BIPM-maintained volt, 483 594 GHz$(h/2e)$	$V_{76\text{-}BI}$	$1 - 7.59(30)\times10^{-6}$	V	
		$= 0.999\ 992\ 41(30)$	V	0.30
BIPM maintained ampere, $A_{BIPM} = V_{76\text{-}BI}/\Omega_{69\text{-}BI}$	A_{BI85}	$1 - 6.03(30)\times10^{-6}$	A	
		$= 0.999\ 993\ 97(30)$	A	0.30
X-ray standards				
Cu x-unit: $\lambda(CuK\alpha_1) \equiv 1537.400$ xu	xu$(CuK\alpha_1)$	1.002 077 89(70)	10^{-13} m	0.70
Mo x-unit: $\lambda(MoK\alpha_1) \equiv 707.831$ xu	xu$(MoK\alpha_1)$	1.002 099 38(45)	10^{-13} m	0.45
Å*: $\lambda(WK\alpha_1) \equiv 0.209\ 100$ Å*	Å*	1.000 014 81(92)	10^{-10} m	0.92
Lattice spacing of Si (in vacuum, 22.5 °C)[b]	a	0.543 101 96(11)	nm	0.21
$d_{220} = a/\sqrt{8}$	d_{220}	0.192 015 540(40)	nm	0.21
Molar volume of Si, $M(Si)/\rho(Si) = N_A a^3/8$	$V_m(Si)$	12.058 817 9(89)	cm^3 mol^{-1}	0.74

Digits in parentheses indicate the standard deviation uncertainty in the last digits of the given value.
Quantities in braces, such as $\{e\}$, refer to the numerical value only.
[a] The scalar magnitude of the neutron moment is listed here. The neutron magnetic dipole is directed oppositely to that of the proton, and corresponds to the dipole associated with a spinning negative charge distribution. The vector sum $\mu_d = \mu_p + \mu_n$ is approximately satisfied.
[b] The lattice spacing of single-crystal Si can vary by parts in 10^7 depending on the preparation process. Measurements at the Physikalische-Technische Bundesanstalt in the Federal Republic of Germany indicate also the possibility of distortions from exact cubic symmetry of the order of 0.2 ppm.

Source: E. Richard Cohen and Barry N. Taylor, "The Fundamental Physical Constants," *Physics Today*, **42**, 8, Part 2, American Institute of Physics, New York, copyright © 1989. Reproduced with permission.

PERMISSIONS AND PHOTO CREDITS

COVER

The cover photograph of Dr. N. S. Nogar's Resonance Ionization Mass Spectroscopy Laboratory is used courtesy of the Los Alamos National Laboratory. Photographer: Henry Ortega.

The pulses of 537 nm light are produced by a dye laser pumped by a 308 mn XeCl excimer laser. The vacuum chamber is a diffusion-pumped time-of-flight spectrometer, with a flight tube 0.4 m long. Ions are generated by the laser light in the source region of the mass spectrometer and then are accelerated to the detector, a channel electron multiplier.

INTRODUCTION

Figures I.2 to I.15: Daryl W. Preston, *Experiments in Physics: A Laboratory Manual for Scientists and Engineers*, Wiley, New York 1985. Copyright © 1985, John Wiley & Sons, Inc. Reprinted by permission.

EXPERIMENT 2

Figures 2.3 and 2.5: Charles P. Poole, Jr., *Electron Spin Resonance: A Comprehensive Treatise of Experimental Techniques*, Wiley, New York, 1983. Copyright © 1983, John Wiley & Sons, Inc. Reprinted by permission.

Figure 2.7: Edward L. Ginzton, *Microwave Measurements*, McGraw-Hill, New York, 1957. Copyright © McGraw-Hill, 1957, Figure 1.17. Reproduced with permission.

Figure 2.15: Photographer: Terry M. Smith, Instructional Media Center, California State University, Hayward.

EXPERIMENT 3

Figure 3.10: Charles P. Poole, Jr., *Electron Spin Resonance: A Comprehensive Treatise of Experimental Techniques*, Wiley, New York, 1983. Copyright © 1983, John Wiley & Sons, Inc. Reprinted by permission.

EXPERIMENT 4

Figure 4.1: Reproduced with permission of Gaertner Scientific Corporation. Photographer: Terry M. Smith, Instructional Media Center, California State University, Hayward.

EXPERIMENT 6

Figure 6.3: Lewis C. Eggebrecht, *Interfacing to the IBM Personal Computer* (No. 22027), Howard W. Sams & Co., 1983. Copyright © 1983, Howard W. Sams & Co. Reproduced by permission of the publisher.

EXPERIMENT 17

Figures 17.3b and 17.4: F. E. Martin and P. H. Sidles, "Low-Cost Cryostat for Laboratory Instruction in Physics," *Am. J. Phys.* **41**, 103, American Institute of Physics, New York, 1973. From F. E. Martin and P. H. Sidles, *Am. J. Phys.* **41**, 103 (1973). Reprinted by permission of the American Institute of Physics.

EXPERIMENT 18

Figures 18.3 and 18.4: Robley D. Evans, *The Atomic Nucleus*, McGraw-Hill, New York, 1972. Copyright ©, McGraw-Hill, 1972, Figures 1-1 and 1-6. Reproduced with permission.

Figure 18.6: C. M. Davisson and R. D. Evans, "Gamma-Ray Absorption Coefficients," American Physical Society, 1952. C. M. Davisson and R. D. Evans, *Rev. Modern Phys.* **24**, 79 (1952). Reprinted by permission.

EXPERIMENT 19

Figure 19.5: Photographer: Terry M. Smith, Instructional Media Center, California State University, Hayward.

EXPERIMENT 20

Figures 20.2 and 20.5: E. Bleuler and G. J. Goldsmith, *Experimental Nucleonics*, Holt, Rinehart & Winston, Inc., New York, 1952, Figures 19.1 and 19.2. From *Experimental Nucleonics* by E. Bleuler and G. J. Goldsmith, copyright © 1952, Holt, Rinehart & Winston, Inc., and renewed 1980 by E. Bleuler and G. J. Goldsmith. Reprinted by permission.

Figure 20.6a: F. S. Stephens, *Nuclear Spectroscopy*, Part A, Academic Press, Orlando, FL, 1960. From F. S. Stephens in *Nuclear Spectroscopy*, Part A. Reprinted by permission of Academic Press.

Figure 20.6b: *Laboratory Manual A: Semiconductor Detectors and Associated Electronics*, EG + G Ortec. Oakridge, TN, 1968. Reprinted by permission of EG + G Ortec.

EXPERIMENT 21

Figure 21.2a: Paul A. Tipler, *Modern Physics*, Worth, New York, 1978. Copyright © Worth Publishers, Inc., New York, 1978. Reproduced by permission.

Figure 21.2b: Paul. A. Tipler, *Foundations of Modern Physics*, Worth, New York, 1969. Copyright © Worth Publishers, Inc., New York, 1969. Reproduced by permission.

Figure 21.2c: R. Eisberg and R. Resnick, *Quantum Physics of Atoms, Molecules, Solids, Nuclei and Particles*, Wiley, New York, 1985. Copyright © 1985, John Wiley & Sons, Inc. Reprinted by permission.

Figure 21.6: Jearl Walker, "The Amateur Scientist," *Scientific American*, New York, February 1986, p. 116. From "The Amateur Scientist," by Jearl Walker, copyright © 1986, Scientific American, Inc. All rights reserved.

EXPERIMENT 22

Figure 22.2: A. Hudson and R. Nelson, *University Physics*, Harcourt Brace Jovanovich, Orlando, FL, 1982, Figure 32-17. Copyright © 1982, Harcourt Brace Jovanovich, Inc. Reprinted by permission of the publisher.

APPENDIX A

Figure A.2a: A. D. Martin and P. J. Quinn, "Electron Spectroscopy Using a Franck-Hertz Tube," American Association of Physics Teachers, 1984. A. D. Martin and P. J. Quinn, *Am. J. Phys.* **52**, 1114 (1984). Reproduced with permission.

APPENDIX B

Figures B.7 and B.8: P. J. Ouseph, *Introduction to Nuclear Radiation Detectors*, Plenum Press, New York, 1975, from Tables 4.4 and 4.5. Copyright © Plenum Press. Reproduced with permission.

APPENDIX D

Figures D.8, D.9, and D.10: Courtesy of Professor Eugene Y. Wong, Department of Physics, University of California, Los Angeles.

INDEX